CONTINENTS IN MOTION

Other books by Walter Sullivan

Landprints

Black Holes: The Edge of Space,
the End of Time

Quest for a Continent

Assault on the Unknown

We Are Not Alone

CONTINENTS IN MOTION

The New Earth Debate

Second Edition

Walter Sullivan

AIP

American Institute of Physics
New York

Second edition. First edition published by McGraw Hill Book Company
in 1974.

© 1991 by Walter Sullivan
Printed in the United States of America
second printing, 1992

American Institute of Physics
335 East 45th Street
New York, NY 10017-3483

Library of Congress Cataloging-in-Publication Data

Sullivan, Walter.
  Continents in motion: the new earth debate / by Walter Sullivan.
    —2nd ed.
    p.   cm.
  Includes bibliographical references and index.
  ISBN 0-88318-703-5 (case) — ISBN 0-88318-704-3 (pbk.)
  1. Continental drift. I. Title.
QE511.5.S93 1991
551.1'36—dc20   90-22116
                                                          CIP

Among the many who contributed to the revolutionary discoveries described in the pages that follow, none performed a more essential role than those to whose memory this book is respectfully dedicated:

William Maurice Ewing
1906–1974

and

Harry Hammond Hess
1906–1969

Contents

Preface to the Second Edition   viii

Acknowledgments   ix

Introduction   1

1 Birth of a World-Shaping Idea   3

2 Careening or Drifting Poles   20

3 Colliding Worlds and Other Catastrophes   31

4 The Balloon Hypothesis   54

5 The Mantle Controversy   62

6 Magnetic Footprints   87

7 The Doubters   113

8 'Mohole' or 'Nohole'?   135

9 An Epic Voyage into the Past   146

10 Of Seas Turned to Deserts and Lands Drowned or Arisen   163

11 Antarctica—The Crucial Puzzle Piece   186

12 Changing Geography and the Diversity of Life   204

13 The Afar Triangle and the Red Sea Hot Spots   220

14 Of Plumes, Eruptions, and Mid-Ocean Islands   235

15 Wherefore, Then, Were the Mountains Thrust Up?   251

16 The Appalachians—Born of a Vanishing Ocean   267

17 The Search for Universality   285

18 Earthquakes: Prediction or Prevention   307

19 Can the Genie Within Be Tamed?   323

20 Earth's History and Man's Destiny   338

21 "But Still It Moves!"   356

References   379

Credits and Permissions   409

Index   415

About the Author

Preface to the Second Edition

Since the original publication of this book there have been no major changes in the concepts expressed here, but there have been a number of developments that have reinforced or elaborated the theory. These, as much as feasible, have been incorporated into the present edition. New techniques have deepened our knowledge of the planet. The Consortium for Continental Reflection Profiling (COCORP) has used truck convoys, vibrating in unison, to probe deep layers of the earth's crust, confirming some of the extensive overthrusts on both coasts. Data from worldwide earthquake detectors have been analyzed, by a method known as seismic tomography, to learn about the deep interior. Analysis of ancient magnetism, fossils, and regional rock composition have shown that all continents are amalgamations of fragments, or "terranes," some of which, in the past, travelled thousands of miles to reach their present positions. Such terranes, moving north, were plastered against the West Coasts of the United States and Canada, some becoming part of Alaska. Revolutionary discoveries regarding the sea floor were made by the deep-diving submersible, *Alvin*, and other such craft, as well as by unmanned devices. These revealed countless metal-rich geysers erupting from ridges under the oceans. The *Alvin*, in some cases with this writer aboard, was able to study geology of the Cayman Trench, which reaches from south of Cuba to Guatemala and forms the northern edge of the Caribbean crustal plate. The discoveries continue and this book is intended to make them more meaningful.

Acknowledgments

This book, dealing with unusually diverse fields of science and widely separated regions of the world, could not have been written without the generous help of many specialists in the United States and elsewhere. Sir Edward Bullard of the University of Cambridge, Sheldon Judson of Princeton University, and J. Tuzo Wilson of the University of Toronto read and commented on extensive portions of the manuscript. Among those who scrutinized individual chapters special thanks are due to John Bird of Cornell University, Kenneth Hsü of the Federal Institute of Technology in Zurich, John Rodgers of Yale University, and Warren Hamilton of the United States Geological Survey, who reviewed the chapters in several stages of preparation. A number of others provided valuable information or checked portions of the manuscript dealing with their own work.

While it is not possible to name all those who were helpful, I wish to express my indebtedness, apart from those already mentioned, to the following (affiliations refer to the time of the first edition):

Tj. H. van Andel of Oregon State University, Don Anderson, director of the Seismological Laboratory of the California Institute of Technology, Tanya Atwater of the Massachusetts Institute of Technology, Robert Ballard of the Woods Hole Oceanographic Institution, William E. Benson of the National Science Foundation, Lord Blackett (P. M. S. Blackett) of the Royal Society of London, Robert Bostrom of the University of Washington, F. R. Boyd of the Geophysical Laboratory of the Carnegie Institution of Washington, Edwin H. Colbert of the Museum of Northern Arizona, Stirling A. Colgate, president of the New Mexico Institute of Mining and Technology, Allan Cox of Stanford University, Kenneth Deffeyes and R. H. Dicke of Princeton University, William R. Dickinson, Stanford University, Robert S. Dietz of the National Oceanic and Atmospheric Administration, Charles L. Drake of Dartmouth College, Helen Dukas of the Einstein Estate, David H. Elliot of Ohio State University, Maurice Ewing of the Lamont-Doherty Geological Laboratory, and Páll Einarsson of that observatory.

Also Rhodes Fairbridge of Columbia University, Charles H. Hapgood of Keene State Teachers College, Pembroke J. Hart of the Geophysics Research Board of the National Academy of Sciences, John H. Healy of the National Center for Earthquake Research, James R. Heirtzler of the Woods Hole Oceanographic Institution, Lowell S. Hilpert of the U.S. Geological Survey, John T. Hollin of the University of Maine, Sir

Harold Jeffreys of the University of Cambridge, F. K. Jones and V. M. Gray, archivists at Jesus College, Cambridge, H. D. Klemme of Weeks Natural Resources, Inc., Leon Knopoff of the University of California at Los Angeles, Roger L. Larson and Walter C. Pitman III at the Lamont-Doherty Geological Observatory, Xavier Le Pichon of the Centre Océanologique de Bretagne, and John Lyman of the University of North Carolina.

Also Paul S. Martin of the University of Arizona, Thomas R. McGetchin of the Massachusetts Institute of Technology, Charles E. Melton of the University of Georgia, H. W. Menard of the Scripps Institution of Oceanography, A. A. Meyerhoff of the American Association of Petroleum Geologists, Paul A. Mohr of the Smithsonian Astrophysical Observatory, Jason Morgan of Princeton University, Walter H. Munk of the University of California at San Diego, Otto Nathan of the Einstein Estate, Carel Otte of the Union Oil Company of California, Melvin Peterson, Lynn Victoria Allan, and Tom Wiley of the Deep Sea Drilling Project, Troy L. Péwé of Arizona State University, Ruth Piggot (Mrs. Charles S. Piggot), Robert W. Rex of the University of California at Riverside, Peter A. Rona of NOAA, William B. F. Ryan of the Lamont-Doherty Geological Observatory, James M. Schopf of the U.S. Geological Survey, Frederick J. Sawkins of the University of Minnesota, Bobb Schaeffer of the American Museum of Natural History, Sean C. Solomon of the Massachusetts Institute of Technology, Lynn Sykes of the Lamont-Doherty Geological Observatory, Stephen L. Talbott, editor of *Pensée*, Haroun Tazieff of the Centre National de la Recherche Scientifique, Nikolaas Tinbergen of Oxford University, Victor Vacquier of the Scripps Institution of Oceanography, James M. Valentine of the University of California at Davis, Immanuel and Elisheva Velikovsky, Frederick J. Vine of the University of East Anglia, Peter R. Vogt of the U.S. Naval Oceanographic Office, Meyer W. Weisgal of the Weizmann Institute of Science, J. D. H. Wiseman of the British Museum (Natural History), and Isidore Zietz of the U.S. Geological Survey.

A number of the concepts set forth in this book remain controversial and none of those listed above can in any way be held responsible for its content.

I owe a special debt of gratitude to several libraries that were particularly hospitable: the geology libraries of the California Institute of Technology, Columbia University, and Princeton University, as well as the Science and Technology division of the New York Public Library and the science section of that library's Mid-Manhattan Branch. At the Goddard Space Flight Center of the National Aeronautics and Space Administration the Public Affairs Office sought out most of the colored photographs. Their identification and interpretation would have been impossible without the generous help of Paul D. Lowman at that center.

Most of the maps and diagrams were prepared by Andrew Sabbatini, who remained patient through many revisions. Additional credits for other illustrations from previously published material are provided at the back of the volume. Brenda Nicolson did much of the typing. At the McGraw-Hill Book Company Gladys J. Carr and her associates were responsible for the fine appearance of the original edition.

For the second edition, I appreciate the assistance of Irene Aranovich, Andrew Prince, Mara Keire, Tom von Foerster, Rita Lerner, and their colleagues at the American Institute of Physics.

Much of the material in this book was obtained in covering these historic developments for *The New York Times*. It was the wisdom and foresight of *The Times* that made such coverage possible. Finally, my wife Mary not only suffered through the long years of the book's gestation, but edited and re-edited every page of the manuscript. She alone can appreciate the extent of that labor.

A Word About Units of Distance, Weight, and Temperature Used in This Book, as Well as References Cited

Because the United States is engaged in a gradual conversion to the metric system, as all other industrialized nations have already done, metric units are used. However, since many American readers are still unfamiliar with those units, the ones more familiar to them are given in parenthesis where necessary for clarity. Where meters are used, they can be read as "yards," since the meter is only slightly longer than a yard, and no equivalent in feet is given. Likewise, where kilometers are used in a general sense, as in "hundreds of kilometers," no equivalent in miles is given, since one need only remember that a kilometer is slightly more than a half mile. "Billions" are used in the American sense of 1,000 million. To avoid interrupting the text with footnotes, references are listed in the back of the book, chapter by chapter.

Introduction

WHO HAS NOT WONDERED WHAT THRUST UP THE MOUNTAINS AND CARVED OUT the seas? Who has not marveled at the incredible forces that buckled, folded, and distorted the rock layers of our planet—as they lie revealed, for example, where a highway has been newly cut through bedrock? Who, in fact, has not pondered how our planet came to be the way it is, with continents and seas, volcanoes and island chains, all obviously the products of some kind of formative process?

It is the good fortune of those living in the final decades of the twentieth century to be able to witness, at first hand, one of the most profound revolutions in man's understanding of nature. For a concept of the Earth has emerged that explains why its surface is the way it is—why the mountains stand where they do; why volcanoes and earthquakes rim the Pacific; why the coast of California from San Diego to San Francisco is creeping northwest relative to the continent as a whole.

There have been other such flashes of insight in the past, but they have been rare. One occurred when Sir Isaac Newton realized that the laws governing the fall of an apple on Earth also controlled the orbits of the moons of Jupiter, as well as the flight of the Moon around the Earth and of the Earth around the Sun.

His mathematical formulation of gravity as the explanation of these motions represented an enormous philosophical, as well as scientific, revolution. It indicated that the physical laws controlling events on Earth were also applicable throughout the universe. Newton, in the 17th century, first undermined the traditional view that the heavenly bodies answer to some other, ethereal laws beyond our comprehension. It became accepted that there is but one set of laws—one science—for all the universe, but this concept was slow in permeating the consciousness of mankind.

Other great revolutions in understanding have included that for which Charles Darwin was chiefly responsible, explaining the evolution and diversity of life on Earth, and the one inaugurated in this century by the relativity concepts of Albert Einstein. Relativity, however, involves speeds (that of light), enormous masses, and other conditions so far beyond our daily experience that the theory is still remote from the ordinary citizen. The new understanding of the Earth does not lean so heavily on difficult mathematical formulations. It concerns phenomena that surround our daily lives. It answers questions that men have asked since the earliest times. Its development represents one of the great scientific detective stories of all time.

The manner in which this theory has gained acceptance underscores the fallibility of scientists and the fact that fashions prevail in science as they do in clothing or hair styles. Yet it also demonstrates how, to those with vision, seemingly disparate, unrelated discoveries can suddenly be brought together into a theory that not only is plausible but explains a variety of age-old problems.

It is a tale of concepts repeatedly rejected, modified, rejected again, that finally led to a comprehensive picture of our planet Biblical in its grandeur. In essence the theory views the Earth as alive, constantly in flux, its continents in motion with respect to one another, carried by the creeping movements of gigantic plates of the Earth's crust. Where these plates clash—beneath California, across Turkey, along the coasts of Japan, Alaska, Chile, and Peru and in many other areas—there may be devastating quakes and volcanic eruptions.

The theory bears on an extraordinary range of subjects: a former ice age in the Sahara, the presence of fossil "African" reptiles near the South Pole, the global distribution of ore and oil, climate changes past and present, and the circumstances responsible for destroying some species of life and giving birth to others.

It is characteristic of science that rarely, if ever, is a theory the final word in explaining nature. The Copernican view of heavenly motions, falteringly championed by Galileo, was shown by Einstein to be an imperfect description of reality. But, broadly speaking, it was incontrovertible; and there is reason to believe that the new theory of the Earth, broadly speaking, will also stand up to future challenge. Part of nature's beauty is its intrinsic mystery. Just as a shade of mystery helps make a woman beautiful, so nature always seems to hold something back. No matter how deeply we probe its inner sanctum, there is always another door to open.

So it is with the new theory of the Earth. We live in a time of exhilarating discovery about our planet. If the reader wishes to share in this exhilaration he must become familiar with a few terms and ideas that may be new to him. But in this book an effort has been made to make this process as painless as possible. The discoveries set forth in the following pages are one of the glories of our time, in which all should share. It is to be hoped that the reader will emerge with a new appreciation of the majesty that is the Planet Earth.

Chapter 1

Birth of a World-Shaping Idea

IN 1914, AS THE THIRD REGIMENT OF THE QUEEN ELIZABETH GRENADIER GUARDS was spearheading the German advance into Belgium, a young reserve lieutenant named Alfred Wegener was shot in the arm and hospitalized. His recovery was swift, but no sooner had he returned to the front than a wound in the neck put him out of action for the rest of the war.

As he lay recuperating an idea that had enlisted his interest several years earlier became an obsession. As early as 1910 he had been struck by the remarkable "fit" between opposite sides of the South Atlantic—the bulge of Brazil seems to fit almost perfectly into the great embayment of West Africa.

The idea that the two coastlines might once have been in contact seemed preposterous, but a year later, by accident, he had come across a report of geologic and fossil links hinting that the two regions had, in fact, once been joined. In 1912, therefore, he had presented to the Frankfurt Geological Association a proposal that the continents drift with respect to one another. Then an expedition to Greenland and the outbreak of war had cut short his exploration of this possibility.

Now, during a prolonged convalescence, he was able to do his research leading, in 1915, to publication of the first comprehensively developed theory of continental drift. To believers in drift—for many years a small fraternity of scientific outcasts—this book, *The Origin of Continents and Oceans,* was a classic, but to a large part of the scientific community it was a classic example of fuzzy reasoning.

Wegener was not the first to speculate on drift. As soon as exploration had produced reasonably accurate maps of the continents, the striking fits of some opposing coastlines became evident, notably the way in which the bulge of Brazil conforms to the bight of Africa on the opposite side of the Atlantic. It has often been noted that Sir

Francis Bacon, as early as 1620, drew attention to such "conformable instances," although he seems to have been impressed by the similarity in shape between Africa and South America, rather than by the fit of their opposing coastlines.

Geology in the 17th and 18th centuries was dominated by the concept of the Biblical Flood as a major force in shaping the Earth's surface. It was a geology based on a belief that past catastrophes had brought about sudden and radical changes. Theodor Christoph Lilienthal, a German theologian, concluded from Biblical references that there was a tearing apart of the Earth's surface after the Flood and, in 1756, pointed to the coastline fits as evidence for this.

Early in the 19th century Alexander von Humboldt in Germany proposed that the Atlantic had been carved out by catastrophic water action, forming a gigantic "valley" with parallel sides. Closer to the modern view were the ideas of Antonio Snider in 1858, although he still thought in catastrophic terms. At the time of the Flood, he said, there was a great outpouring of material from within the Earth that rifted and pushed the continents apart, forming the Atlantic. He cited the similarities of fossils and certain rock formations on opposing sides of that ocean and published, in Paris, a map that reassembled the Americas, Europe, and Africa like pieces of a jigsaw puzzle (Fig. 1.1). Others explained the opening of the Atlantic in terms of Sir George Darwin's concept that the Moon was thrown off from a fast-spinning Earth. This, it was proposed, left a great cavity where there is now the Pacific Ocean and the flow of material to fill this cavity dragged the Americas away from Europe and Africa.

 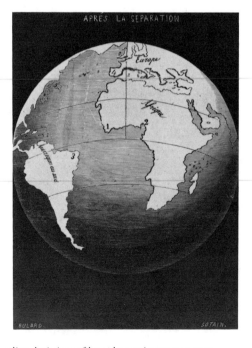

Figure 1.1 Snider's maps of 1858 are some of the earliest depictions of how the continents may once have fit together.

However, by the dawn of this century, the idea of past catastrophes as molders of the Earth's surface had given way to "uniformitarianism." This was the view, still dominant, that all of the forces that can be seen at work on the Earth today, operating over long periods of time, should be adequate to explain what has happened.

Yet, as knowledge of the world's fossils became more complete, it was evident that what has happened was not easy of explanation. There was evidence that polar areas had once been hot and tropical areas had once been covered with ice. In 1908 an American, Frank Bursley Taylor, proposed in a privately printed pamphlet that the Moon was captured by the Earth's gravity during the Cretaceous Period, 100 million years ago, and came so close to the Earth that its tidal pull dragged the continents toward the Equator. Their plowing through the ocean floors, he proposed, wrinkled their Equator-facing fronts to produce such mountains as the Himalayas and Alps. Another American, Howard B. Baker, proposed that a close approach of Venus to this planet tore the Moon from the bosom of the Earth and set the continents in motion.

However, it was Wegener who first tried to muster all the evidence for continental drift. In doing so he anticipated today's dependence on highly diverse fields of science to bring into focus what actually is happening to the Earth's crust. Indeed the magnificence of this tale, as it has finally evolved, rests in part on the breadth of human intellectual endeavors brought to bear on this global problem. The inputs are from such diverse fields as meteorology, geology, oceanography, seismology, geomagnetism, paleontology, evolution—even mountaineering and polar exploration.

Wegener was one of that breed of scientist-explorers who dominated polar exploration early in this century (Fig. 1.2). From what his companions have written of him he was not the swashbuckler one often associates with adventurers. Johannes Georgi, among the last to see him alive, described him as "A man of medium height, slim and wiry, with a face more often serious than smiling, whose most notable features were the forehead and the stern mouth under a powerful, straight nose." Yet, despite the stern mouth, he was a passionately loyal comrade—a quality that finally cost him his life.

Born in Berlin, where his father was a preacher of the Evangelical Church, his early specialization was planetary astronomy and he obtained his doctorate in 1905, having produced as his doctoral thesis a conversion into the decimal system of the Alphonsine Tables of planetary motions. The tables were originally prepared in 13th-century Spain under King Alphonsus X, based on the pre-Copernican concept of the solar system.

Wegener's interest soon shifted to upper air weather observations and he conducted a variety of experiments at the Royal Prussian Aeronautical Observatory at Lindenberg, southeast of Berlin, using kites and balloons. This was the heyday of ballooning, and almost every day the newspapers carried an account of some new venture or some record broken. The year 1906 was marked by widely publicized flights across the Alps, across the English Channel, and from West Point on the Hudson River to Glen Head, Long Island. Late in the year, when Wegener's brother Kurt entered an international competition for the Kaiser's Trophy, a vast multitude assembled around what became Berlin's Tegel Airport to see the 17 balloons lift off. As reported in *The New York Times*, "The adjacent fields were covered with thousands of automobiles and carriages, and fully 100,000 persons had assembled to see the ascent." The balloons—Austrian,

Figure 1.2 Alfred Wegener was a lifelong advocate of the theory of continental drift. To support the theory, Wegener used an interdisciplinary approach. This openness is one of his many legacies to the field of plate tectonics.

Belgian, German, and Swiss—were released at five-minute intervals. Kurt did not place in that race, but earlier in the year the two Wegener brothers broke the world's record for long-duration flight in a free balloon. For 52 hours, from April 5 to 7, they drifted with winds that carried them across Germany, over Denmark, out across the Kattegat—that body of water separating Denmark and Sweden—and back over Germany.

Thus, at the age of 26, Alfred Wegener became established as a man of courage and enterprise. He was also by now a qualified meteorologist and was invited to join a Danish expedition going to Greenland that same year. He spent two winters there, making weather observations and establishing a long, and finally fatal, association with that icy land. On his return to Germany he became a lecturer in astronomy and meteorology at the University of Marburg, and it was there he established what was to be a lifelong relationship with Johannes Georgi. The latter came to hear him lecture and, when Wegener became head of the Meteorological Research Department of the Deutsche Seewarte (German Marine Observatory) in Hamburg, Georgi joined him as a member of its staff.

Meanwhile Wegener's effort to build his case for continental drift had been interrupted, first by a return to Greenland for a memorable crossing of its high-domed ice cap in 1913, and then by World War I. The Greenland crossing was carried out with but a single companion, J. P. Koch, a Dane whom he had met on the 1906 expedition. On this 700-mile trek their five sledges were hauled by ponies, rather than dogs—a tactic that proved successful, in contrast to Captain Robert Falcon Scott's experience with ponies on his dash to the South Pole a year earlier, ending in death for his entire party.

In the years following World War I Wegener continued to seek out evidence for continental drift, several times revising his book, the last edition of which appeared in 1928. He argued that until the Carboniferous, or Coal-Forming, Period, some 300 million years ago, all the continents were united in a single supercontinent: Pangaea. He prepared maps showing them fitted together like puzzle pieces (Fig. 1.3). Then, he said, in successive ruptures they broke apart. Antarctica, Australia, India, and Africa began separating in the Jurassic Period, or "Age of Dinosaurs," some 150 million years ago. In the subsequent Cretaceous Period Africa and South America separated "like pieces of a cracked ice floe." The final separation, Wegener said, was that of Scandinavia, Greenland, and Canada at the start of the ice ages, a million or less years ago.

The Mid-Atlantic Ridge that rises above water to form such islands as Iceland and the Azores was, in his view, composed of continental material left behind when the continents that now flank the Atlantic broke apart. The curved island chains, such as the Antilles that enclose the Caribbean, or Japan and other island arcs of the western Pacific—or even Florida and the Florida Keys—he saw as bits of continental material that broke off and were left behind in the general westward drift of the continents.

The mountains that line the western coasts of the Americas, from the Andes to Alaska, were wrinkles of the Earth's crust, he said, formed by pressure as the great blocks of continental rock pushed west. Likewise the wall of snowy mountains that rises above the steaming New Guinea jungles was pushed up by the northward drive of the Australia–New Guinea block, and the mountains of Asia, from the Himalayas north to the Tian Shan on the borders of Soviet Central Asia, arose from the northward pressure of India.

Such a theory of mountain building was, to Wegener, far more plausible than the traditional idea that shrinkage of the entire planet, like a dried, wrinkled apple, had been responsible. The shrinkage concept went back at least to Sir Isaac Newton. In 1681 he proposed, as a possible explanation for mountains, "ye breaking out of vapours from below before the Earth was well hardned, the settling & shrinking of ye whole globe after ye upper regions or surface began to be hard."

However, it became obvious that mountain building had not been confined to the earliest stage of crustal formation, but had occurred during a succession of episodes throughout the Earth's history. It was therefore a gradual shrinkage, produced by slow cooling of the interior, that became the most popular explanation of what wrinkled the land. It was assumed that the Earth, like a true child of the Sun, had been fiery hot to begin with. How else to explain the internal heat still manifest in volcanic eruptions?

Early in the last century Sir Humphrey Davy, first to isolate a number of the chemical elements, proposed that there are continuous "fires" within the Earth—oxidation processes that provide a continuing source of heat. But further investigation persuaded him that this was implausible.

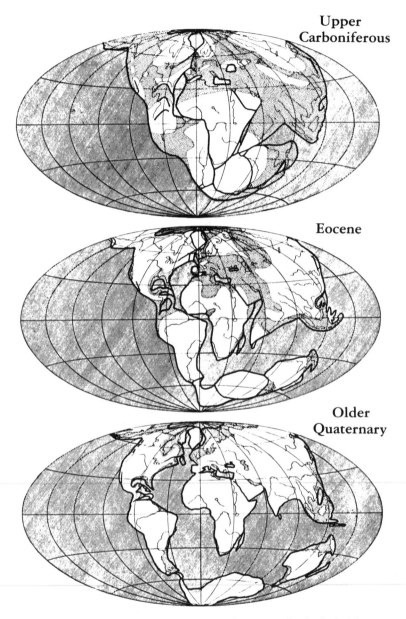

Upper Carboniferous

Eocene

Older Quaternary

*Figure 1.3 **Wegener's breakup of continents** as he envisioned in his book, The Origin of Continents and Oceans.*

By mid-century Lord Kelvin (William Thomson) had calculated that, even if the Earth at its birth had been as hot as the surface of the Sun, it could not be more than a few hundred million years old and still have left as much internal heat as was observed. This argument dismayed the geologists, who needed a much longer time to explain the long and involved history of the Earth indicated by rock formations. It was evident that great mountain ranges had been repeatedly built up and eroded away, and it seemed impossible that this could have occurred in so short a time.

It was, as Wegener pointed out, the discovery of radioactivity at the end of the nineteenth century that resolved the dilemma and gradually eroded the idea that shrinkage produced the mountain ranges. Radioactive elements (chiefly uranium, thorium, and a radioactive form of potassium) are found in many kinds of rock, and their slow decay into other elements generates heat. On the Earth's surface this heat dissipates so rapidly that it is virtually unobservable, but deep in the Earth it accumulates and provides a continuing source of energy.

The discovery of radioactivity has made it possible to estimate how much time has elapsed since various rock formations cooled and solidified. Because radioactive elements decay into other substances at known rates, some of these decay processes can be used as stopwatches to determine the age of a rock formation. From this it is now known that the age of the Earth is measured in thousands of millions of years and it has been possible to establish a timetable of the geologic periods that, in Wegener's hypothesis, were marked by a gradual breakup of the original supercontinent.

Not only did the discovery that the Earth had not cooled a great deal argue against a simple shrinkage of the world, but so did evidence for enormous overthrusts in some mountain areas, particularly the "nappes" of the Alps (Fig. 1.4). Mark Twain is said to have quipped that, if Switzerland were rolled out flat, it would make quite a sizable country. Wegener cited a study indicating that, before its compression, the Alpine landscape was 10 to 12 times wider than now. Yet extensive areas of the world's surface are not wrinkled or overthrust at all. In Wegener's view it was far more likely that such concentration of mountain-building in specific localities derived from the pressures of continental drift than from a uniform, overall shrinkage.

In support of the drift hypothesis he mustered a wide variety of arguments dealing with distributions of living and fossil species, ancient climates, geologic "fits" between lands now widely separated and what he took to be indications of contemporary drift.

Among the problems facing those who studied the distributions of animal and plant species over the globe, he pointed out, were the peculiarly close relationships of species now widely separated. For example, the lemurs, most primitive of the primates, are found only on the island of Madagascar, in nearby East Africa, and across the Indian Ocean in Ceylon, India, and Southeast Asia. Why this distribution on opposite sides of so large an ocean? Furthermore the hippopotamus is found on Madagascar, as well as in Africa. Assuming that those wallowers in muddy streams evolved on the mainland, how could they have swum 250 miles of open sea to reach Madagascar? It therefore had been proposed that a land called Lemuria once linked India, Madagascar, and Africa.

Similarly, Wegener said, Australia's links seemed to be with distant South America, more than with the Sunda Islands off its north coast. Marsupials such as the kangaroo and opossum today are found only in Australia and the Americas. It was strange, he

Figure 1.4 The Engelberg in central Switzerland. Layers of sedimentation are clearly visible, being rocks that formed beneath the sea—at sufficient depth to compact sediments to rock—then were thrust many miles north before being eroded to the landscape shown here. Proponents of continental drift saw these distinctive "nappes" of the Alps as evidence supporting their theory.

wrote, that the life forms of Australia, relative to those of the nearby Sunda Islands, are "like something so foreign as to have come from another world!"

Perhaps the most dramatic puzzle was the distribution of fossil remains of a complex of trees and other plants that flourished during the Carboniferous Period. They were typified by a seed fern, known (because of its tonguelike leaves) as *Glossopteris.* Each region of the Earth, in the past as well as now, has tended to have its characteristic vegetation. A suitably trained botanist, parachuted into a forest with no other clues as to his whereabouts, could in many parts of the Earth tell where he was within a few hundred kilometers by the mixture of plants that he found there. It was therefore surprising to discover that the *Glossopteris* vegetation thrived in as widely separated regions as India, Australia, South America, and South Africa. In fact the Scott and Shackleton expeditions to Antarctica had found it embedded in coal seams in mountains near the South Pole.

How could one form of vegetation have been common to these lands now separated by thousands of kilometers of ocean? Geologists in India had defined an extensive formation, largely of sandstone, with coal, reptilian fossils, and, through extensive layers, the *Glossopteris* flora. When the Austrian geologist Eduard Suess realized that this formation occurred as well in Africa and Madagascar, he proposed that those regions were all linked by a giant continent, much of it now submerged beneath the Indian Ocean. "There is every reason to suppose," he wrote at the turn of the century in his five-volume classic, *The Face of the Earth*, "that the nature of the basement of the Indian Ocean resembles that of the surrounding continents" (a supposition now shown to be erroneous).

He called this great continent Gondwanaland, for a region of India inhabited by an aboriginal tribe, the Gonds, where the *Glossopteris* formation was extensive, and proposed that it extended across Africa and the South Atlantic to include Brazil. The British research ship *Challenger* had discovered a relatively shallow region in the central Atlantic (its true nature as the Mid-Atlantic Ridge could not be determined by the cumbersome sounding methods of the 19th century) and this had given new life to the classic idea of a submerged continent, Atlantis, beneath that ocean.

Suess, therefore, thought of Gondwanaland as a continent, much of which had sunk beneath the waves, rather than as a land that had fractured and drifted apart. Wegener, however, argued that it was physically impossible for such great land areas to sink out of sight into the sea. He cited the newly established principle of isostasy whereby (it was assumed) all parts of the Earth's surface—continents, islands, and ocean floor—are floating on material that, in very slow motion, behaves like a fluid.

This had been shown by the gradual uplift of those areas of Scandinavia and North America heavily laden with ice during the last great glaciation. The Finnish coast along the northern Baltic is rising about one meter every century. At its maximum the ice load pushed down the landscape as much as 300 meters, and, as it melted, finally vanishing some 10,000 years ago, the land rose and still has not fully recovered its original height. In America the north shore of Lake Superior has been rising steadily and a tide gauge on the pier at Fort Churchill, in Hudson Bay, shows an uplifting of two meters per century.

Thus the continents are like great barges. When a barge is loaded, it sinks and water beneath it flows aside to make room for its greater submerged volume. When unloaded, the water presses back in to lift the barge back to a height commensurate with its reduced weight and greater buoyancy.

But if the Earth's crust is, in this sense, afloat, why do high mountains and lofty plateaus not sink to a more modest level? In 1855 British scientists near the Himalayas in northern India conducted experiments with a plumb bob that, they expected, would show a substantial gravitational tug by the world's most massive mountains. Gravity is generated by the entire mass of the Earth, but it was assumed that there was enough mass in the Himalayas to deflect a plumb bob from hanging vertically. Deflections of about one minute of arc (one sixtieth of a degree) were expected, but instead the effect was one sixtieth that amount, or less.

The reason was that the mountains had, in fact, sunk to their appropriate depth. They still stood high in the sky because they were supported by a thick layer of relatively light rock. Similarly, a pinnacled iceberg rises high above the sea because it is supported by an underwater part that is extensive and slightly less dense than the surrounding water.

And what about the crust beneath the oceans? Was it also afloat? Wegener pointed to a striking feature of the Earth's surface. While, in his day, information on the ocean floor was meager, it was known that continents, as a rule, are fringed by a shallow shelf that may extend a few kilometers to a few hundred kilometers off shore, whereupon the bottom slopes steeply down to oceanic depths measured in thousands of meters. The continental margin showing on maps is variable from era to era, being determined by the contemporary sea level. At times sea levels have been higher and the water reached farther inland. During ice ages, when much of the world's water was locked in ice sheets, the sea level was perhaps 50 meters lower, extending the dry land far beyond present shorelines and, for example, forming a land bridge between Alaska and Siberia.

From a structural point of view, therefore, the true margin of the continent is the steep slope that, in most areas, lies under water some kilometers off shore. Beyond this slope, where depths are measured in kilometers, soundings were few, but, Wegener pointed out, it was clear that the surface of the Earth falls into two distinct classes, or provinces. One is the ocean floor, two to five kilometers (one to three miles) below sea level. The other is represented by the continents and their slightly submerged shelves. Very little of the surface lies at intermediate depths.

The answer, Wegener believed, was that the ocean floors represent a deeper layer of the Earth, formed of material in which the continents are "floating." The material of such a layer would have to be denser than continental rocks, just as water is denser than the ice of a berg—otherwise the berg would not float. If the ocean floors were of continental-type rock, such as granite, the force of gravity over the ocean basins should be somewhat weaker than on continents because there would be less material there to produce such a force. But, if the floors were made of very dense rock, this could make up for such a deficiency.

Until the 1920s, gravity measurements at sea were virtually impossible. The most accurate determinations on land were obtained by very precise timing of pendulum swings. A pendulum clock is quite accurate because the pendulum rate is controlled by the force of gravity, which remains rather constant. However, there are regional variations in gravity and these can be determined by highly precise measurements of pendulum rates.

It was not practical, in Wegener's day, to operate a pendulum on a rolling ship, but the Dutch geodesist Felix Andries Vening Meinesz was allowed to ride a submarine of the Royal Dutch Navy, the *K XIII*, from the Netherlands to Java via the Panama Canal, making pendulum measurements while the craft was submerged and stable. The results showed that gravity in the oceans differs little from gravity on land. This implied that the ocean floors are very dense and "floating" at their proper depth and, in Wegener's view, ruled out the possibility that continents had sunk into the oceanic depths.

One way to prove that the oceanic floors are not of continental type would be to retrieve some specimens. "However," Wegener wrote, "it will be impossible for a long time yet to bring up samples of rock outcrop from these depths either by dragnet or other means." Actually, in the years that followed, dredging devices brought up rock fragments that almost invariably were denser than those, like granite, typical of continents. And now, after almost a half-century, there has been the extraordinary achievement of drilling through the entire sequence of ocean floor sediments to sample the rock underneath, providing, as will be seen later, dramatic confirmation of the modern drift theory.

Advocates of the idea that continents sometimes sink to become oceans and that ocean floors rise to become continents had cited the fact that marine fossils are often found in continental rocks, embedded in sediments that clearly were laid down beneath the sea. However, Wegener pointed out that almost invariably these are remains of shallow-water organisms that lived on or above the continental shelf, close to shore. The red clay typical of deep sea sediment is rarely found.

As evidence that South America and Africa were once joined, Wegener, in the later editions of his book, presented the enthusiastic findings of several adherents of continental drift. One wrote of Brazil: "Anyone who knows southern Africa will find the geology of this landscape startling. At every step I was reminded of the formations of Namaland and the Transvaal. The Brazilian strata correspond perfectly in every detail to the strata series of the southern African shield." Diamond-bearing formations (known as pipes) occur in corresponding locations on opposite sides of the ocean and in India as well. The South African geologist Alexander L. Du Toit wrote of a visit to South America: "Indeed, viewed even at short range, I had great difficulty in realising that this was another continent and not some portion of one of the southern districts in the Cape...."

Du Toit, a professor at Johannesburg University, had become an ardent supporter of drift, although his theory differed somewhat from that of Wegener. He envisioned two primordial continents: Laurasia in the north, embracing Europe, Asia, North America, and Greenland (its name derived from the Canadian "Laurentian" and "Eurasia"). The remaining lands were grouped in a southern continent that he called Gondwanaland, although it was quite different from the original concept of a continent much of which has now vanished beneath the sea.

When Gondwanaland broke up, according to Du Toit, two of its fragments—Africa and India—pushed north, crumpling and fracturing the floor of an intermediate body of water, the Tethys Sea, to form the Alps and Himalayas. Tethys, in Greek mythology, was the wife of Oceanus and mother of the seas. If geologists had been prepared to believe in such "preposterous" concepts, this would have explained one of the most dramatic puzzles of their science: the fact that the rocks forming the highest mountains in the world—the Himalayas, now far inland—are crumpled, overthrust, or up-ended slabs of oceanic sediment!

Wegener found matching features on opposing sides of the North Atlantic as well. It had long been observed, for example, that the Caledonian mountain system that runs the length of Norway and crosses northern Scotland, had features in common with the Appalachian System of North America, but it was assumed either that this was true merely because they were formed far apart by similar processes or that they were a single formation linked beneath the sea. Wegener argued that before the Atlantic began to open they abutted one another directly.

Summarizing the evidence from both sides of the ocean, north and south, he wrote: "It is just as if we were to refit the torn pieces of a newspaper by matching their edges and then check whether the lines of print run smoothly across. If they do, there is nothing left but to conclude that the pieces were in fact joined in this way."

Finally he cited the evidence of great climate changes—fossil palm trees in Spitsbergen and coal in Antarctica, for example—showing that those regions were once warm. Deposits of mixed sand, gravel, boulders, and clay typically left by a melting ice sheet show that South Africa was once partially ice-covered. Large gypsum deposits, formed in Iowa, Texas, and Kansas during the Permian Period 250 million years ago, indicate a hot, arid climate in those areas, as do salt deposits laid down at the same time in Kansas and Europe, Wegener said.

The only reasonable answer, he added, was a combination of continental drift and changing of the Earth's spin axis. Early in the century the Hungarian physicist Baron Roland von Eötvös, an extraordinarily meticulous and ingenious experimenter with gravity and centrifugal forces, had calculated that, because the Earth is spinning and its shape bulges at the Equator, there should be a very slight force nudging the continental blocks toward the Equator. This force, which Wegener called the *Polfluchtkraft* or "pole-fleeing force," when combined with a tendency for tidal drag to pull the continents westward, might account for continental movements, Wegener believed, if the spin axis changed from time to time, causing the pliant Earth to adjust its shape to produce a new equatorial bulge and new directions of drift.

Another proposal for the motive force was simple gravity. Reginald A. Daly of Harvard University, in his book, *Our Mobile Earth,* suggested that the planet bulges in some directions, producing enough of a slope so that continents sometimes slide downhill to new positions.

The closest that Wegener came to present-day thinking on the motive force was in a brief discussion of possible convection currents within the Earth. Convection currents arise in any fluid heated on a stove. Hot material rises over the flame, spreads and cools at the top, and sinks down the sides to be reheated and rise again. Wegener cited a proposal that hot rock within the Earth could flow in this manner, rising under the continents, spreading and cooling to sink beneath the oceans. Such a flow, he believed, could have broken up the original supercontinents.

Critics of the drift theory had argued on various grounds that the interior of the Earth is as rigid as steel. Actually, Wegener replied, the interior material is much like steel in that, when hot and subjected to great pressure, it behaves in slow motion like a fluid. Thus, in terms of quick response, as to an earthquake wave, its behavior is rigid, but to long-term forces it responds like a fluid, much like ice in a flowing glacier.

Before the forces responsible for drift disturbed the Earth's crust, Wegener said, flat, continental rocks may have covered the entire globe. In that early period, he added, one great ocean, Panthalassa (from "thalassa," the Greek for "sea"), blanketed the world. Then drift processes compressed, folded, and divided the crust to expose underlying material that became the ocean basins. The first of these basins to form, he believed, was the Pacific.

In 1926, after the third edition of Wegener's book had appeared, the American Association of Petroleum Geologists organized a symposium on his theory as part of its New York meeting. While a few spoke in favor of it, the "big guns" among American geologists fired full salvos in opposition, citing (with some justification) various weaknesses in Wegener's arguments.

The chairman, W. A. J. M. van Waterschoot van der Gracht, a Dutch geologist who was vice-president of the Marland Oil Company in Ponca City, Oklahoma, was one of the few sympathizers. "I am personally convinced that there *is* continental drift," he told the meeting. The proposals, he said, "though revolutionary, are serious and far from having to be summarily dismissed. They are not wild dreams... ."

He harked back to his student days when a fierce debate centered on the idea that great sheets of rock, or "nappes," had been thrust scores of miles across the landscape in some regions, notably the Alps. "Its mere possibility," he said, "was then as firmly denied, as is now the possibility of continental drift." He urged the assembled geologists to look upon the arguments with open minds. "If the opponents of this view, of whom I see many in my audience, and from whom we will presently hear, will only consent to see it in this light, we may hope to make better progress in gradually approaching the truth."

Despite this plea a number of the professors did their best to tear Wegener's arguments apart. Rollin T. Chamberlin, professor of geology at the University of Chicago, said geologists might well ask if theirs could still be regarded as a science when it is "possible for such a theory as this to run wild."

Charles Schuchert, professor emeritus of paleontology and historical geology at Yale, conceded that there might be some slow movement of the continents, but he rejected Wegener's arguments. For example, he described a test of the theory in which tracing paper was laid over a globe, and coasts on the western side of the Atlantic were outlined. Then an attempt was made to fit this outline to shorelines on the eastern side of that ocean and it was found, he said, that this could only be done by greatly elongating the Americas. "It is evident, therefore," he said, "that Wegener has taken extraordinary liberties with the Earth's rigid crust, making it pliable so as to stretch the Americas from north to south about 1500 miles."

Furthermore, he said, animal species on opposite sides of the Atlantic are not as similar as Wegener argued, nor does the geology fit in the detailed manner that had been claimed. In short, he said of Wegener, "he generalizes too easily from other generalizations."

Bailey Willis, professor emeritus at Stanford University, argued that continental movements would have produced great faults and distortions that are not in evidence. Wegener's book, he added, "leaves the impression that it has been written by an advocate rather than by an impartial investigator."

However, the fiercest attack was that of Edward W. Berry, professor of paleontology at The Johns Hopkins University. Wegener's method, he said, "is not scientific, but takes the familiar course of an initial idea, a selective search through the literature for corroborative evidence, ignoring most of the facts that are opposed to the idea, and ending in a state of auto-intoxication in which the subjective idea comes to be considered as an objective fact."

Chester R. Longwell of Yale, while feeling the concept had yet to be proved, was more tolerant. The theory, he said, "shows little respect for time-honored ideas backed by weighty authority.... Its daring and spectacular character appeals to the imagination both of the layman and of the scientist. But," he continued, "an idea that concerns so closely the most fundamental principles of our science must have a sounder basis than imaginative appeal."

Thirty years later, looking back on this symposium, Longwell cast a baleful eye on the ridicule that had been heaped on Wegener's idea. "With such extreme views," he said, "I have little sympathy. We know too little about the Earth and its history to indulge in such final judgments. Why must we—or how can we—make up our minds finally in this matter while we admit great gaps in critical geologic information and in knowledge of geophysical principles?... In my view the hypothesis of continental drift has not failed so utterly that it deserves the death sentence.

"But hypotheses are tools," he added, "and must prove their worth if they are to command our respect. To me the concept of moving continents is *hypothesis,* and as such it must be a target exposed to merciless fire of fact-finding and critical analysis."

After the 1926 symposium, to which he, too, was a contributor, Wegener tried to strengthen his argument with further elaborations and revisions of the book. By then he had what he considered the "first precise astronomical proof of a continental drift" in observations that seemed to show Greenland and the United States drifting west, relative to Europe. He had asked his Danish colleagues of the 1906–1908 expedition to check their longitude determinations in Greenland against earlier observations. The results suggested drift. In 1922 the Danes then undertook more observations, using radio time signals for greater precision, and a variety of comparisons with the earlier positions seemed to show a steady westward drift of about 35 meters a year. Other observations reportedly indicated an annual increase in distance between Paris and Washington of slightly less than a third of a meter. There was, however, great skepticism about these determinations of changing longitude, most of which were made by observing the Moon. Longitudes obtained in this way tend to conceal systematic errors.

In 1936 and 1948—long after Wegener's death—the Danes repeated the Greenland observations and found no evidence for drift. Indeed the expected rates of movement are too slow to become evident through ordinary position-determining methods. Today there are several laser reflectors on the Moon, installed by American astronauts and aboard Soviet unmanned vehicles, or *lunakhods.* These have been used by American, French, Japanese, and Soviet observatories to reflect focused, very intense pulses of laser light back to Earth. From these, and observations on extremely distant celestial sources, by widely separated radio astronomy observatories, extremely accurate determinations of relative positions have become possible, showing that the continents are currently in motion, relative to one another.

The final effort of Wegener's career was an expedition to Greenland launched on the initiative of his old friend and student, Johannes Georgi. Wegener had finally found an academic position at the University of Graz in Austria after a long and futile search for such a post in the German universities. He did not conform to the classic image of a research scientist. His arguments tended to be qualitative, rather than numerical and mathematical, and they ranged across the entire spectrum of physical sciences.

It was regrettable, Georgi wrote later, "that this great scholar, predestined for research and teaching, could not get a regular professorship at one of the many universities and technical high schools [college-level institutes] in Germany. One heard time and again that he had been turned down for a certain chair because he was interested also, and perhaps to a greater degree, in matters that lay outside its terms of reference—as if such a man would not have been worthy of any chair in the wide realm of world science." The post he finally won was broadly conceived, for that time of intense specialization—a professorship in meteorology and geophysics.

During this period, in 1926 and 1927, Georgi had been making weather observations from the far northwest corner of Iceland and had detected a very high and intense flow of air from the direction of Greenland—what has come to be known as the jet stream. Hence he conceived the idea of setting up a winter station on the crown of the Greenland ice sheet to learn about this flow, which clearly helped determine European weather. He wrote in 1928 to his teacher and former chief, "as he was the greatest expert on Greenland," to ask his advice and support.

Of the possibility of establishing an observatory in the middle of Greenland, Wegener replied: "...That is an old plan of Freuchen's, Koch's and mine! If only the War had not happened, it would have been carried out long ago. But meanwhile Freuchen has lost a leg, Koch is in hospital, and I too have had my own trouble and am no longer a young man."

His reference was to Peter Freuchen, whose books told vividly of his life among the Eskimos, and Koch, with whom Wegener had crossed Greenland in 1913. The outcome of Georgi's proposal was that Wegener himself organized the expedition, one of whose tasks, apart from weather studies, would be to measure the thickness of the inland ice, using a new seismic technique in which small explosions were fired and the time recorded for the sound waves to travel to the rock floor beneath the ice and return to the surface. (The ice proved to be 2000 meters thick.)

The chief goal, however, was to set up the station to be known as "Eismitte" (Mid-Ice) to carry out weather observations throughout the polar night. The expedition was outfitted with Icelandic ponies, dogs, and propeller-driven sledges. But their plans went awry. Ice and storm delayed unloading of supplies. By digging and roofing a large pit on the crest of the inland ice a temporary station was set up at Eismitte so that observations could be carried out until arrival of a prefabricated hut and supplies for wintering-over. Desperate, after an agonizing succession of delays, Wegener set out from the coast on September 21, 1930, with 15 sledges carrying 4000 pounds of supplies, including enough to carry the two men at Eismitte through the winter. The party consisted of 13 Greenlanders, Wegener, and Fritz Loewe. The season was late, and after 100 miles of trekking through new snow, driving winds, and extreme cold, all but one of the Greenlanders turned back.

Wegener, however, was determined to make contact with the two men at Eismitte before the polar night isolated them, or help their escape if they had decided to march out. Loewe and a Greenlander named Rasmus Willumsen agreed to accompany him. While the bulk of supplies had to be cached, they continued to haul a minimum load of gasoline, Christmas parcels, a gramophone, and scientific instruments; but even these finally had to be left by the wayside.

Meanwhile morale at Eismitte was sagging. Georgi, who was one of the two men there, wrote in a diary letter to his wife on October 29: "We feel ourselves abandoned here. Wintering here in the middle of Greenland is indeed no trifle. And it is unfortunate that we should have to do it now under such unfavorable circumstances! We are very depressed... ."

The next morning the mood reversed itself. The two men were in sleeping bags since, to conserve meager fuel, they had turned off their heater, when they heard the snow creak overhead. They scrambled out in their underclothes to find a Greenlander, bulky in his heavy clothes, standing beside a dogsled and grinning. He pointed to Wegener and Loewe, who were not far behind.

"A marvelous performance in a temperature of − 65 °!" wrote Georgi. "I am sure no one has ever traveled in Greenland in such a temperature." Loewe's feet were so frostbitten that he could not return. Indeed it was uncertain whether or not they would have to be amputated.

"Wegener wants to start back again with Rasmus very early tomorrow morning," Georgi's letter continued. "With their dogs pretty well worn out it is a race with death." Meanwhile three men were to stay at Eismitte, living on rations that were skimpy for two. They survived, but Wegener and Willumsen never reached the coast.

As their return became more and more delayed those at the coastal base camp became increasingly alarmed. Encamped nearby was the British Arctic Air Route Expedition, led by H. G. ("Gino") Watkins, who himself died in Greenland two years later. Its goal was to learn what mountain ranges and weather phenomena might impede direct, polar air routes from Europe to North America. Watkins offered the use of his two planes to search for the missing men, but winter was upon Greenland and the task was hopeless.

The next summer Wegener's body was found on the trail—apparently he had not frozen to death but died, possibly, of a heart attack (he had just passed his 50th birthday). The Greenlander had buried the body and continued on with the diaries and letters, but was never found. The contents of Georgi's letter to his wife survived because he had kept a carbon copy.

Thus ended the career of one of the most ardent advocates of continental drift—a man who, had he lived to a ripe age, would finally have seen much of his theory vindicated. At the time of his death, however, the main body of geologists (with a few exceptions, notably in the Southern Hemisphere) found the idea that the continents drift about like great barges more than they could stomach. Sir Harold Jeffreys, dean of Earth scientists, published calculations showing that the Eötvös force and tidal drag on which Wegener leaned heavily were both far too weak to drive a continent through the rigid ocean floor. In his definitive work, *The Earth: Its Origin, History and Physical Constitution* (first printed in 1924 and revised several times thereafter), he dismissed continental drift as unfit for serious consideration, citing the words with which a colleague had disposed of the theory that mountain ranges had been formed by gross contraction of a cooling Earth: "It is quantitatively insufficient and qualitatively inapplicable. It is an explanation which explains nothing which we wish to explain."

In the end Wegener himself realized that the weakest element of his theory was its attempt to explain what force could be powerful enough to push continents through the ocean floors. He noted that the behavior of falling bodies and orbiting planets was observed and recorded some time before Newton formulated the laws of gravity that explained their behavior. "The Newton of drift theory," he wrote, "has not yet appeared."

Nevertheless, in retrospect, Wegener's contribution was of major importance. It was he who first mustered the help of every field of science that seemed even remotely applicable. He broke determinedly from the German academic tradition of specialization that had evolved in the previous century. He cited, disparagingly, a geologist who argued that only through geology would the drift debate be resolved and, conversely, a paleontologist who wrote: "It is not my job to worry about geophysical processes...

only the *history of life* on the Earth enables one to grasp the geographical transformations of the past."

In seeking a broader approach, despite detailed flaws in his arguments, Wegener anticipated the mobilization of a dazzling array of varied lines of attack that has finally won over the scientific world.

"Scientists still do not appear to understand sufficiently," he wrote in the fore-word to his last edition, "that all Earth sciences must contribute evidence towards unveiling the state of our planet in earlier times, and that the truth of the matter can only be reached by combining all this evidence."

In the years that followed, the scorn heaped upon this theory became such that to support it was almost as damaging to one's scientific reputation as to argue that "flying saucers" might carry visitors from another world. The questions of climate change and species distribution, past and present, that Wegener and the other "drifters" had sought to answer remained open, but serious scientists were discouraged from pursuing them further and the field was left open for amateurs to promote even more sensational theories.

Chapter 2

Careening or Drifting Poles

PERHAPS BECAUSE DAILY LIFE TENDS TO BE RATHER HUMDRUM THERE IS A HUMAN craving for drama. It draws crowds to a fire; it helps fill theaters. It has made popular the idea that we are under surveillance by "flying saucers" from afar. And it gave strong appeal to catastrophic theories of the Earth's history that flourished in the public mind after the majority of professional scientists lost interest in the possibility of continental drift.

Such catastrophic theories of the Earth are not new to science. Until the mid-19th century they seemed to explain the vast deposits of sand, gravel, silt, and "erratic" boulders—stones unrelated to the local bedrock—that were strewn across parts of northern Europe and America. It was assumed that these had been scattered by one or more great floods, like the one described in *Genesis*, and some believed these occurred when the Earth periodically tumbled from its spin axis, causing the oceans to sweep across the continents.

Such catastrophes, it was thought, could account for the fossil remains of extinct species, from dinosaurs and mammoths to a multitude of smaller creatures. What else, it was asked, could have annihilated so many animals? The extinct species came to be known as antediluvian—meaning "from before the Deluge."

Likewise those who studied geology assumed that only spasmodic convulsions of the Earth could have contorted and sheared the rock layers that, in cliffs, now lie exposed in cross-section, or could have thrust up the sharp peaks of the highest mountains.

One of the great scientists of the 18th century, Georges-Louis Leclerc, Comte de Buffon, is credited with being father of the idea that catastrophic changes have occurred when the spin axis of the Earth was altered. An early speculator on evolution, he was one of the first to wonder how species come and go. He believed some forms can degenerate, as well as advance. Thus he saw the apes as degraded forms of men and the ass as a degraded horse.

Another French naturalist, born when Buffon was an old man, carried the argument further. He was Georges Léopold Chrétien Frédéric Dagobert, Baron Cuvier, who established the study of fossil vertebrates (or back-boned animals) as a science. It had been believed that the giant bones turned up from time to time in England were the remains of elephants brought there by the Romans, but Cuvier showed that they were, in fact, fossil remnants of antediluvian relatives of the modern elephant. His explanation of such remains was that the seas had periodically swept across the land, wiping out vast numbers of animals. One such catastrophe, he concluded, was relatively recent.

It was evident, he said, that "the surface of our globe has been subjected to a vast and sudden revolution, not further back than from five to six thousand years; that this revolution has buried and caused to disappear the countries formerly inhabited by man, and the species of animals now most known; that contrariwise it has left the bottom of the former sea dry, and has formed on it the countries now inhabited; that since the revolution, those few individuals whom it spared have been spread and propagated over the lands newly left dry, and consequently it is only since this epoch that our societies have assumed a progressive march, have formed establishments, raised monuments, collected natural facts, and combined scientific systems.

"But the countries now inhabited, and which the last revolution left dry, had been before inhabited, if not by mankind, at least by land animals; consequently, one preceding revolution, at least, had overwhelmed them with water; and if we may judge by the different orders of animals whose remains we find therein, they had, perhaps, undergone two or three irruptions of the sea."

However in the 1830s a Swiss naturalist named Jean Louis Rodolphe Agassiz began to explore a suggestion that the erratic boulders scattered to the very summits of the Jura, on the Swiss-French border, had been carried there by glaciers. He concluded not only that the Alpine glaciers had once been far more extensive, but that virtually all of Switzerland had been covered by one great ice sheet. In fact, he wrote, "great sheets of ice, resembling those now existing in Greenland, once covered all the countries in which unstratified gravel (boulder drift) is found." When the ice sheets melted, he said, these deposits were the residue. Agassiz eventually joined the Harvard faculty and became a leading figure in American science.

Meanwhile Charles Darwin and others had come forth with a partial explanation of the mass extinctions. From the theory of evolution and the concept that only the fittest individuals and species survive, it became apparent that as the environment changed or more efficient species evolved, some creatures, unable to cope with the new situation, had been driven to extinction.

In geology, too, the idea that the mountains were thrown up in great convulsions gave way to the view that all the processes needed to shape the face of the Earth are presently at work and evident if we look carefully enough. The processes are slow and often buried deep beneath the surface. They may have been more intense during mountain-building episodes of the past—but not in a sudden, catastrophic sense.

This, as noted earlier, was the doctrine of "uniformitarianism," which, to a large extent, has governed scientific thinking ever since. It is so entrenched, in fact, that the revival of dramatic explanations for some developments in the Earth's history has been greeted with great hostility.

The proponent of what is probably the most sensational of the hypotheses to emerge from the post-Wegener period is a little-known electrical engineer named Hugh Auchincloss Brown. In 1974, at the age of 94, he was still at work promoting his theory, whose gestation had begun in Wegener's day.

In 1911 he became interested in reports that mammoths had been found frozen in the Arctic "with buttercups still clenched between their teeth." Perhaps, he thought, an accumulation of ice at one or both poles periodically upsets the equilibrium of the spinning planet, causing it to tumble "like an overloaded canoe." This would cause catastrophically rapid climate changes with the oceans sweeping across the continents in global floods such as the one described in *Genesis*.

As an engineer he knew the bulge of the Earth around the Equator stabilizes its spin, but he decided that the accumulation of Antarctic ice, which in some areas is now two or three miles thick, could be enough to topple the spin despite this stabilizing effect.

In a series of broadsides circulated to congressmen, government leaders, scientists, and journalists over a half century, as well as in a book published at his own expense, Brown argued that the Earth capsizes at intervals of about 8000 years, each time wiping out whatever civilization has managed to emerge. The next one, he says, is overdue, and the Eskimos may be among the few survivors, because the polar areas will be the least subject to catastrophic water action.

To a large extent it was the evidence of the "quick-frozen" mammoths that convinced him—and has influenced others of this school. "The fact that they were found in this condition," he told a *New York Times* reporter in 1948, proves that they were subjected to a sudden "quick-freeze" that killed and preserved them immediately. What could have caused this? Obviously, only some sudden drastic change in the weather such as might have happened if tropical regions were suddenly transported to the icy Arctic.

Furthermore, he said, "tales of sudden floods and the mysterious appearance and disappearances of large land masses are found in the folklore and legends of all races of men." As we shall see, Brown was not the only man of this period who saw evidence for past cataclysms in these legends. In any case, with the birth of the United Nations he urged, in 1946, the formation of a parallel agency, the Global Stabilization Organization, to use atomic bombs or whatever other means necessary to reduce the Antarctic ice sheet.

A particularly ominous omen, according to Brown, was the wobble in the Earth's spin. This he took to be the first sign of an impending tumble. The wobble is a fact and its characteristics continue to challenge scientists seeking to explain it. However, the stability of the Earth's spin, with an equatorial bulge making it 43 kilometers fatter "at its waist" than it is from pole to pole, is so great that it would be difficult to find a scientist who believes the axis is likely to change in any sudden way, even if a lopsided chunk of ice does accumulate in the South Polar area.

Brown's predictions of impending doom have done more to titillate popular fancy than arouse scientific concern. The wobble, however, is a puzzle that goes to the very heart of many problems that relate to the ever-changing planet on which we live. Variations in the spin axis of the Earth fall into two categories. One class is caused by external forces (chiefly the gravity of the Moon and Sun acting on the Earth's equatorial bulge) that alter the aim of the spin axis in space—that is, relative to the stars. The dominant variation in this respect is the slow precession that causes the axis to sweep out a broad circle in the sky every 25,800 years. At present the axis points to Polaris, the North Star.

The other form of variation involves changes of the axis relative to the Earth itself. Such changes alter the geographic positions of those points on the Earth's surface traversed by the spin axis—the North and South Poles. They are largely attributed to influences within the Earth such as changes in the distribution of mass.

If the mass were uniformly and symmetrically distributed about the axis, the Earth would spin in a stable manner. The North and South Poles would remain in precisely the same positions. But the distribution of mass is constantly changing. Continents erode, transferring their material onto the ocean floors. Air and water masses migrate with the seasons. Earthquakes suddenly shift sections of the Earth's crust. Hence it was long suspected that the Earth must have at least a slight wobble.

The situation can be likened to a football in flight or a spinning top. If either spins exactly on the axis of its figure, the spin will be stable. However, if the axis of its figure is not exactly the same as the spin axis, it will wobble. The great Swiss mathematician, Leonard Euler, recognized in the 18th century that, for a given object, spinning at a given speed, the rate of the wobble will be the same, whether it is barely perceptible or radical. He thus calculated the characteristic wobble rate of the Earth to be 305 days, or about 10 months. That is, the North Pole of the spin axis should return to the same spot every 305 days.

Generations of astronomers looked for this effect without success. Finally, in 1891, coordinated and highly precise observations were made from Berlin and from Waikiki in the Hawaiian Islands on the opposite side of the Earth. It was found that when the latitude of Berlin increased, the latitude of Hawaii decreased and vice versa—just what would be expected if the Earth's axis were rocking.

Meanwhile S. C. Chandler, a well-to-do actuary and amateur astronomer in Cambridge, Massachusetts, had gone back through 200 years of astronomical records and, in that same year, declared that the wobble is a combination of two effects. One proved to be an annual cycle apparently caused by seasonal redistributions of oceanic water and atmosphere over the Earth's surface. A region of high atmospheric pressure over Siberia in winter seems to play a dominant role. The other contribution to the wobble has a 14-month cycle and is now known as the Chandler Wobble.

When the 14-month Chandler Wobble became known, Simon Newcomb, another of those who had looked for such an effect, pointed out that, if the Earth's interior were somewhat elastic, this could slow down the wobble from Euler's figure of 10 months to 14. However, this puzzling wobble has challenged all attempts at a full explanation. More than 1700 papers have been written on the subject and the answers still remain elusive. The wobble was long monitored by the International Latitude Service, formed in 1900 to keep a constant watch on the Earth's erratic behavior. It subsequently became the International Polar Motion Service, which operated five or six stations spaced around the world in the same latitude (39 degrees 8 minutes north). Every clear night each observed the same stars, but rocking of the Earth from night to night was so slight that only by combining observations from all the stations was it possible to see the effect. The system has been made obsolete by the new techniques of laser ranging on reflectors left by astronauts on the Moon, on reflectors carried by Earth satellite and by long baseline interferometry, using radio emissions from extremely distant quasars as reference points in the sky. The new system is known as the International Earth Rotation Service.

It has been found that the total drift of the poles, over a 14-month period, is no more than about 22 meters in a wandering, but roughly circular, path. Sometimes the 12-month and 14-month cycles that control the wobble work against one another and the polar motion is minimal. But once every seven years the two effects work together and the drift reaches 15 centimeters (six inches) per day.

It was suggested that earthquake activity (in terms of total energy released throughout the world) seems to reach a peak during this period of maximum polar drift. Internal properties of the Earth gradually reduce, or "dampen," the wobble, which then periodically is restored. It is this reactivation that baffles Earth scientists. Some have proposed that great earthquakes give the Earth's spin a jolt, from time to time, but doubts have been expressed regarding this explanation.

If the sensational warnings of Hugh Auchincloss Brown did little to stir up scientific interest in the possible influence of ice sheets in altering the Earth's spin axis, an article in a science fiction magazine did so. One evening in the early 1950s this article, on "Newcomb's great empirical term," came up for discussion in a chat in the office of Walter Munk at the Scripps Institution of Oceanography at La Jolla, California. With him were the institute director, Roger Revelle, and John Isaacs.

As noted earlier, Simon Newcomb had offered an explanation for the Chandler Wobble. His "great empirical term" derived from extensive and very precise observations of the Moon's position relative to landmarks on Earth. The Moon appeared, at times, to be "out of place," apparently because of changes in the Earth's spin rate that could not readily be accounted for.

Newcomb's findings had puzzled scientists for almost a century. In 1876 Lord Kelvin told a meeting of the British Association for the Advancement of Science (popularly known as the British Association), of which he was president, about his worries on a recent trip abroad:

"Disturbed by Newcomb's suspicions of the Earth's irregularities as a timekeeper," he said, "I could think of nothing but precession and nutation, and tides and monsoons, and settlements of the equatorial regions, and meltings of the polar ice."

It was this last idea that interested Munk, Revelle, and Isaacs. What, they wondered, would be the effect if, from the ice cap covering Greenland, or that which blankets Antarctica—or both—enough ice melted to raise worldwide sea levels a substantial amount? If this occurred sufficiently fast that the crust of the Earth did not have time to sink under the weight of added oceanic water, or rise under the lightened ice load, there would be a significant transfer of material from one or both poles to the oceans, which are more equatorial. This would slow the Earth's spin. The same effect is achieved by the fast-spinning figure skater who puts out her arms to slow her spin.

In describing their analysis, in 1952, Munk and Revelle cited evidence from celestial observations that the rotation rate of the Earth had changed "rather suddenly" on three occasions—roughly in the years 1860, 1900, and 1920. Observations of the spin had long been made, not only by the International Latitude Service, but by other observatories under the auspices of the international time service, or Bureau International de l'Heure, which, in Paris, collects observations that relate both to the wobble and spin rate.

When Munk and Revelle wrote their paper, the largest, seemingly abrupt, shift amounted to a change of about 5.5 thousandths of a second in the length of the day. This may not seem much, but the effect is impressive if one considers that it involved the entire mass of the Earth. The two scientists noted that investigators "have searched for suitable catastrophic events" at the reported times of change but, they added, "none seems to have been forthcoming." In another paper that year, they cited studies of celestial observations recorded by the ancient Greeks, Babylonians, Chinese, and Egyptians from which it had been concluded that, throughout historic time, the day had been lengthening at a rate of about two thousandths of a second per century. This effect (apart from a well-known slowing of the Earth's spin by tidal friction) was, they suspected, the result of a redistribution of weight on the Earth from melting of glacier ice and the consequent rise in sea level.

From their own calculations they found that, if sea levels rose another 10 centimeters (four inches), due entirely to Antarctic melting, the redistribution of mass on the Earth's surface would cause the North Pole to move one meter toward Chicago. If the sea level rise were all due to melting in Greenland, the drift would be nine meters toward Greenland. And if the melting occurred uniformly in both regions, the motion would be two meters toward Newfoundland. Actually, from 1900 to 1925, a combination of these factors moved the pole three meters toward Greenland, they concluded, with little motion of this sort having occurred since then.

So far as changes in the length of day were concerned, shifting weight toward the Equator by melting of polar ice would slow the Earth's spin; shifting it toward the poles (when ice sheets grew) would remove weight from the equatorial region and thus accelerate the spin. But such slow changes could not account for the relatively abrupt changes in the Earth's spin. Munk and Revelle therefore proposed that the answer might lie deep within the Earth. They pointed out that some sort of dynamic change appeared to be taking place within the liquid portion of the Earth's core, manifested on the surface by a relatively rapid drift of the magnetic field. This field, which controls the aim of a compass needle, can be likened to the effect that would be produced at the Earth's surface by a powerful bar magnet in the core of the Earth—not that such a magnet really exists. The main portion of the field is presumably generated by the combined action of the Earth's spin and the heat-generated flow, or convection currents, in the hot, molten outer core. While this flow probably has a certain symmetry about the spin axis, the resulting magnetic field (defining the North and South Magnetic poles) is not exactly symmetric to that axis. The two magnetic poles are therefore somewhat displaced from the geographic poles, which are defined by the spin axis.

What was significant to Munk and Revelle, however, was the fact that this offset magnetic axis is slowly drifting westward around the Earth. Returning to the analogy of a bar magnet in the core, it is as though the magnet were currently displaced from the Earth's center toward Brazil and were moving west fast enough so that, in two or three centuries, it will be under Peru.

This means that something is changing rather rapidly in the core of the Earth and, while there is little mechanical linkage between the liquid core and the enveloping mantle, there are probably electromagnetic links that can be "switched on" and "switched off" rather abruptly, accounting for the observed changes in spin rate.

The two authors concluded their report in *Monthly Notices of the Royal Astronomical Society* with a sly reference to the customary demand that a scientific hypothesis make predictions that can be verified. One "advantage" of their theory, they said, is that the core is "even less accessible than the Antarctic."

Subsequently, in 1960, Munk and Gordon J. F. MacDonald of the University of California in Los Angeles, who later became one of the three initial members of President Nixon's Council on Environmental Quality, published a book-length study of the Earth's rotation in which they explored the possibility of more extensive polar wandering. By then there was magnetic evidence from the rocks of various continents suggesting that the world's geography may not always have been as it is today and that the North Magnetic Pole wandered around a large part of the Pacific Ocean before it reached its present position in the Arctic. It was argued that this was coupled to a wandering of the spin axis. To Munk and MacDonald, however, the manipulations of geography that seemed necessary to accommodate the magnetic argument made the latter's significance questionable. For example, they said, if one accepted this line of reasoning it was necessary to rotate and relocate England relative to North America, Spain relative to France, and Scotland relative to England.

"It is usually a bad omen," they wrote, when the number of necessary manipulations grows "at the same rate as the number of independent determinations." They scoffed at the arguments for polar wandering, stating that the "usual starting point" of any discussion of the subject, "is to presume the Earth to be in equilibrium until suddenly disturbed by some implausible rearrangement of matter. The ensuing motion of the pole is then computed for an Earth made of material that can be modeled by a combination of springs and dashpots [shock-absorbing devices]. Finally, the computed polar path is found to be in agreement with a bewildering array of paleontological, and more recently, paleomagnetic evidence." They could not rule out polar wandering, they said, but they were highly skeptical of it.

One proposal that they examined was an argument by Thomas Gold, then at the Royal Greenwich Observatory, that the Earth's equatorial bulge does not prevent such wandering if the Earth's interior is plastic. He suggested that a slow drift in the spin axis could be induced by accumulations of matter, such as the rapid building up of an ice sheet.

He and Herman Bondi (like others before them) had estimated the extent of the Earth's inner plasticity from its effect in retarding the Chandler Wobble. They had also developed a "steady state" concept of the universe—arrived at independently by Sir Fred Hoyle—which views the universe as eternal and (despite its expansion) unchanging in its essential nature.

Writing in the British journal *Nature,* Gold pointed out that a perfect sphere, when spinning, "would possess no stability of its axis of rotation at all; the smallest beetle walking over it would be able to change the axis of rotation relative to markings on the sphere by an arbitrarily large angle; the axis of rotation in space would change by a small angle only."

If the Earth were entirely rigid, Gold argued, the equatorial bulge would seriously inhibit any change in the axis. But, he pointed out, the interior is plastic—indeed it is because of this plasticity, this ability to flow, that the bulge exists as a by-product of the Earth's spin. The extent of the bulge is almost exactly what it would be if the Earth were, in fact, fluid. Therefore if a mass of ice accumulated on a continent midway between the Equator and one of the poles, this would exert a substantial disruptive effect on the spin. The axis would slowly change as the equatorial bulge adjusted toward the new spin axis.

Gold proposed that, from time to time, there are changes in the distribution of matter on or within the Earth sufficient to induce a drift of the poles. The speed of such drift, he added, depends on the rate at which material within the Earth can flow, but he estimated the duration of each shift at between a hundred thousand and a million years. This, to him, seemed a more plausible explanation of ancient climate changes than continental drift. The polar shifts, he said, apparently have occurred at intervals of about 50 million years, bringing about the great changes in climate and evolution of species that define the periods into which geologic time is divided.

"The occurrence of continental drift over great distances," Gold wrote, "would imply new and surprising data about the construction of the Earth and in particular its crust; while the occurrence of wandering of the poles over great distances would fit in well with all that is known about the Earth, and would reaffirm what can already be inferred from other data."

Actually the idea of polar drift had a venerable tradition. In 1876 Lord Kelvin supported the idea:

> ...we may not merely admit, but assert as highly probable, that the axis of maximum inertia and axis of rotation, always very near one another, may have been in ancient times very far from their present geographical position, and have gradually shifted though 10, 20, 30, 40, or more degrees without at any time any perceptible sudden disturbance of either land or water.

Wegener, who also believed in such polar wandering, said, however, that it occurred sufficiently fast that the Equatorial bulge could not adjust to a new spin axis before there was extensive flooding. The oceans inundated Equatorial lands, providing a watery bulge around the Equator until the shape of the Earth adjusted to the new spin.

Another who believed the poles may not always have been where they are today was Paul Siple, one of America's most experienced polar explorers. As a Boy Scout selected by national competition, he accompanied Admiral Richard E. Byrd on his first Antarctic expedition in 1929 and made numerous trips south in subsequent years, serving as leader of the Little America and South Pole outposts.

In 1946 he and I voyaged to Antarctica together, and during the six-week journey we spent long hours in his cabin going over maps of the world on which he picked out what he thought were the locations of former poles. His candidates were points surrounded by circular features, such as the curved island arcs of Indonesia or the Aleutians. These, he thought, were formed by the centrifugal effect of the Earth's spin around ancient poles.

As will be seen, a very different explanation for the island arcs has now emerged, but the close-up pictures of Mars sent to Earth by Mariner 9 in 1971 and 1972 revealed a series of concentric features around both Martian poles. They were evident nowhere else on Mars. Bruce C. Murray and Michael C. Malin at the California Institute of Technology pointed out that these features were not exactly centered on the present poles and proposed that they may manifest past changes in the Martian spin axis. The Mariner photographs show gigantic volcanoes near the Martian equator, one of them large enough to cover the East Coast of the United States from Washington, D.C., to New York City. These and a mammoth rift valley along part of the Equator indicate movements of material within Mars that could have altered the planet's spin.

What was probably the most ambitious exploration of the polar wandering hypothesis ever undertaken was that of two men who, like Siple, were not members of the scientific "establishment." They had set out to explore the frightening predictions of Hugh Auchincloss Brown, although they soon laid aside the idea of any sudden lurch of the Earth's spin. Their study, by two "amateurs," might have been totally ignored, had it not been for the sympathetic support of Albert Einstein, father of relativity theory, whose own openmindedness had made possible one of the great conceptual revolutions of all time. One of the two men, Charles H. Hapgood, was a New Hampshire history teacher who, after being graduated from Harvard, studied at the University of Freiburg, specialized in the history of science, and became an associate professor of history and anthropology at Keene Teachers College in New Hampshire. His coworker was James H. Campbell, an engineer who had helped develop the Sperry gyroscopic compass and hence was familiar with the laws governing spin stability. He was 84 years old in 1958 when an account by Hapgood of their joint effort was published under the title *Earth's Shifting Crust*.

In an introduction to the book Einstein, by then in his mid-70s, wrote:

> I frequently receive communications from people who wish to consult me concerning their unpublished ideas. It goes without saying that these ideas are very seldom possessed of scientific validity. The very first communication, however, that I received from Mr. Hapgood electrified me. His idea is original, of great simplicity, and—if it continues to prove itself—of great importance to everything that is related to the history of the Earth's surface.

The great physicist not only corresponded extensively with Hapgood, making a variety of suggestions and comments, but proposed that Hapgood, to facilitate his further research, be appointed to the Institute for Advanced Study in Princeton, then headed by J. Robert Oppenheimer. This was turned down, whereupon Einstein proposed Hapgood for a Guggenheim Fellowship—a suggestion which was also rejected.

While the starting point of the Hapgood-Campbell investigation was Brown's concept of a catastrophic tumbling of the Earth, the two men, as already noted, soon decided that changes in the spin axis must be slow, requiring thousands of years, and that it was only the crust that moved, like the loose skin of an orange slipping intact over the fruit inside. Such motion of the rigid crust over the nonspherical interior, they concluded, would produce a network of ruptures in the crust, and they cited as evidence for this a dramatic discovery of the early 1950s. It had been found that the network of ridges bisecting the oceans are, for thousands of miles, cleft by deep rifts. The global pattern was reminiscent of the cracked shell of a hardboiled egg. This discovery, which proved to be an initial step in the emergence of a comprehensive, generally accepted theory to explain salient features of the Earth's crust, will be described more fully later.

To Hapgood this network was a consequence of the crustal slippage envisioned in his theory. He pointed out that the ice covering Antarctica, a sheet larger than Europe and thousands of meters thick, is not centered on the South Pole. He calculated its center of gravity to be 555 kilometers (345 miles) from the polar axis. While he recognized that the crust of the Earth sinks under an ice load and that such subsidence should eventually lower the center of gravity until it is in equilibrium with other parts of the spinning Earth, he believed the ice is accumulating so fast that such slow adjustment could not keep pace with it.

The most recent migrations of the poles took place 8000 and 18,000 years ago, he said, and the next one is likely to come in from 10,000 to 15,000 years. However, he proposed that the displacement of the Antarctic's icy center of gravity toward eastern India is already exerting sufficient pressure in that direction to have caused two devastating earthquakes there in 1897 and 1950.

On January 27, 1955, Hapgood and Campbell presented their hypothesis before a group of specialists assembled for that purpose at the American Museum of Natural History in New York.

Walter H. Bucher, professor of geology at Columbia and former president of the Geological Society of America, acting as chairman of the group, questioned one critical assumption of the theory, namely, that snow is accumulating in Antarctica and adding to its ice sheet faster than that sheet is being depleted by various factors including the breaking off of icebergs. He pointed out that measurements of snow accumulation near the coasts could not be taken as an indication of total accretion (it was subsequently found that snowfall along the coast is often ten times what it is inland).

There was also evidence that the ice sheet was once thicker than it is today, at least in some areas, for scars presumably produced by a flowing ice sheet showed on rocky summits high above the present level of ice. Furthermore, Bucher said he found it hard to believe that the deep valleys with U-shaped cross sections typical of those gouged by glaciers could have been carved in the brief time (8000 years) since the last shift of poles in Hapgood's timetable. Hapgood's reply was that such evidence of past ice action could have been produced in another period when Antarctica experienced an earlier sojourn in one of the polar regions.

Bucher's objection was raised by other critics of Hapgood's hypothesis and subsequent findings in the Antarctic gradually made it clear that, while the ice sheet there has grown and shrunk a number of times, it has existed for at least five million years. It is to Hapgood's credit that, in his revised presentation, *The Path of the Pole*, published in 1970, he shelved the idea that growth of the Antarctic ice cap was a likely cause for the polar movements that he envisioned. However, he continued to argue for gross changes in locations of the poles during the relatively recent ice ages, placing the North Pole, for example, in the Hudson Bay area only 20,000 to 50,000 years ago.

Although today the concept of continental drift—more properly of independently moving plates that carry continents with them—predominates, the idea of gross changes in the spin axis, relative to the Earth's crust, is far from dead. They are seen by some as an integral part of the Earth's ever-changing geography. But the movement is far slower than that envisioned by Hapgood and, most recently, has been in a direction opposite to what he proposed (that is, from Siberia to its present position, rather than from Canada).

Chapter 3

Colliding Worlds and Other Catastrophes

OF ALL CONTEMPORARY THEORIES TO EXPLAIN THE PRESENT STATE OF OUR PLANET and its more recent history, none has probably had so popular an appeal—at least in the United States—as that of Immanuel Velikovsky. Nor has any invoked such wrathful indignation from the scientific community. When his book, *Worlds in Collision,* appeared in 1950 it produced upheavals in the publishing world almost as dramatic as the theories promulgated by the book itself. So furious was the inveighing of scientists against it that, like a show "banned in Boston," it quickly became a best-seller.

Velikovsky, by any mode of measurement, was an extraordinary man. He was born in Vitebsk, Russia, in 1895, and decided early in life to study medicine. This was difficult, for he was Jewish and in Czarist Russia medical education for a Jew was hard to come by. He went briefly to France to study and, after a visit to Palestine, continued his premedical education in Edinburgh. On a visit to Russia in 1914 he was trapped by the outbreak of World War I and resumed his medical training there, receiving his degree from the University of Moscow in 1921.

His heart, however, was almost as much with Jewish culture as with medicine. While he was doing his postgraduate studies in Berlin, from 1921 to 1923, he met Chaim Weizmann, later to become Israel's first president, who had undertaken the establishment of a Hebrew University in Jerusalem. Velikovsky joined in this effort by co-editing two series of volumes, the *Scripta Universitatis,* containing articles by Jewish scholars and published on behalf of the burgeoning university in Jerusalem. The mathematical-physical section of the series was edited by Albert Einstein.

Weizmann, it seems, asked Velikovsky to start setting up the university in Jerusalem, but he turned down the proposal, apparently not relishing the prospect of intensive fund-raising and administration that this would entail. However, late in 1923 he and his wife moved to Palestine, where he practiced medicine and began studying psychoanalysis. He met and corresponded with Freud and contributed an article to *Imago* (the psychoanalytic journal that Freud published in Vienna) which was later published in English under the title "Tolstoy's Kreutzer Sonata and Unconscious Homosexuality." Velikovsky even decided to analyze Freud himself, so to speak, writing on "The Dreams That Freud Dreamed." He also was the first—or at least one of the first—to recognize the importance of an electroencephalogram (EEG), or print-out of electrical impulses from various parts of the brain, in diagnosing epilepsy. When he read Hans Berger's pioneering paper of 1929 on the monitoring of electrical emissions from the brain, Velikovsky saw its application to epileptic attacks, whose "lightning-like" onset he compared to the effects of an electric short-circuit. In a paper prepared in 1930 and published the following year he urged the study of epileptic seizures with an EEG and suggested the possibility of diverting, from the brain, the rapid electrical fluctuations that, he suspected, were involved.

A major change in the course of his career occurred when he and his family moved to New York in 1939 to further his research on a book dealing with Freud's three heroes: Oedipus, Akhnaton, and Moses. As Velikovsky began delving into ancient Egyptian, as well as Hebrew texts, some of the Biblical catastrophes described in *Exodus*—such as the rain of fire, plague of darkness, and parting of the Red Sea—seemed also to be reflected in the Egyptian writings. Might these, he asked himself, have been worldwide events of some terrible sort? There followed a research undertaking of formidable dimensions. He examined ancient chronicles from pre-Columbian America, China, India, Iran, Babylon, Israel, Egypt, Iceland, Finland, Greece, and Rome.

In many of them he found accounts of catastrophes that he decided had occurred coincidentally throughout the world. Finally, to explain them, he devised an admittedly extraordinary theory: that the Earth, during the period covered by these traditional accounts, had gone through a succession of cataclysmic encounters with comets and planets. The chief villain was Venus, which, he concluded, had been thrown off by Jupiter in the form of a comet that then flew an eccentric orbit, twice bringing it near the Earth. It was an idea that had features in common with one proposed by Howard B. Baker, in a series of articles beginning in 1911. As noted earlier, Baker had suggested that the Earth and Venus came close enough for the gravity of Venus to tear the Moon from what became the Pacific Basin, setting the continents in motion to open up the Atlantic Basin. His hypothetical encounter with Venus came in the Miocene Period, some 20 million years ago, whereas Velikovsky saw it as much more recent, accounting for many of the Biblical catastrophes. Hydrocarbons in the form of naphtha from the "comet tail" of Venus, he said, fell on the Earth, causing a rain of fire; the Earth's spin axis tumbled so that the Sun seemed to stand still in the sky, as recounted in *Joshua*; seas swept the lands and the Red Sea was drained briefly. Venus, at one point, collided with Mars, he said, which also repeatedly came near the Earth before the planets settled into their present orbits.

Velikovsky interwove his interpretations of ancient history with recent scientific discoveries. The fact that the Earth's magnetic polarity occasionally has reversed itself—a phenomenon that would cause the north-pointing needle of a compass to point south—he saw as evidence that when comets or planets almost collided with this planet, electric discharges took place between the two bodies, producing lightning bursts sufficient to reverse the Earth's magnetic polarity. Like Hugh Brown, he argued that the spin axis has repeatedly swung to new positions, shifting both geographic locations of the poles and the aim of the axis toward the stellar constellations.

In the works of Herodotus, Velikovsky found an account which he cited as evidence that, on occasion, the world has changed its direction of spin. Herodotus said that the Egyptians and their priests, in recounting their history, told him, "that the time from the first king to that priest of Hephaestus, who was the last, covered three hundred and forty-one generations of men....Four times in this period (so they told me) the sun rose contrary to his wont; twice he rose where he now sets, and twice he set where now he rises..."

(However, in citing this passage, Velikovsky omitted the following statement, which was inconsistent with the calamities that he associated with sudden changes of spin. Herodotus wrote: "yet Egypt at these times underwent no change, neither in the produce of the river and the land, nor in the matter of sickness and death.")

Like other advocates of spin-axis tumbling, Velikovsky said the presence of coal in the polar regions and the evidence for former ice sheets in the tropics testified to radical climate changes in the past. He pointed out that recent probing of mid-ocean sediments had shown them surprisingly thin—not the deep accumulations one would expect had the oceans lain peacefully in place for millions or billions of years.

Velikovsky told his tale with a prose style that was Biblical in its sweep. Summarizing his vision of the past, in his book, *Earth in Upheaval,* he wrote of

> hurricanes of global magnitude, of forests burning and swept away, of dust, stones, fire and ashes falling from the sky, of mountains melting like wax, of lava flowing from riven ground, of boiling seas, of bituminous rain, of shaking ground and destroyed cities, of humans seeking refuge in caverns and fissures of the rock in the mountains, of oceans upheaved and falling on the land, of tidal waves moving toward the poles and back....

In these upheavals new mountains were thrust up, he said, and the flat-lying strata of rock were compressed into the contorted folds visible in cliffs; the sand, gravel, silt, and erratic rocks attributed to past ice ages were strewn across the world by great floods;

> ...and animals were swept to the far north and thrown into heaps and were soaked by bituminous outpourings; and broken bones and torn ligaments and the skins of animals of living species and of extinct were smashed together with splintered forests into huge piles; and whales were cast out of the oceans onto mountains; and rocks from disintegrating mountain ridges were carried over vast stretches of land, from Norway to the Carpathians, and into the Harz Mountains, and into Scotland, and from Mount Blanc to the Juras, and from Labrador to the Poconos; and the Rocky Mountains moved many leagues from their place, and the Alps traveled a hundred miles northward....

Velikovsky cited a puzzling astronomical observation, made in 1922, that "a considerable amount of heat" is emitted by the night side of Venus—a conclusion drawn from observations of the cloud tops, since the unbroken cloud cover of that planet does not permit direct viewing of its surface. From other observations it was assumed that Venus rotates very slowly, if at all, relative to the sun so that for prolonged periods one side is in sunlight and the other side is in darkness. Why, then, should the night side appear almost as warm as the sunlit side?

"The night side of Venus radiates heat," said Velikovsky, "because Venus is hot." Petroleum fires "must be burning there," he said, and he also proposed that the planet, beneath its clouds, gives off heat as an aftermath of its close passages of the Sun and heat-generating encounters with Earth, Mars, and probably Jupiter. Although Velikovsky apparently was unaware of it, a decade earlier Rupert Wildt, then at Princeton University, had proposed that the surface temperature of Venus must be above the boiling point of water. He assumed there was sufficient carbon dioxide in the Venus air to act like the glass of a greenhouse, permitting sunlight to reach the surface but impeding the escape of heat radiated at longer wavelengths. Wildt's argument was seized upon and publicized, at the time, by Sir Harold Spencer Jones, Britain's Astronomer Royal.

Velikovsky realized that the wild trajectories of Venus and Mars postulated in his theory were difficult to reconcile with the laws of planetary motion set forth by Sir Isaac Newton. According to these laws the planetary orbits around the Sun are essentially under control of solar gravity. Velikovsky, however, argued that magnetic fields powerful enough to change such orbits exist within the solar system.

He believed, for example, that the Earth's magnetic field, which on the Earth's surface is capable of moving only a delicately balanced compass needle, becomes strong enough above the atmosphere to reach across space and influence the librations, or rocking motions, of the Moon.

As early as 1946 John J. O'Neill, science editor of the *New York Herald Tribune,* published an account of Velikovsky's thesis. It was derived, O'Neill said, from "a magnificent piece of scholarly historical research" which showed a striking correspondence among cataclysmic events described in Sumerian, Chaldean, Hindu, Chinese, Mayan, Aztec, Icelandic, Egyptian, and Hebrew records. He described it as "a stupendous panorama of terrestrial and human history which will stand as a challenge to scientists... ." However, book publishers were frightened by the sensational character of Velikovsky's theme and it was not until 1950 that the first book-length exposition, *Worlds in Collision,* was published by the highly respected firm of Macmillan and Company. The book came out shortly after *Harper's Magazine* had printed an article on his theory by Eric Larrabee, one of the magazine's staff writers, entitled "The Day the Sun Stood Still."

The popular enthusiasm for Velikovsky's reconstruction of past events was greeted by the scientific community with almost unanimous dismay. Typical was the comment by Chester Longwell, the Yale geology professor, following a review of the book in *The American Journal of Science,* of which he was editor:

And why, readers may well ask, should a scientific journal give the least attention to such patent nonsense? Frankly, our chief concern is to focus attention on the publisher, rather than on the book or the author. But doesn't a publishing house have the right to print any kind of literature that comes within legal bounds? Yes, so long as the literature is represented for what it is. The Macmillan spring catalogue lists the Velikovsky book under the heading "Science" along with four other books whose titles suggest that they may be properly classified. The four authors must feel much flattered at finding themselves in company so distinguished!

We are given advance notice that *Worlds in Collision* is to be followed shortly by more of the same kind. Under what appropriate heading shall we advise that the publisher classify, in future, works that do not rise to the level of good science fiction, and are best described as burlesque of both science and history? If we are to judge from the tactics employed in advertising the subject of this review, its publisher is concerned chiefly with the classification "best seller."

As more and more scientists—particularly the astronomers—heaped ridicule on the book and made ominous threats against Macmillan (whose textbook sales were a lucrative source of income), the publisher became increasingly alarmed. Looking back on this episode, Harold Strong Latham, editor-in-chief of Macmillan at the time, wrote: "To the hundreds, perhaps thousands, of letters the publisher received, replies were sent out stressing the fact that never for a moment had *Worlds in Collision* been promoted as a contribution to science but rather that its entertainment values, its readability, had always been stressed." Latham said he liked to think that the book appealed to readers of the day as had the works of H. G. Wells and Jules Verne—"much, too, as many publications of a later time were to find wide acceptance as science fiction."

However, as noted by the indignant Professor Longwell, Macmillan had listed the book in its catalogue under "Science."

The more it sold, Latham said, "the more vigorously these objectors fought it. It was described as an insult to science and its suppression was demanded.... Professors who headed science departments in great universities refused to consider the adoption for college use of textbooks issued by the publisher of Velikovsky, 'We just could not have confidence in their scientific worth,' was their hue and cry.... Ultimately the Company's business end injected itself into the argument with the suggestion that the Trade Department explore the possibility of transferring the work to another publisher which did not have a textbook department. We could no longer jeopardize the standing of one of the most profitable departments of the business.

"After some little persuasion Dr. Velikovsky agreed that a transfer might be made, and arrangements were soon completed with Doubleday to take over the book. This had not been difficult to accomplish, for *Worlds in Collision* was a 'going' item; indeed, at the time of the transfer, it was at the top of the best-sellers list in nonfiction.

"Of course the excitement in the Macmillan offices during these days of controversy was high, and to put a stop to all the speculation and gossip I called the Trade staff together and informed its members that as of that day Velikovsky was no longer published by Macmillan. If anyone asked why, the answer was to be 'We know nothing!'"

Meanwhile, one man on the fringes of the astronomical community made the mistake of taking Velikovsky seriously. He was Gordon Atwater, head of the Hayden Planetarium at the American Museum of Natural History in New York, where Hapgood's theory of crustal slippage also came up for scrutiny. Atwater was impressed by *Worlds in Collision,* even though he did not agree with everything in it, and he went so far as to plan a planetarium show depicting the dramatic sequence of celestial events described in the book. The show was never produced. Apparently he was already in trouble with his superiors and this was the last straw. He was summarily dismissed.

It was inevitable that there should be a "backlash." A handful of physical scientists stepped forward and said Velikovsky merited at least a polite hearing. But a much larger group of non-scientists vented their spleen against the "Scientific Establishment" by accusing it of arrogance, misrepresentation, and tyranny. Virtually one entire issue of a magazine called *The American Behavioral Scientist* was devoted to the subject. Velikovsky's proponents argued that the predictions of his theory were being confirmed right and left. He had said that Venus must be hot, and radio observations from Earth and, subsequently, space probes showed that its surface temperature is 400 to 500 degrees Centigrade (750 to 930 degrees Fahrenheit). He had said that, since the atmosphere of Venus was a residue of the material that rained fire on the Earth, it must be "rich with petroleum gases and hydrocarbon dust." Whereupon, it was reported (unofficially) that a space probe had confirmed the existence of such an atmosphere. The planet Jupiter "is cold, yet its gases are in motion." Velikovsky told an audience at Princeton in 1953, "It appears probable to me that it sends out radio noises as do the sun and stars." Two years later pulses of radio emission from Jupiter were detected. He said there should be great magnetic fields in space and they, too, allegedly were discovered.

The reaction of most scientists to the theory was seen, by *The American Behavioral Scientist,* as manifesting panic on the part of the Establishment, lest doctrines in which leaders of that Establishment had a vested interest be overthrown. Velikovsky had arrived at his conclusions, the magazine said, through techniques of the social scientist—the study of historical records. And the message was clear: his method had led him to the truth, whereas the physical scientists had become wedded to stale, outmoded hypotheses.

Alfred de Grazia, professor of government at New York University and editor of the magazine, wrote that "The associations of science are still among the primitive and puerile mechanisms of modern life." After describing how scientists had trampled on Velikovsky's arguments, he said, without qualification:

> He has shown that the present order of the solar system is quite new and that unaccounted forces help govern it. He has struck at a great part of the Darwinian explanation of evolution. He has upset several major theories of geology and offered substitutes therefor. He found space a vacuum and made of it a plenum. While his ideas are not at all beyond criticism, as a cosmogonist he appears in the company of Plato, Aquinas, Bruno, Descartes, Newton and Kant.

The *Bulletin of the Atomic Scientists,* in its April 1964 issue, counterreacted with outrage. "The pages of the *Behavioral Scientist* report," it said, "spill over with anger and passion, and good judgment quickly falls by the way."

Among the few "legitimate" scientists who stood up for Velikovsky's right to be given a hearing was Lloyd Motz, an astronomy professor at Columbia University. He wrote to *Harper's Magazine* in October 1963: "I do not support Velikovsky's theory but I do support his right to present his ideas and to have these ideas considered by responsible scholars and scientists as the creation of a serious and dedicated investigator and not the concoctions of a charlatan seeking notoriety...."

Einstein, who was enthusiastic about Hapgood's polar wandering hypothesis, was indignant at Velikovsky's treatment by the scientific community. Apparently Velikovsky, aware of the help given Hapgood, hoped to receive some form of endorsement from Einstein, a fellow Zionist and neighbor in Princeton.

Velikovsky gave him *Worlds in Collision* and, shortly before Einstein's death, a copy of *Ages in Chaos.* In a gracious note of thanks for the latter gift Einstein pointed out that the book did not tread on the toes of theoretical physicists like himself:

> I look forward with pleasure to reading the historical book that does not imperil the corns of my guild. How it stands with the corns of people in other scientific fields, I do not as yet know. I think of that

touching prayer: Holy Saint Florian spare my house, set fire to other ones! [Saint Florian was said to offer protection against fire] I have already read carefully the first volume of the memoirs to "Worlds in Collision" and have added some easily erased marginal notes. I admire your dramatic talent....

Einstein hated to give offense, but soon after sending his letter of thanks to Velikovsky—and only a couple of weeks before his death—Einstein gave a long interview to I. Bernard Cohen, a Harvard science historian, an account of which was published in *Scientific American.*

"The subject of controversies over scientific work led Einstein to take up the subject of unorthodox ideas," Cohen wrote. "He mentioned a fairly recent and controversial book, of which he had found the non-scientific part—dealing with comparative mythology and folklore—interesting. 'You know,' he said to me, 'it is not a bad book. No, it really isn't a bad book. The only trouble with it is, it is crazy.' This was followed by a loud burst of laughter."

Einstein deplored the pressure brought to bear on the book's publisher by some scientists. He felt, wrote Cohen, that such a book "really could not do any harm, and was therefore not really bad. Left to itself, it would have its moment, public interest would die away and that would be the end of it. The author of such a book might be 'crazy' but not 'bad,' just as the book was not 'bad.' Einstein expressed himself on this point with great passion."

It was, however, Harry H. Hess, one of the nation's leading geologists and head of the geology department at Princeton University, who persisted in demanding that Velikovsky be given a fair hearing. Hess had come to know him after Velikovsky moved to Princeton in 1952 and began using the library of the university's geology department in his research for *Earth in Upheaval.* In 1956 Velikovsky proposed several inquiries for the forthcoming International Geophysical Year of 1957–1958 that might lend support to his ideas, such as whether the magnetic field of the Earth is stronger above the atmosphere than on the Earth's surface, whether the rocking motions of the Moon are linked to relative movements of the Earth's magnetic poles, and whether there is evidence for reversals of the Earth's magnetic field in Biblical times. Hess agreed to pass on these suggestions to the I.G.Y. leaders.

In 1963, when Hess had become chairman of the Space Science Board of the National Academy of Sciences, he again agreed to pass along some of Velikovsky's proposals for space experiments. However, in that same year he wrote to Velikovsky: "We are philosophically miles apart because basically we do not accept each other's form of reasoning—logic. I am of course quite convinced of your sincerity and I also admire the vast fund of information which you have painstakingly acquired over the years...." He cited some of Velikovsky's seemingly successful predictions, but said: "I am not about to be converted to your form of reasoning though it certainly has had successes."

Hess obviously was impressed with Velikovsky as a man of deep sincerity, with an extraordinary command of history, legend, and the scriptures of many faiths. He realized that Velikovsky's ideas could serve as a challenge to sharpen the analytical wits of students and in 1965 he inaugurated a discussion group, known as Cosmos and Chronos, to study such subjects as "catastrophes and Earth history." The idea spread to other campuses. The Carnegie Institute of Technology and the University of Pittsburgh jointly invited Velikovsky to speak. As a symbol of revolt against scientific and academic orthodoxy, he became a popular campus lecturer. Eventually he was even asked to speak at one of the country's leading space centers—the Ames Research Center of the National Aeronautics and Space Administration near the south end of San Francisco Bay. But not all his appearances were without controversy. In 1967, at the invitation of the Rittenhouse Astronomical Society of Philadelphia, he was scheduled to speak in the auditorium of the Franklin Institute in that city. However the meeting place was shifted abruptly to the nearby auditorium of the Philadelphia Free Library. The Franklin Institute, with its links to that father of American science, Benjamin Franklin, apparently had decided it wanted to have nothing to do with the controversial Velikovsky.

When he was invited to speak by the Princeton Section of the American Institute of Aeronautics and Astronautics in 1966, this writer went to listen. Addressing an overflow audience in the auditorium of the university's Woodrow Wilson School of Public and International Affairs he thundered like an Old Testament prophet. In a sonorous, slow, grim manner he warned that, while the planets of the solar system have now settled into stable orbits, there is no stability on Earth. Man harbors in his collective soul the memory of horrendous events in the past and seems to be trying to match them with the development of weapons that can bring a new holocaust.

From his studies with Freud and other psychologists he had concluded that mankind, through the mental trait known as denial (in which a person avoids thoughts of unpleasant things, such as one's own death), had pushed the memory of past cataclysms into hidden chambers of the human soul—"collective amnesia" he called it.

The Velikovsky controversy simmered into the 1970s. A magazine called *Pensée,* published by the Student Academic Freedom Forum in Portland, Oregon, devoted one entire issue in 1972 to the subject (it was largely sympathetic to Velikovsky) and promised to continue the discussion through 10 issues, into the 1973–1974 academic year. The first article, written by David Stove, a lecturer in philosophy at the University of Sydney, and originally published five years earlier, was entitled "The Scientific Mafia."

Editorially the magazine accused the scientific community of "libel and character assassination" against Velikovsky. It cited "the rude failure to acknowledge his correct prediction of 'surprising scientific discoveries,' " and said his work had been subject to "farcical criticism." These "black marks," it added, "cast a disconcerting pall over the achievements of modern science."

Thus the affair was fueled by contemporary hostility to science in general, and the scientific community had to share some of the blame, for to most scientists the arguments of Velikovsky were too absurd to merit detailed refutation. To nonscientists unfamiliar with those details, his "predictions" seemed convincing, although on closer examination they proved less impressive.

For example, from what is now known of the Venus atmosphere it would be most remarkable if its surface temperature were not oven hot, for the planet is covered with a massive blanket of carbon dioxide, producing a surface air pressure about 100 times that on Earth. This gas blanket traps solar heat and keeps both the day and nights extremely hot—even hotter than predicted, from such a "greenhouse effect," by Rupert Wildt in 1940.

While Jupiter does produce radio emissions, they have no relationship to emissions of the type generated by the Sun and other stars, as proposed by Velikovsky. The surprising discovery, with regard to magnetic fields in space, was not that they exist, but that the Earth's magnetic field, far from becoming more intense above the atmosphere, as Velikovsky expected, terminates in a well-defined boundary, the "magnetopause," about one quarter of the way to the orbit of the Moon except directly opposite the Sun, where it streams out into a long tail. This magnetic envelope is blown into an extremely elongated teardrop by thin, high-velocity gas flying outward from the Sun—the "solar wind." This wind carries with it ever-changing magnetic fields so weak only the more sophisticated spacecraft instruments can detect them.

The report of a hydrocarbon atmosphere on Venus was also spurious. The Soviet probes found the Venusian air to be formed of almost pure carbon dioxide. Velikovsky's citation of evidence in the rocks that the magnetic poles of the Earth have reversed position has little bearing on his hypothesis of catastrophic events in historical times, for, as will be seen, these reversals took place long before his hypothetical catastrophes.

His prediction that the Moon's surface, less than 3000 years ago, was "repeatedly molten" has been disproven by the various analyses performed on samples from all the lunar landing sites. In no case was a rock found that had been molten within the past three billion years (despite Velikovsky's unconvincing efforts to refute those findings). If the Moon has been unaffected by external catastrophe for so long a period (apart from meteorite impacts), the changes that have occurred to the Earth's landscape must be almost entirely of internal origin.

Another pillar of his case, and that of other adherents of cataclysmic lurches of the Earth, has been the argument that "quick-freezing" of animals, large and small, in the Arctic testified to an abrupt climate change. While Hapgood favored a slower shift of the axis, he too made much of the discovery of such things as a mammoth, frozen intact, with a last mouthful of buttercups between its teeth.

Velikovsky cited, in particular, the remains of extinct creatures uncovered in frozen muck during placer mining in central Alaska. In these operations—later largely suspended—an entire valley floor was stripped of its frozen soil to expose the gold-bearing gravel underneath. This was done with water jets so powerful they could demolish a house in seconds. During the summer months the top few inches of thawed muck were swept away by jets playing back and forth across one area. Then that sector was left to thaw for a few days while another part of the valley floor was attacked.

In this process, the remains of animals, trees and plants, long buried in the frozen muck, were exposed. They included mammoth, mastodon, lion, camel, super-bison, horse, and two extinct forms of musk ox (bootherium and symbos), all of which had vanished from the Americas when Columbus arrived, plus moose, lynx, caribou, and musk ox.

"Under what conditions did this great slaughter take place," Velikovsky asked, "in which millions upon millions of animals were torn limb from limb and mingled with uprooted trees?" He concluded that the Pacific and Arctic Oceans had repeatedly swept across the land. But, he asked, what could have caused these waters to leap from their basins, "and wash away forests with all their animal population and throw the entire mingled mass in great heaps scattered all over Alaska?"

Hapgood was not quite so melodramatic, but, in his book, *Earth's Shifting Crust,* he accepted the view that a sudden climate change must have occurred. He said, "...the discovery of complete bodies of mammoths and other animals in Siberia, so well preserved in the frozen ground as to be in some cases still edible, seems to argue a cataclysmic change."

In particular he cited the Berezovka Mammoth, the largely intact remains of which, stuffed and partially restored, repose in the Museum of the Zoological Institute of the Soviet Academy of Sciences in Leningrad. It was discovered in 1901 in the eroding banks of the Berezovka River in northeast Siberia and an expedition was dispatched by the Imperial Academy of Sciences to study it. When the expedition reached the remote site, exposed parts of the animal had allegedly been eaten by wolves, although some scientists suspect either that the putrefied flesh tended to fall apart as it thawed, or that the eating was done by ice-age wolves before the monster was fully buried. The Russians built fires to thaw the remainder of the carcass and, as the thaw proceeded, the stench was almost unbearable.

The reports state that a mixture of grass and sedge was in the animal's mouth and that its stomach contained a wide variety of plant remains, many of them typical of an Arctic grassland. They included, among other things, a form of buttercup. The state of flowering and fruiting of these plants suggested that the animal's last mouthful was cropped in late July or early August.

Hapgood's thesis, in *Earth's Shifting Crust,* was that volcanic eruptions, as a byproduct of slow slippage of the Earth's crust, suddenly clouded the skies and led to a sharp drop in temperature. The Berezovka Mammoth, he wrote, "is feeding quietly in the grassy meadow, and he has just swallowed a mouthful of buttercups, and has gathered up, with his trunk, a new mouthful of wild beans. The temperature is warm, and there is no sign of what is about to occur." Then, in Hapgood's hypothetical account, with sunlight cut off abruptly by volcanic ash in the sky, the weather quickly chilled and the animal froze to death.

In *The Path of the Pole,* published 12 years later, Hapgood was not so melodramatic. Indeed the concept of widespread quick freezing of animals does not stand up in the light of what is now known. The chaotic jumbles of bones and trees uncovered by placer-mining can be explained without resort to oceanic invasions of the hinterland. As soon as those devastating water jets work loose a fragment of bone or wood it is hurled across the surface until, often, it ends up in a tangle of branches and other debris in some sheltered spot.

Furthermore Troy Péwé, who, at the University of Alaska, studied those processes that altered ice age landscapes, notes that the "elephant graveyards" where many bones of extinct members of the elephant family are found lie where streams converged and presumably carried down, from the heights, a variety of debris, including animal remains and fragments of vegetation.

But the most powerful evidence against any sudden quick freeze lies in the recently obtained ages of those frozen remnants collected during the period when the placer mines of central Alaska were active.

In the summer of 1935, I worked in Alaska for New York's American Museum of Natural History, combing the mines around Fairbanks for bones and other ice-age relics uncovered by the thawing operations. Our base of operations was the University of Alaska at College, and we tried to visit each of the mines at least once every week or two, hoping in that way not to miss too many of the bones exposed by each removal of a new muck layer.

These valleys were heavy with a pungent, rather evil, odor from the resumption of decay in vegetable matter frozen for thousands of years. In no case can I remember finding even the partially intact skeleton of a large animal. The bones were dismembered and scattered across the thin veneer of mud overlying the still-frozen ground below.

My trophy of the summer was the lower part of a super-bison leg with its fur, tendons, hoof, and some flesh intact. I boasted later that we could have made a "gamey stew" from it, but I doubt that the world's greatest chef could have made it palatable.

In some areas excavation of the valley floor had exposed prehistoric stream beds with the turf along their banks still intact (although black, matted, and malodorous). Here and there, on these banks, were circular patches, the size of a dinner plate, which, when cut around the edge with a hunting knife, could be lifted to expose a ground squirrel nest. Inside, almost always, we found a squirrel family curled up for its winter nap. A tiny tunnel led uphill to a chamber full of grain. Another tiny tunnel led downhill to a latrine. And in a number of cases there were still bits of fur, even flesh, clinging to the skeletons.

My guess at the time was that during the last ice age some winter was simply too severe for these little creatures. They obviously had not starved with grain still on hand. Péwé believes that debris, washed down slopes by snowmelt, mudslides, or accumulations of melt-water, entombed or drowned these animals in their nests. Even some of the mammoths, he thinks, were buried and frozen in this way. Others may have fallen into pits of muddy water that eventually froze; some died in summer, but in the cool climate decayed only in part (or were partially devoured by scavengers) before mudslides and winter finally enveloped them to repose in nature's deep freeze until, thousands of years later, mining operations or river erosion (as along the banks of the Berezovka River) thawed the ground and exposed them to view.

Central Alaska has not produced almost intact mammoths such as those found in Siberia, although in 1948 the remains of a baby mammoth were found near Fairbanks, its head, neck, trunk, and one almost hairless foreleg still intact. A few years later the skull of an enormous adult was uncovered, with some hair preserved, as well as tusks four meters long.

In Siberia the finds have included numerous partially preserved ice-age rhinoceroses—the animals apparently never crossed the land bridge of Alaska, possibly because pasturage along the route was unsuitable. But it was the great herds of mammoths that dominated the ice-age scene. In the past 250 years bones, tusks, and other remains of some 117,000 of these monsters have been found in Siberia alone. Thousands of mammoth teeth have been dredged up off Norfolk, England, and giant teeth are also commonly brought up by trawlers off the East Coast of the United States—relics of a time when much of the world's water was locked in ice sheets, sea levels were scores of meters lower, and herds of quadrupeds wandered what is now submerged continental shelf.

Velikovsky assumed that these animals perished in one or more cataclysmic floods. Those found in Alaska, he said, died "in rather recent times"—either at the end of the ice age or in the millennia that came soon thereafter. But he recognized that when they died was critical with regard to his hypothesis. "A problem the archaeologists will have to solve," he said, "is that of clarifying whether the extermination of life in these regions of northwest America and northeast Asia, resulting in the death of mammoths, took place in the seventh and eighth or 15th century before the present era (or earlier)—in other words, whether the herds of mammoths were annihilated in the days of Isaiah or in the days of the Exodus."

It has been more than 20 years since he formulated this question, and there have now been many age determinations, using the decay of a radioactive form of carbon as a stopwatch, much as the decay of other elements is used to date rocks. It is known, roughly, how much of this radioactive carbon, called carbon-14, was mixed with ordinary carbon when the animal died. Carbon-14 is manufactured at a more or less constant rate in the high atmosphere by radiation from space, and, on the average, it decays into nitrogen at a fixed rate. Its abundance in the air remains uniform because of its constant replenishment. But once it is incorporated into flesh or bone, this replenishment ends, and the extent to which it subsequently has decayed, relative to the other stable forms of carbon, can be used to estimate how much time has elapsed since the animal lived.

In this way the Russians have estimated that the Berezovka mammoth cropped its last mouthful some 45,000 years ago, placing its death in the early part of the last, or Wisconsin, Ice Age (thought to have extended from 73,000 years to 10,000 years ago). The ages of frozen remains found in Central Alaska are scattered rather uniformly through the last 20,000 years of the Wisconsin Ice Age. There is no hint of one or two cataclysmic events and all ages are far greater than those proposed by Velikovsky.

While such data undermine his argument, the fact remains that, from the point of view of large mammals, something rather catastrophic did happen toward the end of the last ice age. An extraordinary number of them, in all parts of the world, became extinct in a relatively short time. In the Americas, for example, 49 species whose individuals weighed more than 45 kilograms (110 pounds) vanished, including elephant-sized sloths capable of pulling down treetops and nibbling the crowns, camels with giraffelike necks likewise able to browse the upper branches of American forests, a variety of horses (some zebralike), a rodent the size of a small rhino, a herbivore as large as a hippopotamus, a giant beaver, diverse *Proboscidea* (mammoths and mastodons)—in fact, an entire bestiary of odd-looking creatures. It is estimated that extinction of the great herds of super-bison, mammoth, and horse reduced the total weight of large animals on the land by more than 90 percent.

They had survived a succession of ice ages. Why did they succumb during the terminal phase of the last one? This remains one of the most tantalizing questions in all natural history. Speculation has included epidemics, a change in climate so rapid the animals had no time either to migrate or adapt—though not at the catastrophic speed envisioned by Velikovsky and Hapgood—swift alterations of flora, and "human overkill."

The last hypothesis was advanced by Paul S. Martin of the University of Arizona. He pointed out that, beginning in Africa more than 40,000 years ago, the extinctions of large animals seem to have spread into Europe, northern Asia, across the Bering land bridge to the Americas, and, 8000 to 13,000 years ago, even to isolated Australia. He postulated that this represented the spread of big game hunting technology whose entry into the Americas, where no animal had ever had to contend with human hunters of any sort, was catastrophic. In particular, Martin pointed to the development about 11,000 years ago of the so-called Clovis points, flaked with great skill and provided with a fluted shaft to fit into a split spear pole. (They take their name from the Clovis archeological site in New Mexico.)

Were Clovis points and similar spears the first weapons of mass destruction? Were they responsible for the first catastrophic impact of man upon his environment? Martin's "human overkill" idea is disputed by a number of other scientists. Some argue that man reached the Americas long before the extinctions. Others ask how a few thousand inhabitants, scattered across the whole of North America, could have wiped out millions of large mammals? Why did not a few survive—enough to keep the species alive? Perhaps, some say, man was responsible in that he applied the *coup de grâce,* disposing of the few survivors of some other catastrophe.

Apparently on at least some occasions early man was a mass killer. At Solutré, the French archeological site that gave its name to the Solutrean Period of prehistory, the remains of an estimated 100,000 horses have been found in a single deposit. K. K. Vereshchagin in the Soviet Union has reported the discovery of "many hundreds" of skeletons left, apparently, where stone-age hunters had driven the animals into narrow canyons and slaughtered them with clubs, spears, and other weapons. On the pastures of the Dzhugut–Kala plateau herds of various grazing animals were driven over cliffs to perish on the rocks below. And another weapon of early man was the fire drive which could maim or kill thousands of animals. Such fires, on occasion, also destroyed much of the grassland on which a species was dependent.

Paul Martin argued that the big game hunters, entering North America via the Bering land bridge, were able to penetrate the continent when the main Canadian ice sheet retreated eastward toward Hudson Bay and the ice flowing off the coastal mountains of the West also diminished, opening an ice-free corridor in between. This occurred about 11,500 years ago, and the hunters found themselves in a world whose huge mammals had never seen a two-legged predator and fell easy prey to spears tipped with those masterpieces of stone chipping, the Clovis points (see Fig. 3.1).

These hunters, according to Martin's hypothesis, advanced some 15 miles each year, proliferating (since the food supply was virtually unlimited). In the advancing frontal zone, which was about 100 miles deep, the population density was considerable—about one person per square mile. In a decade or two, by the time this zone had passed a region, most of the large species had been annihilated. Behind the advancing front, therefore, the fauna lay wasted and within 1000 years the entire continent had been swept clean. So swift was the passage of the hunters that they left little evidence in the form of artifacts or kill sites.

Was this the way that large-animal extinctions occurred, not only in the Americas, but in other parts of the world? The search for an acceptable explanation continues, with careful study of pollen entombed with the remains and even of the dung left in caves by some of these creatures (which can be studied to learn the nature of their changing diets). It has already been found that Northrotheriops, a pony-sized ground

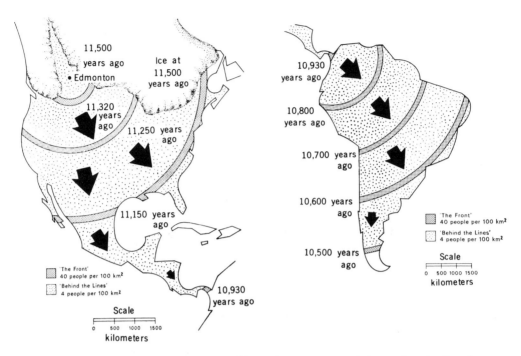

Figure 3.1 Progressive extinction of large ice-age mammals *by an advancing line of hunters, as envisioned by Paul S. Martin.*

sloth that lived in Arizona, Nevada, and New Mexico, ate such desert shrubs as mormon tea and Joshua tree. Did they thrive on these plants—so unappetizing to other browsers? In doing so were they able to convert into meat a form of vegetation that now "goes to waste" over vast regions of North America? Or was this diet a desperate effort to survive in a deteriorating environment? The answers to this question and to man's role in the great extinctions remain to be found, but they are important, for they bear, as well, on the future. It may be that man's extinction of other forms has a peculiar momentum. By now we have eliminated a large part of the world's wildlife habitats to accommodate our agriculture, our housing, our industries, and transportation networks. Is this to continue until no wild species are left? Is this to become virtually a "one-crop" planet—the single crop being *Homo sapiens*, exposed to all the epidemic threats that endanger one-crop regimes? Insofar as archeology and paleontology can throw light on man's past role, in this respect, those fields of research take on new relevance.

It is a striking feature of history that the tides of opinion have a cyclic character, whether in politics, economics, or science. Despite the outraged reactions to Velikovsky's theories, the revolt against all forms of catastrophism that was born in the 19th century had almost run its course by the 1950s, and sober scientists began to consider that on occasion rather dramatic events may have changed the face of the Earth. It was, perhaps, not surprising that one of the most sensational of these proposals was developed in the open-minded atmosphere of Harry Hess's geology department at Princeton. John T. Hollin, a British graduate student there who had been stationed at a

United States research camp on the coast of Wilkes Land in Antarctica, explored a proposition whereby swift floods almost as catastrophic as those envisioned by Velikovsky could be generated. He had read a 1964 proposal by A. T. Wilson of Victoria University in New Zealand that the ice ages might be caused by periodic "surges" of the Antarctic ice sheet.

The annual snowfall over the South Polar region adds continuously to the bulk of the ice sheet covering that continent-sized area and, while the ice flows toward the coast and breaks off into icebergs, Wilson argued (as had Brown and Hapgood) that it accumulates faster than it can be shed. As the ice becomes thicker, it increasingly bottles up the heat rising from the Earth's interior until finally the floor of the ice warms enough to produce a water layer. This lubricates the underside of the ice and, it was proposed, has periodically allowed large sections of the ice sheet to slip, all at once, into the sea.

It is known, from areas where the ice pushes far out from the coast as an apron of "shelf ice," that, no matter how thick it is on land, once water-borne the sheet spreads from 1000 meters thickness as it first becomes water-borne to 300 meters as it breaks off into the open sea as icebergs. This is the thickness of all the aprons of shelf ice around the rim of Antarctica. Hence, according to Wilson, when the ice sheet, which now averages about two kilometers thick over an area larger than Europe, slips into the sea, it too would spread across the frictionless water to a uniform thickness of 200 or 300 meters.

This vast, snow-clad crust of ice on the southern seas, when combined with the snowy landscape of Antarctica itself, would constitute a brilliant-white surface comparable in extent to all of Asia. It would reflect so much sunlight back into space that the entire atmosphere of the Earth would be cooled, starting an ice age.

Hollin pointed out that such a "surge" would produce a very rapid rise in sea level. If the entire ice sheet slipped off, he reported at a 1972 conference on possible ice age causes, the seas would rise 70 meters; but he proposed as more likely a partial slippage, producing a 20-meter rise. If this occurred, whether it took days or a few years, the results would be disastrous for most major cities and for those many food-producing areas that lie close to sea level.

He pointed out that surges of individual glaciers have been well documented. In such a situation the upper sector of a glacier, where most of its snow accumulation occurs, reaches a critical volume and then flows rapidly down its valley. The Kutiah Glacier of northern India is said to have advanced as much as 360 feet a day, and in 1934 Kenneth Mason presented to Britain's Royal Geographical Society a rather incredible account of two old ladies being overtaken by the Garumbar Glacier in the Himalayas.

The closest parallel to the hypothetical Antarctic surge that Hollin found occurred in the Bråsvellbreen, part of the Spitsbergen ice cap, which, sometime between 1935 and 1938, advanced up to 21 kilometers (13 miles) on a front 32 kilometers (20 miles) wide.

Evidence on the Antarctic coastal mountains of glacial action above the present ice level shows that the ice there has been thicker in the past, but Hollin argued that such evidence reflects short-term fluctuations along the coast. It is the great mass of ice in the hinterland that he believed may have been accumulating steadily since a great surge initiated the most recent ice age, about 73,000 years ago. He pointed out that the drilling of a hole through the ice sheet at Byrd Station, in the Antarctic hinterland, penetrated 2200 meters of ice to the rock floor and showed that, in fact, there is a water layer at the bottom. More recently probing of the ice sheet by airborne radar has even disclosed extensive "lakes" beneath it.

As support for the plausibility of surges Hollin cited the London "brickearth," once considered dramatic evidence for the Biblical flood. The bricks from which older London was built derived largely from brickearth dug along the Thames estuary. Much of it was laid down in thin layers separated by traces of vegetative remains, clearly representing a cycle of flooding and regrowth of plant life. But there was one deposit, about 10 meters thick, that seemed to have been formed all at once. In it, during the last century, were found the bones of numerous ice age animals. In one pit, alone, were the remains of more than 84 extinct elephants of every age and size, and it was reasonable at the time to assume they were victims of the Biblical deluge. It was noted by British scientists of the last century that the more fleet-footed animals, such as deer and carnivores, were scarce in these deposits, and one geologist, R. P. Cotton, suggested "that they...were generally able to avoid being submerged with their less active contemporaries by escaping to the hills...."

Unfortunately this brickearth is no longer accessible. Either the pits have been worked out or the formation has been built over in the general development of the Thames estuary.

However, Hollin pointed to the exposure, by the building boom in Washington, D.C., of the remains of a cypress swamp seven to 20 meters above present sea level that flourished during the warm period between the last two ice ages, then was submerged by what may have been a sudden rise of the sea to at least 20 meters above its present level. This would have flooded most of the area now occupied by downtown Washington, including the present site of the White House.

At the 1972 conference it was reported, from the study of oxygen samples extracted from deep within the Greenland ice, that some 89,500 years ago the climate suddenly shifted from one warmer than it is today to one characteristic of a full-fledged ice age. The climate did not remain frigid, however, and within 1000 years was warm again. Not until 73,000 years ago did the last ice age begin, according to this study. Evidence from sediments in the floor of the Gulf of Mexico also has indicated an abrupt cooling some 90,000 years ago. While this could have been caused by an Antarctic surge, it was pointed out that a similar climate change would be induced by prolonged volcanic activity that clouded the skies and cut off solar energy.

One bizarre by-product of the surge hypothesis was a fear that some reckless government, with nuclear energy at its disposal, might set such a surge in motion. This would destroy most of the urban and food-producing areas of coastal nations—meaning virtually all of the industrialized countries of the world. In the years immediately following presentation of the Wilson-Hollin hypothesis, the danger was apparently discussed at high levels of government in Washington. In 1968 Gordon J. F. MacDonald, who in that year became vice chancellor for research and graduate affairs at the University of California in Santa Barbara, pointed to the malevolent use of a surge in an analysis of ways in which Earth science could be misused for catastrophic changes of the environment:

> If the speculative theory...is correct (and there are many attractive features to it), then a mechanism does exist for catastrophically altering the Earth's climate. The release of thermal energy, perhaps through nuclear explosions along the base of an ice sheet, could initiate outward sliding of the ice sheet which would then be sustained by gravitational energy....
>
> What would be the consequences of such an operation? The immediate effect of this vast quantity of ice surging into the water, if velocities of one hundred meters per day are appropriate, would be to create massive tsunamis (tidal waves) that would completely wreck coastal regions even in the Northern Hemisphere. There would then follow marked changes in climate brought about by the suddenly changed reflectivity of the Earth. At a rate of one hundred meters per day, the center of the ice sheet would reach the land's edge in forty years.

Who would stand to benefit from such application? The logical candidate would be a landlocked equatorial country. An extended glacial period would ensure near-Arctic conditions over much of the temperate zone, but temperate climate with abundant rainfall would be the rule in the present tropical regions.

That such ideas should be taken seriously by so sober a geophysicist as Gordon MacDonald is significant in that he was at the time executive vice president of the Institute for Defense Analyses in Washington as well as a member of President Lyndon B. Johnson's Science Advisory Committee.

While the occurrence of Antarctic surges remains controversial, one phenomenon that Velikovsky regarded as evidence of past catastrophes has now been widely recognized, although contrary to his contention, it seems unrelated to sudden changes in the Earth's spin. Velikovsky had pointed to a report on the Swedish Deep-Sea Expedition of 1947–1948 and other evidence that sudden changes have altered the ocean floor. During the 1940s, when most of the world was at war, the Swedes developed a device that could extract long cores from the bottom of the sea. Such cores were obtained by driving a tube down into the sediment, thus extracting a cross section of layers laid down by thousands of years of oceanic history. Every day some of the tiny organisms that live in the sea die and their remains, in the form of tiny seashells, fall to the ocean floor like a continuous, gentle snow. As climate and sea temperature change, so do the populations of these organisms, and thus the sediment layers embody a history of oceanic climate, as well as of evolution.

Since the history of this evolution has been partially reconstructed, it is possible to identify how many thousands or millions of years ago a layer was laid down through examination of the tiny shells found within it. Furthermore, because the accumulation typically is no more than a centimeter per 1000 years, a core of sediment layers a few meters long may extend back hundreds of thousands of years.

Driving the tube as deeply as possible was a challenge. The late Charles S. Piggot of the Carnegie Institution's Geophysical Laboratory tried using explosive charges that detonated on contact with the bottom, driving the coring tube into the sediment. But handling the explosives was hazardous and the method met with only limited success. One problem, in all coring efforts, was that the tube tended to push the sediment down and to the sides, rather than draw it in. Börje Kullenberg of Sweden designed a "piston corer" that worked somewhat like a hypodermic needle in that a piston was pulled out of the coring tube, as the tube pushed down into the bottom, to help suck in the sediment. Early models of the device were not much more successful than the explosion-driven corer, and it was not until after World War II that an effective version came into general use, ultimately making possible, under suitable circumstances, the extraction of cores as much as 30 meters long.

The Swedish expedition of 1947–1948 not only had a piston corer but was also equipped with depth charges developed by Sweden's Bofors armament works (which had also provided the combatants of World War II with weapons). These charges generated shock waves that not only echoed off the bottom, indicating water depth, but probed the thickness of the sediment blanketing the bottom, sometimes to depths of a kilometer or more.

Armed with a special winch to lower the one-and-a-half-ton corer and other deep-sea research tools, the Swedish expedition set forth in the four-masted training ship *Albatross* for a 15-month journey around the world. On its return Hans Pettersson, director of the Oceanographic Institute at Göteborg, Sweden, wrote a colorful account of the findings, which was quoted at some length by Velikovsky. Some 200 cores were obtained, totaling more than a mile in length. They provided, Pettersson said, "evidence of great catastrophes that have altered the face of the Earth." These included not only the ice ages and periods of widespread volcanic eruption but events that "raised or lowered the ocean bottom hundreds and even thousands of feet, spreading huge 'tidal' waves which destroyed plant and animal life on the coastal plains."

The evidence for such seemingly catastrophic events included the discovery that parts of the midocean floor are carpeted with sand and silt similar to that produced by wave action and erosion along a coast. In 1947 such material, typical of New England beaches, was found by the 142-foot ketch *Atlantis* of the Woods Hole Oceanographic Institution in a deep basin of the western Atlantic—a region one would expect to be covered with the ooze characteristic of midocean sediment.

The other indication of extraordinary occurrences in the ocean basins was the existence of canyons cutting deep into the rims of continental shelves and into the slope, beyond the shelf, that descends to the deep-sea floor. One of the best-known of these canyons lies off New York, having been observed as early as the 1840s in a survey of the approaches to that port. It is aligned as though it were an extension of the Hudson River, and its Upper Gorge, with depths greater than a kilometer, would be as awesome as the Grand Canyon, if the oceans were dry. The canyon extends from the shelf 200 kilometers (125 miles) downslope to the abyssal plain, whose smooth surface suggests carpeting with sediment carried down the canyon.

Many such canyons, it was found, lie off river mouths with aprons of sediment spread across the deep-sea floor at their lower ends, as one would expect where a river empties into a shallow sea. But the lower end of the Hudson Canyon, for example, lies more than 4.5 kilometers (2.8 miles) under water. Was it possible that the sea levels were once lowered by four kilometers until only shallow lakes remained in the deep basins? Or could the sea floor have once been that much higher and then subsided, Atlantis-like? How else, one might ask, could canyons have been carved through solid rock and coastal sand have been spread into mid-ocean?

While Velikovsky felt he could explain such phenomena by lurches of the Earth, others were on the track of more plausible explanations. The initial hint came, not from ocean studies, but from the first detailed surveys of the two great lakes bordering Switzerland: Lake Constance and Lake Geneva. The Upper Rhine flows into the former, the Upper Rhone into the latter. In both cases, it was found, deep gorges cut into the deltas that extended far into the lake from the river mouths. In 1883 F. A. Forel reported to the French Academy of Sciences that, in the case of Lake Geneva, the gorge ran six kilometers (four miles) out into a very deep part of the lake, and he tried to explain how the seemingly placid flow of water, once the river entered the lake, could account for this.

His answer was entry into the lake, during spring, of a deep, invisible torrent of frigid, silt-laden (and thus very heavy) water that scoured the bottom, even though the surface of the lake gave no hint of the fierce activity a few score meters below.

The Upper Rhone is fed by glaciers and, in spring, by melting snowfields. It is chalky with "rock flour" ground from the mountains by flowing ice, and one cubic meter of the water, Forel reported, carries 130 grams of such material. Fresh-water rivers emptying into oceans would not carve such channels, he said, since the fresh water is lighter than sea water and would remain near the surface. In this respect, however, it seems he was wrong. One must try to imagine what the Hudson River was like 10,000 years ago, when the last great ice sheet was melting but had not retreated far enough north to clear a drainage route down the St. Lawrence River. Vast amounts of icy, silt-laden water flowed down the Hudson Valley and out across the continental shelf (then partly exposed by the lowering of sea level). This, it would seem, could have carved a mighty underwater canyon, particularly if massive discharge of fresh water from the melting ice sheet reduced the salinity of coastal waters, allowing the coldest, heaviest torrent to hug the sea floor.

The first attempt to relate the phenomenon seen in Lake Geneva to the carving of canyons like that off New York was in 1936 by Reginald Daly of Harvard. The process, now known as a turbidity current, seemed too far-fetched for acceptance at that time, although since then it has proved to be a major factor in shaping that vast region of the Earth that lies under water. And the scars left by turbidity currents in ancient seas have now been found, thrown high into mountain systems like the Alps, as sandstones that provide clues to the past history of this planet.

Although no human eye has ever gazed on such an occurrence, it is an event of almost unbelievable magnitude. In a matter of hours, a turbidity current can carpet thousands of square kilometers of the deep-sea floor with sand, gravel, and silt transported from hundreds of kilometers away. The only comparable occurrences that occur on dry land are a powder-snow avalanche—the most dread occurrence of the high mountains—and a volcanic *nuée ardente* such as that which, within seconds in 1902, killed all but one of the 30,000 inhabitants of Saint Pierre on the island of Martinique. (The sole survivor was a felon, relatively protected in his dungeon.) In both the avalanche and volcanic eruption material becomes mixed with air producing a form of "superheavy air" that sweeps down the slope. Being almost free from restraint by friction, the rate of descent may match that of a diving airplane. The powder-snow avalanche occurs under dry-snow conditions. It not only descends a slope but may race across a valley and a short distance up the opposite slope, wreaking havoc in its path. Such, apparently, was the avalanche that, on February 10, 1970, struck the ski resort of Val d'Isère in the French Alps. The dining hall of a relatively new hostel, operated by the Union of Fresh Air Centers, was filled with young people at breakfast when they heard a distant roar that grew louder and louder. The avalanche descended the far side of the valley, crossed the Isére River and National Highway N202, and smashed through the windows of the concrete building. Some of the occupants were swept out windows on the opposite side; others were buried under hard-packed snow; and 38 died of suffocation. In a *nuèe ardente* it is volcanic ash and incandescent gas from an explosive eruption, rather than snow, that are mixed with air.

Few, if any, took seriously the proposition of Daly that such things happen underwater until Philip Kuenen at the University of Groningen in the Netherlands performed a series of experiments using a tank more than 30 meters long that could be tilted to simulate the conditions under which turbidity currents might originate. His results led Bruce C. Heezen and Maurice Ewing, at what was then the Lamont Geological Observatory of Columbia University (it is now called the Lamont–Doherty Geological Observatory), to examine the effects of an earthquake that had badly shaken the Grand Banks, south of Newfoundland, on November 18, 1929. The quake occurred in a crossroad region of submarine cables and broke at least 12 of them. Since the cables were in use, the time each was severed was a matter of record (Fig. 3.2).

It was found that the six cables within 95 kilometers (59 miles) of the quake center broke at the time of the quake. These were:

New York to St. John's, Newfoundland, "New York No. 1" (Commercial Cable Co.)
Canso, Nova Scotia, to Fayal, Azores, "Main No. 4" (Commercial Cable Co.)
Halifax, Nova Scotia, to Harbour Grace, Newfoundland (Imperial Cable Co.)
New York to St. John's, Newfoundland, "New York No. 2" (Commercial Cable Co.)
Canso, Nova Scotia, to Fayal, Azores, "Main No. 6" (Commercial Cable Co.)
New York to Bay Roberts, Newfoundland, "No. 1" (Western Union Cable Co.)

Another cable, linking Cape Cod with the French islands of St. Pierre and Miquelon, south of Newfoundland (operated by the French Cable Co.), held up for 13 minutes before breaking. However, it was the timing of breaks in five additional cables, much farther out to sea, that testified to an avalanche-type occurrence of great speed and scope.

"For more than 13 hours after the earthquake," Heezen wrote later, "cables farther and farther to the south of the [quake] epicenter went on breaking one by one in a regular succession. Each break was downslope from the one before, and the last took place in the deep ocean basin 300 miles [480 kilometers] from the epicenter."

The cables severed in this manner were:

Halifax to Fayal (Imperial Cable Co.)
Cape Cod to Brest, France (French Cable Co.)
New York to Fayal (French Cable Co.)
New York to Bay Roberts, Newfoundland, "No. 2" (Western Union Cable Co.)
New York to Horta, Azores, "No. 1" (Western Union Cable Co.)

From the timetable of breaks, Heezen and Ewing deduced that, in its early stages, the torrent of turbid water that caused them was moving at more than 83 kilometers

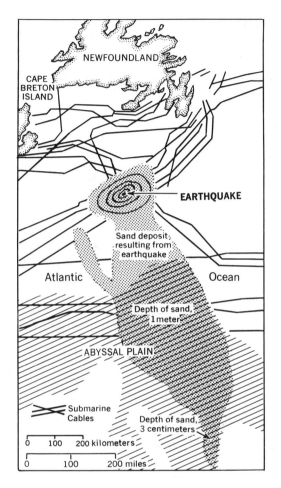

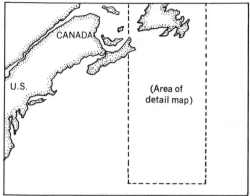

Figure 3.2 Submarine avalanche
On November 18, 1929, at least a dozen submarine cables were severed following an intense earthquake on the Grand Banks. In 1951, Bruce Heezen and Maurice Ewing deduced, from the timing of these breaks, that an underwater avalanche, or "turbidity current," had swept far out into the mid-Atlantic spreading a vast apron of sediment over the sea floor.

(52 miles) an hour on an extremely broad front. One section of cable 200 kilometers (125 miles) long was swept away.

From later sampling of the sea floor it appears that at least 100,000 square kilometers (40,000 square miles), extending some 800 kilometers (500 miles) from the earthquake site, were carpeted with material laid down by this event, much of it to a depth of one meter. According to Kuenen, the volume of this sediment must have been about 100 cubic kilometers (24 cubic miles)—"enough to load a row of tankers 20 ships wide running round the Equator." That it all came down in one great torrent is shown by its grading—a feature typical of turbidity current deposits. In such grading the pebbles drop out of the fast-moving cloud first, forming a layer directly on top of the preexisting sea floor of typical, midocean ooze. Next, the heavy sand grains sink from the cloud, then smaller grains, and at last the finest-grained silt, producing a characteristic bed of sediment grading continuously upward from the coarsest layer at the bottom to the finest at the top.

While, in this case, the sediment was thrust into water suspension by an earthquake and by quake-induced landslides on the steep continental slope, such landslides may also occur periodically off the mouth of a river where sediment accumulates on a slope until it is unstable and breaks loose. In this regard Africa's Congo River has probably caused as much grief to a cable company as any in the world. During the late 19th century a cable was laid along the coast, keeping far enough off shore to come no closer than 110 kilometers (65 miles) from the river mouth. This was considered a safe distance, but within seven years the cable had broken five times.

The breaks occurred where it crossed a canyon reminiscent of that off New York—one, it would seem, that had been cut, or at least was kept open, by periodic turbidity currents from the river mouth. Since the floor of this canyon was two kilometers under water, fishing up and repairing the cable each time was a major undertaking. In 1893 the cable was therefore relaid in a detour to bring it no closer than 200 kilometers (125 miles) from the river mouth. Instead of improving matters, this made them worse. In four years it broke nine times.

The cable company decided to try laying the cable close to the river, in shallow water, so at least repairs would be easier. But 22 times in the next 40 years it had to be spliced and in 1937 the entire route was abandoned.

Detailed surveys of the Congo canyon by the Lamont Observatory's research schooner *Vema* showed it to be flanked, on both sides, by levees about 100 meters high. Similar levees sit on the flanks of the Hudson Canyon and, on a smaller scale, flank the gorge on the floor of Lake Geneva. They are apparently formed of material that settles out of a turbidity current when it overflows the canyon banks.

The sudden manner in which accumulated sediment slumps to start a turbidity current is illustrated by the frustrating effort to maintain two long jetties on either side of the entrance to South America's Magdalena River, flowing from Colombia into the Caribbean. They have repeatedly had to be rebuilt when the bottom slumped and swept out to sea. On one occasion 250 meters of a jetty vanished overnight, and when engineers sought to rebuild it they found the water depth to be 40 meters. When a cable 24 kilometers (15 miles) off shore was fished up from a depth of 1.5 kilometers (one mile) for repair, it was wrapped with green river grass.

While the manner in which the great submarine canyons were carved is still debated, many now believe some, at least, were cut by material from the melting ice sheets that accumulated as sediment on the rim of the continental shelf and then, periodically, raced down the slope as a turbidity current.

In any case it is evident that sudden cataclysms take place under the sea as well as on land. Their occurrence, however, does not violate the concept of uniformitarianism, if the latter is broadly interpreted. There is no need to call upon unique—still less super-natural—events to explain the shaping of our planet's surface, above and below water, even though some of them may have been episodic and catastrophic.

In the early 19th century, religion and geology were, to some extent, wedded to one another. Thus William Buckland, in his *Reliquiae Diluvianae,* wrote in 1823 that geology was "the efficient Auxilliary and Handmaid of Religion." His view was dominated by the concept of a "general flood which swept away the quadrupeds from the continents, tore up the solid strata, and reduced the surface to a state of ruin."

The uniformitarianism that has replaced such catastrophism, as noted recently by Stephen Jay Gould of Harvard University, has split into two schools: There are those who believe in "uniformity of rates or material conditions" (leading to skepticism about occasional catastrophes); and there are those who believe simply that the laws of nature remain constant, both in space and time, even though such laws account, occasionally, for dramatic events. It was the latter view, according to Gould, that enabled Sir Charles Lyell, who in the last century helped lay the foundations of modern geology, "to exclude the miraculous from geologic explanation."

It is noteworthy that Harry Hess, Velikovsky's most distinguished champion (on humanitarian, more than on scientific grounds), was to help frame a theory of the Earth that, in slow motion, was no less dramatic than that of Velikovsky. This was the concept that the Earth's surface is shaped by contesting crustal plates, growing in some areas and being swallowed into the depths of the Earth in others. While its early protagonists did not experience ridicule as sharp as that which slashed at Velikovsky through much of his career, for a number of years they had to endure the scorn of their colleagues.

However, the emergence of this theory was preceded by one that seemed ideal in its simplicity, for it saw the Earth simply as an expanding balloon.

Chapter 4

The Balloon Hypothesis

ON A PARTIALLY INFLATED BALLOON IT IS POSSIBLE TO FIT TOGETHER CUT-OUT REP-resentations of the Continents so that they cover most of the surface. When the balloon is inflated further, these "continents" draw apart and "ocean basins" appear between them.

Such an explanation of continental drift, through gross expansion of the Earth, has been considered by a number of scientists and the idea, in modified form, was championed into the 1970s.

Probably its most ardent and eloquent advocate has been S. Warren Carey, professor of geology at the University of Tasmania in Hobart. In 1958 he organized an international conference there to assess the drift hypothesis, and he invited Chester Longwell of Yale as "principal guest."

Longwell, in an introductory address, made it clear that he leaned more to polar wandering than to changes in relative positions of the continents. In particular he frowned on Wegener's idea that "ships" of continental rock are "plowing through a sea" of oceanic crust (which he considered to be more rigid). "Can a ship plow through a sea stronger than its own prow?" Longwell asked.

When it came his turn to speak, Carey said the answer lies in expansion of the Earth. This would tear apart the continents, forming ocean basins, and no plowing through crust would be necessary. The young Earth, he said, was less than half its present diameter and its surface, with an area less than a quarter what it is today, was completely covered with continental crust. Then steady expansion not only pulled the continents apart but thrust material up under the continental plates, producing mountain ranges.

The problem, he conceded, was to explain what could cause such gross expansion. If the Earth were cool to begin with and then heated, this would produce swelling. Conversion of some internal material to a gaseous form could blow up the balloon some more, as could other such "phase changes." But none of these seemed adequate to double the Earth's diameter.

Carey therefore challenged physicists to provide an explanation. It was his duty as a geologist, he said, to "follow fearlessly" where his observations led him. "It might lead ultimately to an absurdity whereupon I would abandon it, but it has not done so yet. On the contrary it prompts me to ask the physicist to seek an Earth model, which will expand at an increasing rate with time."

Some physicists have, in fact, come up with theories that call for all celestial bodies to expand as time passes, although the reaction of the great physicist Niels Bohr to the first of these was somewhat typical. One day in 1937 George Gamow, the Russian-born cosmologist, was sitting in his room at Bohr's institute in Copenhagen, where Gamow was a guest, when Bohr came in. Waving the latest issue of the British journal *Nature* he exclaimed: "Look what happens to people when they get married!"

He was referring to the British theorist, P. A. M. Dirac, who had just wed the sister of Eugene Wigner, the Hungarian physicist. Dirac had proposed that the relationship between gravity and the electrical force is slowly changing and that, in effect, gravity—the one force that acts over long distances in the universe, controlling all orbits and trajectories (as well as holding us on the Earth)—is becoming weaker. To him the ratio between the electrical force and gravity (which is now a very large number, gravity being so weak) was originally a one-to-one relationship. Gravity has been weakening ever since, the ratio today, according to the Dirac hypothesis, being determined by the age of the universe. If gravity is becoming more feeble, then all material pressing on the Earth's interior is becoming lighter and the Earth must be expanding.

Although Dirac was acknowledged to be almost in a class with Einstein as a theorist (one of his predictions anticipated the discovery of anti-matter), his idea was not warmly received. Edward Teller, the Hungarian-born physicist who later came to be known as "father" of the hydrogen bomb, pointed out that, if gravity were stronger in the past, the Sun would have generated more energy and the Earth would have been too hot for life. The theory, he said, would lead us to expect a temperature on Earth "near the boiling point of water" from 200 to 300 million years ago. This, to put it mildly, would have been uncomfortable for the dinosaurs. It would, in fact, have ruled out the existence of life until relatively recently.

Teller, however, used estimates of the age of the universe prevalent in the 1940s. It was shown subsequently that the universe is at least three times older than he had assumed, and hence the decline in gravity would have been much more gradual.

Dirac's idea was elaborated by Pascual Jordan, professor of physics at the University of Hamburg, and the proposal also received support from Laszlo Egyed, professor at the Geophysical Institute of Eötvös-University in Budapest. From what was known of past inundations of the continents, Egyed pointed out, it was evident that the areas covered by water had declined in intermittent but persistent fashion for the past 400 million years. The reason, he felt, was obvious: to begin with, the entire surface of the Earth was continental, but gradually the planet has expanded, forming ocean basins that could accommodate more and more water and thus freeing increasing areas of continental land from submergence.

Meanwhile, in the United States, Robert H. Dicke of Princeton and his former student, Carl H. Brans, also came to the conclusion that the Earth was expanding, but for quite different reasons. They had undertaken a careful scrutiny of Einstein's gravitation theory (the "general theory" of relativity) and had concluded that it should be modified in a manner that predicted a slow drop in the force of gravity.

In the traditional view, as summarized by Dicke, "Distant matter of the universe, spherically distributed about the Earth, is without a noticeable effect on the solar system. There are no locally induced consequences of the expansion of the universe." But, he asked, is it really true that we are so independent of the universe that surrounds us? He and Brans concluded that as the universe expands, gravity weakens—a somewhat different concept from that of Dirac. (In 1971 Sir Fred Hoyle proposed a still different hypothesis whereby gravity weakens and, at the same time, the masses of all atomic particles become greater.)

In 1957 Dicke calculated, from current estimates of the rate at which the universe has been expanding, that gravity has weakened enough over the past 3.25 billion years to cause the Earth's volume to swell some 15 percent. This would increase its circumference only 1800 kilometers (1100 miles)—far from enough to account for the ocean basins. However, he wrote, the expansion "would amount to some 50 cracks 20 miles wide distributed over the ocean bottom." Another possibility, he said, was that it had produced many small cracks that soon filled with sediment. Or it could have been masked by "creep" of the ocean floor.

On the Moon, where, unlike on the Earth, there is little alteration of the landscape by sedimentation or other such processes, the surface is split, here and there, by "rilles" like the one examined by the Apollo 15 astronauts. Furthermore, in 1972 the Mariner 9 spacecraft in orbit around Mars transmitted pictures showing remarkable cracks in the Martian surface. These were seized upon as encouraging by believers in a slow weakening of gravity.

However, the most dramatic evidence seemed to lie in a discovery whose full impact came on the heels of Dicke's 1957 prediction of sea-floor cracks. The discovery was the product of an intensive postwar effort to learn the nature of the ocean floor, an effort of which the 1947 voyages of the *Albatross* and *Atlantis* were but a preview. The oceans cover 71 percent of the planet's surface, and relatively little was known of this region. As Maurice Ewing, head of the Lamont Geological Observatory quipped, trying to understand the Earth with such meager knowledge of the sea bed was "like trying to describe a football after being given a look at a piece of the lacing."

After World War II a host of new instruments made it possible, figuratively, to roll back the waters and see what lay beneath. These included, in addition to piston corers of the type used by the *Albatross* and *Atlantis* and explosive charges for probing sediment depths, such devices as underwater cameras, instruments that could measure the flow of heat from the depths of the Earth up through the ocean floor, and magnetometers of such sensitivity that when towed along the surface, they could record subtle variations in the magnetic properties of rock buried beneath sediments lying under several kilometers of water.

The use of these devices in some cases was funded generously by the United States Navy which, in the early postwar years, still had ample funds available and found itself one of the few government agencies in a position to support basic research. Certainly it was an advantage to missile-carrying submarines to know the precise positions of landmarks on the sea floor, or for anti-submarine technicians to know the sound-propagating nature of deep-sea layers. But the bulk of the research was designed simply to learn more about that little-known arena where the Navy operates.

The instrument that laid bare the seabed topography for the first time was the echo sounder. Midocean soundings with the traditional leadline of the sailor were hopeless, for the water is far too deep. Only by heroic efforts could such soundings be made. On its memorable cruise of 111,000 kilometers (68,900 miles) across the Atlantic and Pacific, from 1872 to 1876, the British research corvette *Challenger* repeatedly lowered a 200-pound weight on a hemp line. Each sounding took hours, but for the first time it was shown that the mid-ocean basins reach depths below sea level greater than the height above the sea of the loftiest mountains. It was also found that there is a broad ridge bisecting the Atlantic, although with the soundings spaced 160 kilometers (100 miles) apart it was not clear whether it was simply a hump, or a more mountainous feature.

The laying of the first trans-Atlantic cables, in the latter part of the 19th century, encouraged such laborious soundings, but it was the invention of sounders which determine depths by echoing off the bottom that revolutionized mapping of the sea floor. In a typical such device a sound pulse is transmitted and, at the same moment, a needle electrically inscribes a mark on the left margin of a moving scroll of paper. The needle then travels slowly across the paper until an echo returns from the bottom, whereupon a second mark is inscribed. The spacing between the mark on the margin and the second mark displays the depth. As the scroll moves down, the process is continuously repeated, producing a succession of marks that indicate variations in depth.

In this way a ship can automatically chart a profile of the ocean floor without stopping. The first echo-sounding profile across the Atlantic was recorded by an American vessel in the early 1920s, although the fully automatic sounders did not come into general use until World War II, and even then most were not designed for mid-ocean depths. Their purpose was to keep ships from running aground. However, from 1925 to 1927 the German research ship *Meteor* used echo-sounding for an extensive survey that, for the first time, revealed the ruggedness of the Mid-Atlantic Ridge. It has been said that there are more mountains between New York and London than between New York and San Francisco. Furthermore, in the South Atlantic, the *Meteor's* measurements of deep water salinity and temperature showed that some kind of barrier separates bottom waters on the two sides of the ocean—the first hint that a ridge bisects the entire Atlantic from north to south.

Not long after the *Meteor* voyages an expedition financed by the Danish brewing industry's Carlsberg Foundation set forth around the world in the ship *Dana* and discovered an undersea barrier bisecting the western Indian Ocean. It was named the Carlsberg Ridge in honor of the expedition's benefactors, placing the brand name of a famous beer permanently on the map. The discovery that anticipated what was to become a key factor in the debate over Earth expansion was then made, in the mid-1930s, by a British expedition that explored the floor of the Indian Ocean from the ship *John Murray.* R. B. Seymour Sewell was expedition leader with John D. H. Wiseman as geologist. Using an early form of echo-sounding device, they found that a deep gully splits much of the Carlsberg Ridge, which extends southeast into the Indian Ocean from the island of Socotra, at the mouth of the Gulf of Aden.

The British discovered another ridge nearby in the Arabian Sea which they named the Murray Ridge, and it too was split by a rift valley. The Carlsberg rift was about 300 meters deep, relative to the ridge crests that flanked it. Wiseman and Sewell, in their report, noted the similarity between these rifts and those extending from the Jordan Valley down through the Red Sea and Gulf of Aden. The submarine canyons, as well, seemed related to the Rift Valleys that extend the Red Sea cleavage down into Africa.

The two scientists noted that maps of worldwide earthquake activity (though still somewhat crude at that time), showed a zone of quakes along the course of the Carlsberg Ridge and also along the Mid-Atlantic Ridge. "While providing us with an additional point of similarity between the Rift Valley and the Carlsberg Ridge," they wrote, "the presence of these earthquake belts indicates the lines along which the Earth's crust is in a condition of instability...."

According to Wiseman, now at the British Museum (Natural History), he and his colleagues suspected, in the light of their findings and the evidence for Mid-Atlantic earthquakes, that a rift valley also split the Mid-Atlantic Ridge. They planned an expedition to investigate this possibility in 1940, but World War II intervened.

After the war a steady improvement in earthquake recorders (stimulated in part by American efforts to discriminate between natural quakes and underground nuclear weapons tests) led to more precise charts of worldwide quake activity. Not only were more sophisticated instruments installed at existing stations, but new ones began to operate in regions where none had been before. Networks of seismic stations were set up in India, South Africa, and elsewhere.

As a result it became evident that lines of frequent earthquake occurrence formed a far-flung pattern. At a meeting of the Royal Society of London in 1953, the head of the International Bureau of Seismology in Strasbourg, J. P. Rothé, displayed a map of earthquake activity over the previous 30 years (Fig. 4.1). From it, he said, "there appears more and more clearly to be an almost *continuous* seismic zone extending for a length of more than 30,000 kilometers, splitting into two parts the Indian Ocean as well as the Atlantic Ocean." Pointing out that these zones joined one another midway between Africa and Antarctica, he said they "present an almost perfect continuity from the Gulf of Aden to Spitzbergen" (see his map of Fig. 4.1 and the more recent map of worldwide earthquake activity, Fig. 6.7).

It was, however, the efforts of researchers at the Lamont Geological Observatory that drew attention to the full scope of this feature. Early in the 1950s Marie Tharp, a studious young surveyor's daughter on the observatory staff, began work on a map of the Atlantic floor, using the echo soundings that were beginning to accumulate in considerable quantity. The observatory had been set up for Columbia University by Maurice Ewing at what had been the summer estate of Thomas W. Lamont, the financier, overlooking the Hudson River at Palisades, New York. Unfortunately the soundings made by the *Meteor* on its Atlantic voyages had been lost in the Allied bombing of Berlin, but Miss Tharp began plotting the ocean-floor profiles obtained more recently by research ships. She was struck by a feature in each line of soundings across the Mid-Atlantic Ridge. Wherever a ship traversed it, riding the waves one or two kilometers

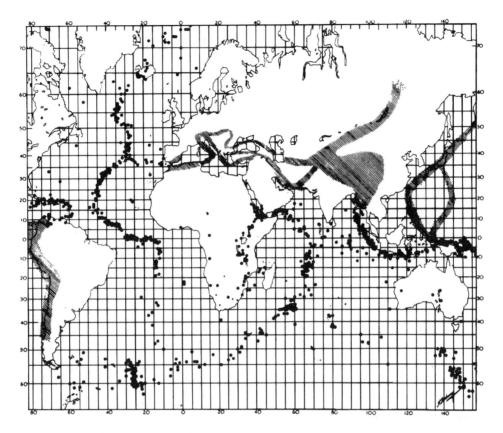

Figure 4.1 Earthquake Activity, 1920–50. With this map, J. P. Rothé demonstrated in 1953 that the Atlantic and Indian oceans are split in two by sharply defined lines of earthquake activity. The shaded areas mark other major seismic zones.

above, the ridge was split by a deep cleft. Thus it began to look as though the prewar prediction of Wiseman and Sewell of a rift valley down the Mid-Atlantic Ridge was correct.

Only a half-dozen crossings were available, and the observation did not produce much excitement at the observatory until Miss Tharp began working on another assignment. The Bell Telephone Laboratories had asked for help in determining whether midocean earthquakes were responsible for a rash of breaks in trans-Atlantic cables. Miss Tharp's job was to plot the earthquake locations and see to what extent they coincided with the breaks. She found that virtually all of the quakes had occurred at shallow depths beneath the floor of the rift valley bisecting the Mid-Atlantic Ridge.

Bruce Heezen, her boss, noticed that this line of earthquake activity not only extended around Africa into the Indian Ocean but was global, forming a network reaching across the Arctic Ocean to Siberia and from the Indian Ocean, south of Australia, into the Pacific. Could this entire network consist of oceanic ridges split by rift valleys?

Knowledge of the vast Pacific floor was still only fragmentary. Scattered soundings at the turn of this century had shown a hump in the Pacific floor off Mexico; and when, in 1946, the United States Navy expedition, *Operation Highjump* (with this writer aboard), sailed southward across the Pacific, from Panama to the South Polar pack ice, the 11 ships were ordered to follow tracks at least 50 miles (80 kilometers) apart, operating their echo-sounders so that, collectively, they would sweep a broad band of the ocean floor. This confirmed the existence of a great north–south ridge, poking out of the sea at Easter Island and, at its northern end, skirting the West Coast of Mexico. Now known as the East Pacific Rise, it is much less steep-sided and rugged than the Mid-Atlantic Ridge and is not cleft by a central rift. But, like the other ridges of the active, global network, its entire length is punctuated by periodic earthquakes.

The suggestion, in the maps of worldwide earthquake centers, that they mark a network of submarine ridges manifesting some intense form of ocean floor activity, came at a propitious time, for during the International Geophysical Year of 1957–1958 the maritime nations launched the most ambitious program of oceanic study ever undertaken up to that time. A number of research vessels—including the *Vema* and *Conrad* of the Lamont Geological Observatory, the *Argo, Horizon,* and *Spencer Baird* of the Scripps Institution of Oceanography in California, and two British ships, the *Owen* and *Discovery*—ran lines of soundings across earthquake-implied routes of the ridge network and showed that a global system does, in fact, exist.

In tracing this ridge system there was no simple way to confirm the existence of that part of it indicated by a line of earthquakes across the Arctic Ocean, since pack ice prevents ships from reaching there. A ridge had been identified by soundings from Soviet and American stations on the drifting floes. Known as the Lomonossov Ridge, it lay closer to the Pacific side of the Arctic Ocean than the line of earthquakes and was itself free of such activity. A few spot soundings through the ice suggested that there might be a second, active ridge along this line of quakes, and confirmation came from the journeys of the nuclear submarines *Nautilus* and *Skate* under the North Polar ice in the late 1950s. The submarines made soundings of the ocean floor below and also recorded the bottom profile of the ice above them.

When the worldwide nature of the ridge-and-rift network had been confirmed, Heezen, Tharp, and Ewing declared that it represents the most extensive geographic feature on Earth—a submarine mountain system 65,000 kilometers (40,000 miles) long, extending into all the oceans and, in many sectors, cleft by a central rift. Heezen pointed out that the Grand Canyon of the Colorado, along its most majestic 80 to 100 kilometers (50 to 60 miles), averages not much more than 1000 meters deep and is from 6 to 30 kilometers (4 to 19 miles) wide. The Mid-Atlantic Rift, he said, although hidden from human eyes, is twice as deep and, for hundreds of kilometers, is 12 to 50 kilometers (7 to 30 miles) wide. It was this succession of deep submarine gorges, forever plunged in darkness, that American and French oceanographers decided to explore, by deep submersible, in the mid-1970s.

Heezen, as noted, concluded that the rifts arose from gross expansion of the Earth, much as envisioned by Carey. A somewhat more modest expansion was proposed by J. Tuzo Wilson of the University of Toronto, who later, as president of the International Union of Geodesy and Geophysics, was to play a major role in "selling" the drift hypothesis to a skeptical scientific world. He pointed out that the parallel mountain chains constituting the oceanic ridges covered an area of the Earth's surface roughly equal to the increase in area to be expected if the expansion envisioned by Dicke had taken place. The implication, he said, was that the ridges were a by-product of that expansion.

Dicke himself at first welcomed what seemed a confirmation of his theory, but by 1962 enough magnetic data were on hand to suggest strongly that the two sides of the Atlantic were close to one another as recently as 200 million years ago (a relatively short time, geologically speaking). The expansion rate needed to produce the ocean in so limited a time, he pointed out, was 300 times greater than that suggested by his theory. Furthermore, if the Earth had been expanding that fast, this would have slowed its spin, lengthening the day an appreciable amount over the last 2000 years. Yet he cited evidence that there has been no such change. He concluded, therefore, that the midocean ridges and rifts could not be attributed to Earth expansion. If gravity has weakened, causing expansion, the process has been so slow that any resulting fissures have been erased by those more fast-acting processes that continuously stretch, squeeze, and blanket the Earth's surface.

While Carey has kept the expansion idea alive, it now has few adherents. Meanwhile, new and old discoveries concerning the ocean floor had presented to those seeking to understand the Earth a great challenge. It was evident that the oceans and continents have little in common:

The continents are characterized by folded mountains, by great masses of granite, by rocks of all ages (back at least three billion years) and by very complicated structures.

The ocean basins, on the other hand, feature only volcanic mountains. They are floored by aprons of lava, by basalt rather than granite. Their rocks in all cases are quite young and with structures that are relatively simple. The challenge was to integrate these features into a plausible theory of the Earth.

Chapter 5

The Mantle Controversy

ON THE EVENING OF MARCH 26, 1957, STUDENTS AND FACULTY AT PRINCETON gathered in Guyot Hall to hear Bruce Heezen tell about the newly discovered rifts. When he was through, Harry H. Hess, Chairman of the university's geology department, stood up to comment. He said, in essence: "You have shaken the foundations of geology."

To Hess discovery of the rifts was of enormous significance, and it was to play a key role in the geologic revolution that, within three years, he would help launch (Fig. 5.1). Unlike some proponents of the view that great changes are taking place in geography, Harry Hess was no more flamboyant than the suggestion of a moustache that he wore. Both physically and scientifically his movements were deliberate—even cautious. Yet his chain smoking gave hints of volcanic activity within. So sage were his views, in the eyes of Washington bureaucracy, that he was named chairman of the Space Science Board, responsible for guidance of the nation's space program. Yet it was in the opposite direction—down rather than up—that his chief interest lay.

Throughout his career he was concerned with what lay beneath the oceans. In the early 1930s, as a doctoral student at Princeton following his graduation from Yale, he accompanied Vening Meinesz, the Dutch gravity expert, on a submarine voyage to measure gravity near trenches in the sea floor off the West Indies. Beginning in the 1920s, as noted earlier, Vening Meinesz had made several long submarine voyages, swinging his pendulums in the stable environment of a submerged vessel to record local gravity variations. Following his journey to the East Indies in a Dutch submarine, Meinesz explored the ocean floor off the West Indies aboard the *S-21* and *S-48* of the United States Navy. In the East Indies he found that a zone of markedly weakened

Figure 5.1 Harry Hammond Hess, 1906–1969

gravity 160 kilometers (100 miles) wide and 8000 kilometers (5000 miles) long follows the sequence of trenches that parallels the island arc of the Indies on their southern flank, curving north with the trend of the chain to pass west of New Guinea and then up the east coast of the Philippines. A similar gravity "anomaly" was discovered, following the trenches on the oceanic side of Japan and in the West Indies.

By the 1950s the trenches had emerged as one of the most remarkable features of the Earth's surface, and the strangely weak gravity above them was to provide Hess and others with one of their strongest clues as to what is going on beneath the oceans. The trenches are scattered around the rim of the Pacific basin like great long gashes in the Earth, some of them extending farther below sea level than Everest rises above it by a margin of a kilometer or more (Fig. 5.2). Typically they lie to seaward of an arc of volcanic islands, or a coastal mountain range with volcanic activity, like the Andes. It was evident that upthrust of the islands or mountains and downthrust of the trenches is all part of the same process.

Since water pressure on the trench floors is about one ton per square centimeter (eight tons per square inch), it may be some time before wheeled sea-bottom vehicles are able to journey along these sunken highways of the deep, but such a journey will be an eerie one. It will be in total darkness except for the luminous appendages of the sea creatures that live there, and the vehicle will have to climb over a succession of saddles that separate the sections of deepest trench. If the walls could be seen, they would be awesome, for some trenches are comparable to seven Grand Canyons piled one atop the other and extending from New York to Kansas City.

In the mid-1950s only three ships had winches and tapered cables capable of lowering sampling devices to the floors of the trenches. One was the *Albatross* of the

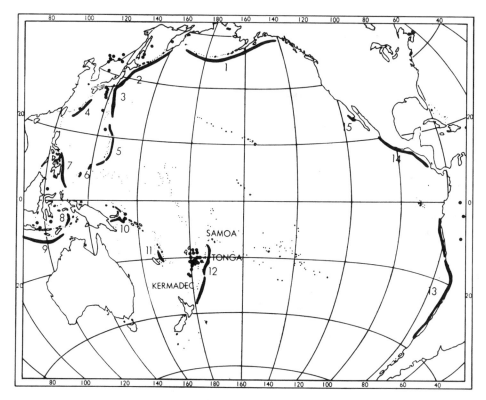

Figure 5.2 Pacific trenches. *The rim of the Pacific basin is marked by a succession of trenches that cut into its floor like deep gashes. They are: (1) the Aleutian Trench, (2) the Kurile Trench, (3) the Japan Trench, (4) the Nansei Shoto Trench, (5) the Marianas Trench, (6) the Palau Trench, (7) the Philippine Trench, (8) the Weber Trough, (9) the Java Trough, (10) the New Britain Trench, (11) the New Hebrides Trench, (12) the Tonga-Kermadec Trench, (13) the Peru-Chile Trench, (14) the Acapulco-Guatemala Trench, and (15) the Cedros Trough.*

1947–1948 Swedish expedition (later used by the Danes on the *Galathea* expedition of 1950–1952); another was the *Spencer F. Baird* of the Scripps Institution of Oceanography in La Jolla, California; and the third was the Soviet research vessel *Vityaz*. Probing the trenches is particularly difficult because they are so narrow that a ship is apt to drift from its position directly over the trench floor while lowering its dredge or other device.

Perhaps the most remarkable discovery in this sampling work was the nature of sediments brought up from these extraordinary depths. As noted in 1955 by Roger Revelle and his colleague, Robert L. Fisher, samples of material recovered from even the deepest trenches "resemble in many ways deposits laid down in shallow water." Revelle, who later became director of the Center for Population Studies at Harvard, ran the Scripps Institution of Oceanography at La Jolla during the years when it pioneered in exploration of the Pacific floor, much as the Lamont and Woods Hole institutions on the East coast led the way in exploring the floor of the Atlantic. Revelle was a man of impressive physical dimension, hearty voice, and great warmth. It is typical of the light-hearted spirit at Scripps that the chief corridor of its building, perched on a bluff overhanging the Pacific, is lined with choice photographs of its most illustrious associates. They include one of Sir Edward Bullard, sitting on a rock, clothed only in a sun hat, and another of Revelle, on a ship's deck, stripped to the waist and wrestling with an enveloping tangle of cable like some modern Laocoön.

To Revelle and Fisher the gashes in the ocean floor were of extraordinary interest. "The size and peculiar shape of the Pacific trenches," they wrote, "stir our sense of wonder. What implacable forces could have caused such large-scale distortions of the sea floor? Why are they so narrow, so long and so deep? What has become of the displaced material? Are they young or old, and what is the significance of the fact that they lie along the Pacific 'ring of fire'—the zone of active volcanoes and violent earthquakes that encircles the vast ocean?"

They pointed out that, in contrast to the volcanoes of Hawaii, whose lava is highly fluid and whose eruptions are relatively well-behaved, the volcanoes around much of the Pacific rim are explosive. The outer slopes of Hawaiian volcanoes, such as Mauna Loa, are very gradual because the lava swiftly flows away from the point of eruption, producing a mountain that, while high, is shieldlike in profile. This is in sharp contrast to the steep cones of such volcanoes as Fujiyama in Japan, Rainier in Washington, Shishaldin in Alaska, Popocatepetl in Mexico, and other circum-Pacific sites.

The explosive nature of some volcanoes around this rim is all too familiar to those living near them. In 1883 Krakatoa blew its top in the East Indies, throwing some 20 cubic kilometers (five cubic miles) of fragmented rock and dust to a maximum height of 80 kilometers. The explosion was heard as far away as Australia, 3500 kilometers away. In 1912 the supposedly extinct Katmai volcano on the Alaska peninsula exploded and Kodiak Island, 160 kilometers away, was buried under a third of a meter of ash. For a time darkness reigned over much of Alaska. In 1956 Bezymyannyy Volcano in Kamchatka blew off its entire summit, throwing dust 43 kilometers aloft, tearing the bark from trees 20 kilometers away and producing torrents of mixed ash and snow-slush that swept 80 kilometers down nearby valleys. The 1963 explosive eruption of Agung Volcano in Bali was reported to have killed 1100 people and injected so much dust into the stratosphere that for many months there were dramatically red sunsets over the United States and many other lands. Taal volcano in the Philippines has undergone a succession of explosive eruptions. One, in 1965, took almost 200 lives. And more recently the eruption of Mount St. Helens in 1980 blew off its side and killed 60 people (Fig. 17.4).

Such activity has helped manufacture several of the best harbors of the Pacific. Deception Island, off the Antarctic Peninsula, is a giant crater flooded by the sea, but it is far from dead. Once, when it was a favorite haven for whaling fleets, water in the harbor boiled, peeling bottom paint off ships that failed to exit in time. An eruption occurred in 1967, but no one was there to witness it.

This "ring of fire" around the Pacific is obviously related to the trenches that parallel the inner side of the ring. Whatever is pulling the ocean floor down into the trenches must also be responsible for the volcanic activity. Revelle and Fisher noted that the deficiencies in gravity over the trenches detected by Vening Meinesz aboard his submarines were among the largest known on Earth. "The trench-producing forces must be acting against the force of gravity," they said, "to pull the crust under the trenches downward."

By this time Hugo Benioff of the California Institute of Technology had identified another striking feature associated with the trenches and the volcanic zones alongside them. Using improved methods for pinpointing the place within the Earth where an earthquake has occurred, he showed that there is a sloping zone of intense earthquake activity that begins at shallow depth below a trench and extends down and inland, under an island arc, as in the Aleutians, or under a continental rim, as in Chile.

He plotted all earthquakes that had occurred in South America from 1906 to 1942 and found that they delineated such a sloping zone 4500 kilometers (2800 miles) long, following the western edge of the continent. It began at relatively shallow depth below the trench that skirts the coast; but under the Andes and farther east it extended down to more than 600 kilometers (400 miles)—one tenth of the way to the center of the Earth. Volcanic eruptions occur over that part of the slope where quakes are taking place at depths close to 100 kilometers (60 miles).

These sloping regions, now known as Benioff zones (see Fig. 5.6), dip down under the continents at about 45 degrees (although the slopes vary from place to place). It was tempting to speculate that the quakes result from friction between a descending ocean floor and the continental block which is being under-ridden. When the ocean floor material reaches a critical depth (roughly 100 kilometers) the heat and pressure melt out of it the most volatile components that then push upward as lava.

There is a striking similarity to the chemical composition of these outpourings around the Pacific perimeter. The resulting rock is called andesite, since it is typical of the Andes. There is a clearcut "andesite line" that separates the islands of the western Pacific, formed by this process and displaying this typical rock, from those of the central Pacific, such as Hawaii.

Two other striking features of the trenches had been noted. One was that the most prominent ones are all about the same depth—10 kilometers (six miles) below sea level—suggesting they were all products of a similar, uniform process. The other was that the continental side of a trench tends to be steeper than the ocean side, the latter curving down in a manner suggestive of material being sucked under.

Pursuing his effort to understand the forces that have shaped the ocean basins, Harry Hess pondered these clues and recalled a discovery he himself had made during World War II. In the course of his prewar research aboard submarines he had joined the Naval Reserve and was called to active duty the day after the Pearl Harbor attack. By the time of the island-hopping invasions of the Western Pacific he had become commander of a troop transport, the *U.S.S. Cape Johnson*, and, as the ship shuttled back and forth carrying troops to landings in the Marianas Islands, the Philippines and Iwo Jima, he insisted that the quartermaster on watch keep the echo-sounder running continuously. (He had seen to it that his ship was fitted with a sounder capable of recording deep-sea depths as well as coastal shoals.)

On a number of occasions, when he looked at its moving scroll, he saw that, far beneath the ship, the bottom had abruptly risen to a flat-topped summit resembling a volcano sliced off at the top. When the war was over, he reported the discovery of 160 of these "drowned ancient islands of the Pacific basin," rising from 3000 to 4000 meters above the ocean floor, but rarely coming closer than 1000 meters to the water's surface. He named the flat-topped ones "guyots" in honor of Arnold Guyot, Princeton's first geology professor (Fig. 5.3).

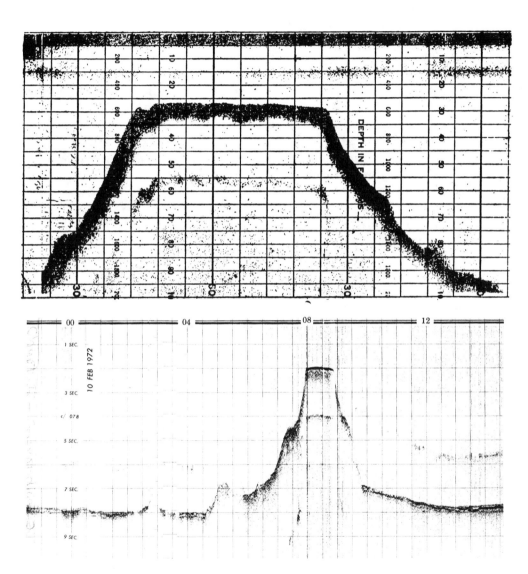

Figure 5.3 Traces of Pacific floor. *Above is a trace of the Pacific floor obtained by Harry Hess with the depth recorder on his Navy transport during World War II, showing one of the flat-topped features, or table mounts, that he called "guyots." Its summit is 1000 meters below the sea surface and rises some 3000 meters from the ocean floor. Below is the trace of a guyot at almost the same site as the one above, recorded some 600 kilometers (400 miles) east of Saipan in the Western Pacific by a "sparker" sounding device aboard the* USNS LYNCH *in 1972. It took the ship fifty minutes to cross the flat-topped summit, indicating a width of fifteen kilometers (nine miles). The vertical scale is exaggerated.*

It seemed almost certain that the guyots were volcanoes that once rose above sea level and were flattened by water action. But why had they not produced atolls? One of Charles Darwin's most memorable accomplishments had been to show that the atolls of the Pacific were ancient volcanic islands that sank below the sea but, as they did so, provided a waterline platform on which tiny marine organisms could build coral reefs. As the islands sank, the coral grew, and, when the island itself vanished entirely beneath the water, a necklace of coral islets remained, enclosing a shallow lagoon—one of the typical atolls.

Hess at first believed that the guyots were formed some time in the Precambrian Era, more than 600 million years ago, before the coral-forming organisms had evolved and when the seas were far shallower than today. Then, in 1956, it was found that sediments retrieved from the tops of some Mid-Pacific guyots contained fossils of shallow-water species dating from the relatively recent Cretaceous Period, 100 million years ago. What, Hess wondered, could possibly have submerged these mountains to such great depths in so short a time?

It was also discovered that some of the sea mounts are tilted on the edges of oceanic trenches as though, having been formed upright, they were now riding a moving carpet down into the trench. One, on the seaward rim of the Tonga-Kermadec Trench, rises 8200 meters from the ocean floor, coming to within 370 meters of the water surface. It is one of the highest mountains in the world, but—if anyone could see it—the mountain would appear to be leaning, like the Tower of Pisa. This strongly suggested to Hess that such sea mounts are being slowly drawn into the abyss. As he put it, they "ride down into the jaw-crusher."

Finally, in formulating his explanation, he had before him some astonishing data on the flow of heat up through the ocean floor. Until the 1950s there was no information at all on oceanic heat flow. On land, European miners, in the 16th and 17th centuries, found that, as they dug deeper, the rock became warmer. It was apparently Robert Boyle, discoverer of the law that bears his name, relating volume and pressure in gases, who first speculated on the cause of this. He made no measurements himself, finding the poorly ventilated mines of the 17th century unbearable. Because he was "particularly subject to be offended by any thing that hinders a full freedom of Respiration," he wrote in 1671, "I was not solicitous to goe down into the deep mines." Instead, he took the word of others regarding a steady rise in temperature with depth and proposed that this was because, very deep down, there are actual fires, or "magazines of hypogeall heat," and that this heat is carried upward either by conduction—"as when the upper part of an Oven is remisly heated by the same Agents that produce an intense heat in the Cavity," or by an upward flow of hot material or vapor. Such an upward flow is typical of a so-called convection current, seen in a churning pot on the stove, where hot material rises over the flame.

It was not until the late 19th century that systematic attempts were made to measure the flow of heat outward from the Earth's interior, and they were all done on land (except for one measurement in a mine shaft extending a short distance under the Irish Sea). When heat is flowing in a uniform manner up through the Earth and escaping at its surface, the extent of this flow can be calculated by recording the drop in temperature within the Earth as the heat moves toward the surface. The requirements are: temperature measurements at two depths a known distance apart and a determination of the efficiency of the material between those points in conducting the heat upward.

On the continents such measurements are made in deep wells or shafts that penetrate below the depth subject to seasonal temperature fluctuations. The chief problem is not recording the temperatures but accurately determining the conductivity upon which accuracy of the results is heavily dependent. Hence, in 1935, the British Association for the Advancement of Science set up a committee to collect information on the conductivity of rock in mines and wells where temperature measurements had been made, or were likely to be made, and an energetic young geophysicist at Cambridge University, then in his late 20s, was assigned as one of the two experimenters. He was Edward C. Bullard and, on the eve of World War II, he began wondering how such measurements might be made in the deep ocean floor.

It was assumed that heat flow there should be far less than on the continents, since it was known that continental rocks, such as granites, generate heat through their relatively rich inclusions of radioactive elements (uranium, thorium, and the radioactive form of potassium). Dredgings in the deep sea had brought up virtually no native granite, and it was suspected that the ocean floors were of basalt—dark, dense, lavalike rock with little radioactive material—and, below that, "basic" rock from which the basalt had been extracted and which was even less radioactive.

Bullard saw that it would not be necessary to penetrate more than a short distance into the sea floor to make a heat flow measurement, since there are no seasonal changes in water temperature at such depths and the temperature situation there must be extremely stable.

After the war (during which he became Assistant Director of Naval Operational Research in Britain), he sought in vain to find a sponsor for his heat-flow experiment and a ship to carry it. Finally, in 1948, the Scripps Institution of Oceanography came to his rescue, and he teamed up there with a graduate student named Arthur E. Maxwell. Since the workshop at Scripps was too busy to build the device, Bullard and Maxwell did so themselves. After some sea trials that produced no results Bullard left to become director of the National Physical Laboratory in England, but Maxwell continued working on the device with his colleagues at Scripps.

Meanwhile the Swedes aboard the *Albatross* had attempted some measurements, but the results apparently were inconclusive, and by 1950 Maxwell and Revelle, the director at Scripps, were ready for a full-scale attempt. The device was a hollow steel spear, three meters long and 4.2 centimeters (1.7 inches) in diameter, fitted with a water-tight, pressure-resistant chamber at its upper end to house the recording devices. It was lowered to within a short distance of the bottom and held there until it had cooled to the water temperature at that depth (which is very cold). Then it was allowed to fall and penetrate the sediment. For 30 to 40 minutes it was left in place, allowing time for most of the heat generated by the friction of its penetration to dissipate.

Two temperature sensors, one near the point at the bottom of the spear and the other near the top, transmitted readings to recording devices in the waterproof chamber so that the temperature differences could later be determined. At the same site a coring device was lowered to sample a cross section of the sediment. This was sent, sealed to preserve its initial moisture, to Bullard's laboratory in England for a conductivity analysis.

In this way heat flow measurements were made from the Scripps research ship *Horizon* at six sites in the deep Pacific, and at all but one the flow of heat from beneath the sea was comparable to that on the continents. Maxwell and Revelle pointed out that, if the oceanic crust, which is only five to nine kilometers (three to six miles) thick, were made of ocean-type basalt, this could explain only 15 percent of the observed flow. And the upward conduction of heat attributable to cooling of the deep interior could only explain 20 percent of what was recorded. Hence, they said, either there is a remarkably high concentration of radioactive material immediately under the oceanic crust (in the upper part of the very thick region, known as the mantle, that separates the crust from the liquid core), or material must be rising through the mantle in the form of a convection current, carrying heat with it.

There were major objections to both ideas. It seemed hard to believe the mantle under the oceans would be much more radioactive than that below the continents. And grave doubts had been raised as to the plausibility of convective flow in the mantle. From observations of earthquake waves that had traveled through the deepest parts of the mantle it seemed that it was as rigid as steel down to the liquid core, 2900 kilometers (1800 miles) below the surface.

Commenting on the Maxwell–Revelle paper in *Nature,* Bullard said it "gives a result which is completely unexpected, and demonstrates again how little we know of submarine geology." Bullard soon had his own device in operation and measurements were made in the Atlantic from the British research ship *Discovery II,* one of them in the Mid-Atlantic rift valley. They showed that the flow of heat through the Atlantic floor was also remarkably high and quite variable from place to place, suggesting that something hitherto unsuspected was going on there.

The flow, Bullard reported, "is too large to be produced by any likely chemical process." For example, he said, to sustain this heat flow over the millions of years that its generation presumably had been maintained would have required the burning of a coal seam 300 meters thick. He noted that some scientists had postulated convection currents rising through the mantle beneath the continents as a force for mountain-building. Perhaps, Bullard suggested, such currents actually rise beneath the oceans and sink under the continents.

Hess, in seeking to reconcile all of these findings concerning the sea floor, was mindful of an elaboration by his old mentor, Vening Meinesz, of the convection current hypothesis. That grand old man of Dutch science was known as much for his fortitude in squeezing his huge frame into the cramped quarters of small submarines for long gravity measurement voyages as for his scientific achievements. As early as 1930, confronted with his findings of remarkably weak gravity over the trenches in front of island arcs, both in the Caribbean and East Indies, he had begun considering the possibility of convection currents moving slowly in the deep Earth and, with such currents, he eventually sought to explain the distribution of continents over the Earth's surface.

In the churning pot, hot material rises directly over the flame, spreads and cools on the surface, thus becoming heavier, and sinks down the sides to be drawn in over the flame at the bottom and heated again. Such a circulation pattern is a "convection cell." It occurs in the flow of air in a room heated by a radiator. And, Vening Meinesz said, it may also occur at a very slow rate in the "solid" Earth under our feet.

Specifically Vening Meinesz considered the material forming the mantle to be sufficiently hot and plastic to flow, much as seemingly rigid ice flows in a glacier. By then it was known, from earthquake studies, that the Earth has an inner core, probably of nickel and iron, surrounded by an outer core of similar material in the molten state (see Fig. 5.4). Outside of that is the "rigid" mantle, constituting the bulk of the Earth's interior, much as the pulp forms most of the inside of an apple. The crust of the Earth is relatively thin—comparable to the skin of an apple. The existence of the liquid outer core had been deduced from study of shear waves generated by earthquakes. In such waves particles along the path move at right angles to the direction of wave travel, as in the waves generated when one wiggles a rope secured at the far end. Since liquids, being non-rigid, have no shear strength, these waves cannot pass through them, in contrast to sound waves (that is, pressure waves), in which motion of the particles is

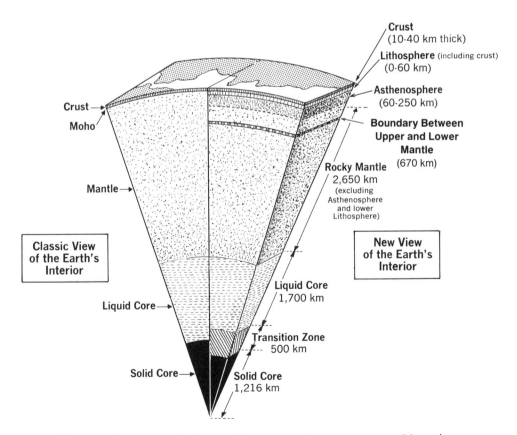

Figure 5.4 Interior of the Earth. Old and revised views of the structure of the earth.

along the travel path of the wave. It was found that a shadow zone exists within the Earth which cannot be traversed by shear waves and it was deduced that this must be the liquid part of the core.

When the Earth was young, Vening Meinesz said, the convection pattern took its simplest form, with all the cold material sinking in the same region (Fig. 5.5). As it did so, its lighter components resisted sinking and formed the counterpart of a "froth" on the surface. In this way a great patch of light rock (chiefly granite) accumulated, while the heaviest material (chiefly iron and nickel) settled into the core.

Once the core had grown to substantial size, there was less room for the circulation pattern and it subdivided into several convection cells. In this new situation some of the rising currents came up under the froth layer—the original supercontinent—and tore it apart. Spreading of this rising material continued to push the fragments away from one another, forming an ocean basin. Along the centerline of this basin, the rising material produced a ridge, its central rift valley torn open by tension generated by the spreading.

As the core continued to grow, Vening Meinesz said, the convection patterns were further constrained. New ones evolved. The patterns of continental drift changed and new mid-ocean ridges evolved.

There were, however, objections to the existence of such massive convection cells. To many geophysicists the Earth's interior was far too rigid for such movements. They noted that, while mountains and continents have sunk to their appropriate depths on the largest scale, they have not done so on scales measured in miles rather than hundreds of miles. The "rind" of the Earth is clearly rigid. The dawn of the space age

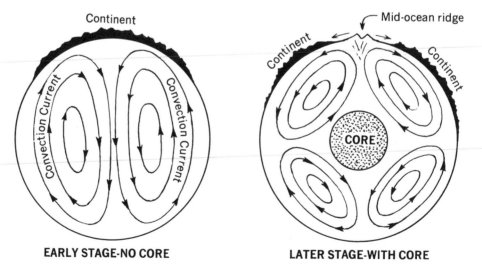

EARLY STAGE-NO CORE   **LATER STAGE-WITH CORE**

Figure 5.5 Convection currents. *Vening Meinesz speculated that before the formation of the earth's core, there was a single system of flow, or "convection cell," within the earth. Then as light material accumulated on the surface to form a primordial continent, heavy material sank to form a core. The growing volume of the core broke the circulation pattern into separate cells whose flow tore the continent apart and formed a midocean ridge along the line of rupture.*

provided low-flying Earth satellites whose orbits were highly responsive to regional variations in gravity. These, too, indicated a "lumpiness" in the shape of the Earth that seemed to imply great internal strength (although adherents of convection argued that the lumps were over rising currents of mantle material).

Furthermore the overall shape of the Earth, with its flattening at the poles and its equatorial bulge, was interpreted as evidence of internal resistance to flow. Gordon J. F. MacDonald, the young Earth scientist who, with Walter Munk at Scripps, had studied the Earth's rotation and its internal response to stress, argued that the polar flattening was too great to represent an immediate response to the Earth's present spin rate. Rather, the flattening was that which, had the Earth been completely liquid and able to adjust instantaneously to a changing spin rate, would have been produced when the world was spinning faster, some 10 million years ago. In other words, there is a 10-million-year lag in adjustment of the Earth's shape to conform to the slower spin.

This implied a very rigid interior, MacDonald said, and therefore large temperature differences would be needed to set it in motion. Heat-flow measurements on the East Pacific Rise tended to be high, relative to the ocean floor as a whole, and those that had been made in the Acapulco Trench, off Mexico, were unusually low, as one would expect if hot material were rising under the ridge and flowing away from it to sink, after cooling, down into the trench. Even so, MacDonald contended, to keep the mantle churning in the proposed manner would require far greater temperature differences, between ridge and trench, than those observed. Taking "realistic values" for the plasticity of the interior, he said, "we see that the thermal requirements of convection in the mantle are so severe that the hypothesis can be dismissed."

Indeed there are many today who still doubt that there can be a turn-over of material through the full depth of the mantle. However Bullard, Maxwell, and Revelle argued that conditions in the mantle must permit convection currents simply because there seemed no other explanation for the amount of heat coming up through the ocean floors. While this flow was not great enough to meet MacDonald's objections, it was sufficiently high to be inexplicable without convection. On continents the radioactivity of granite accounted nicely for the observed heat. The ocean floors were assumed to be made of basalt, which is only a third as radioactive, yet their production of heat was similar to that of the continents.

Bullard and his colleagues argued further that, if the heat under the oceans is largely brought up by rising convection currents, "there should be corresponding regions of low heat flow in places where the material is sinking." The Acapulco Trench area, they said, "may be an example of such a region."

Too often, in scientific findings, the wish is father to the discovery. In this case the authors were refreshingly candid. "In brief," they wrote, "it seems not impossible that the objections to the existence of convection currents in the mantle can be overcome by reasonable assumptions about the properties of the material of which it is composed. That the mantle does really possess these properties is not self-evident. In fact, the main reason for supposing that it does is the desire to have convection currents to account for the oceanic heat flow."

The synthesis, by Harry Hess, of all this evidence into a relatively comprehensive picture of the great movements taking place on the sea floor did what Wegener had failed to achieve. It set the stage for what finally became general acceptance of the modern theory of mobile plates. His initial presentation was published informally by Princeton in 1960 as the draft of a chapter for a forthcoming book on the oceans and, at the time, was not widely circulated. (It was not until 1962 that his proposal appeared in the general scientific literature.) "Whole realms of previously unrelated facts," Hess wrote excitedly, "fall into a regular pattern which is highly suggestive that close approach to satisfactory theory is being attained." Although he considered himself a "dedicated" uniformitarian he said he believed in one "great catastrophe" which was a single great convective overturn like that envisaged by Vening Meinesz for the initial step in formation of the Earth's interior. In this process the heaviest stuff sank to form the Earth's core and the lightest stuff remained on the surface over the descending convection current to form the lighter rocks of a great proto-continent.

As the core grew, the convection currents became more constrained, rising under the mid-ocean ridges, flowing away from them toward the continents, and then vanishing under the trenches. Since the ridges were elongated features, these patterns of flow must be banana-shaped, he said. He cited the newly found rift valleys along the ridge centerlines as evidence of such flow. It would also explain another puzzle: The blanket of sediments on most of the ocean floors is so shallow that, assuming any reasonable rate of sedimentation, the accumulation could not have gone on for more than 200 or 300 million years at the most. Yet, if the ocean basins were as old as the continents, the accumulation time should have been 10 or 20 times longer. Hess cited a report by Maurice Ewing and his brother John that there was almost no sediment on the Mid-Atlantic Ridge and what there was tended to lie in the valleys. A similar report had come from the East Pacific Rise, indicating recent birth. And the fossil remnants of sea life, hauled up by piston corers and other devices, were rarely if ever older than the Mesozoic Era, which began some 225 million years ago. Yet the fossils on land showed that oceanic life had existed 10 times longer than that. How could the ocean basins be permanent features if they had no really old fossils?

Most important of all, Hess concluded that the rock of the ocean floor is not really part of the Earth's crust at all. It is essentially the top of the mantle. Ever since his student days he had been obsessed with curiosity about the composition of that vast and stubbornly inaccessible region known as the mantle and comprising 84 percent of the volume of the Earth's interior. In fact Hess, when he promulgated his theory, was playing a central role in the so-called Mohole Project, whose goal was to retrieve samples of mantle and determine unequivocally what the bulk of the Earth is made of.

He strongly suspected that fragments of the mantle were already in geological collections—notably specimens from St. Paul Rocks, rising from the Mid-Atlantic Ridge abreast of Brazil. These, clearly, had been thrust up from the Earth's interior recently. He concluded that the ocean floors are formed of mantle rock chiefly of a type known as peridotite, altered by chemical reaction with water into serpentine.

The idea that the continents are resting on mantle material that, as a lower layer of the Earth, constitutes the ocean floor was like that of Wegener—but with a basic difference. In the process envisioned by Hess there was no need for the continents to plow through the ocean floors like ships through a frozen sea. He saw mantle material rising as a convection current through the ridges and forming a new ocean floor that spreads away from the ridges in both directions. Thus the ocean floor is constantly moving

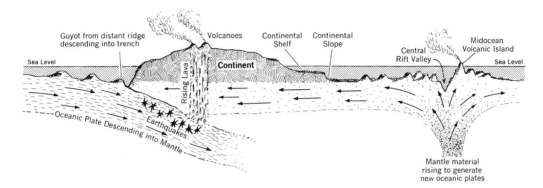

Figure 5.6 Continental drift. *The concept of sea-floor spreading as envisioned by Harry Hess is shown schematically above. New ocean floor is formed along midocean ridges, while old sea floor descends beneath continents or island arcs, generating, as it does so, a down-sloping zone of earthquakes, or "Benioff zone," and volcanic activity. More recently it has been argued that, where the descending slab is very ancient and therefore cool, the lava originates in the mantle above it.*

away from the ridges toward regions where it descends into the depths, as in the Pacific trenches (see Fig. 5.6).

"This is not exactly the same as continental drift," he said. "The continents do not plow through oceanic crust impelled by unknown forces, rather they ride passively on mantle material as it comes to the surface at the crest of the ridge and then moves laterally away from it."

He concluded that the Pacific guyots and deeply submerged sea mounts were volcanic mountains that had formed on the crest of a Pacific ridge, rising to the surface of the sea. The ridge, he proposed, was active some 150 million years ago and is now extinct. The flow of newly forming sea floor away from this ridge slowly carried these mountains into deeper and deeper water. The present flow, away from a younger ridge, is, in some cases, carrying them down into the "jaw-crushers" of ocean-rim trenches. Where continents meet a down-turning of the ocean floor, as colder mantle material sinks back to the depths, the pressure against the continental front produces major deformations (such as the Andes). Likewise sediments and sea mounts riding the descending floor are swept up, crushed, and often are "welded" onto the continents, accounting for some of the complexities of coastal geology.

In conclusion Hess said the Earth is a "dynamic body" with its surface constantly changing: "The Atlantic, Indian, and Arctic Oceans are surrounded by the trailing edges of continents moving away from them, whereas the Pacific Ocean is faced by the leading edges of continents moving toward the island arcs and representing downward-flowing limbs of mantle convection cells..." (see Fig. 5.6).

The active lifetime of midocean ridges is only as long as that of the convection current that sustains them—200 or 300 million years. "The whole ocean is virtually swept clean (replaced by new mantle material) every 300 to 400 million years," he wrote, which would account for the thin veneer of oceanic sediments. In short, he said, "The ocean basins are impermanent features, and the continents are permanent although they may be torn apart or welded together and their margins deformed."

Hess said he considered his paper "an essay in geopoetry"—a free flight of speculation. While his theory will need revisions, he added, "it is hoped that the framework with necessary patching and repair may eventually form the basis for a new and sounder structure."

Hess had discussed these ideas with another explorer of the sea floor, Robert S. Dietz of the Navy Electronics Laboratory in San Diego, and Dietz elaborated the Hess theory, giving it its name: "sea-floor spreading." He had been a Fulbright Fellow in Japan during the mid-50s and had accompanied various navy oceanographic expeditions into the Pacific. Like Hess, he had concluded that the sea mounts in the western part of that ocean are being carried into the jaws of the Marianas Trench.

Dietz called attention to the discovery that sections of the Pacific floor off California have been transported enormous distances, relative to other sections of the floor. A series of widely separated bands of mountainous terrain had been found extending westward across the Pacific from North America. There was evidence of volcanic activity along these so-called fracture zones, and one was marked by an escarpment a kilometer or more in height. A detailed magnetic survey off the West Coast had revealed a north-south striped pattern in which the stripes, alternately, represent bands of sea floor that seemed slightly more or slightly less magnetic than the average for the sea floor as a whole. While the reason for this pattern had not yet been deduced, the stripes were broken at the fracture zones in a manner indicating extraordinary transverse displacements. It was as though a newspaper had been cut vertically and the right side displaced so that its lines no longer matched the lines on the left half. Like the printed lines, the sea-floor pattern made it possible to determine how great the displacement had been. For the various fracture zones it ranged from 200 kilometers (125 miles) or less to as much, in the case of the Mendocino Fracture Zone, as 1200 kilometers (750 miles). "This last displacement," according to Henry Menard of Scripps, "is by far the largest that has been measured on Earth...A similar displacement would tear a continent in half." Furthermore, he said, whatever had torn the ocean floor asunder had not, apparently, affected the adjacent continent to any great degree—a situation difficult to understand (see Fig. 5.7).

Dietz took the fracture zones to be evidence that sections of the sea floor have long been in motion at different relative speeds, like trains moving down parallel tracks in the same direction, but at separate velocities. The continents drift on the flowing mantle, he said, until one descending convection flow meets another descending one. When a continent reaches such an area, it comes to rest, like a floating cake of soap over the whirlpool of a draining tub, and remains there until the flow pattern changes (see Fig. 5.8). In this way, he said, continents tend to be compressed by converging currents, "which accounts for alpine folding, thrust faulting, and similar compressional effects so characteristic of the continents."

Dietz believed that the pattern of magnetic stripes lay at right angles to the direction of sea-floor spreading and represented some form of "stress pattern," but, as will be seen in the next chapter, the stripes soon proved to be far more significant than that. In any case, Dietz recognized that the offsets of the fracture zones indicated great mobility. "While the thought of a highly mobile sea floor may seem alarming at first," he wrote, "it does little violence to geological history."

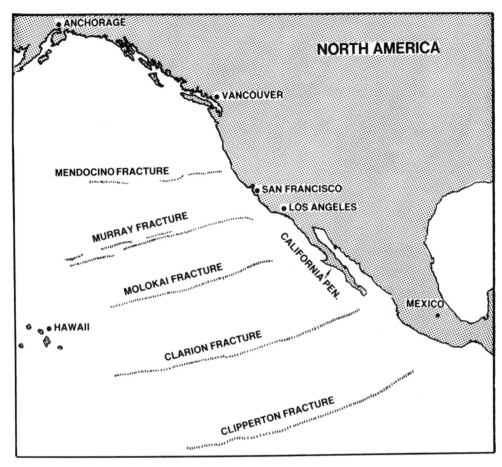

Figure 5.7 Fracture zones. *Earth scientists were baffled by discovery of a series of east–west fracture zones cutting the east Pacific floor. Displacement along the Mendocino Fracture Zone, said Henry Menard, "is by far the largest that has been measured on earth."*

Thus the stage was set for a remarkably elegant confirmation of this hypothesis through surveys of sea-floor magnetism. Enthusiasm for the Hess-Dietz hypothesis was far from universal, however. Arthur A. Meyerhoff of the American Association of Petroleum Geologists wrote that the Dietz argument was "open to serious doubt" and chided Dietz and Hess for allegedly failing to seek out and give credit to earlier proposals along similar lines.

"Failure to do this simple type of research properly," he wrote, "is, in the minds of some, a reflection on one's ability as a scientist and even leaves room for doubt on the validity and accuracy of the published results of new research."

Meyerhoff cited, in particular, the hypothesis published in 1931 by Arthur Holmes of the University of Edinburgh (an earlier version appeared in 1929) that hot rock rises beneath a mid-ocean ridge and flows away from it, carrying continental blocks on its back. The rock then descends into oceanic trenches. In other words, Holmes postulated convection currents. Hess had mentioned Holmes's contribution

PERMIAN -225 million years ago

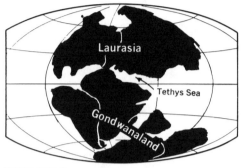

TRIASSIC -200 million years ago

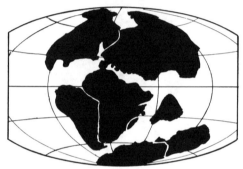

JURASSIC -135 million years ago

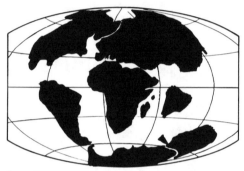

CRETACEOUS -65 million years ago

Figure 5.8 Timetable for continental breakup as envisioned by Robert S. Dietz and John C. Holden.

CENOZOIC -Present

in his 1960 paper. In fact, Wegener himself had mentioned convection currents as a possible force for moving continents in the final edition of his book in 1928. Hess and Dietz, in their reply to Meyerhoff's charge, pointed to a number of differences between the Holmes concept and their own, including the use in the new hypothesis of a mass of recently acquired information on the ocean floors.

Holmes himself felt that, from the meager information then available, his proposal had almost no observational support. "Purely speculative ideas of this kind," he wrote modestly in 1945, "specially invented to match the requirements, can have no scientific value until they acquire support from *independent* evidence."

Meyerhoff argued that others, too, were being credited improperly with inventing "new" ideas that were old. He said that although the convection current hypothesis was being attributed to Vening Meinesz and the Earth-expansion idea to Carey, Dicke, and Egyed, such proposals had been made long before. However, having rapped the knuckles of his geological colleagues, Meyerhoff confessed that he, too, could not "state categorically" that the sea-floor spreading hypothesis was original with Holmes.

He was wise not to do so, for the idea, to a large extent, had been spelled out a half century earlier by an English parson. This relatively little-known prophet was the Reverend Osmond Fisher who, at the age of 18, entered Jesus College at Cambridge University after preparing at Eton and King's College, London. He was one of those who first applied mathematical techniques to geology, studying the formations of Dorsetshire, East Anglia, and the Isle of Wight. But it was a grander arena—the shaping of the entire crust of the Earth—on which he finally focused. His work was facilitated by his selection to be rector of Harlton, six miles from Cambridge, by Jesus College, which controlled the patronage of this "living." The latter provided a home and an annual income of about £ 140.

It was Fisher's book, *Physics of the Earth's Crust,* published in successive editions from 1881 to 1891, that anticipated many of the arguments presented as "new" in the 1960s. Although the shrunken-apple concept of the Earth was much in vogue, he concluded that no plausible shrinkage could explain the great compression of some regions, pointing out that layers of crustal rock in Scandinavia and the Alps have been thrust over other layers for scores of miles. The crushing of the east coast of North America to produce the Appalachians had compressed 160 kilometers (100 miles) of terrain to a width of only 90 kilometers (65 miles).

"The more one considers the instability of the Earth's crust, and the magnitude, according to our ordinary standards, of the movements which have occurred," he wrote, "the more one is lost in amazement. And it is impossible to avoid using the kind of exaggerated language which is the opprobrium of popular geology. Rocks of a single geological period, thicker than the heights of the highest mountains, have been deposited over certain areas, sunk, been re-elevated, crumpled and denuded, and still stand as lofty mountains." He cited one such formation in Wales that is 7000 meters thick and Appalachian sedimentary deposits estimated to be 13 kilometers (eight miles) deep.

While conceding that settling of the crust because the Earth shrank within its skin could induce horizontal pressures "enough to crumple up and distort any rocks," he calculated that the wrinkling to be expected from such shrinkage would produce elevations of only a little more than six feet (two meters).

He therefore proposed that flowing cells of liquid material circulating beneath the rigid crust—he called them "convection currents"—are responsible for the compression. Molten rock rises under the oceans, he said, spreads toward the coasts, and sinks beneath the continents. This subterranean flow pushes against the coasts, producing the mountain ranges that parallel the edges of continents.

The upwelling of lava occurs chiefly under the mid-ocean ridges, which he called plateaux. "It is on these plateaux that the volcanic islands of mid-ocean are based," he wrote, "and it is obvious that the condition of the molten substratum with upward currents pressing against the underside of the crust is exactly that, which would tend to rupture it, and open fissures, and originate volcanic vents."

Such volcanic islands occur here and there along the centerline of the Atlantic from Iceland in the north (where open fissures are much in evidence) to Tristan da Cunha close to the South Polar seas.

Where the molten rock nears a continent and begins sinking back into the depths, frightful stresses can develop within the Earth, Fisher said. "Now it is known that most of the earthquakes which disturb the much shaken islands of Japan originate beneath the sea on that side; which shows that the sub-oceanic crust in that region is in a very unstable condition, as it would be if it were thus sinking."

Fisher believed the down-current pulls some of the crust with it. He described how a crust forms on the lava lakes of Kilauea Volcano in Hawaii, remains buoyant for a while, cools, and then sinks, citing the perceptive observations of Mark Twain, who watched this process on his visit to Hawaii in 1866: "Here was a yawning pit upon whose floor the armies of Russia could camp, and have room to spare," Twain wrote. The greater part of the crater floor, he said, was inky black, "but over a mile square of it was ringed and streaked and striped with a thousand branching streams of liquid and gorgeously brilliant fire! It looked like a colossal railroad map of the state of Massachusetts done in chain lightning on a midnight sky...Occasionally the molten lava flowing under the superincumbent crust broke through—split a dazzling streak, from five hundred to a thousand feet long, like a sudden flash of lightning, and then acre after acre of the cold lava parted into fragments, turned up edgewise like cakes of ice when a great river breaks up, plunged downward, and were swallowed in the crimson caldron."

Similarly, Fisher believed, the drifting crust of the Earth under the oceans becomes so cool and dense that it sinks back into the depths when it reaches the continental margins.

Unlike the true drifters—those who believe the continents to be in constant (or at least intermittent) motion with respect to one another—Fisher argued for a single episode of drifting, which he tied to Sir George Darwin's proposal that the moon was torn from a fast-spinning Earth, removing a large section of its crust. It was this event, leaving the Pacific basin as a giant scar, that set in motion the great convection currents within the Earth, Fisher said.

Sir George, son of Charles Darwin of evolution fame, postulated ejection of the Moon some 50 million years ago—a time, he said, when the Earth was spinning once every five hours. The centrifugal force of this spin was not enough, by itself, to throw off part of the planet, but he proposed that the spin rate matched the Earth's natural oscillation frequency like the critical speed that can set a car vibrating severely. With the Earth's natural oscillation frequency the same as its spin rate of five hours, Sir George argued, solar gravity would act on the Earth in tempo with this oscillation, producing ever higher tides in the semi-molten Earth until a huge chunk was thrown off and became the Moon.

Molten material then flowed in from the interior of the Earth to fill the cavity, and it was this flow, Fisher believed, that pulled the Americas loose from Europe and Africa, carrying them toward the Pacific cavity and producing the Atlantic. The fluid flow thus set in motion within the Earth then persisted as convection currents beneath the oceans. As evidence for such fluid motion he cited the changing orientation of the Earth's magnetic field. The magnetism, he argued, is generated by this flow and changes direction as the movements of molten material within the Earth are altered.

The idea that the Earth's interior, below a crust some 35 kilometers (22 miles) thick, is formed of hot, fluid lava was the accepted view, but this was shown to be impossible by William Thomson, later Lord Kelvin, who had graduated from Cambridge four years after Fisher. Kelvin pointed out that, if the Earth were almost entirely fluid, it would respond to the gravitational pull of Sun and Moon much as the oceans do, producing twice-daily tides of the land as well as the sea. With both land and sea responding together, there would be no visible tidal action along the coasts of the world. Kelvin concluded, therefore, that the bulk of the Earth's interior must be as rigid as steel and Fisher's idea of convection currents soon was forgotten. While his proposal that the magnetism of the Earth is generated by fluid motions beneath the crust was invalid, it anticipated the modern view that it is produced by such flow in the liquid core.

It can be argued that even Benjamin Franklin had glimmerings of the modern concept of surface plates driven by the flow of material underneath them. In a letter to Abbé Soulavie on September 22, 1782 (read subsequently to a meeting of the American Philosophical Society), he wrote:

> Such changes in the superficial parts of the globe seemed to me unlikely to happen if the Earth were solid to the centre. I therefore imagined that the internal parts might be a fluid more dense, and of greater specific gravity than any of the solids we are acquainted with; which therefore might swim in or upon that fluid. Thus the surface of the globe would be a shell, capable of being broken and disordered by the violent movements of the fluid on which it rested... If [these thoughts] occasion any new inquiries and produce a better hypothesis, they will not be quite useless. You see I have given a loose to imagination; but I approve much more your method of philosophizing, which proceeds upon actual observation, makes a collection of facts, and concludes no farther than those facts will warrant.

As the volume of scientific findings and proposals swells to ponderous proportions, it becomes more and more difficult for a scientist to determine if his work is original and to avoid repeating earlier efforts. But, as the history of the sea-floor spreading theory shows, the problem of learning what others have done and proposed existed before the deluge of scientific papers that has burgeoned in the final decades of this century.

A development soon after Hess and Dietz put forth the concept of sea-floor spreading made that theory seem more plausible. This was confirmation of an old and discredited idea that a relatively soft layer of rock lies between 60 and 250 kilometers (40 and 150 miles) below the surface. Early in the century some scientists had proposed such a layer as a source of the lava that erupts through volcanoes. However, it became evident that the crust of the Earth is underlain by denser rock, assumed to be the uppermost part of the mantle. After an earthquake in 1909 Andrija Mohorovičić, at the University of Zagreb in what is now Yugoslavia, observed that seismographs less than 800 kilometers (500 miles) from the quake had recorded two sets of tremors, as though there had been two quakes. More significant, the farther an instrument was from the site, the greater the separation of the first arrivals of these two sets of waves. From this he deduced that the waves had traveled via two layers, the lower, denser, and more rigid of which carried the waves much faster than the upper layer.

The demarcation between these two layers, assumed to mark the boundary between crust and mantle, came to be known as the Mohorovičić discontinuity, or "Moho." On the continents it is some 40 kilometers (25 miles) below the surface and at sea from five to 10 kilometers (three to six miles) below the ocean floor.

As early as 1926, four years before he left the University of Frankfurt in his native Germany for the California Institute of Technology, Beno Gutenberg reported evidence for a deeper layer, centered at a depth of about 150 kilometers (93 miles), in which earthquake waves travel six percent more slowly than immediately below the crust. This, he thought, could mean that at that depth the mantle material, under the existing conditions of heat and pressure, changed from a crystalline to a "vitreous" state, more amenable to flowing. Gutenberg therefore came to believe that continental drift was a possibility—not a very popular viewpoint at that time.

It was in large measure the nuclear weapons race that led to confirmation of Gutenberg's proposal. An underground nuclear explosion produces a sharp crack, compared to the "rumble" of a natural earthquake, and this made for cleaner earthquake data. Furthermore, in anticipation of a treaty that might ban such tests, an intensive effort was launched to develop instruments and techniques capable of discriminating between natural quakes and clandestine weapons test explosions. Typical of this effort was the construction of a seismic array consisting of more than 500 stations spread over an area of southeast Montana half again as large as Massachusetts, their readings channeled into a central computer facility.

At the same time an international effort had been launched as a sequel to the International Geophysical Year of 1957–1958. Known as the Upper Mantle Project, it was initiated by Vladimir V. Beloussov, who had represented the Soviet Academy of Sciences in the directorship of the I.G.Y. As justification for the project it was stressed that processes in the upper mantle are largely responsible for such phenomena, of vital importance to human welfare, as volcanic eruptions, earthquakes, mountain-building, and the formation of ore bodies.

By the time this program ended, in 1970 (and by the time a treaty banning aboveground tests had been signed), the more sophisticated earthquake observations had made it clear that there is, in fact, a layer of less rigidity in the region from 60 to 250 kilometers (40 to 150 miles) below the surface. It is apparently not formed of glassy basalt, as Gutenberg had thought, but of crystalline rock which, because of the combination of pressure and temperature at that depth—and a small but critical admixture of free water—can flow. Some described it as a "crystal slush," although that gave an exaggerated idea of its fluidity. The earthquake studies showed that most quakes associated with volcanic eruptions originate at depths between 60 and 200 kilometers, supporting the view that the lava comes, originally, from that "slushy" layer, although subject to various kinds of alteration en route.

One of the most violent earthquakes of the century, that which shook Chile on May 22, 1960, helped convince the scientific world that the soft layer was real. The quake imparted so sharp a blow to the Earth that the entire planet vibrated as a unit, like the ringing of a bell. The complex task of analyzing these vibrations, as recorded by detectors tuned to very long-period earthquake waves, was undertaken by Frank Press and Don L. Anderson, then his student at the California Institute of Technology. The results were best explained by the existence of the soft layer.

Early in the century an American geologist, Joseph Barrell, had proposed that the Earth is rigid only to a depth of about 100 kilometers (60 miles), below which it is plastic. He called the rigid layer the "lithosphere," from *lithos,* the Greek for "stone." Beneath it, he said, was the "asthenosphere," from *asthenes,* the Greek for "weak." These terms are now in wide use to describe the drifting plates, formed of the rigid lithosphere (including the topmost part of the mantle, as well as the crust above it) and the plastic asthenosphere on which the plates ride. The term "crust" arose in the days when the Earth's surface was thought of as a layer of slag on top of a molten interior. The nature of the moho—the transition zone between crust and upper mantle—is still uncertain, but it is believed to be a change either in chemical composition or in molecular structure from a less dense phase, above, to a denser phase below (see Fig. 5.4).

Discovery of the plastic asthenosphere began winning new and important converts to drift. At the University of Toronto Tuzo Wilson saw in it a "lubricating channel" upon which the continents and sea floor could move. He pointed out that, if oceanic islands and guyots were formed on a ridge and then rode away from it on the conveyor belt of moving ocean floor, then islands near the ridges should be relatively young and those farther away correspondingly old.

Wilson put his students to work searching the scientific literature for clues to the ages of some 40 islands and, from the limited information available, it appeared that the farther an island was from the mid-ocean ridge the longer ago it was formed. Furthermore, some sites on the Mid-Atlantic Ridge have apparently been generous erupters for millions of years, their output of lava being carried both to the east and west by the flow of ocean floor away from the ridge. One such source of material seems to lie under Tristan da Cunha Island in the South Atlantic. Submarine mountain chains reach from it toward both South America and Africa (although they are not at right angles to the mid-ocean ridge, as one would expect). Similar undersea features reach from Iceland, on the Mid-Atlantic Ridge, toward both Europe and Greenland (see Figs. 14.1 and 10.4).

Also conforming to this hypothesis are the Hawaiian Islands, each of which becomes older from southeast to northwest (Fig. 5.9). The big island of Hawaii, at the southeast and youngest end of the chain, is the only one with fully active volcanoes (Fig. 5.10). Its nearest neighbor, to the northwest, is capped by a dormant volcano, Haleakala. The period of activity that built the remaining islands becomes more and more ancient as one follows the chain out 2640 kilometers (1640 miles) to Kure, an atoll whose parent volcano has entirely vanished beneath the sea. Thus Hawaii seems to rest, today, over the spot in the mantle from which the island-building material is erupting, but, like all its predecessors in the chain, it is destined to be carried northwest, making room for a new island to grow over the deep-seated "island-making machine." Such an island, Loihi Sea Mount, is already evolving southeast of Hawaii.

Wilson also looked for clues to a former marriage between North America and Europe. One of the most striking geographical features of the British Isles is the Great Glen Fault. It cuts Scotland in two, running diagonally southwest from Inverness on the

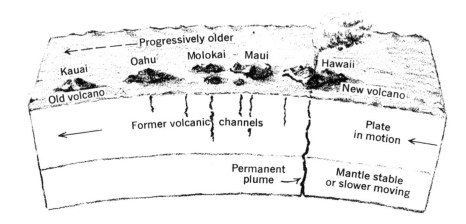

Figure 5.9 Formation of the Hawaiian Islands as envisioned by J. T. Wilson. Pacific plate moves over a plume of hot rock. Older islands are carried north and west, sinking and eroding as they go.

North Sea through Loch Ness and Loch Linnhe to the Isle of Mull on the Atlantic side, and was said to represent a slippage of northern Scotland 105 kilometers (65 miles) southwest, relative to the remainder of Britain.

There is a similar cleavage in North America, Wilson said, though it is not so striking. It bisects Newfoundland, separates Nova Scotia from New Brunswick, and then follows the coasts of Maine and Massachusetts as far as Rhode Island. He called it the Cabot Fault. Both the Cabot Fault and Great Glen Fault, he said, were formed in late Devonian and early Carboniferous times, about 350 million years ago, when Europe and America should have been snugly fitted together. In fact, he proposed they composed a single, continuous fault. In other words, one of the salamanderlike inhabitants of that time could have made its way along one deep valley from what is now Boston to the site of Inverness.

Wilson also concluded that there was an earlier Atlantic Ocean—a "Lower Paleozoic Atlantic"—which existed several hundred million years ago and then was snuffed out by an earlier phase of continental drift. He pointed to a remarkable feature of fossil remains of coastal marine organisms at that early time: fossils from Scotland and the Norwegian coastal zone were far more closely related to American species of that period than to those of Europe. Likewise, some fossils from parts of the North American seaboard had European, rather than American affinities.

Wilson's explanation was that, after the earlier Atlantic was closed out and a great continent was formed by welding together Europe, Africa, and the Americas, this continent again split apart to form a new Atlantic Ocean; but part of what had been Europe and Africa stuck to North America, and part of what had been North America stuck to Europe. In other words, the cleavage did not occur along exactly the same line as the previous closure. A portion of continental crust bearing fossils of an earlier European way of life clung to America and vice versa.

Wilson, who had been a doubter, had thus become fully converted to drift theory, arguing for a complex history of movements in which one continent married another,

Figure 5.10 Lava fountain at Pu'u O'o on the island of Hawaii. The fountain height reached approximately 800 feet (243.9 meters).

divorced it, and drifted off to wed some other piece of real estate. His early efforts regarding drift were not received with unbridled enthusiasm. He recalls how, in 1963, his paper proposing assembly-line formation of the Hawaiian Islands was turned down by the *Journal of Geophysical Research*. The specialist asked by the journal to comment on his submission "recommended rejection on the grounds that the paper contained no new data, had no mathematics in it and disagreed with current views."

"Instead of arguing about the matter," Wilson said, "I slapped it into the *Canadian Journal of Physics*."

After assuming the prestigious post of president of the International Union of Geodesy and Geophysics, he became one of the most eloquent protagonists of drift. When the union held its General Assembly at Berkeley in 1963 Tuzo Wilson threw down the gauntlet before his colleagues. "The hypothesis of drift," he said, "has only been accepted by a few geologists, mostly scattered in the southern hemisphere, and is now being looked upon with favor by a few oceanographers and other geophysicists...Some, like myself, have of course changed from one view to another.

"It will be difficult for most of us to accept that large amounts of what we have written and taught has been erroneous. It will be embarrassing to admit that because traces of new petroleum are only found in tropical regions it is likely that older deposits not now found in the tropics have drifted away from them."

Other sciences, he said, have become bogged down in blind alleys from which they could escape only through a revolution in thinking: "I believe that Earth science is ripe for such a revolution, that in a lesser way its present situation is like that of astronomers before the ideas of Copernicus and Galileo were accepted, like that of chemistry before the ideas of atoms and molecules were introduced, like that of biology before evolution, like that of physics before quantum mechanics. Before each revolution all the pegs seemed square and all the holes round. In each case it was not until it was realized that one had to discard the whole frame of reference and seek another that answers came in a flood. If Earth scientists have been trying to fit the history of an Earth which has in truth been mobile into the framework of a rigid and fixed pattern of continents, then it is not surprising that it has been impossible to answer the major questions. It is not our methods nor our observations that have been wrong, but our whole attitude.

"All of these major revolutions cast their shadows a long way before them. Greek astronomers foreshadowed Copernicus by 2000 years; Darwin got ideas about evolution from his grandfather. It should not surprise us that Wegener's and Holmes's ideas have been before us for some time without acceptance."

A potent test of the drift theory, Wilson said, would be to find out if, in their movements, the continents had left evidence of their motion on the sea floor. "No sailor wishing to know if his ship is moving looks in the lounge, he looks at navigating instruments or observes the wake."

Wilson thought that, in the chains of submarine mountains linking such Mid-Atlantic islands as Tristan da Cunha with the continents on either side, he had found such a "wake" of continental drift. But a far more convincing "wake" was imprinted on the floors of all oceans. Although Wilson did not realize it at the time, the first samples had already been discovered in the striped magnetic patterns of the Pacific floor off North America—patterns that came within a few miles of the auditorium at Berkeley where his audience was seated.

Chapter 6

Magnetic Footprints

ONE WOULD HARDLY EXPECT THAT ETRUSCAN VASES HAVE A TALE TO TELL OF THE past history of the Earth's core, but that, in fact, seems to be the case. Late in the 19th century, from studies of magnetism it was realized that, when certain substances, including potting clays, cool after high heating, they become imprinted with the magnetic field of the Earth. The force lines of this field, as observed at the surface of the Earth, resemble those that would be generated by a bar magnet near the center of the Earth. The shape of such a magnetic field is often demonstrated in the schoolroom, two-dimensionally, by holding a bar magnet under a sheet of paper sprinkled with iron filings. The filings instantly arrange themselves in a beautifully symmetrical pattern of curved lines that tend to radiate from points nearest the two poles of the magnet.

On the Earth's surface the force lines are vertical at the magnetic poles and horizontal at the magnetic equator. Since the compass needle is oriented solely by the horizontal component of the Earth's field, at the magnetic poles (where there is no horizontal component) the needle swings aimlessly, whereas at the magnetic equator (which is quite close to the geographic equator) the force controlling a compass needle is strongest (although still several hundred times weaker than that between the poles of a toy horseshoe magnet).

One of the tools of the polar explorer is a dip needle, which is similar to a compass except that it swings in a vertical, instead of horizontal, plane. The dip of the needle indicates roughly his distance from the magnetic pole. When it becomes vertical, the explorer knows he is at that pole.

Pots, urns, and vases, old and new, can serve as frozen dip needles—frozen at the time of their firing, because, when they are very hot, the magnetic particles within them become liberated and can orient themselves with the force lines of the Earth's magnetic field. When the vase cools, the particles are locked into place and the orientation of the magnetic field, measurable by laboratory methods, is preserved for posterity.

By 1907 magnetic analyses had been done on Etruscan vases of the eighth century B.C., Greek vases of the seventh century B.C., neolithic pots from about 1500 B.C., and

87

various volcanic lavas that also had captured the local magnetic field when they cooled. It was assumed that the pots had all been baked upright and the remanent magnetism within them, therefore, indicated the local dip of the field at the time of their firing. It was thus shown that the local magnetic field, in the area where each of these objects was fired, had changed.

Ever since the 17th century navigators have known that there is a slight year-to-year change in the magnetic field, requiring small corrections in compass headings. These changes have local peculiarities, presumably related to the internal composition of the Earth. The overall axis of the field drifts slowly westward, taking about 10,000 years to make one complete journey around the Earth's spin axis, and the strength of the field, at any one point, may vary by 50 percent within several thousand years. It has long been assumed that these changes manifest some form of activity deep within the planet. As Christopher Hansteen put it, early in the last century, "The Earth speaks of its internal movements through the silent voice of the magnetic needle."

Another way in which the Earth's magnetic field becomes imprinted for future reference is in the formation of underwater sediments. As the material settles to the bottom, or before loose, wet sediment is compacted and finally turns to rock, the magnetic grains within it are free to align themselves with the Earth's field, but from then on they are locked in place.

In the 1940s scientists from the Department of Terrestrial Magnetism of the Carnegie Institution of Washington painstakingly studied a long succession of ice age varved clays from New England. A varved clay is one that has been laid down, layer upon layer, by annual cycles of sedimentation, leaving a record of the past somewhat analogous to that formed by rings in a tree. From the clays it was found that between 15,000 B.C. and 9000 B.C. magnetic "north," as seen from New England, drifted back and forth to either side of true north.

If clays laid down during the past few millennia carried a record of the Earth's magnetic history, what about rocks formed from similar sediments of much greater age? Was it possible that, within the sedimentary rocks that carpet much of the Earth, there lurk hidden compasses frozen into place millions of years ago? To test this idea John W. Graham, a young doctoral candidate working with the Carnegie group, collected sedimentary rocks from the Hudson Valley, Colorado, Utah, Wyoming, and Montana. Finally, in sediments that had lain in West Virginia for 200 million years, he found convincing evidence that such a magnetic memory had survived intact. For a distance of 50 miles the formation showed a consistent direction for the magnetic pole—one very different from that of today.

Even more powerful evidence came from lava flows. In some regions of the Earth, such as Iceland, California, Oregon, the Aleutians, Hawaii, and Japan, there have been intermittent eruptions for many millions of years, laying down lava layer upon lava layer, each of which has trapped the magnetic field in existence when it cooled. When the oxides of iron and titanium within such lavas are heated above a certain level (known as the curie point and usually considerably below the melting temperature) atoms within these substances align themselves with the local magnetic field. Then, once the rock cools below that level they become locked in their orientation.

The resulting "magnetic memory" of the lava is preserved indefinitely, although the rock may also acquire a more transient or "fluid" form of magnetism that helps mask the permanent form. This fluid magnetism can be "washed" from a specimen by rapidly fluctuating magnetic fields—a process that does not affect the permanent imprint.

From early studies of remanent magnetism in old lava flows and sedimentary rocks it appeared that locations of the magnetic poles in the distant past were very different from those of today. This was taken by some to indicate slow wandering of the Earth's spin axis relative to surface geography. The difficulty in assessing such evidence for polar wandering was the absence of any accepted explanation for the main magnetic field of the Earth and its relationship to the Earth's spin. In 1946 Patrick M. S. Blackett started to investigate magnetism, including the magnetic properties of rocks on the Earth's surface, and the results ultimately led him into the camp of drifters—the first member of that scientific aristocracy, the Nobel Laureates, to become a full-fledged champion of continental drift.

The manner in which rotating bodies that carry no electric charge (like the Earth) generate magnetic fields was of special interest to Blackett because of his search for an explanation of the behavior of cosmic rays. The "rays" are actually atomic nuclei traveling almost at the speed of light. They must be generated in certain regions of the cosmos, yet they rain on the Earth uniformly from all directions (although the Earth's magnetic field alters their paths as they approach). The energy of some of them far exceeds that of any particles generated by processes native to the Earth. If it could be learned where they come from in the sky, that would be a clue to the nature of their origin. The fact that, instead, they arrive from all directions suggested that something in space is warping their paths so that their final approach is no clue to their source.

The obvious explanation for such warping was the existence of far-flung magnetic fields in space. It had just been discovered that some stars are magnetic, but the absence of any well developed theory to explain how such fields are generated focused Blackett's attention on this problem. He tried to solve it by borrowing gold from the Bank of England in order to test a proposal, made some 30 years earlier, that the magnetism of the Earth and Sun might arise simply as a by-product of rotation by a massive object. The idea had intrigued Einstein. Late in 1930 an electrical engineer in Louisiana had written Einstein, arguing for a relationship between the magnetic and gravitational fields of the Earth that, the letter-writer thought, might provide a path to a unified theory for those two basic phenomena: gravity and electromagnetism. Einstein replied: "Your letter has interested me to the utmost, since I have for long been convinced that the magnetic field of the Earth is not produced by magnetic bodies, but by the entire mass of material linked to the Earth's rotation..."

Since, presumably, only objects of considerable mass would produce a significant amount of magnetism in this way, Blackett turned to "the old lady of Threadneedle Street"—the Bank of England—to enable him to construct a sphere of substantial mass, but workable size. Thanks to the trusting cooperation of that institution he was able to set up in his laboratory a cylinder of pure gold weighing 15.2 kilograms (38.5 pounds). The Earth's daily spin caused the cylinder to rotate, relative to the heavens, and Blackett developed an extremely sensitive magnetometer to see if any magnetism was produced. He could detect none. (It was for Edward Bullard, his wartime associate in naval operations research, and Walter Elsasser, a German geophysicist who had come to the United States in the 1930s, to propose what became a widely accepted explanation for the Earth's magnetism, namely that it is generated by a dynamolike process involving the Earth's spin and motions of molten iron within the outer core.)

Armed with his supersensitive magnetic detector, Blackett turned his attention to the magnetic memory within ancient rocks. He saw in the frozen compass needles within rocks of all ages and from all parts of the world a way not only to unravel the history of the Earth's magnetic field, but to assess the possibility of continental drift.

In a 1954 lecture in Jerusalem he said: "Major countries will have to study the magnetism of their own rocks just as they do their own geology. I have no doubt at all that the result of this work will, in the next decade, effectively settle the main facts of land movements, and in so doing will have a profound effect on geophysical studies of the Earth's crust." The actual time interval was somewhat more than a decade, but this was a prophetic statement.

When Blackett first became interested in continental drift he was professor of physics at the University of Manchester, having taken over that post from W. Lawrence Bragg, who, with his father, William Henry Bragg, had won a Nobel Prize in 1915 for their discovery that X-rays could be used to determine molecular structure (by a technique known as X-ray crystallography). It was this development that later made it possible, for example, to decipher the double helix structure of DNA (deoxyribonucleic acid), the code-of-life molecule.

When Blackett (later to become Lord Blackett) arrived in Manchester the geological community there was far from hospitable to drift theory, as his predecessor, Bragg, had discovered. One day in 1919 Bragg went for a walk in the nearby Derbyshire hills with Sydney Chapman, an inveterate walker and cyclist. Chapman, then also on the Manchester faculty, was to become one of the world's leading authorities on the Earth's magnetism and ultimately was to serve as chairman of the committee that organized the International Geophysical Year.

He had just completed a walking tour of Norway in the course of which he had stopped off to see Jakob A. B. Bjerknes, the great Norwegian meteorologist. It turned out that a conference of German and Norwegian weathermen was being held there. Chapman had been invited to sit in on the sessions, at one of which Alfred Wegener expounded his drift theory. Chapman was greatly excited by the idea of moving continents and, on his walk with Bragg, explained the theory at length.

"Chapman's description of it impressed me so much," Bragg wrote later, "that I can remember the exact spot where he began to talk about it and where the idea of the very great movement of the continents involved in the theory dawned on me. I was so thrilled that I wrote to Wegener for an account of his theory, got it translated, and presented it to our Manchester Literary and Philosophical Society." This prestigious organization had been founded a century earlier by John Dalton, father of modern atomic theory. But, Bragg said, its geological members were "furious." Until then, he added, he had never known at first hand what it meant to "froth at the mouth." In fact, he said years later, "words cannot describe their utter scorn of anything so ridiculous as this theory, which has now proved so abundantly to be right."

Blackett was not one to be deterred because a concept was unpopular, particularly when he found that his new field of interest, the remanent magnetism in old rocks, might resolve the controversy. His students fanned out over Britain and far more remote parts of the world to collect specimens and determine, from laboratory analysis, the orientation of magnetic field lines that had passed through them thousands or millions of years earlier. If the placement of the sample was carefully noted before it was extracted from its parent formation, the former orientation of the magnetic force lines within it could be determined, not only in terms of their dip, as in the ancient pots, but three-dimensionally (in terms of north–south and east–west vertical planes and the horizontal plane).

One of the first indications of changed geography derived from such analyses came from three members of Blackett's group: J. A. Clegg, Mary Almond, and Peter H. S. Stubbs. They found that the "compass needles" in rocks laid down in England during the Triassic Period, 200 million years ago, did not point to the magnetic pole of that period deduced from rocks elsewhere. The apparent explanation, they reported in 1954, was that England, in the interim, had rotated 34 degrees clockwise. Similarly from the continent came evidence that Spain had swung away from France, opening out the Bay of Biscay and crushing the hinge area to form the Pyrenees.

The protégé of Blackett who was to become the chief British ball-carrier for drift theory was S. Keith Runcorn, a stocky, red-haired native of Lancashire who seemed to relish controversy. Runcorn studied lava flows in Oregon and then, in 1954, scrambled down into the Grand Canyon, collecting shale samples "from top to bottom." From these and specimens gathered elsewhere it was possible to reconstruct the migration route of the North Magnetic Pole over the past several hundred million years, from the southeast Pacific toward the Philippines (in terms of present geography), then north across China and Siberia into the Arctic Ocean and its present position. Since the work on New England clays indicated that in more recent times the magnetic pole had never wandered very far from the geographic North Pole (which lies on the spin axis of the Earth), it appeared to Runcorn that the spin axis, too, had migrated. "We can only suppose from this," he wrote, "that...the planet has rolled about, changing the location of its geographic poles. Either mountain building or convection currents in the mantle might account for this rolling."

His argument was that the thrusting up of new mountains, particularly in midlatitudes, would disrupt the perfect equilibrium of the Earth's spin, just as would the development of an ice sheet in such an area, as pointed out by Walter Munk of the Scripps Institution of Oceanography and others. Convection currents, likewise, could upset the stability by introducing new material where it had not previously existed, and Runcorn believed they could be the answer. But his papers arguing the point were not always warmly received. He tells of one rejection slip from a scientific journal citing a referee who said, "This is another in a number of papers by Runcorn on convection in the mantle and I hope it will be the last." Runcorn, however, gradually came to the view that the magnetic records from Europe and North America were best explained by relative motion of the continents.

By 1960 it was evident to Blackett's group that radical changes in the magnetic latitude of various regions had occurred. Europe and North America were near the magnetic Equator 300 million years ago, they said. Australia "appears to have moved in latitude in a somewhat complicated way." Some 400 million years ago it was near the Equator, then drifted to within 2300 kilometers (1400 miles) of the South Pole 200 million years ago, and finally north to its present latitude. "India has apparently moved farther and faster than any other continent," they wrote, from a latitude farther south than that of Australia 150 million years ago, across the Equator to its present location.

By this time Runcorn had moved to King's College, at Newcastle upon Tyne, and had formed a powerful team of investigators of rock magnetism. Using a map of the globe with today's geography they plotted positions of the North Magnetic Pole as far back as the Precambrian Period, more than 600 million years ago. What struck them was the difference between the path of the migrating pole as indicated by American rocks and that derived from European rocks (Fig. 6.1). The positions based on European rocks always lay to the east of those calculated from American rocks. Did this mean there were two magnetic poles? It struck Runcorn and his colleagues that, if they shifted Europe up against North America, the two indicated paths of polar motion came into coincidence.

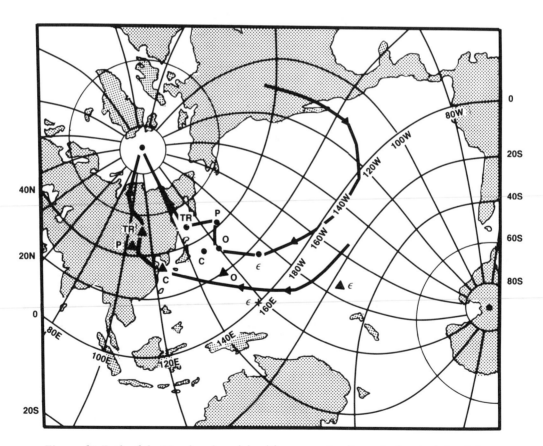

Figure 6.1 Path of the North pole *as deduced from magnetic sediments in America (triangles) and Britain (dots). "TR", "P", "C", "O", and "ε" stand for the Triassic, Permian, Carboniferous, Ordovician, and Cambrian respectively.*

While this, to Runcorn and others, was a strong argument for drift, another aspect of the Earth's magnetism demonstrated sea-floor spreading in a manner that even the most determined skeptics could not ignore. A long-standing puzzle had been the 1909 discovery by Bernard Brunhes that some ancient lava flows in the Massif Central, the mountains that bisect France, are imprinted with a magnetic polarity exactly opposite that of other rocks of somewhat greater or lesser age. Brunhes assumed that this was a local phenomenon. Then, in 1929, a Japanese scientist, Motonori Matuyama, found that in Japan many rocks laid down some 700,000 years ago also were reversely magnetized. Was it possible that the entire magnetic field of the Earth had temporarily flipped over so that a compass needle that normally pointed north would point south? This seemed too preposterous for serious consideration. In the next three decades a number of other formations with reversed polarity were discovered and, in 1950, John Graham of the Carnegie Institution proposed that some rocks, because of their own properties, may become imprinted with reversed polarity. Laboratory tests showed that, in certain rare rock formations, this was the case, but it was found that such "self-reversing" was characteristic of less than one percent of those specimens that, long ago, had acquired reversed magnetic polarity.

Furthermore, rocks laid down over the great span of geologic time seemed about equally divided between formations with normal polarity and those with reversed polarity. This proved to be the case in an extensive series of lava flows laid down in Iceland during the Late Tertiary Period, from one to 20 million years ago. Runcorn found the same pattern in Oregon lavas erupted during the Miocene Epoch of the Tertiary, five to 25 million years ago. Evidence of reversals was found in rocks 400 million years old, and alternate periods of normal and reversed polarity were identified in a timetable that remained consistent throughout a series of South African dikes, or vertical lava intrusions, extending for more than 300 kilometers (185 miles)—the most extensive such formation in the world. This seemed to rule out lightning flashes or other local events as the explanation.

Such evidence led a trio at the laboratories of the United States Geological Survey in Menlo Park, California, to undertake a major investigation. They were Allan Cox, G. Brent Dalrymple, and Richard R. Doell, all three of whom had begun studying the Earth's magnetic history while at the University of California at Berkeley. They realized that the reversals represented a major scientific challenge. "The idea that the Earth's magnetic field reverses at first seems so preposterous," they wrote later, "that one immediately suspects a violation of some basic law of physics, and most investigators working on reversals have sometimes wondered if the reversals are really compatible with the physical theory of magnetism."

They cited the "dynamo" theories of the Earth's magnetic field developed by Bullard and, independently, by Elsasser. In a manmade dynamo the motion of electric wires through a magnetic field generates electricity, whereas within the Earth it is the convective flow of electrically conducting material in the molten, spinning outer core, that it was proposed, generates the main magnetic field. In a power plant some of the generated electricity is used to produce the necessary magnetic field, but to start the process electric power from elsewhere—or permanently magnetic material—is required. What "started" the dynamo within the Earth therefore remains a puzzle and, as

Cox, Doell, and Dalrymple put it, "After centuries of research the Earth's magnetic field remains one of the best-described and least-understood of all planetary phenomena."

What enabled this trio, with contributions from several other groups, to demonstrate that the Earth's field, in fact, does flip over from time to time was the simultaneous occurrence of two processes within rock that is cooling after eruption. One is the freezing of its magnetic "compass needles." The other is the setting in motion of a "stopwatch" that, millions of years later, can be read to determine how much time has elapsed since the cooling occurred and the needles froze into position.

The nature of this stopwatch was first recognized in 1940 by Robley D. Evans of the Massachusetts Institute of Technology. It depends on a radioactive form of potassium (Potassium-40) constituting only 0.012 percent of natural potassium—so little that, while most rocks contain potassium, it does not make them very radioactive. Potassium-40, at a very slow rate, decays into Argon-40, a gas. Because the latter does not take part in any known chemical reactions, it simply accumulates in the crystal structure of the rock. However, if the mineral becomes molten the argon can escape, and in that case the process of accumulation begins again when the rock cools. That is, the stopwatch is reset to zero.

Thus, by measuring the amount of potassium in a specimen it is possible to determine how much Potassium-40 was there when the rock cooled. Then, if the amount of Argon-40 is measured, this indicates how long the accumulation of argon, from Potassium-40 decay, has been going on—that is, the length of time since the rock cooled and captured the magnetic field in existence at that time.

By thus correlating magnetic reversals with age it was possible to show that rocks formed at any given time in the past almost all carried the same magnetic polarity, regardless of where in the world they originated. For some periods the polarity resembled that of today. For others it was reversed.

Early in this research Blackett recognized that the timetable of such reversals could be especially useful to geologists. If, he said, "the reversals have been spaced irregularly in time, then it may prove possible to recognize a 'pattern of reversals' at different places on the globe. If reversals are very infrequent, for instance, only once, say, in 100 million years, then the tracing of such a reversal across the globe should be relatively easy, thus enabling the results to be used for geological dating purposes."

A similar method, based on the irregular spacing between tree rings, is used by archeologists. Year-to-year variations in climate produce a sequence of fat and lean rings unique to all trees growing in a particular area during a given period. In 1946, when I visited the University of Alaska, Louis Giddings, a specialist in this unusual science, known as dendrochronology, told of a journey of about 3000 kilometers (2000 miles) that he and his wife had just completed, riding a skiff down the Peace and Mackenzie Rivers to the Arctic Ocean and then along the coast toward Alaska until ice floes driven against the shore blocked their progress. All along the route he had stopped and, with a hollow-stemmed drill, extracted slender wands from driftwood logs. These wands showed the sequence of ring widths for the entire life of the tree.

In this way Giddings reconstructed the succession of fat and lean years in the Mackenzie Valley for many centuries. Because wood hardly decays at all in the arctic climate, many of the logs were extremely old. Whenever Giddings found one a little older than any heretofore, it was possible to extend the chronology further into the past. Along the Arctic coast were the oldest logs of all—wind polished and gray. Once his chronology had been established, Giddings, with his drill and magnifying glass, could tell almost exactly the year in which a particular tree began growing and when it died, even though this happened centuries ago. Sometimes he could even identify the tributary of the Mackenzie in which it had lived, before drifting down the river, for each valley had its subtleties of local climate. Such chronologies have enabled archaeologists to date with precision prehistoric sites thousands of years old simply by examining wooden objects found there.

If magnetic reversals followed a more or less regular timetable, like that of the 11-year sunspot cycle, they would be of little use, but it eventually became apparent that they were far from regular. The present orientation of the Earth's magnetism was found to be characteristic of rocks laid down during at least most of the past 700,000 years, and this has been named the Brunhes Epoch in honor of the Frenchman who first detected reversed magnetism. Before that, for some 1.8 million years, was the great Matuyama Epoch of reversed polarity, named for the Japanese pioneer in this research. Earlier than that the field was again "normal," as it is today.

Puzzling, at first, was an observation made in lava from the Olduvai Gorge of Tanzania—famous for its relics of man's most ancient ancestry. Two researchers from the University of California at Berkeley, C. S. Grommé and Richard L. Hay, found that lava flows which dated from the Matuyama Epoch and hence should have acquired reversed polarity, showed normal polarity instead.

The validity of this finding was suspect until the trio at Menlo Park—Cox, Dalrymple, and Doell—found that samples from three lava flows in the Pribilof Islands of the Bering Sea, whose age of 1.9 million years was virtually identical to that of the Olduvai specimens, also showed normal polarity, though they had been formed during the Matuyama Epoch. Furthermore, these flows were sandwiched between older and younger flows that showed the typical reversed polarity of the Matuyama.

I. McDougall and Don H. Tarling in Australia, who were also building a timetable of reversals, were skeptical of so brief an "event;" but then, on Reunion Island in the Indian Ocean, McDougall found evidence for one that occurred about a million years ago. "It so happened," reported Cox later, "that Doell and Dalrymple had *also* independently found an event in New Mexico that occurred about a million years ago." It was evident in a lava flow near Jaramillo Creek and was clearly the same event.

By now at least three such flip-flops during the Matuyama Epoch have been identified (in some cases with apparent subdivisions). They have been named, for the localities where first identified, the Jaramillo Event, the Gilsa Event, and the Olduvai Event. There is also one reported for the more recent Brunhes Normal Epoch—the Laschamp Event.

When the American Geophysical Union met in Washington in April 1966, these findings began to fall into place. "I was chairing a section at the meeting," Cox wrote later, "and Neil Opdyke from Lamont Geological Observatory came to me and said he wanted to change his talk because he had some exciting new information. He and a graduate student, Billy Glass, had begun finding magnetic reversals in deep-sea sediments. Amazingly, *they* had found the same event at one million years (the Jaramillo Event). So three groups, working independently, were beginning to find the same fine structure on a global scale."

Soon 28 cross-sections of sediment extracted from the ocean floor with piston corers had been studied—14 of them from south of the Aleutians, eight from the equatorial Pacific, and six from near Antarctica. Opdyke's analyses of the orientation of magnetic particles within these cores showed the same timetable of reversals derived from rock samples on land. By 1967 the Menlo Park group had determined ages for about 135 land samples, covering a span of almost four million years, and they fell into some 16 periods of normal and reversed polarity, ranging in duration from 50,000 years or less to more than a million. Because of this irregularity they were to provide a master identification code for magnetic "tree rings" on the ocean floor (Fig. 6.2). The code could also be applied to deep-sea cores—some of those that Opdyke studied were at least 12 meters long, and it could be shown, from their top-to-bottom changes in magnetic polarity, that they represented some five million years of sedimentation.

A puzzling observation, by James D. Hays of Lamont, was the apparent extinction of certain tiny sea creatures at the times of such reversals. He was studying the distribution, within the cores, of the shells left by one-celled organisms known as radiolaria. He found that at times in the past certain of these species had been abundant; then they vanished abruptly. He reported that, of the eight species that vanished from the cores during the 2.5 million years for which the record was most complete, six disappeared close to the time of a reversal, as recorded in the magnetic particles of the same core. "The correlation between reversal and extinction levels is indeed striking," he wrote. He suggested that a biological susceptibility to weak magnetic fields, during a reversal, might be partly to blame. He noted reports that mud snails, flatworms, and fruitflies can sense magnetism and asked why they should have such sensitivity unless it was of use to them.

He noted that the great extinctions of marine and land animals in the distant past seem to have coincided with times when periods of few reversals gave way to periods of frequent ones. For example, toward the end of the Permian Period some 250 million years ago, almost half the known families of animals, large and small, became extinct. This was the time when the age of amphibians and fern forests gave way to that of reptiles and more modern plants. The next great upheaval came at the end of the Cretaceous Period when the dinosaurs and other reptiles vanished and emergence of the mammals and flowering plants began. While some of the marine reptiles, flying reptiles, and other species had begun to decline earlier, the horned dinosaurs were expanding when the change suddenly came. "Not only did these times of extinction affect a wide variety of animals but they were worldwide in extent," Hays wrote, "affecting both the marine and terrestrial environment."

Harold Urey, who in 1934 won a Nobel Prize for his discovery of heavy hydrogen, proposed that extinctions marking the ends of geologic eras might have been caused by giant meteorite impacts. Then, in 1980, another Nobel Laureate, Luis W. Alvarez, his son Walter and their colleagues at the University of California in Berkeley reported finding in Italy what they took to be evidence of such an impact 65 million years ago, as the Cretaceous Period gave way to the Tertiary and the dinosaurs became

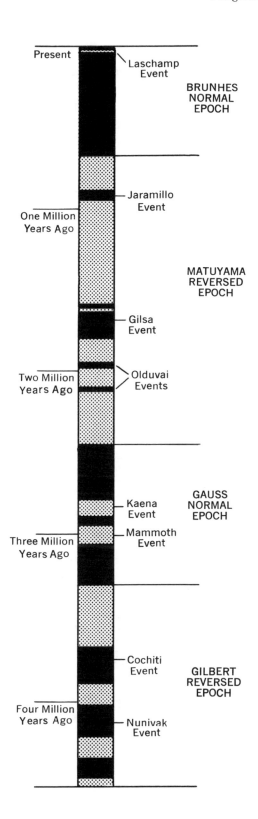

Figure 6.2 Reversals of Earth's magnetic field over the past 4.5 million years. Periods of "normal" polarity, such as that in effect today, are indicated by dark shading and "reversed" polarity by light shading.

extinct. The evidence, they said, was an abundance of iridium in a layer of clay only a few centimeters (a half inch) thick marking the Cretaceous-Tertiary transition. Iridium is rare in the Earth's crust but common in meteorites (and to some extent in some volcanic eruptions).

By 1982 this layer had been found worldwide in 26 sites. Often it was associated with quartz particles that had been shattered, as though by a great explosion. It was argued that debris thrown into the sky could have altered the climate, killing off the dinosaurs and many other species. The impacts, it was argued, could have been by one or more comets thrown out of their orbits on the outskirts of the solar system by a passing star—perhaps one, named Nemesis, in a very large orbit with the Sun that therefore returns periodically.

The argument has not yet been resolved to the satisfaction of all. Some say the iridium and shocked quartz came from volcanic explosions that poured out the lava forming India's Deccan Traps during that period. A number of paleontologists have said the extinctions were gradual. Nevertheless, that there was an impact is now widely accepted and some wonder whether many great turning points in evolution have had such an origin.

The suddenness of extinctions is hard to determine, since one centimeter of sediment may represent 10,000 years or more. The same applies to speeds of magnetic reversal. Presumably the change was not abrupt, as though someone had thrown a switch. But the timing is uncertain since, in terms of crude potassium-argon ages, the flip-flop seems instantaneous.

Needed was a series of lava flows representing closely spaced eruptions that would tell in more detail what happens during a reversal. By 1971 such a sequence had been found near Mount Rainier, documenting a changeover, some 14 million years ago, from reversed to normal polarity. This showed that for a time before any change in orientation of the field—perhaps several thousand years—it steadily weakened to about one tenth its original strength. As recorded in the rock of Mount Rainier National Park, the polarity of the field then began to rock back and forth in an increasingly wild manner. At the peak of this erratic behavior one magnetic pole seemed to swing almost to the opposite end of the Earth and back again. Then the pole that had been in the south began wandering around the north end of the globe and finally settled into a position roughly opposite that from which it started. While the techniques for determining rock ages could not pinpoint the timetable of these motions, they seemed to endure for from 1000 to 4000 years. The field then gradually resumed its original strength. In 1989 Robert S. Coe of the University of California at Santa Cruz and Michel Prévot of the University of Science and Technology in Montpelier, France, reported on analysis of lava in a single six-foot-thick flow on Steen Mountain in southern Oregon. During its cooling, over an estimated two weeks some 15 million years ago, the magnetic field fluctuated wildly, changing direction as much as 90 degrees. Trying to explain this, they said, "truly strains the imagination," and they termed their finding "tentative."

What causes the reversals continues to be debated. Theorists have found that a hypothetical version of the "dynamo" thought to generate the Earth's magnetism might spontaneously reverse its polarity from time to time, but they have found no convincing reason why it should do so. Another proposal is that subtle changes in the spin axis, such as the 14-month Chandler Wobble, might be responsible, or polar wandering fast enough to alter flow patterns in the core.

Bruce Heezen suggested that flipping of the magnetic field may occur when the Earth is struck by meteorites or comet heads. He based this primarily on the discovery that the age of tiny, glassy fragments known as tektites—spread over a large section of the Eastern Hemisphere by some sort of catastrophic event—coincided with the last major reversal, some 700,000 years ago (Fig. 6.3). The streamlined shape of the tektites indicated that they had flown through the atmosphere in a molten state. Many believe they represent splashes of molten rock thrown up by some cataclysmic impact. Several "strewn fields" of tektites have been found in various parts of the Earth, the most extensive of which reaches from the Philippines and Southeast Asia, through Indonesia to Australia and Tasmania. Tiny "microtektites" in sea-floor sediment through much of the Indian Ocean have been attributed to the same shower. Measurements of radioactive decay and radiation scars in the specimens all show that they cooled about 700,000 years ago, but scientists in Australia and a Soviet scientist working on specimens from 120 sites in Vietnam say they occur in formations only 5000 to 10,000 years old. This led, in 1989, to the controversial proposal that they erupted from some other celestial body, were streamlined by flight through its atmosphere, then traveled in space for several hundred thousand years before hitting the Earth. The event responsible for this debris must have been fearsome indeed.

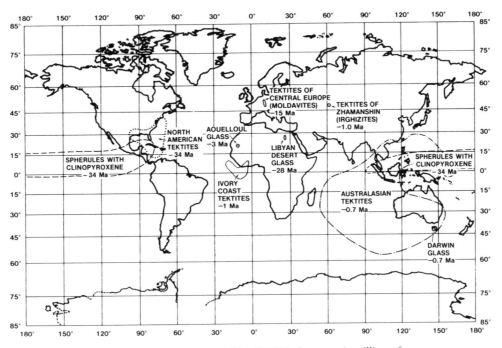

Figure 6.3 Tektite Strewn Fields. "Ma" indicates age in millions of years.

Recently, through two chance discoveries, it has been found that another great strewn field was laid down, some 35 million years ago, from the Caribbean, off the Venezuela coast, as far north as Massachusetts and west to Texas. For a number of years two small strewn fields had been known, in Georgia and Texas. Then, in 1960, John Chase, a zoologist at Ohio Wesleyan University who summers on Martha's Vineyard, off the Massachusetts coast, found an odd-looking piece of green glass that apparently had eroded from the colorful formations of Gay Head on that island. Specialists identified it as a tektite. In 1971 another specimen was found in the cabinet of a recently deceased geology professor at Columbia University, Arie Poldervaart. Its label indicated that it came from Cuba, but there were no details as to the circumstances of its collection. This inspired a close examination of cores extracted from the floor of the Caribbean and, within a layer laid down some 35 million years ago in one such core, about 6000 microtektites were found. It has been determined, in most cases by two independent dating methods, that all of these specimens fell at about the same time. Billy Glass of the University of Delaware, working with a group from the Max Planck Institute for Nuclear Physics at Heidelberg, has estimated the total weight of this fall at 110 billion tons.

A smaller strewn field, whose origin has been debated more than 100 years, lies in Bohemia, near the point where Czechoslovakia, Austria, and Germany meet. It has been reported that the Ries Basin, a circular feature 29 kilometers (18 miles) in diameter, 275 kilometers (170 miles) to the west in Germany, was formed by an impact that occurred some 15 million years ago, at about the same time as the spreading of this strewn field. A similar association has been reported for a strewn field in West Africa. Not long after the discovery of gold in gravel beds of the Ivory Coast, the natives observed the occurrence of strange-looking glass objects in those deposits. They called them "agna" and ascribed to them supernatural powers. It has been found that these objects, and similar ones of microscopic size in cores from the nearby sea floor, fell about 900,000 years ago. A comparable age has been ascribed to an impact that carved out a crater 16 kilometers (10 miles) wide, now occupied by Lake Bosumtwi in Ghana, 240 kilometers (150 miles) to the east of the strewn field. This impact occurred close enough to the time of the Jaramillo Event—a brief flip-flop of the Earth's magnetism—to suggest a possible relationship.

In assessing the role of such impacts, Heezen and Glass (then working together at Lamont) pointed out that small meteorites strike the Earth daily. "A meteorite large enough to produce a small impact crater falls every few thousand years," they said. "A cosmic body sufficiently large to cause a reversal of the geomagnetic field may arrive every few hundred thousand years." But, they proposed, the really big impacts that produce tektite fields and mark turning points in the Earth's history are even rarer: "It seems plausible that the geological clock may strike new periods when even larger bodies hit the Earth every few tens of millions of years."

However, an internal cause for the magnetic flips is probably more plausible. Over the past 2500 years the total field of the Earth has weakened by about 50 percent. This may be the prelude to a new reversal within a few centuries. If one occurs before the cause of such events has been deciphered, the flip may itself provide clues, not only to its cause, but to its possible effects on the Earth's environment and those living within it.

The manner in which magnetic reversals helped resolve the question of sea-floor spreading came from a completely unexpected quarter. During World War II the problem of submarine detection from the air had led to the development of an extremely sensitive device called the Magnetic Airborne Detector, or M.A.D. It was housed in a teardrop capsule that could be lowered from a blimp or scouting plane to trail below and behind the craft, well clear of its magnetic disturbance.

The device was effective for submarines at shallow depth, and after the war geologists looking for oil-bearing formations found it useful for aerial magnetic surveys, since it could delineate, in rough terms, regional distributions of igneous and sedimentary rocks. Might it not be possible to learn something about the ocean floor simply by towing a magnetometer behind a ship? If the towline were, say, 100 meters long, the "bird" being towed would be relatively free of magnetic disturbance by the ship's metal.

In 1952, to test this idea, a magnetometer was towed from Samoa to San Diego, revealing that the sea floor was remarkably variable in its magnetic properties, considering the rather uniform topography along that route. In fact some of the most marked magnetic "features" seemed unrelated to any variations in depth. Lamont Geological Observatory ships began towing magnetometers as a routine measurement, and, although the initial results were not very exciting, they did show a marked peak in magnetic intensity of the sea floor every time a ship crossed the centerline of the Mid-Atlantic Ridge.

Then, in 1955, the United States Coast and Geodetic Survey undertook an intensive deep-water mapping program off the West Coast. The research ship *Pioneer* was to steam back and forth along a succession of east–west tracks spaced eight kilometers (five miles) apart with orders not to deviate more than 150 meters from this assigned pattern—a precision made possible by radio-pulsing stations of the so-called LORAN navigation system. As reported later by Arthur D. Raff, who had developed a towable magnetometer at the Scripps Institution of Oceanography, "The Survey was willing to tow our magnetometer so long as we did not interfere with their work. We almost passed up the opportunity, but we finally decided to give it a try for a couple of months."

After an initial period of monotonous sailing back and forth, the Scripps scientists brought their accumulated data back to the laboratory, where a visiting British scientist, Ronald G. Mason, plotted the variations in magnetic intensity as recorded along successive tracks. He then drew the counterpart of a contour map, showing the hills of high magnetic intensity and the valleys of low intensity (Fig. 6.4). When he was through, Raff wrote, "A single glance was enough to show that we had something quite new in geophysics." Over the whole map were parallel, north–south "hills" and "valleys" of magnetic intensity. Although the intensity variations between these "hills" and "valleys" were only a few percent, the features were unmistakable, varying in width from a few kilometers to a few tens of kilometers. "No dry-land surveys," said Raff, "had ever revealed a lineation that approached this one in uniformity and extent."

And so the Scripps scientists, this time with more enthusiasm, returned to the *Pioneer* and continued to ride it back and forth. By the end of 1956, after roughly a year of surveying, the ship had covered a section of sea floor several hundred kilometers

Figure 6.4 Magnetic patterns *on the Pacific floor charted from data collected on the 1955–1956* Pioneer *survey. Since then it has been found that similar patterns cover most ocean floors. The straight lines indicate what appear to be offsets of the pattern. F. J. Vine and J. T. Wilson interpreted the symmetry on both sides of the Juan de Fuca Ridge as evidence that the sea floor spread away from it. Black areas indicate above-average magnetic intensity, while white areas are below average.*

wide and extending more than 2000 kilometers (1250 miles) from off Mexico to off British Columbia. The pattern of north–south stripes, representing, alternately, above-average and below-average magnetism in the sea floor, covered the entire region. Some of these "magnetic avenues" with their characteristic widths and spacing could be traced, as in an aerial view of the uptown–downtown avenues of Manhattan, but for 100 times the length of that island—2000 kilometers or more. At some points, where broken by east–west faults, the pattern had been greatly displaced, implying major side-slippages of the sea floor. The extent of slippage could be seen by seeking out matching magnetic patterns north and south of the fault. But at first, along the great Mendocino and Pioneer Fracture Zones, no match was evident. Perhaps, said Victor Vacquier of Scripps, the displacements were so great that the corresponding stripes, or "magnetic anomalies," on those sides of the faults displaced to the west had been carried beyond the coverage of the survey. Consequently the Scripps team went to sea again and traced the magnetic patterns hundreds of kilometers farther west. This showed the Mendocino displacement to have been more than 1100 kilometers (680 miles).

In a 1961 article in *Scientific American* Raff discussed a variety of possible explanations for these remarkable magnetic stripes. One was a complex system of electric currents in the crust, although he said "not a shred" of evidence supported such an idea. The variations in magnetism might be imprinted in a layer of basaltic rock forming the sea floor. This was plausible because it was widely believed that the rock immediately under the sediments is basalt and, it was noted, such rock can become strongly magnetized. But how could it have become magnetized in such an odd pattern? There was, Raff pointed out, the "hot spot" theory. If alternate bands of rock were too hot to become magnetized, this might account for the pattern. But it was hard to believe the ocean floor could have such stripes of hot rock, like the red-hot wires on an electric toaster.

Another proposal was that favored by Robert Dietz, namely, that the anomalies might represent some form of stress. It was suggested, for example, that the pattern had been formed by the same compression that buckled the mountain ranges paralleling the West Coast. Since the pattern lies at right angles to the direction of the Earth's rotation, might it be somehow related to the tidal stresses or an adjustment of the Earth's shape as the spin rate slows?

After running through all these possibilities, Raff revealed that his imagination was nibbling around the edges of the true explanation. "One further line of speculation may be mentioned," he wrote. Citing the recent discovery of a worldwide network of mid-ocean ridges, he said the magnetic bands discovered in the *Pioneer* survey and, by this time, elsewhere in the Pacific, "run generally parallel to the ridges there. It looks as though the two are related."

In any case the magnetic stripes were clearly a major challenge to science. In the following year—1962—the British research frigate *H.M.S. Owen,* which was in the Indian Ocean as a participant in the International Indian Ocean Expedition, at Bullard's initiative, did an intensive magnetic survey of an area that measured 40 by 50 nautical miles (74 by 93 kilometers) spanning the Carlsberg Ridge. The echo sounder showed that the ship was steaming back and forth across a succession of ridges a kilometer or more in height parallel to the main Carlsberg Ridge and its central rift valley. When Drummond H. Matthews of Cambridge University brought home the magnetic measurements made on these crossings of the ridge, he turned them over to a new graduate student named Frederick J. Vine for analysis. "I already believed in continental drift, sea-floor spreading and reversals of the Earth's magnetic field," Vine wrote later to this author, "and was particularly looking for some record of drift and spreading within the ocean basins."

What first struck Vine, as he analyzed the data with the aid of a computer, was that the magnetic "signature" of one of the sea-mounts on the Indian Ocean floor indicated it to be reversely polarized. It occurred to him that this could be explained if the sea-mount were a volcanic feature formed during a period when the magnetic field of the Earth was reversed.

If that were the case, and if the sea-floor spreading recently proposed by Hess were correct, then material erupting along a mid-ocean ridge, as it cooled, would become imprinted with the polarity of the Earth's magnetism in effect at that time. After this ribbon of new sea floor had been split and pushed to either side by additional material erupting along the ridge, this newly forming sea floor would also capture the current field; and, if the field by then had reversed itself, the polarity of the imprinted magnetism would be reversed. The effect of this process, after a succession of reversals, would be to produce a series of sea-floor bands parallel to the ridge and alternating in their magnetic polarity.

There seemed a suggestion of such a pattern in the data obtained over the Carlsberg Ridge. The central rift valley displayed a markedly weak magnetic signal, whereas over zones to either side the magnetism was abnormally high. (This was the opposite of the situation in the Atlantic, where the mid-ocean ridge was a line of intense magnetism, but that was understandable since the geometry of the magnetic field at the two locations was quite different.)

In the issue of *Nature* for September 7, 1963, Vine and his thesis supervisor, "Drum" Matthews, published their explanation for the magnetic stripes. Work on the Indian Ocean magnetic survey, they said, had led them to suggest "that some 50 percent of the oceanic crust might be reversely magnetized." In other words, they proposed that oceans throughout the world are paved with such parallel avenues of rock, divided about equally between anomalies indicating normal and reversed magnetism.

As so often happens in science, when the time is ripe for a discovery, it tends to pop up in more than one place at almost the same instant. At about the time that Vine and Matthews offered their explanation for the magnetic bands, Lawrence W. Morley, chief of the geophysics division of the Geological Survey of Canada, and A. Larochelle came up with a very similar idea, but it was not published until a year later and then in a *Special Publication* of the Royal Society of Canada with limited circulation. The Vine–Matthews proposal appeared in *Nature,* one of the most widely read scientific journals in the world.

Even so, it did not create much of a stir at the outset. As Vine put it later, to suggest that a large part of the Earth's surface is paved with long avenues of alternating magnetic polarity was, "to say the least, a little presumptuous," based as it was primarily on data from a small patch of the Indian Ocean. Indeed a good many Earth scientists, at that time, felt just that way about it.

By 1965, however, the key to confirmation of the Vine–Matthews explanation was in hand, although at first no one realized it. A step toward recognition of that key occurred at Cambridge University, early in that year, when what Vine called "a remarkable coincidence" brought together there some of the most enthusiastic and innovative drifters—Tuzo Wilson from Canada, Hess on sabbatical from Princeton, as well as Vine, Matthews, and their mentor, Sir Edward Bullard. Hess was enthusiastic about the Vine–Matthews idea, since it fitted in so nicely with his concept of sea-floor spreading. If it were correct, he pointed out, it should be possible, once the ages of a few magnetic stripes had been determined by deepsea drilling (Hess was then much involved in the Mohole Project), to determine the age of any other part of the sea floor simply by examination of a magnetic map. All one would have to do was count the stripes from the one of known age, much as one counts tree rings—or, as he put it, the varves of thin clay layers laid down each year on the floor of a lake.

During the conversations of this assemblage, Vine and Wilson realized that, if the long magnetic bands, or anomalies, were being manufactured on a ridge, then split and pushed aside as a new band was formed, the resulting magnetic patterns should be symmetrical to either side of the ridge—that is, the sequence of broad and narrow stripes on one side, generated by long and short magnetic epochs, should be a mirror image of that on the other side. It would be as though two conveyor belts, slowly moving away from each side of the ridge, were being painted at the ridge either black or white, depending on the magnetic polarity in force at the time. The two belts would then carry the same sequence of broad or narrow, black and white bands, but in opposite symmetry.

The only detailed magnetic survey then available (in contrast to the simple profiles across the Carlsberg Ridge) was the original one of the sea floor west of North America. So far as was then known, no active ridge existed in that area, but during his Cambridge visit Wilson had deduced evidence for a short segment of ridge, isolated by extensive side-slippage of the sea floor, off the Strait of Juan de Fuca—the entrance to Puget Sound. His suspicions were strengthened by the charted lines of earthquake activity in that area. He and Vine spread out the magnetic map and studied the vicinity of this feature, which they called the Juan de Fuca Ridge (see Fig. 6.4). They were immediately struck by the symmetry of the magnetic avenues to either side of it.

By this time, Cox, Doell, and Dalrymple—that team of inveterate charters of the Earth's magnetic history—had pieced together a rough timetable of reversals that had occurred about one million, 2.5 million, and 3.4 million years ago, with short flip-flops at 1.9 million and possibly at three million years. Vine and Wilson decided to test the hypothesis that this timetable was inscribed in the relative widths of the magnetic avenues on each side of this ridge.

To carry out their test the two men computed the profile that would be recorded if a magnetometer were towed across a ridge that had been manufacturing zones of rock normally and reversely polarized according to the timetable. They realized that the central avenue—the one down the centerline of the ridge—would be double-width relative to the others, not yet having been split in two by the intrusion of material imprinted with a new magnetic reversal. Since they did not know the rate at which the sea floor was moving away from the ridge, they assumed various rates to see if any produced a profile resembling that actually recorded across the Juan de Fuca Ridge. They took into account the water depth (which would make the magnetic changes seem less abrupt to a sensor traveling the water surface three kilometers [two miles] above) and the angle at which the Earth's magnetic field crossed the ridge (which would introduce a small element of asymmetry into the pattern).

They adopted a proposal of Hess that the magnetism was chiefly embedded in a thin veneer of basalt lying on a deeper layer of serpentine (mantle material altered by reaction with water). The basalt was presumed to represent the most easily melted fraction of the mantle rock that had risen into the ridge. This light, molten fraction had flowed out onto the surface, cooled, and formed a magnetized layer about a kilometer thick, thinly coated with sediment. Basalt, when it cools in a magnetic field, becomes strongly magnetized and thus would provide the sea floor with a thin magnetic "icing."

Wilson and Vine found that, if they assumed a flow rate of two centimeters (one inch) per year away from each side of the ridge, the resulting magnetic profile bore a striking resemblance to the one based on actual observations (Fig. 6.5). A similar result was obtained when this test was applied to a magnetic profile across the East Pacific Rise. Thus, said the two scientists, it was possible to account for the striped patterns "without recourse to improbable structures or lateral changes in petrology (rock composition)."

The following November, after Vine had gone to Princeton as a young geology instructor, he attended a meeting of the Geological Society of America in Kansas City, and it was there that the full significance of the magnetic sea-floor patterns became apparent. At the meeting Cox, Doell, and Dalrymple displayed a more extended and detailed timetable of the reversals, including a sharp new "signal" in the sequence—the Jaramillo Event, when, somewhat less than a million years ago, the reversed field of the Matuyama Epoch briefly reverted to normal.

The timetable (Fig. 6.2) now stood out like a great, coded message, inscribed in "dots" and "dashes" (or broad and narrow tree rings) to represent the succession of brief and extended periods of magnetic polarity reaching back at least four million years. This same message, Vine realized, must be written on the floor of every ocean in the world.

Meanwhile, in 1962, the year that H.M.S. Owen magnetically surveyed the Carlsberg Ridge, James Heirtzler at Lamont had proposed that the United States Navy use its elaborately equipped magnetic survey plane for an intensive study of the Reykjanes Ridge—a section of the Mid-Atlantic Ridge extending south from the Reykjanes Peninsula of Iceland. Lamont ships had done extensive magnetic measurements in their Atlantic journeying, and these had shown a strong magnetic anomaly along the centerline of the ridge, particularly in the Reykjanes sector. In view of the striking, elongated magnetic patterns discovered on the floor of the Northeast Pacific, the Lamont group wanted to find out if there were similar lineations along this ridge.

The area was close enough to Iceland and to the electronic aids to navigation there serving trans-Atlantic air traffic (LORAN A stations) so that an aircraft operating from

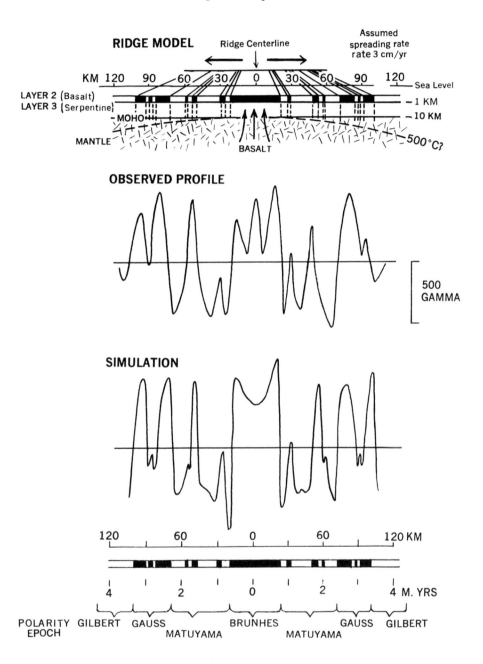

Figure 6.5 Magnetic intensity *as measured by ships crossing the Juan de Fuca Ridge, located on the Pacific floor off the Washington coast. F. J. Vine and J. T. Wilson found the observed profile resembled a computer-simulated profile. They based the construction of the computer simulation on the assumption that the sea floor was manufactured, symmetrically to either side of the ridge, and imprinted with the sequence of normal and reversed polarities of the earth's magnetic field as indicated by the timetable at the bottom of the illustration. While various spreading rates were tried in the simulations, the one that matched best was three centimeters (1.2 inches) a year.*

the air base at Keflavik, on that island, could fly a grid survey over the ridge with great precision. During 1963 the Navy plane flew 58 lines more or less evenly spaced and at right angles to the ridge. Of these, 49 crossed the crest and were used by the Lamont group to map magnetically an area extending 320 kilometers (200 miles) along the ridge and 200 kilometers (125 miles) to either side of it.

What emerged on the plotting table was an extraordinarily symmetrical pattern of magnetic stripes to either side of the ridge (Fig. 6.6). There was a strong anomaly along the centerline, flanked by a succession of narrow stripes. These extended 110 kilometers (68 miles) to either side, whereupon the magnetic bands became considerably broader and weaker in intensity.

If the stronger-than-average magnetism was charted in black, and the weaker-than-average magnetism in white, the map looked like one of those ink-blot patterns formed when a paper is folded, smudging the design on one side to produce a mirror image design.

It was not until two years after the survey flights that this map was published, its appearance almost coinciding with that of the paper by Vine and Wilson explaining the magnetic pattern of the Juan de Fuca Ridge in terms of sea-floor spreading. However, so far as the Reykjanes Ridge was concerned, the Lamont group was skeptical of this explanation. They pointed out that there seemed to be two classes of magnetic avenue: narrow "axial anomalies" that lay close to the centerline of the ridge, and "flank anomalies" that were broader, more irregular, and farther to either side. Because these seemed to be separate phenomena with separate causes, and because the magnetic strength of the features decreased away from the central ridge, the Lamont group concluded that the process proposed by Vine and Matthews three years earlier for production of the Carlsberg Ridge anomalies was "untenable, at least in its present form."

The Vine–Wilson proposal, however, had sought to explain the strength of the centerline anomaly. The central zone of the ridge, they said, was formed entirely of newly erupted rock imprinted with the current magnetism of the Earth. As such rock was pushed away from the ridge and a new magnetic epoch began, the older rock became contaminated with volcanic material that erupted into it. The topography of the ridge flanks showed ample evidence of such volcanic activity. Thus rock carrying the magnetism characteristic of a former era became diluted with lava that erupted through it, cooled, and acquired the contemporary magnetism. This would lessen the magnetic strength of anomalies on the flanks.

Maurice Ewing, the director at Lamont, was slow to be won over, largely because the flat-lying layers of oceanic sediment seemed to him incompatible with such dynamic activity. However, other members of his staff began to be converted, one of the first being Jim Heirtzler. If the theory was correct, the timetable of magnetic reversals, like the dots and dashes of a single, unique coded message, should be inscribed to either side of mid-ocean ridges everywhere. The Lamont scientists applied the timetable to the beautifully symmetrical Reykjanes pattern and found that it fit. They then began to examine magnetic records from all the oceanic surveys they could lay their hands on.

Walter C. Pitman III went to work on data from the South Pacific. Xavier Le Pichon (visiting from France) tackled recordings from the Indian Ocean and Manik

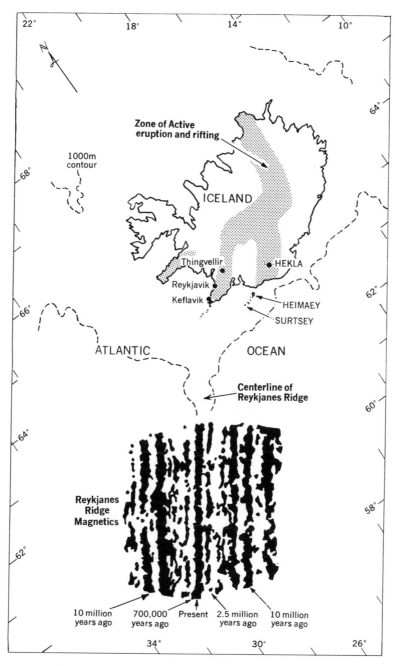

Figure 6.6 Magnetic features of Reykjanes Ridge. *A U. S. Navy plane flew fifty-eight passes perpendicular to the Reykjanes Ridge, located to southwest of Iceland, in order to map the region's magnetic patterns. The data obtained from the flights revealed an extraordinarily symmetrical pattern of magnetic features to either side of that ridge. Their ages, deduced from the timetable of worldwide magnetic field reversals, are shown.*

Talwani of India (who was to succeed Ewing as observatory director) worked on the North Atlantic. They were in a particularly strong position for such analyses, since the magnetic recordings made along tens of thousands of kilometers of oceanic journeying had been catalogued, organized, and stored in a manner amenable to computer analysis. As Warren Hamilton of the United States Geological Survey has put it, when the concept of sea-floor spreading came along, "they were able to print out the profiles on any projection and at any scale, and to play the appropriate orientation and latitude games with them. We would have been decades yet sorting it out without this."

Particularly clean data had been obtained by the United States research ship *Eltanin* on Leg 19 of its survey of the South Pacific, where the East Pacific Rise curves west. At the memorable 1966 meeting of the American Geophysical Union, where Opdyke reported confirmation of the reversal pattern in oceanic sediments, Pitman threw on the screen the magnetic profile obtained by the *Eltanin* as it crossed the rise on Leg 19 of its zigzagging journey. To dramatize the symmetry of this profile to either side of the rise, he displayed above it the same profile in reverse sequence—that is, from southeast to northwest instead of the other way around. Then, below it, he showed a computer calculation of what the profile should look like in terms of the known timetable of reversals and an assumed spreading rate of 4.6 centimeters (1.8 inches) a year from each side of the rise (which was at the center of each profile).

All three profiles were strikingly alike. As the session chairman, Allan Cox, commented later, "The profile was beautifully symmetrical on either side of the oceanic rise. I hadn't really believed in sea-floor spreading up until then because the magnetic data hadn't been very symmetrical. But suddenly there was the incredible symmetry of the *Eltanin*-19 profile. I remember my reaction: 'Good grief! Vine is right after all.'"

Vine himself was there with copies of a new and comprehensive paper on the implications of the magnetic patterns. The following December 16 this report, by a lowly, newly appointed geology instructor at Princeton, was published, in revised form, as the lead article in *Science*. "Magnetic anomalies," said its subtitle, "may record histories of the ocean basins and Earth's magnetic field for 2×10^8 (200 million) years."

Vine pointed out that the observed spreading rate for the North Atlantic was sufficient to account for the entire opening of that ocean according to the time scale indicated by geologic evidence. For example, lava flows on the islands of Mull and Skye off the Atlantic coast of Scotland, erupted as the opening of the North Atlantic began, had been found, from potassium-argon dating, to have cooled some 60 million years ago—right on schedule for the spreading timetable. The magnetic lineations along the centerline of the Red Sea, said Vine, testify to its being an "embryonic ocean floor."

Meanwhile magnetic data were being collected from the Indian Ocean south and southwest of Australia and from the South Atlantic. The icebreaker *U.S.S. Staten Island*, steaming back and forth between New Zealand and the Antarctic, obtained profiles across the southern extremity of the East Pacific Rise. Surveys were already on hand from the North Atlantic and Northeast Pacific. The Russians reported anomalies paralleling the active ridge beneath the Arctic ice floes.

It appears that in all oceans the same pattern prevailed. Marching away from each active ridge was a succession of wide and narrow magnetic stripes conforming to the master timetable of reversals. The same "magnetic message" was recorded every-where, although there were local variations or distortions. While the rate of spreading in a particular region affected the widths of the bands, the *relative* widths in any such sequence followed the timetable.

This made it possible to deduce variations in spreading rate. Alongside the Reyk-janes Ridge the bands are closely spaced and the spreading rate is about a centimeter (or half inch) per year away from the ridge, whereas from parts of the East Pacific Rise it is more than five centimeters (two inches) a year. That is, objects resting on the sea floor on opposite sides of that ridge are moving away from one another at 10 centi-meters a year (the width of a human hand)—an extraordinary speed for any geological movement involving so vast a region as the sea floor.

While there are regional variations in the speed, the rate of flow from any one ridge seems to have remained generally constant for many millions of years. This was evident to Heirtzler and his colleagues when they extended the timetable of reversals all the way from a mid-ocean ridge to the region, near a continental margin, where the magnetic pattern disappears. The chronology derived in this way extended much further back in time than any obtained from lava flows on land. They charted 171 reversals that occurred over the past 76 million years. Because the timetable thus derived from one ocean matched that obtained from magnetic surveys in another

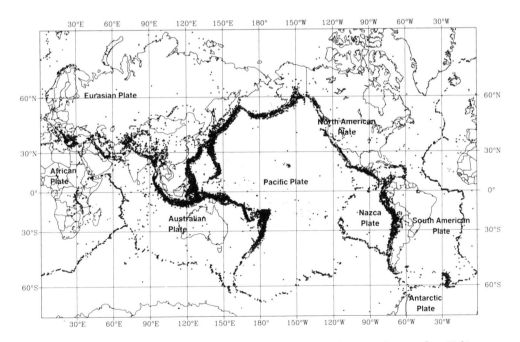

Figure 6.7 Global distribution of earthquakes with magnitudes 5.0 and greater from 1963–88. The charting of earthquake activity has revealed a continuous system of seismically active midocean ridges that are believed to subdivide the earth's surface into a series of gigantic plates. Earthquakes almost never occur within the interior portions of those submarine plates. This pattern influenced Hess' development of his seafloor spreading theory.

ocean, they knew that the spreading rates in both cases had not been subject to major local variations.

By 1968, for roughly half the world's oceanic floor, the main features of spreading patterns had been recorded, with most of the data focused on the more intensively surveyed mid-ocean ridges.

Using, for the six major plates of the Earth's surface, the directions and rates of drift reflected in the sea-floor anomalies, Le Pichon sought to reconstruct the history of continental drift over the past 200 million years (the six plates are delineated in Fig. 6.7). His colleague at Lamont, James Heirtzler, pointed out: "Comprehensive new theories that rationalize large numbers of observations and explain major aspects of the physical world are rare in any field of investigation. Such a synthesis may be within reach in geophysics."

Even to the most ardent protagonist of drift, there had been little to indicate just how the continents were once fitted together and by what routes and what timetables they drifted apart. One might reasonably assume that such information was lost forever. But now there was the history of such motions imprinted on the bottom of the sea.

The spreading of the sea floor south of Australia clearly has been pushing that continent and Antarctica apart. The entire Atlantic, north and south, seems to be growing along the Mid-Atlantic Ridge. But in the North Pacific almost the whole sea floor is moving northwest, sliding along the rim of California and disappearing under Japan and the Aleutians.

"The magnetic lineations in the ocean floor," said Heirtzler, "serve as 'footprints' of the continents, marking their consecutive positions before they reached their present positions." Thus it appeared that, as previously postulated, the present continents originated in two primordial continents: Gondwanaland in the south and Laurasia in the north, themselves children of the supercontinent, Pangaea.

"We found that the slow but steady and prolonged rates of motion," Heirtzler continued, were sufficient to separate South America from Africa—thus creating the entire South Atlantic Ocean—in about 200 million years and to separate Australia from Antarctica in about 40 million years. As more of the sea floor was dated we could establish more exactly just when the various continents separated and how they moved."

These were heady conclusions, but the battle was not over, for there were leading scientists who questioned the magnetic evidence, arguing that the "footprints" were, in fact, figments of wishful thinking.

Chapter 7

The Doubters

IT IS EVIDENT THAT NOT A SINGLE ASPECT OF THE OCEAN-FLOOR SPREADING HYPOTHESIS CAN STAND up to criticism. This hypothesis is based on a hasty generalization of certain data whose significance has been monstrously overestimated. It is replete with distortions of actual phenomena of nature and with raw statements. It brought to the earth sciences an alien rough schematization permeated by total ignorance of the actual properties of the medium.

With these words, published in a 1970 issue of *Tectonophysics,* Vladimir V. Beloussov reiterated his vehement opposition to the sea-floor spreading concept. The British journal *Nature,* taking note of this dissent, described Beloussov as "the Soviet Union's most distinguished geophysicist." Some opponents of the theory, it said in an editorial, "may be identified as diehard reactionaries, adherents to an older order of a world which has passed them by." But, it added, a man like Beloussov "may not be dismissed so lightly."

As early as 1964, when the magnetic evidence was only partially in, the skeptics had begun to muster their arguments. In that year Lord Blackett and Sir Edward Bullard persuaded Howard W. Florey, president of the Royal Society, that the possibility of continental drift had become respectable enough for the Society—one of the most venerable and prestigious scientific academies in the world—to give the subject a hearing. Lord Florey, an Australian pathologist with no particular prejudices regarding drift, agreed, and a two-day symposium was organized by Blackett, Bullard, and Keith Runcorn. The sessions, held on March 19 and 20, 1964, in the lecture theater of London's Royal Institution, brought face-to-face most of the leading protagonists from both sides of the Atlantic.

Gordon MacDonald and Bruce Heezen were there to argue against the convection current hypothesis, Heezen citing the complex ridge system of the Indian Ocean as evidence against it; Tuzo Wilson presented his argument that some strings of islands are formed on a mobile sea floor by a single lava source, like puffs of smoke trailing off from a smokestack; and Runcorn offered evidence for past changes in geography based on the otherwise contradictory positions of the magnetic poles as "seen" through the rocks of different continents.

In opening the conference Lord Blackett indicated his hope that the debate would remain within gentlemanly bounds and cited an episode, memorialized by Bret Harte, concerning a dispute between a man named Jones, who was missing some mules, and his neighbor, Brown, who had dug up some bones. The debate centered on a scientific question: Were the bones those of some rare animal, as claimed by Brown, or were they the remains of the missing mules:

> Now, I hold it not decent for a scientific gent
> To say another is an ass—at least to all intent:
> Nor should the individual who happens to be meant
> Reply by heaving rocks at him to any great extent.
>
> The Abner Dean of Angel's raised a point of order—when
> A chunk of Old Red Sandstone took him in the abdomen;
> And he smiled a kind of sickly smile, and curled up on the floor.
> And the subsequent proceedings interested him no more.

(Old Red Sandstone was familiar to most of the conferees as a widely occurring formation in both Europe and North America.)

The highlight of the meeting was the presentation by Sir Edward and his associates of the results of an experiment using the EDSAC-2 computer of the University of Cambridge Mathematical Laboratory to test the matching of coastlines that, according to the drift theory, had once been joined. As Bullard pointed out, many scientists were not convinced by the claim of close fits between continents. Sir Harold Jeffreys, a leader of the antidrift school, had recently stated: "I simply deny there is any agreement." J. H. Taylor of King's College, London, recalled at the symposium how, in 1958, a scientist (A. H. Voisey) had ridiculed the game of coastline fitting with a demonstration that eastern Australia could be snugly nestled against eastern North America. There even was supporting geologic evidence, he said: the Taconic sequence of rocks in the United States matched the Brisbane schists of Australia.

The purpose of the Cambridge experiment, therefore, was, in Bullard's words, "to put the facts beyond doubt by using the best data available and finding the 'best fits' by objective arithmetic methods." It was evident that a far better fit was obtained if one matched the outer edges of the continental shelves—the so-called continental slopes— as the boundaries, rather than the charted coastlines. This had been recognized by the early protagonists of drift. More recently, with new oceanic soundings in hand, Carey, the Tasmanian proponent of Earth expansion, had produced striking congruences by pushing around plastic cutouts of the continental blocks on a globe. But he had done this "by inspection"—by eye, rather than by mathematical means.

The Cambridge group based its experiment on a dozen charts of the United States Hydrographic Office that covered the entire globe, showing contours for the depths of 100, 500, 1000, and 2000 fathoms (a fathom being six feet or 1.8 meters). The computer was "told" the shapes of the continental rims along these contours in terms of the latitudes and longitudes of points 30 miles (48 kilometers) apart. The computer then tried a series of "fits," based on a succession of "rotation poles."

According to what is known as Euler's theorem, when part of the surface of a sphere moves over that sphere (like a drifting plate of the Earth's crust), its motion can be considered rotation about an axis that cuts the surface of the sphere at a certain point—the rotation pole. If you peel an orange and then slip one section of the skin, representing one fragment of a sphere, around the surface of the orange in a uniform direction, the patch of skin will move in this manner around some "pole" on the orange. Furthermore, the rate at which the skin moves over the orange is not the same everywhere. In the area closest to the pole of rotation the motion is slowest. In fact, if the patch of skin extends as far as that pole, the point at the pole rotates without changing its location.

The Cambridge group assumed that, to determine how the continents were once joined, it was only necessary to identify the rotation pole that generated the best fit when the coastlines were swung back together. By a complex "homing process" the computer zeroed in on the best-fitting poles, using various contour depths to define the continental margins. It was found (as Carey had assumed) that the best fits were produced by the 500-fathom contour, that being the depth at which the rims of the continents, typically, slope most steeply downward.

First, the Cambridge computer fitted South America and Africa together. This was the most obvious match and one that became remarkably snug when the computer had done its job. Then the lands on opposite sides of the North Atlantic were fitted: North America, Greenland, and Europe. "The closeness of these fits exceeded our expectations and fully confirms the work of Carey," Bullard said. He and his colleagues then tried to put the two blocks together: the North Atlantic block and the South Atlantic block. "Here the fit was less good," he reported, although his maps showed striking congruity (see Fig. 7.1).

Some liberties had been taken with geography. Thus features presumably formed since the opening of the Atlantic, such as Iceland, various sea-floor ridges, and the vast delta of sediment deposited in the bight of Africa by the River Niger, were omitted. On the other hand, areas now submerged but suspected as possible continental fragments—notably Rockall Bank, midway between Iceland and Ireland—were figuratively lifted from the water to fill gaps in the jigsaw reconstruction, a somewhat arbitrary step that was later proved correct. Likewise Spain was rotated clockwise to close the Bay of Biscay, bringing Spain and Brittany close together and flattening out the Pyrenees. As noted earlier, there was evidence in the remanent magnetism of European rocks that this had been the original orientation of Spain, and the fit of the French and Spanish coasts, after such realignment, was impressive. Magnetic mapping of the Bay of Biscay later confirmed that its floor had spread, swinging Spain away from France. In Bullard's reconstruction Spain was also moved somewhat west to fill a gap between that land and the continental margin off Newfoundland. In fact some now believe the Pyrenees were formed as much by the eastward sliding of Spain to its present position as by its swing away from Brittany.

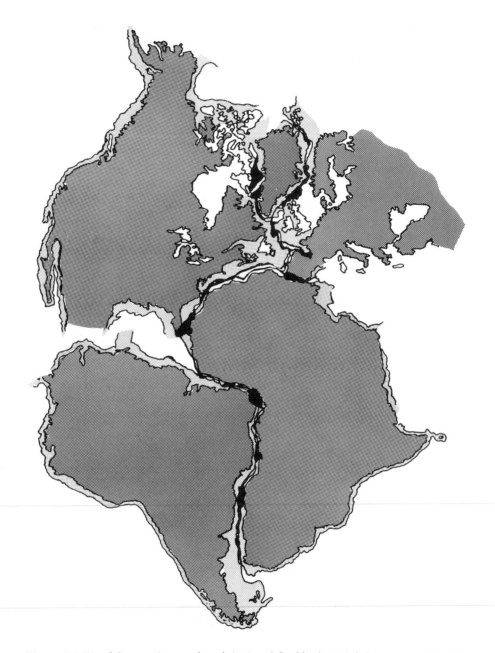

Figure 7.1 Fit of the continents *along their rims, defined by the 500-fathom contour along the edges of the continental shelves, as reconstructed by the computer at Cambridge University. Areas along the rims shaded black represent overlaps.*

"That there may be something anomalous about the position of Spain was noticed by W. H. Auden," said Sir Edward. He quoted from Auden's poem, "Spain 1937," which described it as:

> ... that arid square, that fragment nipped off from hot
> Africa, soldered so crudely to inventive Europe...

In summing up, Sir Edward said there were only two possible explanations for these fits: One was chance similarity, "on a par with the similarity of the coast of Italy to a boot." The other was drift. The new, more precise fitting of the continents now made it possible, he said, to test the hypothesis by checking for detailed geological continuity between coastlines that apparently once were joined.

His presentation, however, did not convince such skeptics as J. Lamar Worzel of the Lamont Geological Observatory. Worzel had followed in the footsteps of Vening Meinesz in making gravity measurements aboard submarines and then had helped develop ways to do so on a rolling ship.

"We have seen the continents authoritatively reconstructed by a computer," Worzel told the meeting caustically. "This seems most convincing except of course it seems necessary to discard Central America, Mexico, the Gulf of Mexico, the Caribbean Sea, and the West Indies, along with their pre-Mesozoic rocks (those pre-dating the alleged opening of the Atlantic)!"

Equally "mysterious," he continued, was what happened to all the rock that had lain on the floor of the Pacific in the path of the alleged westward drift of the Americas. How, he asked, could a crustal region comparable in area to the Atlantic be eliminated in the Pacific while new crust was being formed in the Atlantic?

He listed other objections. No longer was the heat-flow picture as clearcut as it had seemed. Places had been found on the ridges where the flow was moderate. If the floor of the Pacific was old compared to that of the Atlantic, why were their rock floors so similar and why is the average thickness of Atlantic sediment twice that of Pacific sediments? One would expect the reverse to be true. Finally, he said, if the Pacific floor is pushing from the East Pacific Rise (known also as the Easter Island Rise) against the west coast of South America, and material from the Mid-Atlantic Ridge is pushing against the east coast of that continent, a terrible battle must be in the offing: "If continental drift is believed, it must be considered that South America is now, or is soon to be, the scene of a gigantic struggle between the westward drifting forces indicated by the Southern Atlantic midocean ridge and the eastward drifting forces indicated by the Easter Island Ridge of the South Pacific."

Worzel had concentrated on transition zones between continent and ocean. In the 1950s, aboard the submarine *U.S.S. Tusk,* he had made gravity measurements across the continental shelf and out to the deep ocean from Cape Hatteras, Cape May, and Martha's Vineyard. His findings led him to question the sea-floor spreading hypothesis. If trenches mark where sea floor, moving away from a central ridge, descends beneath the continents, where are the trenches on either side of the Atlantic? He concluded that trenches are produced by tension in the Earth's crust and suspected, like Heezen, that the mid-ocean ridges might be a product of the Earth's expansion. If the trenches on the rim of the northwest Pacific are swallowing sea floor manufactured along a mid-ocean ridge, he asked, where is that ridge? The nearest one is 10 or 11 thousand kilometers (six or seven thousand miles) away. The idea of a convection cell that wide was to him incredible. "It must therefore be concluded," he wrote in one of his reports, "that the structure of deep oceanic trenches and the mid-oceanic ridge do not support either the hypothesis of extensive continental drift or the hypothesis of convection currents in the mantle." This forthright conclusion was made, however, before discovery of the "magnetic footprints" that won over most of his colleagues at Lamont.

Much of the Royal Society confrontation focused on the manner in which great sections of the sea floor have slid alongside one another, both in the Atlantic and Pacific. Was this evidence for drift—or against it? Victor Vacquier from the Scripps Institution told of the great displacements revealed by the offset magnetic patterns of the Pacific. If one combined the displacements of the Mendocino and Pioneer Faults, he said, the north side has slid 1400 kilometers (870 miles) westward relative to the south side. His colleague from Scripps, Henry Menard, pointed out that the Clipperton fracture zone can be traced from the East Pacific Rise, off Central America, across the Pacific almost as far as the Line Islands—one quarter of the Earth's circumference (See Fig. 5.7). Likewise, he said, the Mid-Atlantic Ridge, at successive points along its snaking journey from Iceland to the South Atlantic, is offset hundreds, or even thousands of kilometers to one side or the other.

Where the Mid-Atlantic Ridge—or any of the other oceanic ridges—is thus offset, the displaced segments are joined by faults similar in structure to the extremely elongated fracture zones of the Pacific. These faults do not merely join the offset segments of the ridge, but often continue for hundreds of kilometers beyond them in both directions.

What, the conferees were asked, could account for these extraordinary features, one of which encircles a quarter of the Earth? To the conventional geologist the faults linking offset ridge segments seemed simply to mark disruption of the sea floor where the ridge had, in some way, been pulled apart. Such faults, where one side, relative to the other, slips right (or left) its entire length, are well known on land and are called trans-current faults. But why should the ridges be offset in this manner?

One possibility, suggested by Dietz, was that if various sections of the Pacific floor are spreading at different rates, the fracture zones would then mark boundaries between fast-flowing and slow-flowing sectors. In this case or in that of a transcurrent fault, one would expect earthquake activity all along the fault from the resulting friction. Yet Lynn R. Sykes, a seismologist soon to join the Lamont group, told the London meeting that this was not the case. Earthquakes along these faults were limited to those segments joining the ridge offsets.

The solution to this puzzle emerged a year later. The scene was Bullard's laboratories at Madingley Rise, on the outskirts of Cambridge, during the same talks that had led to the final breakthrough in the interpretation of magnetic footprints. As noted, a special feature of the active fault zones in the sea floor—and notably of the San Andreas

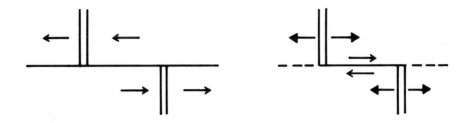

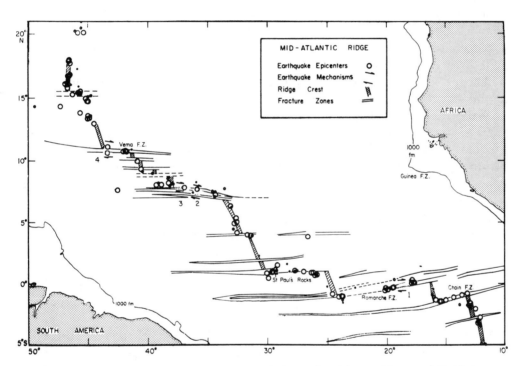

Figure 7.2 Types of Faulting. *In transcurrent faulting (above left), one side moves uniformly in a direction opposite to that of the other side. The mid-ocean ridges were probably offset by transcurrent faulting. In transform faulting (above right) the ridge segments do not move relative to one another, but material flows away from each segment. Along that part of the fault between the offset segments, there is motion in opposite directions, producing earthquakes. Lynn R. Sykes of the Lamont Observatory analyzed data from earthquakes along the Mid-Atlantic Ridge between Africa and South America and along faults marking where the ridge was offset, shown in the lower diagram, published by Sykes in the 1967 issue of the Journal of Geophysical Research. All quakes along those faults were betweeen the offset segments of the ridge, rather than to either side, conforming to the transform fault concept.*

Fault along the western rim of North America—is that they tend to terminate in a ridge, often roughly at right angles to the ridge. The San Andreas, after running up the interior of California, continues along the coast from San Francisco north to Cape Mendocino.

It was Tuzo Wilson who saw the answer. "I got the idea," he says, "while sitting in an office at Madingley Rise and playing with paper models." He realized that, as a consequence of spreading away from the ridges, a form of faulting previously unrecognized by geologists was at work. He called it "transform" faulting because, where the fault met a ridge, in terms of earthquake activity it became transformed into another type of feature—the ridge itself (see Fig. 7.2).

Wilson's suggestion was hard to believe because, in a situation like that in the Mid-Atlantic where such faults link offset segments of the ridge, it would mean that motion along the fault was directly opposite to what one would expect from looking at a map of the ocean floor. Instead of being in a direction pulling the ridge segments apart, it would be reversed.

The reason is that the ridge segments are, in fact, fixed with respect to one another, with the sea-floor spreading away from each of them.

Wilson's theory explained why fracture zones extending far beyond the active part of a transform fault—that is, beyond the sector between the ridge offsets—are free of earthquakes. If the spreading rate of sea floor away from all segments of the ridge was uniform, the only part of the fault where one block rubbed against another would be that between the ridge segments. Elsewhere the motion on both sides would be in the same direction and at the same speed, accounting for the absence of earthquakes there, as reported by the seismologist Lynn Sykes.

The testing of Wilson's hypothesis became possible, in large measure, through the same effort to learn how to distinguish between underground nuclear tests and natural earthquakes that led to rapid advances in the entire field of seismology. As Bruce A. Bolt, head of the University of California's network of seismic stations, put it, "seismology was transformed from a neglected orphan of the physical sciences into a family favorite."

One element of this transformation was the establishment, under subsidy by the United States Coast and Geodetic Survey, of the World-Wide Standardized Seismograph Network with more than 125 stations in cooperating nations around the globe. In addition, Sykes and his Lamont colleagues had access to data from a number of additional stations.

These observatories, by and large, were equipped with identical long-period seismographs. A long-period seismograph is one that records long waves—those that pass a station at intervals measured in minutes, rather than seconds or fractions of a second. These instruments had been developed by Frank Press and Maurice Ewing for the extensive observations of the International Geophysical Year in 1957–1958. The advantage of a worldwide network of identical seismographs was that their output could be processed without correction for response characteristics that—sometimes in indeterminate ways—vary from one instrument design to another.

This network, whose recordings also helped resolve the controversy concerning the plastic properties of the mantle, was capable of recording data of special value in discriminating between bombs and quakes. An important criterion, in that respect, arises from the fact that an explosion pushes the rock out in all directions, whereas a quake arises from slippage of one block against another. The resulting patterns of pressure waves are different. In the case of an explosion, no matter what the direction of the observing station, the first effect it should see in the crust of the Earth is an increase in pressure along the path leading from the event. That is, the first motion of the ground is away from the explosion site. However, because an earthquake is caused by slippage in one particular direction, it pulls on one part of the Earth's crust, as well as pushing on other parts. Stations lying in the direction affected by the pulling action initially will observe a decrease in pressure. The needles on their seismographs will record first motion toward the site—not away from it, as in the case of a bomb.

With a worldwide network it became possible to look at each event in three dimensions, and the direction of slippage could be determined. Furthermore, it was possible to tell which side of the fault had slipped which way by comparing records from stations on one side of the fault with those from a station on the other side.

While Sykes was perplexed by his observation that quakes occurred only along those short segments of oceanic faults linking displaced segments of a ridge, it was not until his colleagues Heirtzler and Pitman did their analysis of the magnetic footprints that he put everything else aside. "I really got steamed up," he recalls, and he began the laborious task of analyzing data on oceanic earthquakes recorded by scores of stations.

By 1966, when a meeting much like the Royal Society symposium was called in New York, he had some preliminary results. The conference was organized by the Institute for Space Studies of the National Aeronautics and Space Administration, and, when its proceedings were published in 1968, his results were even more striking.

By then he had analyzed 30 quakes along the centerlines of mid-ocean ridges, along the faults marking their offsets, and along their extension into Africa. Without exception, every quake consisted of slippage in the manner predicted by Wilson. Along the offset faults it was primarily horizontal and conformed to the "transform" slippage. Along the ridges themselves it was primarily upward from the bowels of the Earth and, to some extent, away from the ridge (see Fig. 7.4). Furthermore, by 1969, he had found that earthquakes near trenches around the Pacific showed the sea floor moving toward the trenches, as was to be expected from sea-floor spreading.

But if earthquake activity occurs only along those segments of the transform faults that join offset portions of a ridge, why do those faults continue, in dormant state, sometimes for thousands of kilometers beyond those limits of activity? This could be explained as follows: Along the active portion, where areas of the sea floor are moving in opposite directions, their rubbing together produces rifts, cliffs, and eruptions. These battle scars then remain after that portion of the floor has passed into the region where it flows parallel with the neighboring region, eliminating further friction, as shown in Fig. 7.2.

When Sykes finished his presentation at the NASA Institute meeting, Don Anderson of the California Institute of Technology put to him the obvious and critical question that emerged from this explanation of the faulting: "What," he asked, "offset the ridges...in the first place?" Much discussion centered on this question. In his original presentation of the transform fault hypothesis Wilson had suggested that the offsets were formed along old lines of weakness when the continents flanking the Atlantic first pulled apart, as shown in Fig. 7.3, taken from his original paper.

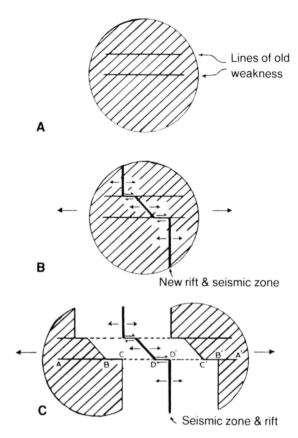

Lines of old weakness

A

B

New rift & seismic zone

C

Seismic zone & rift

*Figure 7.3 **Transform faults.*** T*uzo Wilson proposed that when the two sides of the Atlantic first split apart, the transform faults offsetting the Mid-Atlantic Ridge developed along old lines of weakness. This schematic depiction of Wilson's theory is from his original paper.*

But why the zigzag nature of the ridge-and-fault pattern, which is not evident in the continental margins that presumably spread from those ridges? In fact, as more precise surveys of the ocean ridges were made, it became evident that this zigzag pattern is typical of most, if not all, oceanic ridges, even where shorelines presumably derived from the ridge or rift are quite straight, as in the Gulf of California and the Gulf of Aden. The ridges mark the rifts from which the sea floor is spreading and the transform faults whereby the ridges are offset always seem aligned in the direction of this spreading motion (see Fig. 13.1 and the right-hand map of Fig. 17.6).

On the Mid-Atlantic Ridge this spreading is directly away from the ridge, but in the Gulf of Aden, for example, the motion is oblique to the ridge—an important factor, as will be seen later, in seeking to determine the motion of Arabia relative to Israel and Egypt.

At the NASA meeting Menard proposed, as he had in London, that, while the ridges were originally straight, having been formed by a single great convection cell, the latter broke up into smaller cells that tore the ridge apart into short segments. Vine pointed out that the magnetic patterns on the floor of the Northeast Pacific indicate a complicated history with variations in spreading rates, both in time and location, that could have broken up the ridges. Bruce Heezen said that, at least in the Atlantic, this did not appear valid. The sea floor seems to have spread westward and eastward from the central ridge, pushing the continents apart in a straightforward manner with the east–west faults marking the direction of drift. "This ocean had, as far as I am concerned, offsets built in from the very beginning." he declared. "The ridge was never straight. So the hypothesis that at some time the anomaly had to be in a straight line and was later displaced, can't be entertained as a universal possibility. It could happen somewhere, but in the Atlantic it won't work."

Menard, whose work had been done in the Pacific, spoke up: "Bruce and I," he said, "are always running into difficulties in that our oceans aren't exactly the same." The debate was inconclusive, and, as late as 1972, Sykes was still moved to comment that the offsets of the ridges remain "one of the big, unresolved questions."

By then, however, one geologist had found what he took to be a small-scale replica of ridge offsetting and transform fault generation within a crater at Kilauea—the same frequently erupting Hawaiian volcano whose description by Mark Twain had led Osmond Fisher, in the 19th century, to view it as a living example of the Earth's crustal motions. Wendell A. Duffield of the Hawaiian Volcano Observatory, maintained on the rim of Kilauea by the United States Geological Survey, took a series of photographs of the churning activity inside the crater of Mauna Ulu, on Kilauea's east rift zone, late in 1970 and early in 1971.

He found that the rising lava welled up along lines about one meter wide. As it spread away from those incandescent lines, its surface cooled immediately into dark basalt, forming a crust that was carried away from this line by the flowing lava underneath until suddenly the crust descended into the molten depths. This line of descent, too, was about one meter wide. Furthermore, the incandescent line of rising lava was sometimes offset, as in the transform faults of midocean ridges. The lava surface was 40 meters below the crater rim, forming a pool measuring 90 by 160 meters.

It was, said Duffield, "a naturally occurring miniature version" of the sea-floor spreading process. The lava crust "responded to complex circulation in the underlying melt with plate motions which were apparently controlled by the topography of the top of the lava column and which mimicked motions of Earth's lithospheric plates." The detailed analogy was not perfect, he said, "but the principal plate-bounding structures—rises, sinks, and transform faults—were all clearly displayed on the lava column in an impressive, ever-changing mosaic."

In the early 1970s a variety of laboratory experiments were carried out to seek the causes of transform faulting, using materials that ranged from sheet metal to partially melted wax. One which clearly generated such faults was done with freezing candle wax by Douglas W. Oldenburg and James N. Brune of the University of California at La Jolla (Fig. 7.4).

Molten candle wax (paraffin) was placed in a pan and cooled by a fan until a film had solidified on its surface. A stick embedded in the wax about midway across the pan was then drawn slowly to one side, producing a horizontal motion in the wax. The wax crust that was stretched by this motion split to form the counterpart of a mid-ocean ridge. As with such a ridge, the splits all occurred at right angles to the direction of plate motion, but they were offset because the original splitting occurred along local lines of

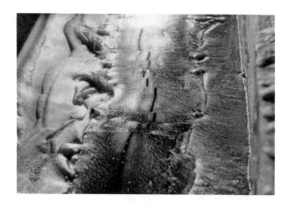

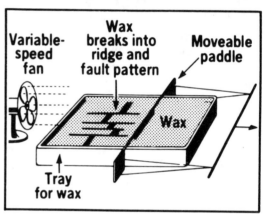

Figure 7.4 Simulation of sea-floor spreading. In an attempt to understand the effects of sea-floor spreading, Douglas W. Oldenburg and James N. Bruno of U.C. La Jolla developed a laboratory experiment simulating this process. They poured molten wax into a tray with a paddle, enabling them to pull it apart when its surface, cooled by a fan, formed a crust. The crust broke in a series of rifts off-set by analogues of transform faults. Molten wax rose to fill the rifts as does lava in sea-floor spreading.

weakness. The spreading then proceeded from each of these offset segments, producing transform faults whose motions were the same as those observed in the oceans. The experimenters concluded that such rifting occurred, not because material was rising into the ridge zone and pushing the plates apart, but through activity that was pulling them away from the ridges.

While much debate focused on the peculiar rifting of the ocean basins, the doubters also drew attention to the puzzling distribution of oceanic sediments. At the Royal Society meeting Worzel cited the abundance of Atlantic sediments, relative to those of the Pacific, as an argument against the alleged youth of the Atlantic floor. Likewise his colleague at the Lamont Geological Observatory (and its director), Maurice Ewing, continued to hold the sea-floor spreading concept at arm's length, even after the team that he mobilized at Lamont had uncovered powerful evidence for the hypothesis. It was the sediments that gave him pause—and it was the more detailed knowledge of their distribution that finally won him over.

Ewing's development of ways to probe the ocean floor played a major role in bringing about the 20th-century revolution in knowledge concerning that vast region of the Earth's solid surface (Fig. 7.5). Ordinary echo sounders had provided profiles of the surface layer, but this told little or nothing about the nature of the structures

Figure 7.5 Maurice Ewing,
1906-1974

underneath. It was in the use of explosions and other means of generating shock waves to probe those structures that Ewing broke new ground.

The use of explosive charges to determine the nature of structures within the Earth had been developed by oil prospectors in the 1920s, although the idea went back to the experiments of Robert Mallet, an Irish engineer, who proposed in 1848 that underground geological formations could be explored by creating small earthquakes with explosives. If one knew the speed with which such waves traveled through various kinds of rocks, this, he said, could be used to determine the nature of deep structures. To find out the speed of shock waves through granite he fired a gunpowder charge in a granite formation and, at some distance on the same formation, he placed a bowl of mercury. By watching through a telescope for the first disturbance of light reflected from the mercury he was able to estimate the speed at which the waves had gone through the rock.

With an archive of such knowledge and more sophisticated seismometers, oil men found they could locate, under flat surface terrain, warped and faulted layers of sandstone that provided traps to which oil could migrate and accumulate. Often there was no clue to these features on the surface.

The application of this technique to the ocean floor owes much to Ewing's experience after graduation from Rice Institute (now Rice University) in Houston. In the summer of 1928, as a paying supplement to his graduate work in oil prospecting with shock waves and gravity measurements, he was employed by the Geophysical Research Corporation in using explosions to search for salt domes beneath the coastal lakes of

Louisiana. He and his colleagues lived in two houseboats and did their field work from launches. It was, Ewing said later, "a valuable and exciting experience in working afloat" and stood him in good stead when he was challenged with a far more ambitious proposal: to survey the underground crustal structures across the boundary zone between North America and the Atlantic Ocean.

Ewing's efforts to land an academic post had brought him to the attention of Richard M. Field of Princeton, who was trying to persuade scientists to find out what the ocean floors are really like (he was soon to take on an international role, in this respect, as chairman of the committees on continental and oceanic structure of the International Union of Geodesy and Geophysics). It was, said Ewing, "Field's personal enthusiasm" that recruited to the endeavor such ultimate leaders of the revolution in Earth Science as Harry Hess, Tuzo Wilson, Sir Edward Bullard, and himself. Field thus helped make Princeton a center of innovative thinking about the ocean floors—as Bullard was to do at Cambridge University in England.

By 1934 Ewing had a job as instructor at Lehigh University in Pennsylvania and, he recalls, "one slightly snowy day in November" Field and William Bowie, Chief of Geodesy for the United States Coast and Geodetic Survey, appeared in his office at the university campus. Bowie was a Lehigh alumnus and made periodic visits there. Their proposal was that Ewing undertake a coastal survey. They believed he could get obsolete equipment from the oil companies and probably money from the Penrose Fund of the Geological Society of America. He did, in fact, get a relatively modest grant of $2000 and was able to modify seismic instruments for work in water depths to 60 meters. In the summer of 1935 he and his colleagues charted the sediment thickness to bed rock and the seismic velocities of both the sediment and underlying rock from inland Virginia to Cape Henry and beyond. Their explosion soundings to 120 kilometers (75 miles) off shore inaugurated the in-depth exploration of the sea floor that was ultimately to reveal the structure of a large part of the Earth's water-covered surface.

The next year Edward Bullard, then in his late 20s, joined Ewing for one of his seismic forays at sea, after which he returned to Britain and tried similar measurements in the Channel and south of Ireland. Meanwhile Ewing, although trained as a continental geologist and oil prospector, had become fully committed to oceanic work. In 1936 he had made a submarine voyage with Harry Hess on the *U.S.S. Barracuda* for gravity measurements, and he began working on methods to determine sediment thickness under the deep-sea floor—beyond the continental shelf where he had already operated. At first this was extremely laborious. Bombs were used to generate shock waves, and recorders were attached at intervals to a two-mile, multi-conductor cable laid out on the ocean bottom. While some improvements were made, it was not until after World War II that Ewing and others were able to develop methods suitable for extensive surveys.

There are two basic types of seismic shooting. One records waves that have traveled a considerable horizontal distance from the explosion site, some of them as they penetrate beneath the land or ocean floor, being bent, or "refracted," to a horizontal path and then bent back upward to a recording station. In such refraction seismology, as it was conducted during the immediate post-war years, one ship sat and listened with its hydrophones while another ship fired a succession of charges at varying distances from the listener. The recordings could then be analyzed to determine layering of the ocean floor and the velocities at which shock waves pass through each layer. These, in turn, provided clues to the nature of the material forming the layers. To penetrate deeply, long-distance observations were necessary and, to generate shock waves strong enough to reach a ship 100 kilometers (60 miles) away, naval depth charges weighing 300 pounds (136 kilograms) were used. However, for shorter distances, charges as small as a half pound (a quarter kilo) were adequate.

The other form of probing requires only one ship and depends on direct echoes off the sea floor and the layers of sediment and rock beneath it. To obtain a continuous profile of the rock basement underneath the sediment, a man stationed aft on the moving ship used to toss grenades over the side at fixed intervals. Such hazardous methods were largely eliminated with the development of non-explosive noise makers—again largely at the initiative of oil prospectors and of Ewing who, by now, had set up the Lamont Geological Observatory. One was a "sparker" towed astern a few meters underwater. Its spark generator produced low-frequency sound pulses that penetrated deep into the bottom. An "air gun" was developed that served a similar purpose. Such devices made sounding the layers beneath the sea floor hardly more difficult than the use of ordinary echo sounders to record a profile of water depths.

It became possible, across the entire width of an ocean, to delineate sediment layers to a depth of one to two kilometers (up to a mile or more). Sparkers and air guns, however, did not deliver enough punch to penetrate as deeply as explosives, which were still used to reach several kilometers down to the Moho. Refraction profiles to the bottom of the oceanic crust were obtained by such ships as the *Atlantis* of the Woods Hole Oceanographic Institution and the Lamont Observatory's *Vema*.

The *Vema,* and Woods Hole's original *Atlantis,* were both research vessels of the old school. The *Vema* had been built in Denmark as a three-masted schooner. She served briefly as a cadet training ship for the United States Merchant Marine Academy, and was then acquired by Lamont. While operating under sail had certain advantages—notably no vibration to disturb sensitive measurements—her rigging was dismantled after 1950 and she operated entirely on her engines.

By 1972, the *Vema* and other Lamont and Woods Hole ships had probed sediment depths along 760,000 kilometers (470,000 miles) of track crisscrossing the North and South Atlantic. The findings differed radically from a prediction made in 1950 by Philip Kuenen of the University of Groningen, the authority on oceanic sedimentation and turbidity currents. Considering the age of the Earth, the erosion of continents and the vast span of time since living organisms first appeared in the seas, he expected that the average accumulation would be about 3000 meters. However, soon after he made this prediction seismic soundings indicated that, at least in some regions, the sediment beds were far thinner. Now the more extensive surveys showed this deficiency to be universal, so far as the deep ocean basins were concerned.

The advocates of sea-floor spreading pointed out that this could be explained if those basins were, in fact, quite young. Ewing, however, was impressed by the flat-lying nature of the sediment layers, like a stack of unruffled blankets. To him these seemed incompatible with so dynamic a process as that envisioned in the spreading hypothesis.

Almost everywhere the soundings revealed three layers. The top one transmitted shock waves slowly and was almost certainly sediment. Beneath it was a layer, from one to one-and-a-half kilometers (a half mile to a mile) thick whose wave-propagating properties suggested that it was basalt, or lava-like rock, that had erupted and spread over the sea floor. Beneath this "Second Layer" was a "Third Layer" of very dense rock.

What struck Ewing and his colleagues was that the surface of the Second Layer was extremely rugged, whereas the sediment on top of it showed no evidence of the deformation that produced those buried mountains. That they were formed of basalt was supported by the fact that rocks dredged up from the deep sea floor are typically basalt and virtually never continental rock, such as granite (unless they have evidently been transported there, as by icebergs). At the time, Ewing thought this rugged, sediment-buried layer might be the primordial crust of the Earth—a most exciting prospect and one that, as will be seen, provided added stimulus for drilling into the ocean floor in the Mohole Project. Harry Hess, however, eventually saw in this discovery evidence for a veneer of basalt spreading across the sea floor—the "icing" that, it turned out, had become imprinted with the magnetic footprints of the continents. Its rough surface would therefore be a continuation of what lies exposed along the active ridges but gradually becomes buried in sediment as the ocean floor carries it away from the ridge.

The distribution of Atlantic sediment has proved to be highly uneven. More than half of it, in volume, lies near the continents and is clearly material eroded from the land and from the continental shelves. Off the mouths of great, muddy rivers like the Niger and Amazon the apron of silt extends hundreds of miles to sea. Most, but by no means all, of the sediment on the deep ocean floor appears to have been deposited there by turbidity currents. These deposits, formed by a succession of turbidity flows, show up in the echo records as thin, clearly distinguishable layers, whereas there is another kind of sediment that displays almost no layering at all.

The latter is believed to be the product of a phenomenon not recognized until the late 1960s. The bottom region of the ocean is murky with clouds of fine particles—a "nepheloid layer" that ranges in thickness from 300 to 2400 meters and is in constant motion with the general flow of bottom water. This deep circulation is on a global scale—like one vast system of convection currents, driven toward warm latitudes by density differences in the water arising from variations in temperature and salinity. In the Atlantic, cold, dense water flows in along the bottom from the Labrador Sea to the north and from the Antarctic to the south. The flow is to some extent diverted westward by the Earth's rotation and is partially blocked by transverse ridges. However, the murky nepheloid layer is so thick that it tends to override such obstacles and may deposit its sediment far from the point of origin. It is suspected that much of this material is stirred into the bottom water by turbidity currents—those catastrophic submarine counterparts of an avalanche—and then is carried great distances in suspension.

It was, however, the variations in sediment thickness across the Mid-Atlantic Ridge that ultimately persuaded Ewing and most of his colleagues that the sea floor has flowed away from that ridge. In 1963, the year before the Royal Society meeting, he and his co-workers reported that a profile of sediment depths showed virtually no accumulation within 10 to 50 kilometers (six to 30 miles) to either side of the ridge centerline, implying that this area was carpeted with newly erupted lava. Then there was an abrupt increase of thickness to 30 to 40 meters and the sediment continued to thicken slightly out to a distance of 100 to 400 kilometers (60 to 250 miles), whereupon there was another abrupt increase in thickness.

To protagonists of sea-floor spreading the symmetrical manner in which the sediments thickened at increasing distances from both sides of the ridge was impressive evidence that the age of the sea floor likewise increases at greater distances from the ridge. However, the abrupt changes in sediment thickness reported by the Lamont group were puzzling. In 1966 Ewing was still counseling caution. Just because the mid-ocean ridge was covered with freshly erupted rock, he said, did not mean the whole sea floor was moving away from the ridge in both directions.

As the magnetic evidence became more persuasive, Ewing was won over and proposed that the sudden increase in sediment thickness, 10 to 50 kilometers on either side of the Mid-Atlantic Ridge, marked the end, some 10 million years ago, of a period of suspended spreading and undisturbed accumulation that had lasted many millions of years. However, further analysis of sediment deposition indicated that such abrupt variations in sediment depth were the result of ocean currents.

Some skeptics argued that the worldwide paucity of oceanic sediments could simply mean that life in the oceans in past eras was meager and sedimentation slow. However, when the timetable of magnetic reversals was applied to remanent magnetism down the full depth of a deep-sea core, it was possible to show that for millions of years there had been no radical change in sedimentation rates (although in some areas layers of sediment are missing). It was found, furthermore, that deposition rates on the ridges are rapid, even though little sediment is found. This rapid deposition is at least in part because the water there is shallow enough for the tiny shells that make up much of the sediment to remain intact. In deeper water, out on the flanks of the ridges and in the ocean basins, the calcium carbonate of the shells becomes soluble in seawater and the shells disintegrate, leaving a red clay typical of the deep ocean floor. The soundings showed these deposits draped over ups and downs of the basins like blanket layers laid loosely over boulders. Presumably this cohesive clay clings to slopes down which other forms of sediment would slide. Furthermore, few if any earthquakes are centered in this region, which is well removed from the ridge, whereas along the ridge quakes are common, causing the sediment to slide into the valleys.

The thinner overall sediments of the South Atlantic have been attributed to the greater extent to which that part of the ocean has opened, relative to the north. Bullard and his colleagues found that South America and Africa have drifted apart in a pattern separate from that of the North Atlantic. The spreading pole, for this drift, turned out to be in the North Atlantic—that is, the two continents swung apart like a gate whose hinge lay north of the Azores.

As noted at the start of this chapter, there have been those who, well into the 1970s, have fought a rear-guard action against the plate theory. On March 29, 1971, at the annual meeting of the American Association of Petroleum Geologists in Houston, a symposium was held on "The New Global Tectonics" (the term "tectonics" refers to that branch of Earth Science dealing with processes that have shaped the Earth's crust).

Arthur A. Meyerhoff, publications manager of the sponsoring association, and his father Howard teamed up to attack the theory on all fronts. The elder Meyerhoff had been a geology professor and had served as executive director of the Scientific Manpower Commission in the 1950s. It was his son who had chided Hess and Dietz for allegedly ignoring the work of earlier scientists.

Their attack was later published as two articles in an issue of the association's *Bulletin* devoted almost entirely to articles for and against the theory. The Meyerhoffs declared the plate theory to be incompatible with what was known of ancient climate zones which, they said, "indicate the constancy of position of the rotational axis, continents and ocean basins for at least 570 m.y. (million years)....A truly devastating fact is that the topological requirements for moving the Americas away from Euroafrica eliminate any possibility of such movement unless the Earth has expanded greatly during the last 150–200 m.y."

They noted that sediment layers within deep-sea trenches, where the ocean floor is supposedly descending into the mantle, show no evidence of disturbance (however, such disturbance has, in fact, been found by deep-sea drilling alongside a trench). They charged that those trying to fit the continental plates together had ignored the fact that the plates are spherical, affixed to the surface of an oblate spheroid—the Earth. "Advocates of the new global tectonics, in short," they said, "have implicitly adopted the geometry and mathematics of a flat Earth in discussions of a spherical body." This was criticism not likely to please Bullard, whose work had been grounded in spherical geometry, or Carey, who used plastic sheets on a globe to explore the continent-to-continent fits.

The two Meyerhoffs also argued that the ice sheets known to have occurred in South Africa during the Carboniferous and Permian periods could not have developed had that region been surrounded by other continents, as assumed for Gondwanaland. In a complaint reminiscent of the outcast state of drifters during an earlier period, they wrote:

> Those who oppose the new global tectonics are subjected to strong criticism. Specifically, opponents are told that they ignore the evidence from the oceans, such as midocean ridges, magnetic anomalies, so-called transform faults, and related features. We ourselves have been told that we select facts which are "minor and unimportant details that in no way affect the overall validity of 'the new global tectonics." This type of accusation suggests insecurity, arrogance, or both.

But the battle that received the most attention was that between two "giants" in the field, for both had served as president of the International Union of Geodesy and Geophysics, the most prestigious position in that family of sciences. One was Tuzo Wilson of Canada, who had become one of the most ardent proponents of the theory. The other was Beloussov, whose caustic comment was cited at the start of this chapter. In 1967 Wilson had delivered an address in Ottawa which was published in the *Canadian Mining and Metallurgical Bulletin.* He sent a copy to Beloussov as an up-to-date statement of his views, and the reaction of the Russian bordered on outrage. Wilson's talk was entitled, "A Revolution in Earth Science," and he made much of the remarkable meshing of three recent discoveries:

1. The imprinting, in successive lava flows, of a timetable of magnetic field reversals extending back millions of years.

2. The discovery that the successive magnetized bands of the sea floor conform, in relative width, to the timetable of reversals in the lava flows.

3. The discovery that this timetable is also found in cores of oceanic sediment—on a vertical scale measured in centimeters, whereas the magnetic bands constitute a horizontal scale reckoned in kilometers.

These three independent, yet interlocking, phenomena are explained, Wilson pointed out, by a single theory—that of sea-floor spreading. "No such accurate correlations and predictions have ever been found before in geology or areal geophysics," he said. "The whole subject of Earth Science has thereby been radically altered." The picture that has emerged, he pointed out, is not identical to that proposed by Wegener, for it does not portray the continents as barges, plowing through the ocean floors. The new theory, he said, "visualizes the continents as being carried about along with the ocean floor like logs frozen in ice."

This is the theory, now widely referred to as "plate tectonics," that, as noted earlier, envisions the Earth's surface as broken up into plates, forming a rigid "lithosphere," that ride on a deeper, more plastic layer, the "asthenosphere." There are about six large plates, plus a certain number of small ones which fill the leftover spaces. The plates, their boundaries marked by lines of earthquake activity, may be partly continental and partly ocean floor. For example, North America (east of the San Andreas Fault) and the North Atlantic west of the mid-ocean ridge constitute a single plate moving away from the ridge. Likewise South America and the western South Atlantic form a plate that is overriding the Pacific floor along the trenches that parallel the west coast of South America. The eastern North Atlantic and Eurasia constitute one huge plate moving away from the Mid-Atlantic Ridge and overriding the Pacific plate along the trenches of the western Pacific.

Worzel had asked the Royal Society symposium how a section of Pacific floor as large as the Atlantic could have vanished from the Earth's surface as the Americas were pushed west and Eurasia was pushed east to open up the Atlantic. According to the plate theory it has all gone down the drain—into the mantle—via trenches.

"It seems that we know what is going on in the Earth," said Wilson. "This could be as important to geology as Harvey's discovery of the circulation of the blood was to physiology or as evolution was to biology. This is the most exciting event in geology for a century and every effort in research should be bent toward it."

If the crustal plates of the Earth have been in continuous motion, creating the surface features that we see all around us, he said, "much of our teaching must have been nonsense. Isn't it time we changed? Is there not evidence of much weakness in methods, both geological and geophysical, which were too feeble to distinguish between drift and nondrift?"

As recently as March 14, 1967, Wilson said, he received a letter from the director of a certain "distinguished geological survey" (which he discreetly did not identify), saying in part: "The opinion of the National Committee is that the subject of continental drift, attractive and stimulating as it is, is not of priority interest to geologists in (our survey)."

A great many geologists and teachers of geology have allowed their science to stagnate, where other sciences have leaped forward, Wilson asserted: "They were so absorbed in improving techniques, in amassing data and in planning computer codes by which to store the information that they forgot that other sciences have simplified their problems by discovering principles." But, he added, geology is now on the move. Textbooks are being rewritten and curricula radically altered.

Beloussov's response, along with Wilson's original talk and his rebuttal to Beloussov, were published together as a "debate about the Earth" in the December 1968 issue of *Geotimes*, the monthly publication of the American Geological Institute. A robust man physically and a genial one socially, Beloussov nevertheless pulled no punches. He cited the once popular view that mountain ranges and other surface features had been formed by contraction of the crust as the Earth cooled.

"The example of the contraction hypothesis," he wrote, "could serve as a warning to all those who are in a hurry to rewrite textbooks. At the end of the last century and even at the beginning of this one, few people questioned the correctness of the contraction hypothesis. And all textbooks were written on the basis of that theory. The fundamental principles of the contraction hypothesis became so much a second nature to geologists that up to the present they have been strongly reflected in the majority of ordinary regional geological papers.

"And yet," he continued, "the foundations of the contraction hypothesis collapsed." They did so, he added, "not only because a new physical phenomenon was discovered—radioactivity—that turned upside down all our ideas on the thermal regime of the Earth. It was undermined also because of its primitive nature; it schematized natural phenomena, reducing them to a state of complete distortion." For example, he said, the contraction theory never succeeded in explaining the gross vertical movements of the crust that, it was evident, had occurred—a failing that he also attributed to the new theory (and one still troublesome in explaining such great uplifts as that of the western plains of the United States).

Beloussov argued against the concept of deep convection within the Earth— "There simply is no foundation to it"—and he tried to show that a global system of internal currents as the moving force for continental drift ran into all sorts of geometric difficulties. "Near the coasts of California you show currents that meet at right angles. Is this possible according to elementary principles of hydrodynamics?" he asked.

The root of the difficulty, said Beloussov (himself primarily a continental geologist), was the focus on oceanic evidence. Knowledge of the ocean floors "remains not only very schematic, but to a very great extent indirect, being founded on interpretation of indirect data," whereas continental geology is known in considerable detail. "In the theory you are so ardently advocating, the geological development of the continents is much more schematized than was done in the contraction hypothesis," he maintained. "The geology of continents is simply and completely annihilated. Could it be that you really want to rewrite the textbooks, and throw overboard a great part of the outstanding achievements in the geology of continents?"

Beloussov acknowledged that the magnetic evidence was Wilson's strongest argument, but he questioned the Canadian's "confident references" to the magnetic bands that parallel the mid-ocean ridges. Are they not premature, he asked, "if we take into account that after more detailed investigations the stretches of anomalies fall apart into rather irregular scattered patches?"

He cited a notorious series of astronomical observations in which the minds of those gazing through telescopes converted patterns of spots into long straight lines. "When making too-general comparisons," he cautioned, "it is always easy to become a victim of illusions: certain groups of patches could appear to look like stretches. Remember the canals on Mars!"

Indeed, in early presentations of the magnetic evidence some (including this writer) were tempted to wonder how much imagination had gone into drawing those beautifully symmetrical bands of alternate black and white, based on magnetic recordings made by ships or planes that had traversed the area at fairly wide intervals. As the data accumulated, the reality of the patterns became more convincing, even though many irregularities and complexities emerged.

An explanation that Beloussov offered later for the symmetry of magnetic bands to either side of an ocean ridge was a series of successively younger and less extensive outpourings of lava from the ridge. Each flow spread a distance that was symmetrical to either side of the ridge. The magnetic bands, or anomalies, would represent the aprons of successively earlier flows protruding from under more recent ones—a scheme, however, that could not explain the conformity of the bandwidths to the timetable of magnetic field reversals.

In his "open letter" to Wilson, Beloussov extolled the advantages of considering many hypotheses, rather than just one. Otherwise, he said,

we will be bitterly reproached (perhaps also ridiculed!) by the coming generations if we call one of such working hypotheses a final theory, if we assert that the truth is at our elbow and that we have only to stretch out our hand and pick the flower. We have dedicated our lives to a difficult science, which, unfortunately, is still assembling fundamental data. We have only just begun to penetrate the secrets of the very shallow interior. It would be most irresponsible of us to tempt young people, saying that all the difficulties are behind us and, instead of leading them along a hard and strenuous path of search and menial labor, inevitable for a scientist, to lull them with delusive hopes and dreams.

Wilson's rebuttal began with a cordial acknowledgment of Beloussov's contributions to international science—his service, for example, on the committee that ran the International Geophysical Year of 1957–1958 and his initiation and leadership of the subsequent International Upper Mantle Project. Wilson agreed on the importance of what was known of continental geology, but he cited much evidence for sea-floor movement that had been disregarded in Beloussov's critique, including the seeming match of the timetable of field reversals with the magnetic patterns. He also drew attention to the new earthquake data of Sykes and others, published since Beloussov wrote his letter.

With regard to the Russian's emphasis on patient collection of new data, Wilson said:

If two groups of scientists both studied whirlpools, and one group held that the water in them did not move they would never understand whirlpools no matter how much data they collected. The other group, which admitted that the water was moving, could understand the nature of whirlpools with little data...

> If indeed the Earth is, in its own slow way, a very dynamic body and we have regarded it as essentially static (he continued), we need to discard most of our old theories and books and start again with a new viewpoint and a new science...If a scientific revolution is in progress in the Earth sciences it provides us all with an exciting opportunity, the prospect of a great revival, and I think we should embrace the change and expect the whole study of the Earth to move rapidly forward.

By 1972 many of Beloussov's countrymen had been won over to the plate theory and he himself, somewhat reluctantly, endorsed a final report of the international committee, of which he was chairman, that ran the Upper Mantle Project. It was during that project, the report said, that "there began the accumulation of an extraordinary assemblage of independent observations—observations individually not convincing but collectively overwhelming, and leading irresistibly to the model of plate tectonics."

Nevertheless what Beloussov had to say about fashions in science should not be taken lightly. And it may be a long time before some of the questions that he raised— notably the nature of the deep currents that are moving the plates—are resolved.

But that is getting ahead of our story. Before the debate between Wilson and Beloussov got under way, a project was launched that many hoped might pierce the crust of the Earth and throw some light on the deeper processes. It was a plan to drill deep enough into the sea floor to sample the mantle.

Chapter 8

'Mohole' or 'Nohole'?

BY SHEER COINCIDENCE THE NAME "CHALLENGER" HAS BEEN ASSOCIATED WITH the two most ambitious efforts—one real and one fictional—to penetrate the crust of the Earth. The real feat, accomplished only after an earlier, ill-fated Mohole Project, was the historic achievement of the drill ship *Glomar Challenger* in drilling over one thousand holes deep into the ocean floor, providing strong confirmation for the hypothesis of sea-floor spreading. The fictional case was that of Professor Challenger, the irascible scientist of Sir Arthur Conan Doyle's tale, "When the World Screamed."

Like so much of science fiction, the story anticipated real developments, although it presented a somewhat bizarre version of them. Professor Challenger had already figured in Conan Doyle's classic, "The Lost World," written in 1911, in which an inaccessible South American plateau is found to harbor surviving monsters of the antediluvian period (it formed the basis for such moving pictures as "King Kong"). In the second tale of the series, Challenger, a huge, black-bearded man, believed the world to be a living organism—a sort of gigantic sea urchin whose shell was the crust of the Earth. "You will recall how a moor or heath resembles the hairy side of a giant animal," he said. "You will then consider the secular rise and fall of land, which indicates the slow respiration of the creature. Finally, you will note the fidgetings and scratchings which appear to our Lilliputian perceptions as earthquakes and convulsions."

To his regret, however, the Earth was not aware of the tiny creatures living on its surface. "That is what I propose to alter," he said. "I propose to let the Earth know that there is at least one person, George Edward Challenger, who calls for attention."

Challenger, according to the story, sank a shaft eight miles into the Earth, breaking through the crust to grayish material, glazed and shiny, that rippled and bubbled, emitting an odor "hardly fit for human lungs." He proposed to announce his presence

135

to Mother Earth by driving a sharp, 100-foot artesian drill into the pulsating gray matter, like a gigantic sting.

Why should he do such a thing, he was asked? His reply was a caricature of scientific motivation. "Away, sir, away" he cried. "Raise your mind above the base mercantile and utilitarian needs of commerce...Science seeks knowledge. Let the knowledge lead us where it will, we still must seek it. To know once for all what we are, why we are, where we are, is that not in itself the greatest of human aspirations? Away, sir, away!"

The drill was injected on his command and there followed a cyclone, an earthquake, a volcano: "Our ears were assailed by the most horrible yell that ever yet was heard...It was a howl in which pain, anger, menace and the outraged majesty of Nature all blended into one hideous shriek...No sound in history has ever equalled the cry of the injured Earth." A spout of "vile treacly substance" shot into the air, and the pit closed like a wound.

Little did Conan Doyle realize the extent to which the Earth is, in fact, "living," churning, and ever-changing. For much of the past century scientists have dreamed of boring into the planet's interior to determine the nature of its inner structure and deep processes. On May 5, 1881, Charles Darwin wrote to Alexander Agassiz, son of Louis, the Swiss naturalist who was among the first to recognize the evidence for past ice ages: "I wish that some doubly-rich millionaire would take it into his head to have borings made in some of the Pacific and Indian (ocean) atolls, and bring home cores for slicing from a depth of 500 or 600 feet."

Darwin hoped, thereby, to confirm his explanation of the atolls—those romantic necklaces of coral reefs and low, palm-covered islets enclosing shallow lagoons in otherwise deep oceanic regions. Darwin believed the atolls rested on sunken volcanoes that once had stood high above the sea. Around their rims, to begin with, were coral reefs that grew higher as the volcanic island subsided. This subsidence formed a lagoon inside the ring of reefs, with the now-diminished volcano rising from the center of that lagoon. A classic example is Bora Bora in the Society Islands. In four years at sea the most dramatic landfall experienced by this writer was his first glimpse of the giant crag of Bora Bora, rising beyond the distant horizon. As we drew nearer, its lower, less steep, vegetated slopes came into view and finally the rim of palm-covered coral islets enclosing its circular lagoon, from which the volcanic remnant rises.

Eventually, according to Darwin's hypothesis, such a volcano sinks entirely beneath the sea, but, as it does so, the fringing coral reef continues to grow upward so that the living portion remains within the habitat of the coral polyps, close to the ocean's surface. Thus, only the necklace reef finally remains, enclosing a shallow central lagoon. It has been said that atoll reefs that have grown upward 1000 meters or more, as the volcano beneath them subsided, represent the largest structures built by any creatures on Earth, including man. But to a number of scientists such extensive sinking of the numerous atoll volcanoes seemed implausible. Darwin hoped that drilling through the coral would extract volcanic rock (basalt) from beneath it, proving his thesis.

Since Darwin's time, broader motives for deep drilling have emerged. One purpose would be to learn something about that part of the Earth's crust that covers two thirds of the planet—the ocean floor. In 1943, at the height of World War II, T. A. Jaggar, president of the Hawaiian Academy of Sciences and founder of the Volcano Observatory on the rim of Hawaii's Kilauea caldera, proposed that 1000 holes be drilled into the floors of the world oceans to obtain core samples to depths greater than 300 meters. Jaggar was a man of grandiose ideas, and this one was embodied in a letter to Richard M. Field, the proponent of ocean-floor exploration at Princeton.

Jaggar pointed out that geology, up to that time, had been based entirely on continental studies and thus was largely a "speculative science." He hoped that the geological community could team up with the oil industry and carry out the project which, he said, would cost "twenty million dollars as a starter." The purpose, he added, would be to "know the whole Earth crust, its thermal gradient, its rock specimens, its inner waters, its physical variables, its resources for future labor, its stimulus for future invention, its topography in comparison with the moon, and its economic minerals in relation to the trivial area of surface that today yields power, iron, copper, oil and aluminum." He noted that, when World War II was over, "thousands" of ships and engineers suddenly would become unemployed. He proposed that some be put to work extracting the 1000 deep-sea cores that, he said, would provide "a century of scientific specialists with materials for chemical, physical and biological analyses in the laboratory."

At the time of this bold proposal the requisite technology was not in sight, but it is just such a reservoir of specimens that was produced, starting in 1968, by the drill ship *Glomar Challenger*. The globe-encircling operations of that vessel and the abortive Mohole Project that preceded it had their roots in Darwin's hope that someone would drill through the coral of an atoll and prove him right. Several attempts had been made and, in 1947, after the first atomic tests at Bikini, holes were drilled to a depth of 779 meters. The drill was still chewing through coral when the hole was abandoned.

The three geologists who participated in this effort were to join in setting afoot the most ambitious drilling project ever planned. They were Harry S. Ladd and Joshua I. Tracey of the United States Geological Survey and Gordon G. Lill of the Office of Naval Research. In the early 1950s, with "basalt or bust" inscribed on the wall of his geology shack, Ladd continued his attempts on islets of the Eniwetok Atoll. He finally brought up basalt from a depth of 1287 meters, confirming Darwin's thesis.

Meanwhile Maurice Ewing, whose schooner *Vema* had been probing the sea floor with seismic soundings and piston corers, was agitating for a program to drill completely through the sediments. The soundings had revealed several tantalizing layers within them, indicative of major changes in the oceans long ago. They also suggested, as noted before, that, if one assumed that the ocean basins were as old as the continents, much of the sediment was missing, unless it lay hidden, solidified in the form of the hard "Second Layer" detected beneath the sediment by seismic probing.

In 1956 an Army scientist, Frank B. Estabrook, outlined in the journal *Science* the importance of boring through the entire crust of the Earth to sample the mantle. The crust is not much thicker, relative to the total volume of the Earth, than the skin of an apple. Yet, he pointed out, the mantle that lies beneath the crush comprising 84 percent of the Earth's total volume, had never been directly sampled. He listed the many scientific problems that could be resolved by obtaining such specimens—some of them, such as the plasticity of the mantle material, its radioactivity, and other properties (although he did not so specify), bearing on the drift controversy.

Apparently little heed was paid to his proposal at the time, but it was only a few months later that the Mohole Project was born. The members of an advisory panel at the National Science Foundation were sitting around a table strewn with requests for money to carry out a variety of studies. NSF, then still a relatively young government agency, was responsible for doling out federal funds for research in the physical and nonmedical biological sciences. One of the panel members, Walter Munk, the authority on waves, tides, and the Earth's rotation from the Scripps Institution of Oceanography, finally broke into the discussion to comment that not a single proposal on the table was grand in scope and aimed at extending knowledge into fundamentally new realms. Would it not be better to set our sights a little higher (or deeper)? he asked.

The problem that most excited another panel member, Harry Hess of Princeton, was the nature of the mantle. And so they began discussing whether or not it would be feasible to drill that deep. Hess and Munk were both members of an offbeat group, the American Miscellaneous Society, whose name symbolized its freewheeling, after-hours character. It was proposed that the membership of this organization, as well as its philosophy, made it particularly suited to explore the possibility of drilling through the crust.

The American Miscellaneous Society had been born within the Office of Naval Research in 1952 when Defense Department funds still were being generously allocated to a wide range of nonmilitary research projects. Some of the requests for such funding were rather unconventional and the scientists responsible for reviewing them decided to create an informal group to discuss the more "miscellaneous," imaginative, and far-out ideas. In its light-hearted moments the society boasted of subdivisions on calamitology, triviology, and etceterology, as well as a Latin motto: "*Illegitimum non carborundum*" (freely translated: "Don't let the bastards grind you down"). AMSOC's members included all three participants in the atoll drilling—Ladd, Lill, and Tracey—as well as others who became prominent in the drift debate such as Maurice Ewing, Roger Revelle, and Arthur Maxwell (the last two having made the Pacific heat flow measurements).

The idea of putting the American Miscellaneous Society to work on the drilling proposal was received with enthusiasm, and it was decided to call a meeting of the group over drinks at the Cosmos Club that evening. William E. Benson, head of the Earth Sciences division in NSF and member of the society, went to phone Gordon Lill, who was its chairman, asking him to meet them at the club later that day.

The discussion there ran late into the night. It was decided to form a group, to be known as the AMSOC Committee, "to investigate the feasibility of drilling to and sampling the mantle." The committee included all the society members except Benson who, because of his government post, had to remain on the sidelines. Thus was the Mohole Project born. While the character of the society had now become more sober, to obtain funds from NSF for a feasibility study it was necessary to have a substantial affiliation. To this end the prestigious National Academy of Sciences agreed to take on the AMSOC committee as one of its affiliates, and Willard Bascom, a specialist in ocean engineering on the academy staff, was appointed staff director for the project. It was Bascom who coined the term "Mohole," since the goal was to penetrate the Moho, or Mohorovičić discontinuity, thought to mark the boundary between the Earth's crust and its underlying mantle. As noted earlier, this lower boundary of the crust had been discovered by a Yugoslav named Mohorovičić who found that, below a certain depth within the Earth, there is a sudden and marked increase in the velocity of earthquake waves. It was assumed that this deeper region was formed of more dense material and it came to be known as the mantle.

Since the Moho lay about 35 kilometers (20 miles) below the continents, drilling to it there was deemed hopeless; but it seemed to lie as little as five kilometers (three miles) under the ocean floors. In some of the early discussions drilling on an atoll was considered, since three members of AMSOC, including its chairman, had participated in such work. But seismic probing on Eniwetok showed it to be underlain by 10 miles of lava rock—the ghost of its parent volcano—and, even if that were penetrated, the region below might not be typical of the mantle. Hence it was decided to explore the possibility of drilling the deep-sea floor—a challenge of formidable dimensions.

While such drilling was eventually to bear directly on the drift controversy, the concept of sea-floor spreading had not yet been proposed when the Mohole Project began to take shape in the late 1950s. Probing with sound waves had revealed a rugged surface buried beneath the sediments that many scientists then believed was the primordial crust of the Earth. They assumed that the geography of the world had been fixed since its formation and that, by drilling through all the sediment, one would extract fossils tracing the evolution of oceanic life from its first appearance on Earth.

It was a heady thought, for most of this early history was missing. The oldest fossil record of any completeness dated back only to the time, about 600 million years ago, when most of the phyla (or basic divisions of the plant and animal kingdoms) had already evolved. In the sediments of the ocean floor, therefore, it was argued that one could hope to find the missing early record of evolution. The successive layers would provide a ladder—a framework of time—on which to hang our fragmentary knowledge of the history of life, of climate, and of the changing orientation of the Earth's magnetic field.

In requesting that NSF in 1958 give AMSOC the modest sum of $30,000 for a feasibility study, the National Academy of Sciences–National Research Council said: "There probably is no project, within the scope of present capabilities, which would give more information concerning the broad picture of the Earth as a planet, than drilling a hole through the sediments, and the so-called basalt layer and finally into the upper mantle. It can be looked upon as a courageous attempt to broaden the base on which the most fundamental of Earth problems rests."

A political motive for the Mohole had also emerged. During the 1957 General Assembly of the International Union of Geodesy and Geophysics in Toronto a resolution was adopted, at the initiative of American proponents of the project, endorsing an attempt to sample the Earth's mantle. The meeting was marked by the first appearance of Soviet scientists at such a conference following the Stalin period, and one of them reported that the U.S.S.R. was also developing a deep-drilling capability. This was the period of intense Cold War rivalry whose climax came when the Soviet Union blazed the way into space, prodding the United States to undertake a "race" to land men on the Moon before the Russians did so—a race in which, it turned out, the Russians did not take part. The "race" to the Moho proved to be of a similar nature.

In a 1959 issue of *Science* Gordon Lill and Arthur Maxwell of AMSOC wrote: "There is intense international competition in science these days which is a kind of substitute for war. It pervades all areas of science." Noting the Soviet deep-drilling program they said: "It would give them great pleasure to beat the United States in this undertaking, since drilling is a field in which we are considered to be proficient." Considering the millions going into the Moon program, "and the billions being spent on atomic bombs," they added, the Mohole cost would not be exorbitant. The American Miscellaneous Society, they said, liked to put it this way: "The ocean's bottom is at least as important as the moon's behind."

From subsequent developments it appears that the Soviet goal was not to reach the Moho but to drill to great depths at sites of special interest within Soviet territory, such as potential oil-producing regions. But, meanwhile, fear that the Russians would demonstrate superiority had helped stimulate a venture that, while not as costly as the Apollo Moon-landing program, still promised to be very expensive in terms of scientific budgets. And, like Project Apollo, it would also depend on new and untried technologies.

That such technologies were not entirely out of reach became evident from reports filtering from the normally secretive oil prospecting fraternity. Methods were being developed for drilling holes three kilometers (almost two miles) into the seafloor beneath 500 meters of water. This was a far cry from drilling in three or four kilometers of water, but it was encouraging. Four oil companies—Continental, Union of California, Shell, and Superior—had banded together, calling themselves the CUSS group, to press this development. They were using a war surplus Navy barge of the largest type, newly converted into a floating rig christened *CUSS I.* It carried a derrick 30 meters high (compared to the standard 43 meter derricks found in oil fields), and it was held over the drill hole by winch-controlled mooring lines attached to large, anchored buoys.

Such a mooring system would not be practicable in the very deep water beyond the continental shelf, but huge, diesel-driven outboard motors had been developed during World War II to propel Navy barges, and Bascom proposed that such motors, each developing 200 horsepower, be installed on the four corners of *CUSS I.* By aiming the motors in suitable directions it should be possible to hold the barge in position over a deep-sea drill hole. Buoys moored to the bottom and held 30 meters below the surface by piano wire would serve as echo-ranging reference points for the man controlling the motors. The *CUSS I* was modified accordingly and, after drilling several test holes in relatively shallow water off San Diego, the barge was towed to an area 65 kilometers east of Guadalupe Island, off Mexico, where the water is 3.5 kilometers deep and where the puzzling "Second Layer" had been detected, by seismic sounding, beneath 300 meters of sediment.

Determining the nature of this Second Layer was itself a challenge worthy of a major effort, and it was one of particular interest to Harry Hess. Seismic exploration of both the Atlantic and Pacific had shown such a layer to be very widespread. The layer above it was clearly sediment, transmitting sound waves slowly—at about 1.8 kilometers per second. The Second Layer transmitted sound much faster—at 4.5 to 5.5 kilometers per second, implying that it was rigid. Below that was the denser Third Layer, taken by many to be the primordial crust of the Earth, with a velocity of seven kilometers per second. Finally, about five kilometers (three miles) down, was the Moho, marking a further increase to 8.2 kilometers per second. It was believed that the Moho represented either a change in chemical composition of the rock, or a sudden shrinkage of molecular structure to a denser, more rigid state (with no change in composition), due to great compression and high temperature at that depth.

By the time *CUSS I* took up its position off Guadalupe in April 1961, Hess had set forth his concept of sea-floor spreading and was arguing that the Third Layer beneath the sea floor is, essentially, the top of the mantle capped by a veneer of basalt (the Second Layer). The Moho, he proposed, simply marks a change in molecular structure caused, perhaps, by high temperature at that level, either currently or at some time in the past.

If one assumed that the oceans date back to the earliest phase of the Earth's history, as noted earlier, most of the sediment beneath the seas seemed to be missing, and hence many suspected that the Second Layer was formed of this sediment, converted to rock. On the other hand, if the sea-floor spreading concept was valid, the sediment was "missing" because the ocean basins had all been formed relatively recently—since the time when reptilian ancestors of modern mammals roamed the drifting land.

When *CUSS I* reached its assigned location, buoy markers were set in place and a succession of holes were drilled into the sea bed more than 3000 meters under the rig's bobbing hull. One of them penetrated 200 meters, and two cores of basalt, one of them three meters long, were extracted. While it was conceivable that the basalt came from a thin layer, with more sediment underneath, Hess doubted this. He concluded that the Second Layer had, in fact, been sampled. The material brought up in cores obtained directly above it was a green-gray clay believed to have been laid down as recently as the Miocene, 25 million years ago—not when oceans presumably first appeared on Earth more than four billion years ago.

In terms of the sea-floor spreading theory, this Second Layer presumably represented a veneer of basalt on top of the mantle. It would be this veneer that was imprinted with magnetic patterns generated by periodic reversals of the Earth's magnetism.

"The success of the drilling in almost 12,000 feet of water near Guadalupe Island," said President Kennedy in a congratulatory message, "and the penetration of the ocean crust down to the volcanic foundations constitute a remarkable achievement and an historic landmark..."

Following the *CUSS I* drilling it was felt that the practicality of reaching the Moho had been demonstrated and AMSOC, as well as its sponsor, the National Academy of Sciences, decided that the big engineering tasks to follow, including construction of a large, specially designed drill ship or platform, should be undertaken by others. The Academy, chartered by Abraham Lincoln as an organization of the nation's leading scientists that could advise the government, was not designed as an operating agency.

It was at this stage that the Mohole Project, for a combination of reasons, began to fall apart. So far it had cost only $1.8 million, most of which had gone to Global Marine

Inc., which operated *CUSS I* for the oil companies. But now the character of the project changed. On July 27, 1961, the National Science Foundation invited bids for design and construction of the ship or platform that would drill a succession of holes into the sea floor, culminating in one reaching the mantle. On August 10 of that year the deadline for bids was extended, and the bidders were asked, among other things, to describe their plans, "if any," to use the AMSOC staff, headed by Bascom. While the scientists of AMSOC wanted to fall back to an advisory capacity, they felt that Bascom and his staff had a knowledge of deep-sea drilling problems that was unique.

A dozen bids were received, and five were selected as finalists. They included the Texas engineering firm of Brown and Root, General Electric, Zapata Off-Shore Co., and two consortiums composed of some of the largest industrial empires of the country: One included Global Marine, Shell Oil, and Aerojet General; the other combined Socony Mobil Oil with Standard Oil of California, Texas Instruments, and General Motors.

A panel of NSF officials, asked to make a preliminary evaluation, reported to Alan T. Waterman, head of the National Science Foundation, on October 20, 1961, that the group that included Mobil was "in a class by itself—outstanding as to every important respect." Fifth place was given to Brown and Root, a firm that had specialized in very large-scale projects.

In the next go-around the Global-Aerojet-Shell combination moved up to first place because, in the words of one NSF official, "they clearly understood their proposal and the problems better than the Mobil team, who had a good writer [for their submission]. Brown and Root also scored well in understanding the odd-ball complexities."

When Brown and Root finally got the contract, there were cries of "political fix." The firm's home town of Houston was represented in Congress by Albert Thomas, chairman of the subcommittee that processed the NSF budget and would have to approve the Mohole appropriation. Daniel S. Greenberg, who reported developments in science politics for *Science* magazine, wrote that the business fortunes of Brown and Root "were alleged to have been closely linked to the Democratic Party, to which it regularly contributed campaign funds." In the Congressional hearings on the controversy that ensued Waterman, as head of NSF, thrust aside such implications. "Brown and Root," he said, "has a very remarkable record in a number of ways. First of all they are quite accustomed to contracts and know how to deal with them. Second, they do them on time and at cost." But the suspicions of a "fix" would not be downed, and furthermore two other controversies emerged, involving cost and how to go about the job.

In 1959 some promoters of the project were saying it could be done at a cost to the government of no more than five million dollars. This was based on the hope that the oil companies and the oil exploration industry, in the expectation of long-term gains in drilling technology, would donate most of the equipment. When Brown and Root came through with its plan in 1963, the cost of building the drill platform and drilling once to the Moho was put at $67.7 million. Other reports put the ultimate cost as high as $125 million. The Pick and Hammer Club, comprising geologists from the Washington area, in its annual spoof show three years earlier, had chosen as its theme "Moho-ho and a barrel of funds." The joke of 1960 had now become a serious matter.

The other controversy, regarding tactics, ripened after the *CUSS I* drilling had dramatized what could be achieved scientifically with holes far less deep, difficult, and costly than the Mohole. The whole history of the ocean basins now lay within reach, using a drill rig mounted on a conventional ship hull. In fact Bascom and his colleagues had already drawn up preliminary designs for such a vessel. They favored gaining experience with an "intermediate" ship before final design of the Mohole drill rig.

Like the other bidders, Brown and Root, possibly sensing that inclusion of the AMSOC staff on its team, at least temporarily, might help them win the contract, had taken Bascom's group on as consultants for a two-month period, but their advice on an intermediate ship was put aside. The contract with NSF did not call for an intermediate ship, since NSF had left the decision on its necessity up to Brown and Root. Moreover, there apparently was friction between Bascom's team and the Texas engineers. Bascom was a dynamic, adventurous, self-made type, once described by *Life* magazine as having "instant charm and a Barnum-like instinct for promotion." In another profile he was termed "as diplomatic as a diamond drill." In any case, Brown and Root put aside the idea of an intermediate ship and proposed construction of a two-deck platform that would ride high above the waves on six columns, supported by two submarine hulls each 113 meters long. The platform itself would be the size of a city block—71 by 76 meters.

Since the submarine hulls would be below the waves and the platform high above them, the craft would be subject to minimal heaving and rolling. It would be propelled by engines in the submarine hulls and would be kept on station by thrusters in the six vertical columns. Whatever trial-and-error development of drilling techniques was needed, it was said, could been done from this platform. Some of the AMSOC technical advisors favored this approach, as did key figures within NSF. Alan Waterman, as NSF director, told the Subcommittee on Oceanography of the House Committee on Merchant Marine and Fisheries that, while drilling by an intermediate ship would be "scientifically exciting," it should not be allowed to interfere with basic Mohole programming and funding. "We believe," he said, "that the sooner a vessel of appropriate capability and capacity is constructed, the sooner both the intermediate and the ultimate objectives of the project will be accomplished." He thus envisaged the intermediate drilling as being done from the ultimate platform.

The idea of drilling to the Moho was a dramatic one that had won considerable support in Congress, and apparently there was a feeling at NSF that any broadening of the goals could jeopardize the whole scheme.

Early in Mohole planning it had appeared that probing the sediments and upper crustal layers would come about as inevitable dividends of the project, and it was this hope that rallied such men as Ewing, Revelle, and Maxwell to the project. However, immediately following the success of *CUSS I* there was a belief among those then active in AMSOC that, while widespread drilling into the sediment and upper crust was now feasible, it should not be part of the Mohole Project. This was reported to Detlev W. Bronk, President of the National Academy of Sciences, on June 14, 1961, by Gordon Lill as chairman of AMSOC. While it was now possible for a drilling vessel "to explore thoroughly the sediments and upper crustal layers of the ocean basins," Lill wrote, "...we must recommend this possible exploration program to you for separate scientific and financial consideration. We expect the initiative for general ocean basin exploration to come from the large oceanographic institutions, and we heartily endorse the importance of the work."

But, six months later, a new man who set high priority on shallower drilling had taken over as AMSOC chairman, and debate over the Mohole goals moved to center stage. The new chairman was Hollis D. Hedberg, vice president for exploration at the Gulf Oil Corporation and also a part-time professor at Princeton. As the scope and cost of the Brown and Root plan became known, it seemed less and less likely (at least to Hedberg and others) that Congress would provide funds for a separate drilling effort, outside the project. Furthermore, the argument that the preliminary holes to be drilled by the giant Brown and Root platform might meet the demands of shallow-hole proponents was quickly dismissed by the latter group. They pointed out that the platform would be too cumbersome for the "one-night-stand" type of drilling necessary for widespread exploration of the ocean basins, and it would be too big to traverse the Panama Canal. Instead, it would probably sit at one site drilling a succession of holes until the contractual requirement of a single penetration of the Moho was achieved. It was estimated by some that three years of round-the-clock drilling would be required to bore that deep.

Thus Bascom and his colleagues, with their plan for an intermediate ship, found an ally in Hedberg (who persuaded Ewing to rejoin AMSOC, from which he had resigned), and in others on the committee with a strong interest in the sediment and upper crust. Hedberg also argued—with considerable persuasiveness—that extracting a single sample from the mantle would be of limited importance because it might not be typical.

Seismic probing of the sea floor in many areas had shown distinct layering, he told a Congressional hearing, "and many individuals have jumped to the conclusion of a nice, systematically layered crust separated from the mantle by a magic surface known as the Moho, and they naively think that a single hole penetrating these layers will give the answer as to what they are. In reality, as I am sure you all realize, the picture is infinitely more complicated and confused. Simplicity in the picture as usual comes only from a little knowledge."

He pointed out that the layering differs from region to region, as do the properties of the material forming the layers, and in some areas no Moho at all is evident: "Realistically, we must recognize that not one hole but many holes of various depths and at various locations must be drilled before we can even begin to intelligently understand the true situation. Does it make sense to strain wildly for a single deep hole to the mantle before we know something about the upper layers of the oceanic crust or before we know where best to drill our mantle holes?" The Mohole Project, Hedberg told the House Subcommittee on Oceanography, "can readily be one of the greatest and most rewarding scientific ventures ever carried out. I must say also that it can just as readily become instead only a foolish and unjustifiably expensive scientific fiasco...There must be insistence that it not be allowed to degenerate into merely another costly publicity stunt."

It was on October 29, 1963, that Hedberg spoke thus, and by then the project was in deep trouble. Because of the escalating costs and controversies, Kermit Gordon,

Director of the Budget, on March 1 of that year had ordered a freeze on further substantial Mohole expenditures until the disputes could be resolved.

Angry voices were raised on Capitol Hill, notably by Senators, such as Thomas Kuchel of California and Gordon Allott of Colorado, from states whose firms had missed out in favor of Brown and Root. Some wags began calling it the "Nohole Project," and Walter Munk, one of those who gave birth to it, asked:

> Were the chances for success enhanced when the project was removed from the authority of an independent group to an inhouse NSF activity? And were the young men who gathered under Bascom's daring and devoted (though sometimes willfull) leadership less qualified than an amorphous engineering firm which had demonstrated no previous scientific interest in the Mohole venture, or as far as I know in any scientific problem? I have never been able to learn on what grounds the direct approach of a few highly skilled people, which served so successfully in Phase I, had to be replaced by the "acres-of-engineers" philosophy.

Thus the project that had been launched on a wave of scientific enthusiasm was foundering in a morasse of political and administrative difficulties. In January 1966 came the death of Representative Thomas of Texas, "Mohole's chief Congressional guardian" (in Greenberg's words). Then reports were published that family associates of Brown and Root had donated $23,000 to a Democratic fund-raising organization.

That was the coup de grâce. In August 1966 Congress took the unusual step of killing the project on its own initiative, cutting off all further funds for it. Mohole was dead, although keels for its submarine hulls had already been laid on the West Coast. (Some of the metal was later donated to Cornell University to serve as radiation shielding in its powerful synchrotron.) Yet from the ashes of Mohole was to emerge the Deep Sea Drilling Project, one of the most successful scientific enterprises of the century.

Chapter 9

An Epic Voyage into the Past

THE PROJECT THAT HAS CONTRIBUTED MORE TO THE HISTORY OF THE OCEANS than any expedition to date began inauspiciously with two years of maneuver and countermaneuver among oceanographic institutions and their leaders on both coasts of the United States. In 1962 it became increasingly evident that, if there was to be any extensive drilling into the oceanic sediments, it would not be within the Mohole Project. On October 10 of that year, for example, C. Don Woodward, Managing Coordinator for Mohole in the National Science Foundation, told the AMSOC Committee that Brown and Root did not plan to include an intermediate-stage ship in the project. However, he said NSF would look favorably on an independent program for drilling to intermediate depths.

Meanwhile the oceanographic institutions had begun sounding out one another on what to do. Cesare Emiliani, of the Institute of Marine Sciences at the University of Miami, and other oceanographers from Lamont, Woods Hole, Scripps, and Princeton, formed a committee called LOCO (for LOng COres) to explore the possibility of chartering a drilling ship. Then Ewing of Lamont, Revelle of Scripps, and J. B. Hersey of Woods Hole established a more formal organization, CORE (Consortium for Oceanic Research and Exploration), which proposed a program similar to that which had been envisioned for the intermediate phase of Mohole. However, Miami was conspicuously omitted from this lineup and NSF, reportedly sensing partisanship, did not come forth with any funds.

Finally, in early 1964, Lamont, Miami, Scripps, and Woods Hole got together and formed JOIDES (for Joint Oceanographic Institutions for Deep Earth Sampling). Under political prodding they were later joined—after drilling had begun—by the University of Washington to increase regional representation. Then, in 1973, when the Soviet Union agreed to join and help finance the project, the Institute of Oceanology of the Soviet Academy of Sciences also was included.

The first JOIDES effort was modest. By 1965 the "dynamic positioning" of drill ships to keep them over a hole in deep water, pioneered by Willard Bascom in the *CUSS I* four years earlier, had been adapted to oil exploration vessels, and one of them, Global Marine's *Caldrill,* was sailing from California to do some exploratory drilling for the Panamerican Petroleum Company off Newfoundland. The oil company agreed to a delay en route and NSF provided funds so that, off Florida, some demonstration holes could be sunk for the oceanographers in water as deep as 1000 meters.

NSF was impressed not only by the drilling but by the organization behind it and so provided further funds, while the JOIDES institutions drew up an ambitious program, with Scripps to direct the project and specialist panels from JOIDES studying possible drill sites and setting the scientific objectives.

By January 1967, with Mohole dead, NSF was prepared to be generous. It signed a $12.6 million contract with Scripps for an 18-month program of drilling in the Atlantic and Pacific. (By 1975 the project had cost some $68 million.) On November 14, 1967, a subcontract for the actual drilling was given to Global Marine, the firm that had operated *CUSS I.*

Only a month earlier the keel had been laid in Orange, Texas, for one of a new class of large drill ships, and plans had been made to adapt it for deep ocean work. The ship (referred to in the previous chapter) was the *Glomar Challenger* (Fig. 9.1), "Glomar" being the first name assigned the newer ships of Global Marine. The name Challenger was derived, not from Conan Doyle's irascible scientist, but from *H.M.S. Challenger,* the British steam corvette that for three and a half years, in the 1870s, had crisscrossed the world's oceans in the first major oceanographic expedition up to that time. The specimens she collected in remote regions and from the deep sea are still used by scientists for reference purposes and, as noted earlier, her soundings were the first to reveal the true nature of the oceanic basins.

The 122-meter hull of the *Glomar Challenger,* like that of its successor, the *JOIDES Resolution,* is reminiscent of a tanker, with her bridge and living quarters at the stern.

Figure 9.1 The Glomar Challenger is a drill ship whose maneuvers are aided by tunnel thrusters in her bow.

Amidships, rising above an open well that gives access to the sea for the drill pipe, is the derrick, its crown 59 meters above the waterline—so tall that even at low tide the ship cannot sail under New York's East River bridges. In many respects her drill rig operates as in an oil field on dry land, although a number of innovations were derived from Mohole preparations before that project's demise. In order to lower a drill pipe several kilometers to the ocean floor it is necessary to screw onto it a succession of 29-meter sections, or "stands," of pipe. These are retrieved, by a special hoist, from a pipe rack on the forward part of the ship and hoisted up inside the derrick. One clamping device close to deck level holds the string of pipe already dangling from the ship while the new "stand" is added. When the drill string is assembled, a motor, called the power sub, is attached to its upper end to turn the entire pipe and the drill bit at its lower end. Once all connections have been made, the clamp is released and the drill begins turning, an enormous block-and-tackle inside the derrick lowering the pipe, as it bores down, retaining enough of the weight to keep the pipe from buckling. The crown block at the top of the derrick remains stationary, whereas the traveling block rides down as the drill penetrates until a new pipe section is needed (Fig. 9.2).

To witness the addition of an added "stand" of pipe, carried out by a well-trained crew with almost military precision, is an awesome spectacle. The clanging and banging of outsized tools, the shouts of the men, the hissing and chugging of engines, and the bewildering speed at which huge pieces of metal fly hither and yon are all sobering. And the drillers themselves know that when something goes wrong, it can be fatal. Once, early in the project, the crew was working through the night, hauling up the pipe, section by section. With 4500 meters of pipe still dangling from the great, floodlit derrick, the topmost stand of pipe was being removed. The procedure ran smoothly until a clamping device took the full weight of the pipe. It had been improperly rigged and one side of the clamp broke free. The clamp tilted so sharply that the pipe was torn apart. William Benson of NSF, one of the scientific leaders on this leg of the expedition, came on deck just in time to hear the shrieking rupture of metal and see the jagged upper end of the pipe sink downward, whirling in wide circles that came alarmingly close to decapitating one of the roustabouts.

"Well, Doc, we wrung her neck this time," commented one of the crew as 4500 meters of pipe sank beyond retrieval to the bottom. However, the ship carries 7300 meters of pipe on her rack, with more in the hold, so she was able to continue without returning to port.

A major difference from drilling on land is the need to provide for heaving of the ship on the bosom of the sea. This motion would lift the drill bit from the rock into which it was chewing, then drop it back onto the rock with damaging force, were it not for shock-absorbing devices called bumper subs. These are telescopic sections of pipe, a short distance above the bit, that allow for a few feet of upward and downward motion and yet, through long, interlocking teeth, continue to impart rotation to the bit.

The bit, 23.5 centimeters (9.25 inches) in diameter, bores a hole sufficiently larger than the five-inch drill pipe so that fluid can be pumped down through the pipe and back up the hole, outside the pipe, to flush out chips of rock as the drilling proceeds. On land a special "mud" is used for such flushing and to provide sufficient

Figure 9.2 The derrick floor *of the* Glomar Challenger *was where the crew drilled for the core samples. In the middle of the photo is the "traveling block" suspended by cables reaching to the "crown block" (not visible) at the top of the derrick. Beyond the derrick is the automated drill pipe racker, which held more than seven kilometers (four miles) of pipe.*

pressure, inside the hole, to keep its walls from collapsing. In ocean drilling, however, sea water normally can be used instead.

Most important of all on these ships is the coring system that retrieves sections of sediment or rock as the drill bores downward. It is from the cores that all the scientific rewards of the expedition have been derived. By the end of the project in 1983 the total length of all *Glomar Challenger* core specimens came to 113 kilometers (70 miles) and many more have been extracted by its successor, the *JOIDES Resolution*. The drill bit has a center section that can be detached and hauled up through the pipe, making it possible to lower a 10-meter core barrel to fit into the resulting hole. The core barrel rides on roller bearings and therefore is not forced to turn as the rotating bit tunnels downward and a central core of sediment or rock, six centimeters (2.4 inches) in diameter, rises up into the barrel. When those on board think the barrel may be full, they haul it up.

The core sample, enclosed in a plastic inner tube, is pushed out of the barrel and, after initial examination, is cut into 1.5-meter lengths which are split lengthwise. Half the split sample is sealed and placed in a refrigerated van below decks. These vans are shipped to one of the two repositories of the project—at Scripps and the Lamont–Doherty observatories—to provide reference material for present and future generations of scientists. The other half is used for immediate study, and what remains is then also relegated to the van.

One of the first tasks, before the core has been cut or split, is to take pinches of material from its two ends for immediate microscopic examination in the core laboratory. If there are tiny fossils in these top and bottom samples, they will tell what span of time is represented (see Fig. 9.3). A single core may represent more than a million years and the evolution of various tiny sea creatures from one form to another.

The deep-sea drilling has greatly enlarged this history of oceanic life. In conventional piston coring the tube is driven no more than 20 or 30 meters into the bottom, producing information, as a rule, on only the more recent periods. But the *Glomar Challenger* has, for example, bored 1300 meters into the floor of the Arabian Sea, beneath 3539 meters of water, the total length of drill pipe thus exceeding 4.8 kilometers (three miles). Such deep penetration has made it possible to fill in great gaps in the history of certain marine organisms—life forms that evolved in areas of the sea or times in the past never before sampled. In the Caribbean, for example, a complete section of sediment from recent times back to the Upper Cretaceous was recovered, spanning about 85 million years. In addition to enlarging our basic knowledge, such information has practical value for geologists in that, for example, it enables them to use microscopic fossils to identify possible oil-bearing formations in sediment laid down beneath ancient seas.

Despite gaps in the fossil record there are usually enough clues to the age of a core for the specialist aboard the *Glomar Challenger*, within minutes, to define the ages of the top and bottom layers and advise whether to core further or reinsert the center bit for drilling at maximum speed. Since drilling must be suspended during each core retrieval, coring is done only when it promises to be productive. Where the sediment lacks fossils or represents a well-documented period of oceanic history, the drilling may

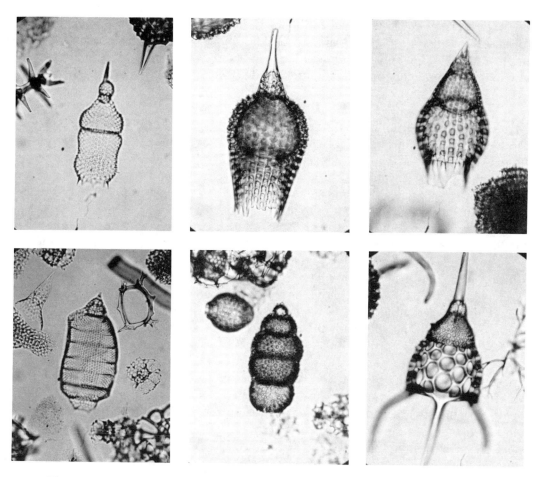

Figure 9.3 Age Indicators. *These microscopic "tests" or shells from sea creatures called radiolarians have been extracted from sediments under the Indian Ocean, tropical Pacific and Caribbean. Each species, during any given period, occurred worldwide and was only characteristic of that one period. All have been magnified 230 times. The top and bottom radiolarians on the left are from the* **Quaternary Period,** *from 0 to 1 million years ago. The two in the middle are from the* **Early Miocene,** *about 24 million years ago. The two on the right are from the* **Middle Eocene,** *about 50 million years ago.*

proceed for scores of meters without a halt for the laborious process of hauling up the center bit, inserting a core barrel, and retrieving a sample.

The drilling and coring would have been impossible were it not for the position-keeping system that enabled the *Glomar Challenger,* even in the deepest waters, to remain directly over the drill hole for a week at a time, despite wind, wave, and current. Through her bow, below the waterline, are two tunnels, each fitted with a propeller that can be used to push the bow to one side or the other. There are two similar tunnels through the stern. In this way the ship can be pushed broadside—or she can be spun, without going fore or aft, by thrusting the bow sideways in one direction and the stern in the opposite direction. When the ship first went through the Panama Canal, the canal pilot was told that her ungainly lines belied her maneuverability. He asked for a demonstration and he, as well as those on a half-dozen nearby ships, watched in amazement as the ship, in about three minutes, made one complete revolution without any forward motion.

As soon as the *Glomar Challenger* reached the desired drill site an acoustic beacon was thrown overboard. Where the water was thousands of meters deep, it might take an hour to sink to the bottom. Once it came to rest, the arrival times of its signals at three of four hydrophones suspended beneath the ship were transmitted to a computer. By comparing those arrival times the computer determined the direction of the beacon and issued appropriate commands to the ship's main engines, as well as to its bow and stern tunnel thrusters, to keep it in its assigned position.

A computerized pilot, however, is no smarter than the computer program that determines "his" actions. During early trials the computer kept the ship within 12 meters of its assigned position for a day at a time, then ran amok, ordering full speed ahead on the ship's twin-screw main engines. On at least one occasion the *Glomar Challenger* dashed 300 meters before the computer could be countermanded. "You can imagine the reaction on board when there were 20,000 feet (six kilometers) of pipe dangling beneath the ship with the end stuck into the ocean bottom," said Arthur Maxwell, later named provost at Woods Hole, after serving as a scientific leader on the ship. The problem, essentially, was "training" the computer to be an effective pilot—programming it, for example, to take into account the ship's sluggish response to engines and thrusters and not "panic" when the response was slow. It turned out, as well, that the sonar beacons first dropped onto the sea floor were not quite loud enough, and the computer would mistake other sounds for the beacon, sending the ship off on what might be called a wild porpoise chase.

The ship's acceptance trials were completed in the Gulf of Mexico on August 11, 1968, three weeks after she departed her native shipyard in Orange, Texas, and without returning to port she began the succession of drilling legs that, by 1973, had carried her into all of the world's oceans. The first hole was bored into the Gulf floor beneath 2832 meters of water. It penetrated more than 762 meters without getting to the bottom of the enormous apron of glacier-ground rock flour and other debris hosed onto the northern floor of the Gulf when the melt-water of the North American ice sheets was racing down the Mississippi Valley, tens of thousands of years ago. This was chiefly a dress rehearsal, but the second site was chosen to test a proposal of Maurice Ewing that the Sigsbee Knolls, a series of peculiar humps in the deepest part of the Gulf floor, were salt domes of the type associated with oil accumulations. Ewing and his colleague, J. Lamar Worzel, were scientific leaders on this initial leg, which ended in Hoboken, New Jersey, on September 23.

The first of these humps had been discovered by the Lamont Observatory's *Vema* in 1954, and in 1961 the same ship, using a subbottom profiler, found many more buried under the sediment. Zigzagging back and forth across the knoll area that runs in a broad north–south band across the deepest part of the Gulf, the *Vema* identified more than 150 knolls, and there probably are many more. If, like the salt domes of the Gulf

Coast, they marked oil reservoirs, this would be a resource of major importance, but many oil men were skeptical.

Salt domes are great, fingerlike columns of salt that, typically, have thrust upward through several kilometers of overlying sedimentary rock. They arise where a thick deposit of evaporites—salts left where seawater has dried up—has been blanketed by a deposit of heavy sediment. The salt, being lighter, tries to force its way up through the sediment, like a cork that is moving slowly up through thick soup. In a sense the salt is "squeezed" up by the weight of material on top of it. Unless that overburden is thick enough, no salt domes arise. (See the salt-dome illustrations, Fig. 9.4.)

In its upward movement the salt dome pierces and bends upward the sediment layers through which it is passing and, where those layers are porous (as in sandstone), droplets of oil derived (presumably) from ancient sea life begin to migrate into the upwarped part of the layer (since oil is even lighter in weight than water). The result, over millions of years, is accumulation of oil and gas in porous, upwarped layers alongside or over the dome. The rising oil and gas may generate considerable pressure in such a reservoir and, in drilling on land, its penetration produces a "gusher." The significance of such an oil trap on the flanks of a salt dome was first demonstrated in 1901 when Captain A. F. Lucas was prospecting for oil at Spindletop, Texas. An eruption of oil and gas blew completely out of control, flooding the countryside with oil. It was the first great Texas gusher, giving birth to the flourishing oil industry of that region.

Until the *Glomar Challenger* began revolutionizing many aspects of marine geology it had been assumed that salt domes occur only where shallow seas, whose floors were close to sea level, evaporated periodically, depositing their salt on the sea floor. Such evaporite layers may become thousands of feet thick. When the sea again fills with water, sediment is laid down over the evaporite and the stage is set for salt-dome formation.

But, the skeptics asked, how could this process occur on a sea floor that is now 3700 meters below the water surface? Whatever the explanation, Ewing believed the Sigsbee Knolls were salt domes, in part because they seemed to be an extension of shallow-water formations known to lie farther north, off Texas and Louisiana, and to the south, off the Campeche area of Mexico.

Hence Hole 2 was targeted for a knoll that protruded above the sea floor, and officials in Washington began to worry that Ewing might be right and the ship might puncture a high pressure reservoir of gas and oil. Some of the whispered fears were rather dramatic: The pipe would be blown up through the bottom of the ship; or a giant bubble of gas would rise to the surface, reducing buoyancy of the sea and allowing the ship to plunge to the bottom. Four years earlier, for example, the *C. P. Baker,* a catamaran-type drill rig, was drilling in the Gulf when it punctured a gas reservoir and became engulfed in frothy water. The rig sank with a loss of 22 lives. Those of a less alarmist bent pointed out that this had occurred in only about 120 meters of water and that such a concentrated effect, in the immediate area of the drill ship, was far less likely where the bottom was thousands of meters below the keel. An additional fear was that a blowout at such depth would be almost impossible to plug and would pollute a large

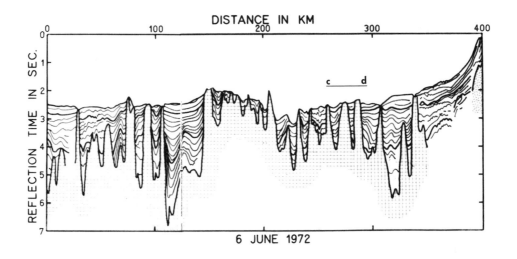

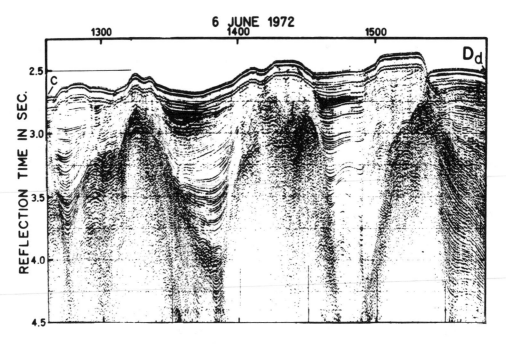

Figure 9.4 Oil Traps. *These intrusive features, or "diapirs," off Angola on the west coast of Africa, are typical of the salt domes that have pushed up through many layers of sediment to form "traps," reservoirs where oil and gas may collect. The top diagram shows the profile of such features recorded in 1972 by the* Atlantis II *of Woods Hole Oceanographic Institution when it sailed from point A to B on the detailed map, next page. Three of these diapirs, located between points c and d on the top diagram, are highlighted in the bottom diagram which was derived from readings of a sub-bottom sounding device.*

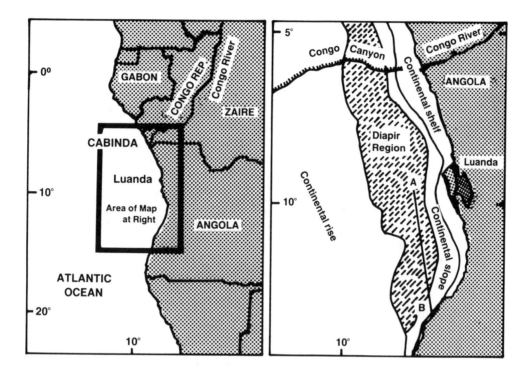

part of the Gulf, producing a worldwide outcry that could force termination of the entire project.

As Ewing came alongside the ship in a launch, prior to the drilling attempt into what has come to be called Challenger Knoll, a colleague on deck shouted (in so many words): "They aren't going to let you drill the knoll!" When Ewing got on board, he found no such dictum in the message file (those responsible for the project in Washington said later they were, in fact, convinced by the structure of the dome that, even if it had produced an oil reservoir, a blowout was unlikely).

Nevertheless, Ewing pressed for an immediate attempt before someone said "no." Cores extracted from the knoll began displaying characteristics of a salt dome, and finally one core was brought up that, in Ewing's words, was "dripping with oil." Prudently the drilling was then halted, and the hole was plugged with cement—a precaution taken at the end of drilling on all 14 holes drilled in the Gulf on a subsequent leg (unless equipment failure made this impractical). The oil in the bottom core from Hole 2 was sent by the American Petroleum Institute to the laboratories of several oil companies who found it to be strikingly similar to that extracted from the heavily exploited oil fields of the Gulf Coast. Furthermore, analysis of cores from this site revealed traces of anhydrite, a saltwater deposit typically formed in a shallow, sun-baked sea that has evaporated—a finding to be repeated, under equally surprising circumstances, when the ship began drilling into the deepest parts of the Mediterranean. More recent surveys of the northwest part of the Gulf by Britain's *GLORIA* wide-span sonar have shown that the complex topography of that region is largely controlled by up-thrusting fingers of salt. The Sigsbee Escarpment marks where this region intrudes into the fan of ice age deposits from the Mississippi.

Despite this early success, the difficulties encountered in the next hole were discouraging. Time after time the core barrel came up empty, or with only a few centimeters of material. The rest had leaked or been washed out the bottom. An improved device for closing the bottom end helped solve that one. Furthermore, the drill bit kept encountering layers of extremely hard rock, known as chert. It typically is derived from the silica-rich shells of sponges, diatoms, and Radiolaria—all creatures of the sea. But, as our early ancestors discovered, the resulting rock, which in its more glassy form is known as flint, is extremely hard. Although a wide variety of drill bits were tried, some studded with 500 to 800 karats of tiny diamonds, the chert wore them to scrap metal within a few feet and the flinty chips jammed the telescoping bumper subs until, with every passing wave, the bit pounded mercilessly on the rock at the bottom of the hole, causing more damage.

On land the drill string can be hauled up, stand by stand, and the drill bit changed. One kind of bit is used for soft sediment and another for hard rock. But at this stage the ship had no way to reenter a hole in the sea floor, once the bit had been hauled out. Later a method was developed whereby a huge funnel—its top five meters in diameter—was installed in the hole so that reentry became possible and, on its first use to achieve a scientific goal, it enabled the ship to obtain a complete cross section of Caribbean sediments over the past 85 million years. Also, by 1972, bits had been found that could cope with the chert, and the time-consuming reentry was avoided wherever possible.

The project records of early drilling attempts in the Atlantic tersely reflect the frustrations:

Site 4 (Hole 4): "Below about 600 feet penetration, the drill with increasing frequency encountered resistant and abrasive chert beds which wore out the bit, and caused operations at this hole to be terminated. The roller bit used here was completely destroyed by the chert."

Site 4 (Hole 4-A): [A drag bit was tried.] "After the pipe was pulled it was found that the drag bit had been destroyed by the chert."

Site 9: "Continuous coring of the acoustically laminated sediment yielded very poor recovery and was abandoned after six attempts."

Site 11: "Recovery was very poor, and only 22 feet (7 meters) of sediment was obtained from nine attempts."

Site 12: "Drilling ceased when an unsampled hard layer destroyed the bit."

Sometimes the drill stuck in the bottom, and, when the drill string was finally pulled free and hauled up (it takes many hours to do so), it was found that the entire down-hole assembly had been lost—bit, coring barrel, bumper subs, and the heavy sections of pipe used to load adequate weight on the drill. By 1972 some 33 such down-hole assemblies had been lost, but most of the difficulties, typical birth pains of a new technology, were worked out early in the project.

Hence, on the second and third drilling legs of the *Glomar Challenger*—from New York to Dakar in Africa, and from Dakar to Rio de Janeiro in South America during

1968 and 1969—it became possible to tackle what, by then, had become the most fundamental problem in the Earth Sciences: the sea-floor-spreading hypothesis. The plan was to drill completely through the sediment to bedrock at lines of sites spanning the Mid-Atlantic Ridge in both the North and South Atlantic. If the theory was valid, at each site the age of the earliest sediment, laid down directly on top of the basement rock, should conform to the timetable of spreading away from the ridge.

One impediment was the thinness of sediment near the mid-ocean ridge itself. Coaxing the drill bit into the bottom until it is deep enough to be stable is known as "spudding in." Drilling into the deep sea-floor can be likened to dangling a sharp-ended wire from the top of the Empire State Building and trying to bore into the sidewalk below. If the sidewalk were covered by mud that was deep enough, the rotating wire would drill its way down until there was enough mud around it to keep it from bending when it reached the hard pavement. But if the mud were only a few centimeters thick, the wire would bend and bow and never bite into the pavement. This was the nature of the problem in the North Atlantic within 75 to 150 kilometers (45 to 95 miles) to either side of the ridge centerline, forcing the scientists to pick sites so far from it that they were beyond the symmetrical, well-defined magnetic anomalies that, it was known from the dating of lava flows on land, had been laid down in the past few million years.

Ideally, the spreading theory would have been tested by determining the age of crustal rock brought up from beneath the sediment at each site. Canadian scientists had produced what they considered evidence for movement of the sea floor away from this ridge by determining the ages of volcanic glass and chunks of basalt dredged from the bottom near the ridge axis. The farther from the centerline, the older they were. The age of the glass was determined from the number of scars, or fission tracks, left in it by radiation. The basalt was dated by extracting, from the core of each specimen, material that presumably had been protected from contact with sea water, which introduces errors into ages determined by the standard potassium–argon method. (For fear of such errors it was decided not to attempt dating the basalt bedrock brought up by the *Glomar Challenger*—at least not until she was able to drill very deep into such a formation.)

Another way to determine age might have been to trace the succession of magnetic reversals down the full length of the hole, starting at the top with the present orientation of the Earth's magnetic field and following its known schedule of flips, back and forth, as recorded in the sediment. This would establish the time when the lowest layer was laid down. But such a method also was impractical. Cores were retrieved only intermittently, and the sediment in them tended to be so distorted by the coring process that the original magnetic orientation of the particles was lost.

The dating, therefore, was done by examination of the microscopic fossils found immediately above the basement rock. Although, on the New York to Dakar Leg in 1968, chert repeatedly kept the drill from reaching basalt at the base of the sediment, such rock was sampled at three sites, and the ages of the fossils found above it in each case conformed, broadly speaking, to the predicted schedule of sea-floor spreading. It was concluded that the sea-floor at Latitude 30° North has been moving away from the Mid-Atlantic Ridge at a rate of 1.2 centimeters (0.5 inches) a year, at least since the late

Cretaceous, some 80 million years ago. However, it was on the next leg, from Dakar to Rio, that the most powerful evidence was obtained. Seven holes, spanning the ridge in Latitude 30° South, reached basalt despite repeated crises.

During the drilling at Site 15, on Magnetic Anomaly 6 (the sixth magnetically defined zone west of the ridge), the wind built up to 30 knots, and the automatic position-keeping system, even with some human intervention, had difficulty holding the ship over the hole. For a time it looked as though the drill string would have to be pulled, but the wind finally subsided. At Site 16, over Anomaly 5 (a magnetically positive anomaly west of the ridge), the current was so strong that, for the main engines to hold the ship on station, she had to be kept headed into the current, even though on that heading wind and waves buffeted her broadside. This set the ship rolling heavily. The bit began pounding on basalt, and torque on the drill string became so severe that rotation sometimes halted completely. When the bit finally was hauled out it was partially broken, but a chip of rock in its wreckage told the story—that the basalt floor beneath the sediment had been reached. At Site 17 it took three holes to reach basalt and also obtain sediment from immediately above it for dating. On arrival at Site 20 a cold front was encountered with winds to 30 knots. When the homing beacon was thrown overboard in five kilometers (three miles) of water, it took off on the shoulders of a deep current, and the ship had to chase it, lest it be carried beyond acoustic range.

For a valid test of the spreading theory it was necessary to be sure that the drill hole had reached rock laid down when that part of the sea floor was formed. There was no way to be sure that the basalt that had been sampled was not a layer laid down over older sediment, but, since ages determined from the deepest sediment layers penetrated by the drill conformed closely to the proposed timetable of spreading, it seemed unlikely that any important part of the sedimentary record had been missed. Furthermore, the timetable showed a remarkably uniform rate of sea-floor movement away from the ridge in both directions—eastward toward South Africa and westward toward Brazil. The distances of these drill holes from the ridge axis ranged from 250 to 1300 kilometers (150 to 800 miles) and the age, at each site, was close to that predicted from the magnetic field reversals, imprinted on the sea-floor with particular clarity in this area (see Fig. 9.5).

Doubters of the spreading concept had pointed out that, from seismic soundings, it appeared that the sediment did not become appreciably thicker at greater distances from the ridge in this area. If the sea floor more distant from the ridge were older, one would expect more sediment to have accumulated there. The drilling confirmed this peculiarity of the sediment thickness, but it was deduced that most of the accumulation occurred early in the history of a sea-floor section, when it was at relatively shallow depths near the ridge, and that later accumulation was so small that, in subsequent years, it added little to the total. "Overall," said the report of the scientists aboard the *Glomar Challenger*, "it is concluded that the sea floor of the South Atlantic has been spreading at an essentially constant rate for the past 67 million years." The rate was such that points on opposite sides of the ridge (and hence, presumably, Africa and South America) have been moving apart four centimeters (1.6 inches) per year throughout that time.

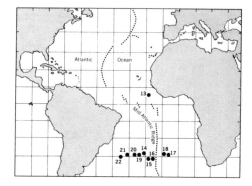

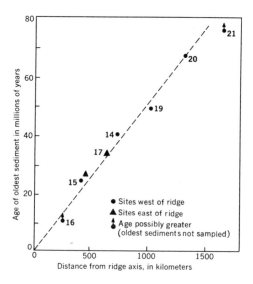

Figure 9.5 Ages of rocks *are a function of distance from the Mid-Atlantic ridge. On both sides of the Mid-Atlantic ridge, the* Glomar Challenger *obtained cores from holes drilled to basement rock at the numbered drill sites shown in the top map. Sediments from the cores revealed that the farther away a site was from the ridge, the greater the age of the deepest sediments. When the ages were plotted on a graph in terms of their distance from the ridge centerline, the plotted points lay in a nearly straight line, indicating a uniform rate of movement away from the ridge both to the east and west.*

Drilling into the Atlantic floor on these and subsequent legs has not only confirmed the spreading concept but has begun to paint a picture of the awesome events that accompanied the birth of that ocean. This cleavage of continents, witnessed by our reptilian ancestors, was not a peaceful process. From the sediments laid down soon after the Atlantic was born, it seems to have been Wagnerian in its eruptions and upheavals—a series of events suited to the vivid and romantic pen of Immanuel Velikovsky. Today the active volcanoes that rise from the Mid-Atlantic Ridge are few and far between—those of Jan Mayen, Iceland, the Azores, and Tristan da Cunha—but, as the Atlantic was being born, this activity must have been on a formidable scale, with new volcanoes thrusting from the sea, darkening the skies, and exploding under cataclysmic steam pressure when rupture of their walls allowed the sea to pour in. Layers of volcanic ash have been found in the sediments at many sites. For example, at Site 10, among abyssal hills west of the ridge in the North Atlantic, more than 30 meters of sediment, immediately above the basalt, contained volcanic minerals. One hole, drilled near the foot of the continental rise off New York, penetrated a succession of ash layers in the last 30 centimeters (12 inches) before reaching the basalt, or hardened-lava basement, at a below-bottom depth of 621 meters. This hole was in the area where the deep-sea floor begins sloping up toward the continental margin. Thus it was relatively near the edge of the American continental block that broke away from northwest Africa at the "grand opening" of the Atlantic. The deepest fossils from this hole were about 155 million years old, and it was estimated, therefore, that the last time land animals could walk directly east from New York to the Sahara was between 170 and 180 million years ago.

The ash beds at the bottom were interlayered with limestone whose lowest levels had been transformed by intense heat to brilliant red and green. These layers were laid down when the Atlantic was relatively shallow and apparently little wider than the Red Sea. The lava erupted, spreading beneath the bottom sediment layers and altering them with its heat. At some sites it was found that the bottom layer of limestone had been converted into marble by the heat, pressure, and disruption of lava intrusions. The swirling, fractured, contorted patterns in some of our decorative marbles bear witness to the agonies of sea-floor sediments during such invasions.

Not only were the birth and growth of the Atlantic marked by extensive volcanism but by activity that carpeted the new ocean floor with remarkably rich layers of metallic sediment. The process may, in fact, still be going on. It has been observed in the extraordinary hot pools of the Red Sea and has produced similar layers in the Pacific. Their importance, for the future of an industrial world with limited metal resources, is obvious. One of the first indications of such far-flung deposits came from drilling at Site 9 on the northeast flank of the Bermuda Rise. For the final 60 meters of this hole the drill cut through clay that, the scientific team reported, underwent a "remarkable change in color," becoming more and more red as the iron content increased. Just above the basalt at the bottom of the hole was a brick-red shale (solidified clay) containing "abundant iron and other metal oxides." There were no fossils in this layer, but immediately above it were shells from the Late Cretaceous Period, 70 to 100 million years ago. Analysis of a shale sample by X-ray diffraction showed 11.7 percent iron oxide. Since then, drilling through the sediments of the world's oceans has revealed metal-rich layers immediately above the basement rock in many (though not all) areas. At one site off the West Coast of North America the drill sampled six meters of virtually pure oxides of manganese and iron, separated from the basement by nine meters of this material mixed with fossil-bearing sediment.

The iron-rich shale at Site 9 lay under about 800 meters of sediment; but extensive metallic deposits have been found on the surface on the East Pacific Rise—the Pacific counterpart of the Mid-Atlantic Ridge—and on the floor of the Red Sea, where geyser-like eruptions seem to be extracting metals from the hot rock rising beneath the rift valley and depositing them on the surface. If this process has occurred—or is still occurring—in the rift valleys of other mid-ocean ridges, such metal deposits could be expected to become more and more deeply buried beneath oceanic ooze as they were transported farther from the ridge. This would explain the thick accumulation of sediment over the metallic layers off New York. Deposits of copper, iron, and zinc were found there at levels more than 240 meters below the bottom. Then, in the summer of 1972, drilling into the Madagascar and Mozambique Basins, on the ship's 25th operational leg, showed that such metallic deposits also lie under the sediments of the

Indian Ocean. All told these accumulations of metal just above the basement rock of the world oceans may be the most extensive on Earth. Evidence from earthquakes and bottom photographs of fresh lava formations, indicating that eruptions are still occurring along the rift valley bisecting the Mid-Atlantic Ridge, led to one of the boldest scientific enterprises of the century, reminiscent of Jules Verne's adventurous tales of submarine journeys. It was FAMOUS, the French–American Mid-Ocean Undersea Study, a four-year project involving exploration of the Mid-Atlantic Rift Valley by deep submersible. One of the primary tasks of the craft, on their hazardous prowling of the black canyons beneath two to three kilometers (one to two miles) of water, was to learn the nature of the eruptions responsible for these metallic deposits and whether or not they are currently taking place. The project is described more fully in the last chapter.

Of equal importance with the metallic carpet beneath the ocean sediments has been the discovery of further evidence for formations that may harbor accumulations of gas and oil. Cores extracted from depths greater than a few hundred feet in three holes near the continental margin off Georgia and Florida were impregnated with methane (the chief constituent of natural gas). The hole off New York penetrated a black layer with carbonaceous material capable of burning. And an account, ringing with excitement rare in a scientific report, came from the scientists on Leg 2, between New York and Dakar—M. N. A. Peterson, N. T. Edgar, and C. C. von der Borch of Scripps (plus Robert W. Rex of the University of California at Riverside, who later did some of the analytical work). From deep in a hole at the foot of the African continental slope near the Cape Verde Islands they extracted material that, they deduced, had come from a vast basin of evaporites (the residue of evaporated seawater) closer to the African shore. If the Atlantic was once a north–south series of narrow, shallow seas, like the chain that now includes the Red Sea, Dead Sea, and Lake Tiberias (the Sea of Galilee), those bodies of water, periodically drying up, would have formed salt beds that later became overlain by sediment suitable for oil accumulations. As the Atlantic grew, the scientists reasoned, these deposits must have been torn in two, one part drifting east with Europe and Africa, the other moving west with the Americas. They pointed out that salt domes of the Senegal–Gambia Basin, off the western extremity of Africa, are known to extend well down the continental slope and, when the continents are fitted back together, this area lies against a bed of evaporites off Cuba and the Bahamas. Likewise, they said, a salt basin off Gabon and the Congo, in the bight of Africa that fits the bulge of Brazil, matches the Sergipe Basin off that South American country.

Meanwhile the belief that extensive oil reservoirs might have formed when the Atlantic was young and narrow received further encouragement from French oceanographers who had studied a number of areas of that ocean's floor. In 1970 they reported that their research ship, *Jean Charcot,* had detected formations resembling salt domes buried in the sediment off Labrador, Newfoundland, Morocco, Portugal, and in the Bay of Biscay. Combining these discoveries with those off Africa, Ireland, and elsewhere (by 1972 such structures also had been found off Angola in Africa and Honduras in Central America), they proposed that both the eastern and western margins of the Atlantic basin are underlain with vast salt deposits.

The Leg 2 scientists from the *Glomar Challenger* summarized their conclusions thus:

These observations suggest that the Mesozoic and lower Tertiary Atlantic [170 to fifty million years ago] may have been very similar to the Persian Gulf and Red Seas today. It is suggested that initially a Persian Gulf type of evaporite sea occurred with the Gulf of Mexico with its Luan Salt being deposited in one of the great evaporite embayments. As the Atlantic started rifting in Jurassic and Cretaceous time, several evaporite basins were formed along the shelves of the rifting ancestral basin and then split apart. By lower Tertiary time a series of peripheral Atlantic evaporite basins including the Senegal–Gambia, Gabon–Congo, the Brazilian, and the Carribbean had formed and were down warped and buried.

The above speculations are intended to advance the hypothesis that the Atlantic, the Red Sea and the Persian Gulf started out as similar features, but that the Atlantic has spread extensively, the Red Sea only slightly, and the Persian Gulf has only subsided a few hundred feet. This hypothesis suggests that additional drilling should be undertaken to outline those areas of old "Atlantic Gulf" that survive in the Atlantic Ocean. The possibility of a deep water petroleum province as large or larger than the Persian Gulf is too important to discard without evaluation.

By the 1990's such optimism had been somewhat tempered by the results of drill holes off the East Coast of North America. For a time there was hope that a region of domelike structures buried in sediment under 4300 meters of water 650 kilometers (400 miles) off the West African coast might indicate the presence of salt domes even on the floor of a deep mid-ocean basin. While the nearby Cape Verde Islands are volcanic, the sounding of sediment layers showed that they had been pushed upward by these intrusive bodies in the manner observed for the salt domes of an oil field. The discovery of such a field 400 miles at sea would have been sensational, but when one of these domes was pierced beneath 250 meters of sediment, on Leg 14 of the *Glomar Challenger's* back-and-forth journeyings across the world oceans, it was found that the intrusive body was a volcanic plug. This disappointment, however, was overshadowed by another series of discoveries—probably the most dramatic to come out of the Deep Sea Drilling Project—made on the previous leg, in which the *Glomar Challenger* ventured into the Mediterranean.

Chapter 10

Of Seas Turned to Deserts and Lands Drowned or Arisen

LITTLE DID THOSE RIDING THE *GLOMAR CHALLENGER* IN THE SUMMER OF 1970, AS SHE sailed past Gibraltar into the Mediterranean, suspect that beneath them there may once have thundered a waterfall grander than any on Earth today. Yet among the surprises awaiting scientists assigned to this drilling leg was evidence that suggested the existence of such a natural wonder.

One of the puzzles they hoped to resolve had been the discovery, beginning in 1959, of a widespread layer deep under the sediments of the Mediterranean floor that was a strong reflector of seismic signals. Further surveys, in 1961, had disclosed dome-like structures rising from this layer beneath the Balearic Basin, an abyssal plain of mid-ocean depth between Italy's Sardinia and Spain's Balearic Islands. This hinted at an evaporite deposit, as did beds of salt that extended onto dry land around the edge of the Mediterranean—in Spain, Italy, Turkey, Israel, and North Africa. But, as in the case of the Sigsbee Knolls in the Gulf of Mexico, it was hard to believe basins as deep as those in the Mediterranean ever could have become dry.

Of the 27 holes drilled in the Mediterranean by the *Glomar Challenger,* 10 were in or near the Balearic Basin. The first was sunk on August 24, 1970, and that evening the drill crew, after hauling up the pipe, brought the bit to the scientific team. "Stuck between its teeth," they reported, "were bits of bedded gypsum, which was the first confirmation we had that an evaporite deposit is present under the Mediterranean."

The third hole showed that the strong-reflecting layer was, in fact, a farflung series of evaporite beds. Some 60 meters of this formation were penetrated and, the report said, "we recovered many cores of anhydritic sediments." Anhydrite, like gypsum, is a calcium sulfate, and both are used in making plaster. But what sent shivers down the spines of those aboard the drill ship was their knowledge that such layers are precipitated from shallow, salty lakes and seas, like the *sabkhas* of the Persian Gulf and Syria. In the Arab countries a *sabkha* is the term for a sun-baked desert, flat and usually near the coast where it is occasionally flooded at high tide and dries repeatedly to form a salt deposit. Anhydrite precipitates out of highly saline water at temperatures above 35 degrees C (95 degrees F). This is a situation typical of desertlike salt beds—not of a deep-ocean floor. The sedimentary structures in the anhydrite cores, the scientists wrote, "are similar in all respects to those of the recent and ancient sabkha deposits." Such precipitates, they added, could not have formed in cool, deep water; yet this material had been extracted from a hole reaching 3148 meters below the present surface of the Mediterranean.

The scientific team on this leg was strongly international, as was appropriate to an expedition into the Mediterranean. The leaders were William B. F. Ryan of the Lamont Observatory in Palisades, New York, and Kenneth J. Hsü of the Federal Institute of Technology in Zurich. Others came from institutions in Vienna, Austria; Cambridge, England; Paris and Brest, France; Catania and Milan, Italy; Bucharest, Rumania; and Berne, Switzerland. In fact, beginning with Leg 2 there had been a determined effort to impart an international flavor to the program. The report on the Mediterranean drilling was prepared by the two leaders, Hsü and Ryan, plus Maria B. Cita of the Institute of Geology and Paleontology in Milan, a specialist in foraminifera (an order of tiny, prolific inhabitants of the sea that provide important clues in determining sediment ages). Some of the Mediterranean cores contained layers with abundant freshwater or brackish-water diatoms that presumably lived when the basins of that sea, while cut off from the world oceans, were occasionally flooded with fresh water from continental rivers. Although the drill was able to penetrate only a few dozen meters of salt beneath the Balearic Basin, the seismic soundings, as well as the outcroppings of this evaporite bed in Sicily, indicated that its thickness might reach "a few kilometers," said the report by Hsü, Ryan, and Cita. Yet they calculated that a single drying of the Mediterranean would deposit salts only about 25 meters thick, of which approximately 19 meters would be contributed by rock salt (sodium chloride).

The explanation, the scientists decided, was that, over a period of about a million years, the sea had dried repeatedly. In some cores Maria Cita found marine oozes between the layers of evaporite, indicating that the basin had been refilled a number of times. This could account for the extreme thickness of the salt beds—too great to have been laid down by the desiccation of a single "tubful" of Mediterranean sea water (some believe such deep accumulations may also occur where salt has slid down from shallower areas). As the ship sailed from one drill site to another, the scientists debated what could account for these findings. They considered the possibility that the salt beds were formed near present sea level and that a catastrophic subsidence, within a few million years, had lowered them some three kilometers (two miles). But the thickness of the salt beds was more satisfactorily accounted for, they thought, by the repeated evaporation of deep basins than by a similar process in shallow lakes. The explanation finally proposed by Hsü, Ryan, and Cita was that the Strait of Gibraltar had acted as a "floodgate" that opened and closed repeatedly during the late Miocene, some six to eight million years ago.

Moreover the retrieval of open-sea sediments from earlier levels beneath the Mediterranean indicated that there had been an open connection between the Mediterranean and the Indian Ocean, permitting cold bottom water free passage between the Atlantic and Indian Oceans. There is ample evidence that a great body of water, the Tethys Sea, once lay between Africa and Eurasia and that the pressure of Africa northward against Eurasia not only reduced that sea to such remnants as the eastern Mediterranean, but generated the mountain ranges from the Alps to the Caucasus of the Soviet Union and the Zagros of Iran. This continuing pressure is presumably responsible for the volcanic and earthquake activity that extends from Portugal through Italy, Yugoslavia, Greece, Turkey, and Iran.

This process, it was proposed, cut the Mediterranean link to the Indian Ocean (now traversed only by the Suez Canal) and pushed the Gibraltar area into a configuration that made it susceptible to repeated opening and closing. When closed, the inflow of water from rivers debouching into the Mediterranean was insufficient to make up for evaporation, and the sea dried up almost to the bottoms of its three-kilometer-deep basins. Then, at each opening, the basins rapidly filled. Even today, Hsü has pointed out, rainfall and the flow of rivers into the Mediterranean are insufficient to make up for evaporation from its surface. If the Gibraltar Strait were closed, that sea could conceivably dry up in about 1000 years.

"When the gate was closed," the three scientists wrote, "the abyssal plain was a playa lake (one that dries up intermittently), but it would become deeply submerged as soon as the gate was temporarily opened...At the beginning of the Pliocene (5.5 million years ago), the gate was crushed and all barriers were removed, and the Mediterranean topographic depression was to remain a part of the world's ocean system to the present day."

The most dramatic feature of the hypothesis was the manner in which that basin filled for the last time—what Hsü, Ryan, and Cita referred to as "the final deluge." East of Gibraltar there is a sharp drop in the sea floor—a scarp of great height that would have been exposed when the sea was empty. Over this, they believe, poured a waterfall such as no modern man has ever seen. According to Hsü the flow was probably 100 times that of today's mightiest—Victoria Falls on the Zambezi River in Central Africa—and 1000 times that of Niagara Falls. It may be that our slope-browed ancestors gazed upon this thundering spectacle. It took 100 to 1000 years for the sea to fill, according to the estimates of the drilling scientists, and for a time a deep-water link with the Atlantic, perhaps carved by the waterfall itself, permitted cold bottom water to enrich the marine life of that sea. But then the strait once again became shallow and today the Mediterranean is too isolated, warm, and salty for many of those earlier species.

This concept of Mediterranean desiccation, followed by a catastrophic flooding in which that sea, like a gigantic bathtub, was again filled, sounded too much like Velikovsky and other champions of calamitous events for a considerable number of scientists. "The idea that an ocean basin of the size of the Mediterranean could actually dry up and leave behind a big hole thousands of meters below worldwide sea level seems preposterous indeed," wrote Hsü, Ryan, and Cita in defense. "We were reluctant to adopt such an outrageous hypothesis until we were overwhelmed by many different lines of evidence. When we first publicized our idea in press conferences at Paris and New York, we were greeted with much disbelief. We recall that our colleagues were vehement in their negative reactions: the idea was thought to be physically impossible."

It was argued, for example, that an abrupt filling of the Mediterranean, adding the weight of some 3.3 million cubic kilometers (close to a million cubic miles) of water to that region, would so alter the distribution of mass on the Earth that it might have upset the stability of the planet's spin. To meet this challenge, the three scientists noted that the eight million cubic kilometers of ice estimated to have accumulated on northwest Europe in the last ice age obviously did not produce any radical changes in the Earth's spin.

However, it was one of the press conferences to which the three scientists referred—the one in New York—that set in motion a chain of events destined to bring support to their hypothesis. Following that conference Ryan received a letter from I. S. Chumakov of the Geological Institute of the Soviet Academy of Sciences which read, in part:

> Last October there had been a brief note in the Soviet press about the end of your expedition on *Glomar Challenger* to study the Mediterranean. On my request the TASS correspondent, Mme. E. L. Shields has kindly sent me a clipping of the article "Drillings Indicate Mediterranean Is Smaller" from The New York Times (October 10, 1970). This article attracted my attention inasmuch as for some time I have been studying the development of the Mediterranean during the late Miocene–Early Pliocene [4 to 7 million years ago] and your preliminary deductions coincide (admittedly, not all of them!) or come very close to my conclusions obtained on the basis of work in some countries in the southern part of the Mediterranean.

His research had included the study of 15 holes drilled into the bottom of the Nile River just south of the Aswan High Dam (see Fig. 10.1). Soviet engineers were helping build the new dam, and the holes were sunk to determine the depths to bedrock and the nature of the sediment on top of it. This revealed that under the relatively flat bottom of the present river the bedrock forms a canyon some 290 meters deep, now

Figure 10.1 Cross section of the Upper Nile. *Before building the Aswan High Dam, Soviet scientists drilled across the relatively flat Nile Valley to bedrock. Their drilling revealed a deep, sediment-filled valley with a narrow gorge at its base. Ancient marine sediment filled the gorge, while the upper valley sediments were freshwater deposits.*

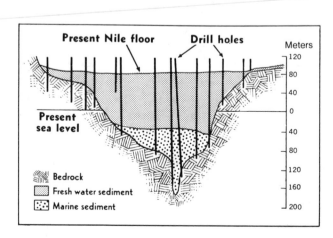

filled with sediment. The lowest part is a narrow gorge with almost vertical walls. Most remarkable, Chumakov found, is the nature of the sediment in the bottom 150 meters of this canyon. It proved to be filled with oceanic fossils of the Pliocene.

In other words this canyon, 1200 kilometers (750 miles) up the Nile, was once flooded by a sudden incursion of the sea some 5.5 million years ago. The most likely explanation, Chumakov believed, was that the canyon was carved when the Nile flowed into a Mediterranean Sea 1000 to 1500 meters lower than today. Then rapid filling of the sea sent salt water up the canyon, and only when the latter silted up higher than the existing sea level did the accumulating fossils change to fresh water forms. The proposal that the Nile might once have flowed into an empty Mediterranean basin via a deep canyon was supported by soundings of the sea floor by various ships, showing traces of ancient river beds. Moreover, Ryan learned of a geologist named Frank T. Barr, with the Oasis Oil Company, who had made a proposition that his colleagues thought absurd at the time.

Beneath the desert sands of Libya's Sirte Basin, 960 kilometers (600 miles) west of the Nile, oil prospectors had found a system of buried channels cut deeply into the bedrock, their floors lying 400 meters below present sea level. Barr believed they were relics of an ancient river system formed when the level of the Mediterranean suddenly dropped in the Late Miocene—the same period deduced by the trio of scientists who had drilled the Mediterranean floor. Nobody believed him, they wrote of Barr's thesis, and he could not get it accepted by a scientific journal until they included it in their report to the Deep Sea Drilling Project.

Further evidence had been recorded earlier by French scientists who had long debated the explanation for a situation along the lower Rhone much like that discovered by Chumakov under the Nile. Deep wells as far north as Lyon, 300 kilometers (almost 200 miles) upstream from where the river empties into the Mediterranean, penetrated marine deposits at the bottom of an ancient, buried fjord. In the Rhone delta this buried channel is 900 meters below the surface. The favorite explanation was a worldwide drop in sea level, although in 1952 one Frenchman, G. Denizot, proposed that isolation and extensive drying of the Mediterranean was the answer.

The argument that a deep basin like the Mediterranean could not dry up and turn into a desert because nothing like that is to be seen on the Earth today was branded by Hsü, Ryan, and Cita as "substantive uniformitarianism" founded on the premise that conditions on the Earth have remained relatively constant throughout geologic history. Such an interpretation of uniformitarianism, disallowing events and processes in the past to which we are not witness today, is considered unacceptable by many of today's geologists because, they feel, it looks at the world from the very limited perspective of human history. A broader interpretation—one that appealed, for example, to Harry Hess—sees the laws of nature and some basic processes, such as slow convection within the Earth, as enduring, but allows for rather dramatic events at scattered points in the Earth's history. The implications of a Mediterranean converted into a great desert basin were many. Such a development could explain the radical change in central European climate in the late Miocene when, as Hsü put it, "the Vienna woods were changed into steppes and when palms grew in Switzerland."

In the mid-19th century Sir Charles Lyell, one of the fathers of modern geology, took note of a change in marine fossils found in various Italian deposits, representing some form of revolution in the environment about six million years ago. He designated this as the end of the Miocene and the start of the Pliocene—the last epoch before the ice ages. It now appears that what altered the environment of the Mediterranean was its desiccation, killing off species with ancestral roots in both the Atlantic and Indian Oceans, followed by opening of the Gibraltar gates and repopulation of the sea with purely Atlantic types.

If basins two or three kilometers deep dried up in the Mediterranean, forming thick salt deposits, this might also represent the process that took place when the Atlantic was largely closed and the Gulf of Mexico isolated, accounting for the salt domes that lie beneath its deepest parts. However, some Earth scientists are still skeptical of "Hsü's waterfall," finding more plausible a gross subsidence of the sea floor following the salt deposition. They argue that, while shallow-water salt deposits were being laid down on the floors of what are now deep basins, similar deposits were forming in areas close to sea level around the rim of the Mediterranean. The most likely explanation, they think, is that the entire region was at that level, but the central part then sank, the Nile and Rhone canyons being cut during this subsidence.

From a practical viewpoint, the most important outcome of the Mediterranean drilling was its demonstration that major oil reservoirs may lie beneath the floor of that sea. While it was tempting to drill into one of the apparent salt domes pressing up through the sediment, the domes were ruled out of bounds by the project planners lest there be a blowout. This would have been particularly disastrous in the confined waters of that sea. Also it was expected that the Mediterranean nations would not take kindly to such drilling in or near their territorial waters.

Nevertheless, there was a "show" of oil in a core from Hole 134 in the Balearic Basin, indicating "the potential of a great reserve," according to the report, and 11 of the holes were plugged with concrete to insure against a blowout. While the technology for exploiting such deep-sea oil deposits has not yet matured, it is under intensive development, and it seems inevitable that, within a few decades, such operations will be under way—in a manner that, one hopes, minimizes the dangers of blowouts and oceanic pollution.

Despite efforts to give the project an international flavor and keep politics out of it, the Mediterranean was not the only area where tempting drill sites were passed up because of political considerations, and some oceanographers began to worry that politics might become a major impediment to scientific exploration, particularly on the continental shelves. Among the sites abandoned under diplomatic pressure was one in an area of promising domes in what the United States considered international waters off the Atlantic coast of Canada. The Canadians had become concerned about activity that could pollute the waters off their coasts—particularly at the prospect of giant tankers hauling out Alaskan oil through the relatively ice-free passages within the Canadian Arctic Archipelago, along the route newly blazed by the specially built tanker *Manhattan.* When they learned of the drilling project off their Atlantic coast, they let it be known that a formal protest would be lodged if the plan were carried out. Likewise, on the Atlantic Leg that followed the Mediterranean drilling, two sites on opposite sides of the North Brazilian Ridge had to be abandoned. Since the United States did not recognize Brazil's claim to a 200-mile zone off its coast, Washington would not intervene, and more direct and informal attempts to obtain clearance from Brazil were unsuccessful.

While the drilling showed that Gibraltar's role in "controlling" the Mediterranean long antedated its modern strategic role, of even more far-reaching consequence was the part played in oceanic history by the Isthmus of Panama. It was known that the isthmus began rising from the sea about 10 million years ago and that for a long time before that there was free passage for water between the Equatorial Atlantic and Pacific. But why did the isthmus rise to form a dam between those oceans, and what effect did it have? To explore these questions the *Glomar Challenger* sank holes on both sides of the isthmus. The Caribbean hole nearest the Panama Canal showed that the isthmus had been the scene of heavy volcanic activity some five to seven million years ago, presumably when Central America was rising from the sea, but it was on the Pacific side that a proposed explanation for that uplift was tested. This was the suggestion that it had been caused by sea-floor spreading from the Galapagos Rift Zone—a series of east–west rifts north of the Galapagos Islands. It had been pointed out that there are ridges parallel to that rift zone, both north and south of it, as though they had once formed a single ridge, along the rift, and then been split apart by the flow of sea-floor generated within the rift and spreading away from it. This would have carried the northern fraction of the ridge to the north, forming what is now the Cocos Ridge, offset by a series of fracture zones (Fig. 10.2). The Carnegie Ridge to the south would be the other fraction of the original feature.

According to this theory the northward moving section of sea-floor descended into an extension of the Middle America Trench along the coasts of Costa Rica and Panama, burrowing under that part of the crust and causing its uplift. This process, that

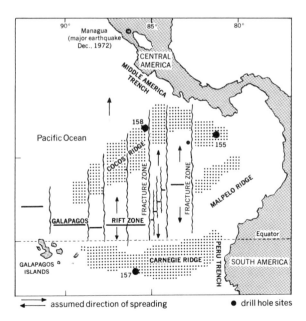

→ assumed direction of spreading ● drill hole sites

Figure 10.2 Spreading in the Equatorial Pacific. To test the proposal that spreading from the Galapagos Rift Zone split an ancient ridge to form two now widely separated ridges, the Cocos Ridge and Carnegie Ridge, holes were drilled at sites 155, 157, and 158. The oldest sediments, found at the northern sites, indicated that they were laid down farther south, near the equator. Geologists believe that the northward motion drove the oceanic plate under the Middle America Trench, thereby lifting Central America from the sea. This plate movement may also be partially responsible for the earthquakes that periodically shake Central America, including the Managuan earthquake of December 1972. (See also Fig. 14.2).

apparently lifted the isthmus from the sea, is still continuing, and periodic surveys of Costa Rican railways show annual uplifts of several centimeters along the Pacific Coast. The movement has filled the trench in that sector and today the Middle America Trench follows the coast from abreast of Mexico City only as far as Costa Rica, where it ends against the northwest foot of the Cocos Ridge. The earthquake that struck the Nicaraguan capital of Managua, farther up the coast, in 1972 apparently involved subsidence, rather than uplift, and its relationship to plate motions in the area is still unclear.

When, in 1971, the *Glomar Challenger* drilled to test the hypothesis that spreading from the Galapagos Rift Zone had closed the Panamian flood gate, the scientific leaders on board were two proponents of the theory, Tjeerd van Andel and G. Ross Heath, both from Oregon State University. Their plan was to drill into the two ridges—the Cocos Ridge that hypothetically had drifted north from the rift zone to cooler latitudes, and the Carnegie Ridge that had remained close to the Equator. If such movement had taken place, then on both ridges the deepest sediments, lying directly on basement rock, should be similar and should reflect the rapid, fossil-rich sedimentation that occurs in the equatorial zone. The southern ridge would show the same kind of sediment from bottom to top, having remained in equatorial waters. But on the north-moving ridge, at shallower, more recent depths, the sediment should change progressively to the type found beneath waters that are cooler and less biologically productive.

This, in fact, proved to be the case. Furthermore the upper parts of the two holes on the northern ridge (at Sites 155 and 158) were interbedded with ash presumably deposited as that part of the sea-floor came closer and closer to the volcanoes of the emerging isthmus.

From the drilling on both sides of the isthmus the following timetable has been deduced for the Central American "flood gate" by various of the participating scientists:

1. Prior to the end of the Cretaceous, the period of dinosaur extinction some 70 million years ago, it was closed by a land connection between North and South America.

2. During the Eocene, 40 to 50 million years ago, when dog-sized horses were roaming the United States, it was open wide with water flowing freely between the oceans. The result in the Caribbean was rapid accumulation of sediment rich in silica shells (largely from Radiolaria).

3. In the Miocene, some 10 million years ago, the "gate" began closing, ending with complete closure in the Late Pliocene, four million years ago. The new land bridge enabled the mammals, which now dominated the land, to migrate between the Americas.

There is a suspicion that the progressive closure of the isthmus, with its alteration of oceanic currents, helped set in motion the ice ages, at least in the north. Prior to the closure a great equatorial current could flow around much of the Earth's circumference—across the full width of the Atlantic and Pacific. Now separate circular systems began to flow in the two oceans, the Gulf Stream being part of the new Atlantic pattern.

This would have altered storm paths, delivering more warm, humid air to central Canada and northern Europe. In winter this would mean heavier snows and, where the accumulation was greater than the summer melt, this would initiate an ice age.

Oceanic drilling has proved a powerful tool for determining the onset of ice-sheet formation because the resulting icebergs, before they melted, carried pebbles and other bits of continental rock far to sea. Such "ice-rafting" lays down a deposit easily recognized in the cores. In the Bering Sea, ice-rafted material from the late Miocene, six million years ago, was found in small quantities and became abundant after the Mid-Pliocene, four million years ago, a time assumed to be the start of "important ice formation" in Alaska. On the Atlantic side of the continent, at two sites in the Labrador Sea, the first ice-rafted material appeared only three million years ago, preceded, chronologically, by a rich population of tiny subtropical plants and animals, presumably carried there by the Gulf Stream on one of its earlier routes.

Early in 1973 the *Glomar Challenger* ventured into the southernmost waters of the world—the Ross Sea of Antarctica—to drill holes within a few dozen kilometers of the Great Ice Barrier, the dazzling cliffs of the Ross Ice Shelf that barred the early explorers from their dream of reaching the South Pole by ship (see Fig. 10.3).

These, and the holes drilled somewhat farther north, showed that the Antarctic continent began shedding icebergs more than 20 million years ago—long before the onset of glaciation in the north. In recent years considerable attention has been given to ancient maps that, it is alleged, show Antarctica free of ice early in human history. Notable among these is a map attributed to Piri Re'is, a Turkish admiral, drawn in 1513 but supposedly based on long-lost information from early voyagers. It purportedly shows, in rough form, the outline of the Antarctic continent as it would appear if free of ice.

Actually, in the early days of exploration, it was generally assumed that a temperate, inhabited land lay far to the south. When first discovered, New Zealand and Tierra del Fuego were considered northern outposts of that continent, which appeared on old maps in various fanciful forms. However, coring and drilling in Antarctic seas seem to have laid to rest any likelihood that the region has been free of ice at any time in human history. In fact, from the *Glomar Challenger* results, it appears that, some five million years ago, there was a great expansion of the South Polar ice. Today the Ross Ice Shelf is an apron of the continental ice sheet that has pushed out over the southern part of the Ross Sea. Once waterborne on that sea, its flow is free of friction, and the ice spreads to an equilibrium thickness of about 300 meters or almost 1000 feet. It is this that accounts for the striking uniformity of the cliffs along its northern front—the Great Ice Barrier of early exploration.

But the drill holes north of that barrier showed that, five million years ago, it became so thick that it scraped the ocean floor at depths as great as 450 and 600 meters below present sea level. This scouring of the sea floor showed up in the Ross Sea drill holes and it could be traced, as a layer echoing sound waves, 300 to 500 kilometers (200 to 300 miles) north of the Barrier. Further evidence for such a remarkable increase in the flow of ice off Antarctica at that time was found in signs of increased ice rafting of continental material from the five-million-year depth in cores extracted from

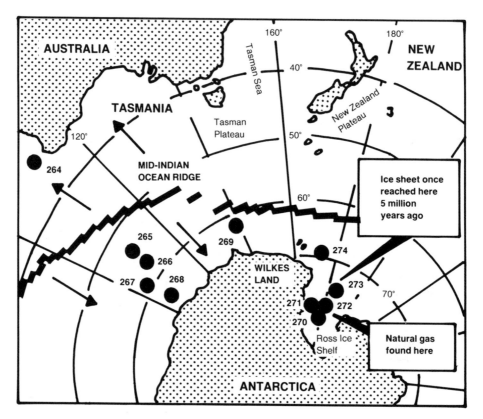

Figure 10.3 Australia and Antarctica Drifting Apart, *as shown by increasing ages of rock extracted from drill holes at greater distance from the ridge. At site 265 it was 13 million years. At site 266 it was 24 million years. At site 267, 42 million years, and at site 268, 50 million years.*

holes between the Mid-Indian Ocean Ridge and the Wilkes Land sector of Antarctica. That ridge bisects the sea separating Australia from Antarctica. Four holes were sunk between the ridge and the Antarctic coast to see if the oldest sediments from each site supported the assumed timetable of spreading from the ridge. They did so, the ages, in order of increasing distance from the ridge, being 13, 24, 42, and 50 million years. It was therefore deduced that this spreading had split Australia from Antarctica about 55 million years ago. It is suspected that, as the separation of Australia and other continents from Antarctica made possible the free flow of oceanic currents around the South Polar continent, the effect, for a time, was greatly increased snowfall on the South Pole area, producing the heavy flow of ice that occurred five million years ago.

Not only did drilling in the Ross Sea show that the Antarctic ice sheet began forming 20 million years ago, but cores brought up from just above bedrock at sites near the Barrier showed the sea-floor there was at or above sea level before the continent became burdened with its heavy ice load. The finding supports the view that the great depth below sea level of the Antarctic continental shelf—far greater than that of any other continent—is a consequence of that region's long-enduring ice burden. If the burden were removed, the continent would rise from the sea and become far more extensive than today.

Taking the *Glomar Challenger*, with no special reinforcement of her hull, into Antarctic waters was a bold venture—and not without its excitement. At times there were as many as 49 icebergs that showed on the radar screen as being less than 16 kilometers (10 miles) away. A radar plot of all nearby icebergs was maintained, and it was decided that, if any came within three miles, during drilling, the ship would have to haul up its drill.

It was known that icebergs, with their great mass and deep draft, tend to move with the ocean current, rather than with the wind, and a current meter was lowered from the ship to record the direction and speed of water flow. Actually, it was found, the bergs moved in a rather zigzag manner apparently because their angular, underwater surfaces were acted upon by the current like sails. At Site 267, one of those chosen to determine the nature of spreading from the Mid-Indian Ocean Ridge, an iceberg bore down on the ship and actually sailed across its position after the drill had been hauled up, enabling the *Glomar Challenger* to escape.

Despite special clothing, the drilling crew became badly chilled at this site, and special diesel-flame heaters were fabricated for the rig floor, using steel pipe and other items on board. After completion of this hole, the ship groped her way through a fleet of icebergs with visibility reduced by fog to 50 meters. "All available marine and drilling personnel were utilized as ice lookouts from various places on the ship," said the operations resume.

At Site 273 in the Ross Sea, three bergs were visible when the station-keeping beacon was dropped overboard, the closest being 13 kilometers (8 miles) away. Since the water was only 505 meters deep, the ship had to be kept within 14 meters of a point directly over the hole to avoid excessive bending of the drill string—a challenge the positioning system normally was able to meet. However a gyro error made manual control necessary, using an iceberg as reference point for the ship's heading, while the drill was hauled up.

The gyro was repaired and the drill lowered to sink a new hole. "Then, two of three icebergs which had been plotted from radar since our arrival on site," said the report, "continued moving towards the *Challenger* as if pulled by a magnet. The first to arrive was a large bergy bit. When the bergy bit was within 3 miles (5 kilometers) of the site, the *Burton Island* (a United States Coast Guard icebreaker) started pushing it and moved it approximately one half mile to our port side. Because of a change in current direction or other reason, the bergy bit once more headed straight for *Challenger*. The hole was filled with heavy mud (traces of gas had been noticed in the last core) and the drill pipe was retrieved until the bit was within 20 meters of the mudline."

The "mud" used in such drilling operations is a special preparation one of whose functions is to keep the walls of the hole from collapsing inward. With the bit just below the sea floor, the ship was in a position to make a quick getaway.

The above-water part of the bergy bit (a small iceberg) was about half the size of the icebreaker, but since most of the ice was submerged, the chunk actually was about twice the size of the ship. The *Burton Island* attacked again and, as it pushed, the bergy bit rolled over and over, at one time smashing the heavy metal of the icebreaker's bow. No sooner was this piece of ice clear than the larger berg began bearing down on the drill ship. This, one, however, was more stable and was readily pushed clear by the icebreaker, whereupon drilling was resumed. When the two ships headed through heavy seas for the next site the icebreaker began taking water through her damaged bow, but the flooding was not serious.

Methane gas and traces of ethane were observed in some of the Ross Sea cores. Since oil was being exploited, or sought for, on the continental shelves of neighboring New Zealand and Australia, as well as those of South America and Africa (all once adjoining Antarctica), it was proposed that similar reserves might lie off the coast of that continent as well.

On this and the next leg the *Glomar Challenger* ran into some of the worst weather of her career. With no land to obstruct them, the circumpolar winds reach 130 kilometers (80 miles) an hour, the seas were 12 meters high, and at times the ship, rolling heavily, was slowed to a walking pace. At one site, south of New Zealand, she was blown 20 kilometers (12 miles) from her station, but, after the storm abated, was able to retrace her steps and find the acoustic beacon, still transmitting on the sea floor.

Thanks to the ship's equipment for printing out radio-transmitted weather maps, as well as photographs of the Earth's cloud cover transmitted by satellites overhead, she was able to avoid the worst of some storms. For example, four to nine pictures were received daily from the ESSA 8 satellite and others from the Nimbus 4 satellite, as well as weather maps from Australia, the Soviet Antarctic station, Molodezhnaya, and the American Antarctic outpost at McMurdo Sound.

In addition to exploring the Panamanian floodgate the *Glomar Challenger* sought to elucidate the nature of the Caribbean itself. While that sea was thought by some to be a section of continent that has sunk, this now seems unlikely, if only because Sir Edward Bullard's reconstruction of the original fit of continents bordering the Atlantic puts part of South America directly where the Mediterranean now lies. Thus the Caribbean would appear to be formed either of new oceanic crust or a patch of the Pacific floor by-passed as the Americas drove westward over the Pacific Plate. Some of the many volcanoes along the island arc that separates the Caribbean from the Atlantic exude lava that forms the rock andesite, typical of the Andes and other volcanic areas around the Pacific. This has led some geologists to include the Caribbean volcanoes in the Pacific "ring of fire." The Caribbean Plate is separated from North America by a series of rifts from the Windward Passage, between Cuba and Hispaniola, through the Cayman Trough and across Guatemala. It is separated from the Pacific by a break in the Earth's crust along the Middle America Trench, where the Pacific floor (the Cocos Plate) is being forced under by eastward motion from the East Pacific Rise. Hence the Caribbean may be paved with relatively new crust, formed as North and South America swung apart. Some of those who took part in the Caribbean drilling have proposed that the volcanic arc of islands enclosing that sea—the Antilles—exists because, while the Caribbean Plate is moving west in the embrace of the Americas, it is not moving in that direction as fast as the Atlantic Plate which, therefore, is descending under the Caribbean, producing the islands and their volcanic activity. While most of the volcanoes from the Virgin Islands to the Grenadines are dead, they are not all so, as witness the terrible eruption of Mount Pelée on Martinique that, in 1902, sent a *nuée ardente* racing down the mountainside to wipe out the town of Saint Pierre. Drilling in this area demonstrated a particularly dramatic uplift of the Earth's crust. More than a century ago Christian Gottfried Ehrenberg, among the first to use microscopic fossils in rocks as a geologic tool, had studied the island of Barbados, which lies in the Atlantic 160 kilometers (100 miles) to the east of the Antilles. The island is an outrider of that chain as the latter overrides the westward-moving Atlantic crust. In the mountains of Barbados Ehrenberg had found sedimentary rocks with a sequence of excellently preserved and diverse open ocean microfossils. They came to be known as the "Oceanic Formation" and had long been a subject of controversy. Could they actually have formed beneath deep ocean and, if so, how did they get where they are?

Drilling 475 meters into a deep basin 400 kilometers (250 miles) northeast of the island, the *Glomar Challenger* extracted a sequence of fossil layers virtually identical to that on Barbados, showing that, in the past 45 million years, the island has risen more than 5 kilometers (three miles). Such uplifting has been observed at several points where the sea floor is approaching the zone where it will descend beneath a continent or island arc. If a sheet of metal is bent downward, it humps upward where it starts to bend down, and the same effect, it is suspected, may act on a crustal plate.

The drilling also revealed dramatic sinkings, including one that submerged a section of continent larger than the combined areas of New Jersey, Connecticut, and Massachusetts beneath the North Atlantic in a manner reminiscent of legendary Atlantis. Plato wrote of Atlantis as though it were a huge island beyond the Pillars of Hercules (The Strait of Gibraltar) in an ocean bounded on all sides by continents (although the Americas were unknown in Ancient Greece). His account was obviously garbled and mixed with legend, for he spoke of Atlantis as the home of a hostile nation that attacked Greece (highly unlikely for a prehistoric island in the Atlantic):

> This power came forth out of the Atlantic Ocean (he wrote), for in those days the Atlantic was navigable; and there was an island situated in front of the straits which are by you called the Pillars of Hercules; the island was larger than Libya and Asia put together, and was the way to other islands, and from these you might pass to the whole of the opposite continent which surrounded the true ocean; for this sea (the Mediterranean) which is within the Straits of Hercules is only a harbour, having a narrow entrance, but that other is a real sea, and the surrounding land may be most truly called a boundless continent.

The empire that ruled this island, said Plato, extended its control as far as Egypt and sought to subdue Athens, but was defeated.

> But afterwards (he continued) there occurred violent earthquakes and floods; and in a single day and night of misfortune all your warlike men in a body sank into the Earth, and the island of Atlantis in like manner disappeared in the depths of the sea.

Scholars in recent years have debated the basis of this tale. It has long been known that a catastrophic eruption buried in ash an ancient community on the island of Thera, 120 kilometers (75 miles) north of Crete, although the nature of that community and the scope of the eruption have only recently become evident. The buried city came to light more than a century ago when the ash, which makes waterproof cement of high quality, was excavated for construction of the Suez Canal. In parts of the island the ash is 300 meters deep.

When the Swedish expedition of 1947–1948 sailed through the Mediterranean aboard the *Albatross,* it found such ash on the sea floor and piston coring by the *Vema* revealed its further extent, as did its discovery on a number of islands (see Fig. 15.4). It is now evident that the ash blanket reaches far across the Mediterranean, toward Egypt, testifying to an explosive eruption in about 1630 B.C. four or five times more powerful than that of Krakatoa in 1883. Ash from the latter reddened sunsets throughout the world for a year. In 1967 the Boston Museum of Fine Arts reported that excavation of the ash-buried town of Thera had shown it to be a Minoan counterpart of Pompeii, with frescoes and various household objects intact. Unlike the victims of Pompeii, most inhabitants seem to have escaped, although the remains of a young man and woman were found and, from radioactive carbon dating, it appears they died at the time of the eruption.

Thus the suspicion has grown that Atlantis was really the islands of Thera and its neighbors in the Santorini group that were shattered and partially submerged by the eruption of 1630 B.C. Thera lies in the zone of volcanism along the north side of the Mediterranean that includes Vesuvius, Etna, and Stromboli. The eruptions are now thought to be a by-product of thrusting by the African Plate under the European Plate.

Plato's detailed description of the government and customs of Atlantis—which he used and apparently elaborated upon as an object lesson in political science—ascribes a role to sacred bulls strongly reminiscent of the Minoan civilization:

> There were bulls who had the range of the temple of Poseidon [Plato wrote]; and the ten kings, being left alone in the temple, after they had offered prayers to the god that they might capture the victim which was acceptable to him, hunted the bulls, without weapons, but with staves and nooses; and the bull which they caught they led up to the pillar and cut its throat over the top of it so that the blood fell upon the sacred inscription.

In the light of the recent discoveries it has been proposed that the eruption of 1630 B.C. generated earthquakes, tidal waves, and clouds of ash that destroyed Minoan cities and towns on the Aegean Islands, carpeted the fields, and killed the cattle on which the Minoan empire depended for food, bringing it to an abrupt end much in the manner described by Plato, 11 centuries later. It has even been postulated that this ash cloud reached Egypt, darkening the sky for three days, as described in *Exodus,* and forming a basis for the account of the plagues in that book of the Bible.

In any case, Plato's Atlantis almost certainly did not lie beyond the Pillars of Hercules, as indicated in his account, since, from what is now known of the Atlantic floor west of Gibraltar, it appears that there has not been a large island there since that ocean was formed. Nor does it seem plausible, in that early time, that a civilization from as great a distance as the Atlantic would attack Athens. Hence the suspicion falls on the Minoan culture of Thera and Crete.

The land that was shown by drilling to have sunk beneath the Atlantic did not do so in a day—or even a million years—and it was far from the Pillars of Hercules. At the time that birds and mammals were evolving among the dinosaurs, it was probably part of the European–American land mass and now shows on maps of the Atlantic as a broad submarine feature, Rockall Bank, midway between Ireland and Iceland, poking above the sea as Rockall Islet. Because Rockall Bank seemed to fill an otherwise empty space when Europe, Greenland, and North America were fitted back together, it had been suspected that it was a submerged block of continental rock, rather than a volcanic structure like the Mid-Atlantic Ridge. A visit to Rockall Islet then confirmed that it was largely formed of granite—typically continental. Sir Edward Bullard has noted that this raises an interesting philosophical point. "If one does what looks like an arbitrary thing to make it fit," he said recently, "one weakens one's case—but if then one finds the arbitrary change can be demonstrated to be all right, one has greatly strengthened one's case. For example the inclusion of Rockall (in fitting together the northern lands) was quite arbitrary, but we then went there and showed it genuinely *was* continental— we had brought off a prediction." When the *Glomar Challenger* sank two bore holes into the plateau, the cores showed not only that it had once been dry land, but that there had been three periods of rapid sinking, separated by two "intermissions" of five to ten million years each in which there was little or no sinking at all. The total subsidence has been at least 1400 meters (see Fig. 10.4).

From the timetable of magnetic patterns on the sea floor it was deduced that when the parting of continents to form the North Atlantic began, this "lost" patch of continent at first remained with that part of the original jigsaw puzzle that became Greenland. Then, some 55 million years ago, as spreading began away from the Reykjanes Ridge, the bank was separated from Greenland and began sinking, the periods of rapid subsidence being 55, 40, and 15 million years ago.

The rock extracted from the bottom of one hole on the plateau had actually been above water before it began its long descent into the depths. The fossils found in sediment layers above that rock basement—first of shallow water varieties, then of life characteristic of intermediate depths, and finally of deep ocean species—told the story of subsidence. From subsequent dating of various ancient intrusions of molten rock it appears that the rifting of Rockall Bank from Greenland was accompanied by extensive tension in the crust across the North Atlantic that led to eruptions in Scotland as well as intrusions into the bank itself and elsewhere.

One indicator of the water depth at which a particular layer of sediment was deposited, widely used by the drilling teams, was the presence or absence of remains of protozoa and those microscopic marine plants whose shell-like "tests" are formed of calcium carbonate. If the seas were drained of water, the high ground of the basins would be white with these tests, much like snow-covered highlands. The low ground, on the other hand, would be reddish-brown, because in deep water the calcium carbonate is dissolved away. Boring into the flanks of the Mid-Atlantic Ridge has produced intact fossils from very deep in a hole, presumably because that area was once in shallower water near the ridge crest. But such shells have also been found at depths within the cores laid down 10 to 15 million years ago, when the area was not on the ridge crest, implying a lowering of this "snow line" at that time—or an uplift of the sea floor. Some oceanographers have taken this to mean that a "rejuvenation" in upwelling of molten rock into the Mid-Atlantic Ridge thrust the entire central Atlantic upward some 2000 meters. This was a time of rapid mountain-building along the Pacific coasts of North

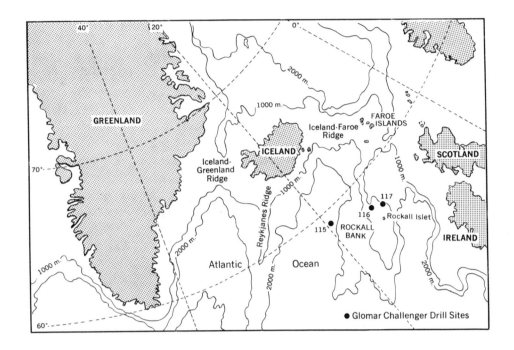

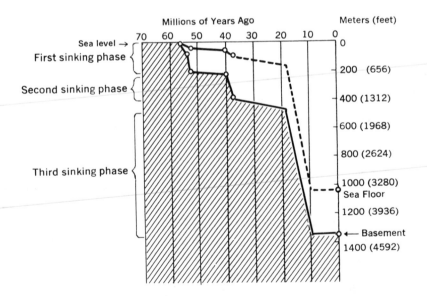

Figure 10.4 Sinking of a continental fragment. *Results from drilling to basement rock at sites 116 and 117 showed that Rockall Bank was once dry land that sank in three successive episodes to deepsea depths. Rockall Bank is apparently a fragment of the ancient continent, Laurasia, left behind when the opening of the North Atlantic pushed apart Greenland and the British Isles. In the lower diagram, the dashed line represents the depth of the sea floor from drilling at site 116, while the solid line represents the depth of the basement rock determined from drilling at site 117.*

and South America, as though there were an intensification of the westward movement of those continents (although no increase in the rate of sea-floor spreading is evident for that time). Those who believe, instead, that the "snow line" moved down argue that variations in the composition and temperature of the sea can modify the solubility of calcium carbonate sufficiently to bring about radical changes in the water depths at which the demarcation line occurs.

However, the fossil evidence for changes in water depth, based on variations in the population of the sea as shown by fossils in the cores, is considered unarguable, and by this means other scenes of gross subsidence were found. For example, a submarine ridge paralleling the Brazilian coast for 1000 kilometers (600 miles) contained coral reefs, now buried under sediment, that were at the surface some 30 million years ago. For a time, the growth rate of the coral kept pace with the sinking of the ridge and it flourished in the tidal zone necessary for its survival. Then for some reason, such as a chilling of the sea or an increased sinking rate, the coral perished, and the ancient reefs now lie more than 3.5 kilometers (two miles) below the surface.

While the drilling provided strong support for the sea-floor spreading concept of Harry Hess, it showed that one of his proposals had to be abandoned. To explain the guyots—those flat-topped mountains submerged deep beneath the sea—in the Western Pacific Hess had proposed that a now-extinct ridge, which he called the Darwin Rise, ran from southeast to northwest across that region. The guyots would have formed in shallow water along this ridge and then been carried to greater depths, whereupon the ridge itself foundered, leaving chains of extinct volcanoes, sinking atolls, and guyots. However, the drilling team on Leg 6 found, from a succession of holes, that the sea floor across the whole central Pacific, instead of originating in a Darwin Rise, had marched all the way from where the western United States now lies (Fig. 10.5). Thus, it was found, the ages of the Pacific floor increase systematically, as one sails west, until one reaches a north–south line following the Marianas Trench and Bonin Trench to the vicinity of Japan. A hole drilled near this line, but short of it, showed that as early as the late Jurassic Period, 150 million years ago, this area was receiving deep-sea sediment. "This region appears to be within the oldest part of the Pacific Ocean," wrote Bruce Heezen and A. G. Fischer, co-leaders of the scientific team, "and is probably the largest remnant of Mesozoic ocean floor left in existence. It may well have been formed by the Mesozoic ancestor of the present East Pacific Ridge." (The last part of the Mesozoic Era constituted the Cretaceous and Jurassic Periods between 70 and 180 million years ago.)

Thus it appeared that the North Pacific floor, for more than 150 million years, has been pushing west from the East Pacific Rise, part of which has now been wiped out by the westward motion of the United States. In the Philippine Sea, separated from the Central Pacific by the Marianas, the crust proved to be much younger, a discovery which at first puzzled some of the drillers. "It certainly does not fit into present theories," said one of them. The explanation seems to be that the great Pacific Plate is descending under the Marianas, Bonin, and Japan trenches so that the Pacific floor west of that line is much younger, having been generated in some other manner.

The southern boundary of this great northwestward moving plate that covers most of the northern Pacific lies along the Caroline Islands, the best-known of which is Truk,

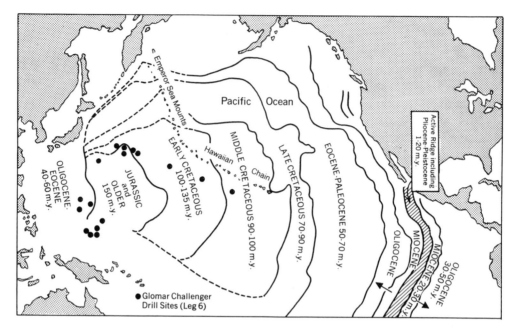

Figure 10.5 Sea-floor ages in Northwest Pacific *indicate that the floor has been pushing west from the East Pacific Rise. The Rise is now partially wiped out by the westward motion of the United States. Ages on the diagram are in millions of years.*

a major Japanese stronghold in World War II. In this region, on the northern flank of the ridge from which those islands are outcrops, the drilling revealed, immediately beneath Oligocene sediments (some 30 million years old), what appeared to be a great sheet of basalt like that which, millions of years ago, spread as a succession of lava flows to blanket much of the northwestern United States. Such great basalt plateaus have been likened to the dark, relatively level seas on the moon. However, the formation north of the Carolines either could represent the original crust of that area, or it could have spread over an older crust at a later date. On this point the 10-man scientific team was noncommittal.

It was on this leg that the drillers had their first brush with a typhoon. Early in the drilling program they had taken care to avoid waters where typhoons, hurricanes, or other maritime hazards were apt to generate a crisis. But gradually they had become bolder, venturing into the foggy, stormy North Atlantic, where icebergs occasionally were encountered, and then they steamed into the breeding ground of West Pacific typhoons (their dodging of Antarctic icebergs coming later).

The danger in drilling in a storm-prone area is not only the length of time, reckoned in hours, that it takes the ship to haul up its drill pipe, but the fact that the pipe sometimes sticks in the bottom and has to be "shot off"—itself an uncertain procedure which involves blasting the pipe in two by lowering an explosive charge to a point directly above where it is jammed.

One of the most agonizing battles to free a pipe—fortunately not under threat of a typhoon—took place during 1971 near the Bonin Trench, southeast of Japan. The drill, beneath 6151 meters of water, had become stuck in flinty rock, and a wide range of tactics was used in an effort to free the costly downhole assembly. First, a lubricating cocktail was pumped down the pipe, consisting of 35 barrels of used engine oil, 20 barrels of diesel oil, and 200 barrels of a gel-and-water mixture. This did not work, so an explosive charge was lowered to within one meter of where the drill entered the bottom. It failed to fire. Then a special explosive severing tool was three times lowered down the pipe, but each time it, too, failed to detonate. The wiggling method used to pull a stake stuck in the ground was then tried—except in this case the "stake" was more than six kilometers (almost four miles) long. To provide enough play in the pipe to make slow-motion wiggling possible, 146 meters of extra pipe were added to the string. Then the ship was driven, first back and forth as much as 600 meters, then in a circle of 230 meters radius, still to no effect. Finally the severing tool was modified and again lowered down the pipe. This time it exploded properly, and the pipe was shot off where it penetrated the bottom, thus salvaging all but the embedded part.

Fortunately, in its first encounter with a typhoon, the drill of the *Glomar Challenger* was not so thoroughly stuck. To cope with its challenging assignment the ship had been equipped to process transmissions from both navigation, and weather satellites. By tuning to one of four United States Navy satellites in polar orbits, electronic devices could pinpoint the ship's location anywhere in the world to within 180 meters, regardless of clouds and other impediments to sighting on Sun and stars. Such precision was important in leading the ship to previously discovered features beneath the trackless ocean. On July 22, 1969, however, it was the weather satellite equipment that sounded the alarm southwest of Guam. A photograph of that region of the Earth, transmitted by ESSA 8 as it passed overhead, displayed to Charles L. Green, forecaster assigned to the expedition by the National Weather Service, the swirling cloud pattern typical of a newly forming typhoon. The position of the ship was close to its center. With the drill pipe buried deep in the ocean floor, miles below, it would take many hours to haul it up and make an escape. Green wrote in his log:

"10 a.m.—Tropical depression right on top of us. Our (atmospheric) pressure dropped 4 millibars in 3 hours. I informed Captain, showed him the ESSA 8 picture which clearly shows it on top of us...Our winds now steady 30 knots with gusts to 40. Seas up to about 8 feet. I hope it moves to the west soon."

Then, later in the day: "They pulled the pipe up and we are running. Wind steady at 40 knots, gusts to 50. Seas up to 12 to 15 feet." The storm turned into "supertyphoon" Viola, with winds of 240 kilometers (150 miles) an hour. Its death toll on Taiwan and the Philippines was at least 20.

One of the earliest tests proposed for the idea of a moving Pacific floor was to trace the drift of the equatorial band of rapid sedimentation. Converging current systems of the Northern and Southern hemispheres bring to the surface, along the Equator, water that is extremely rich in nutrients. In the warm, sunbathed tropics, this leads to a rich bloom of oceanic life and the result is heavy sedimentation. For example, drilling on Leg 9 close to the Equator and far from any continental source of sediment penetrated a sediment layer 4.8 kilometers (three miles) thick. These sediments contain much chalk and other calcium-rich residues of sea life. Hence the zone has been called the equatorial "chalk line." If the sea floor has been drifting, it must have carried with it those "chalk lines" laid down in earlier periods.

To seek this evidence six holes were drilled between Latitudes 42° North and 12° South along a north–south line in Longitude 140° West, and chalk-line evidence was also sought in a number of other holes, including four drilled in an area in the same latitude as Japan and several hundred kilometers to the east of the Japan Trench. The upper layers in these four holes were interbedded with volcanic ash, bearing witness to the fact that this section of sea floor is gradually nearing the volcanoes of Japan—a process that will eventually carry the crust down under the Japan Trench. Even there, more than 3200 kilometers (2000 miles) north of the Equator, the telltale chalk was found, dating back 150 million years. It was evident that the northwest movement of the Pacific floor must be reckoned in many thousands of kilometers. The equatorial sites also showed that the band of rapid deposition was particularly wide 40 to 50 million years ago, and it was proposed that this may have been due to a broad equatorial current far more intense than that of today because the Panama floodgate was open.

The interpretation of mid-ocean cores is complicated by the fact that the material in them may have been swept to the drill site by turbidity currents—those submarine counterparts of avalanches. In the Mid-Atlantic off Brazil cores contained such gemstones as topaz, tourmaline, and aquamarine, not because they were thrown overboard by some profligate cruise passenger (they are generally microscopic in size), but because a turbidity current swept them 1400 kilometers (850 miles) from the mouth of the Amazon. These minerals are typical of the Amazon Valley, and the cores in which they were found were interbedded with layers of plant remains one or two centimeters (an inch or more) thick, apparently generated by Amazonian floods.

This material had accumulated in one of the larger canyons, known as the Vema Fracture Zone, that mark offsets of the Mid-Atlantic Ridge. Since the fracture was presumed to be young, the great depth of sediment detected there by echo-sounding had been, as noted by the drilling scientists, "an enigma to marine geologists for some time." It was now explained by the massive turbidity-current delivery of Amazonian debris. In fact, the rate at which the rift was being filled, deduced from the cores, was so rapid that it could not be much more than a million years old. In some cases the sediment revealed what had happened on nearby land areas. Drill holes off Africa produced thick layers of sediment generated 20 to 25 million years ago as the Atlas Mountains were being thrust up and subjected to rapid erosion. Likewise in the Bering Sea evidence was found for "a time of prodigious continental and island arc erosion" some 12 million years ago. Presumably this was a time of rapid mountain growth on the shores of that sea, bounded by Alaska, the Aleutians, and Siberia. (It was also, as noted earlier, a time when, some believe, there was intensified sea-floor spreading in the Atlantic.) The deluge of sediment onto the Bering Sea floor was followed by an "explosion" of the diatom population as revealed by the plenitude of intricately geometric diatom shells in the cores.

Among the great fans of river sediment probed by the drillers was that spread over the Bay of Bengal by the Ganges and Brahmaputra rivers—the most extensive in the world—and the Astoria Fan, formed of Columbia River sediment swept hundreds of kilometers out over the Pacific floor. Samples of pollen and spores from trees, shrubs, and grasses, extracted from the Astoria Fan, showed that, during the past two million years, Washington and Oregon have never been warmer than today.

In seeking to understand the manner in which sea-floor plates are thrust under continents, scientists have wondered what happens to the heap of sediment that rides atop such a plate. Where the plate bends down to descend under another plate of the Earth's crust, there is typically a trench in the sea floor, an example being the one that parallels the Aleutian Islands on the Pacific side. Skeptics of the sea-floor spreading concept had pointed to soundings that showed flat-lying sediment layers in some trench troughs, suggesting a peaceful, undisturbed history. A drill hole on the landward side of the Aleutian Trench produced tightly compacted and squeezed sediments, tending to confirm that the flat-lying layers of the Pacific had been thrown up against the continental block like bedding crammed against a wall.

Among the puzzling aspects of drilling in many ocean areas were the encounters with chert. Layers of this flinty silica rock within the ocean sediments had proved one of the earliest obstacles to deep drilling. But why should the oceans, at certain well-defined depths, be paved with "tile floors" of this material?

Generations of geologists have debated the origin of chert beds found in former marine deposits now on land. The discovery of such beds in their place of origin beneath the sea now opens the way for more informed speculation; yet the debate continues.

The extent and general continuity of a layer which turned out to be the chert deposits in the North Atlantic was revealed, before the drilling began, by the *Vema* in its sonic profiling of layers below the bottom. This had shown a strong reflecting layer at depths of 100 to 600 meters below much of the North Atlantic, except near the mid-ocean ridge, where sediment accumulation was meager and apparently of more recent date than formation of the "tile." This widespread layer came to be known as Horizon A, and, although at the time its nature was unknown, the drilling has shown it to be one or more layers of chert in a bed some nine meters (and occasionally almost 60 meters) thick, laid down during the Eocene, 40 million years ago. Equally surprising has been the discovery that at many sites there is a major gap in the fossil record immediately below this layer, as though some catastrophic development killed off most or much marine life. Eocene chert was also interbedded in the southeastern Arabian Sea and in the Pacific, where it was found that chert layers had been laid down, as well, during the Jurassic and Cretaceous periods, from 180 to 70 million years ago. It therefore appears that the change in oceanic environment responsible for their formation has occurred more than once.

The chief gap in the fossil record of ocean life was for the period 50 to 70 million years ago in both the Atlantic and Pacific. In some cores the gap represented a longer interval; and in sediment drilled at five sites from the floor of the Tasman Sea, between New Zealand and Australia, and seven sites in the Indian Ocean, it was later—from 20 to 50 million years ago.

The Indian Ocean drilling, on Leg 25 in 1972, was directed by E. S. W. Simpson of the University of Cape Town and Roland Schlich of France's Geophysical Observatory. "Why these accumulations are missing," they said afterward, "is at present a mystery." Some of the gaps, they added, represent "vast intervals of geologic time"—as much as 50 million years. If such a gap were found in only a few locations, it could be explained by bottom currents that had washed away the sediment. The more widespread occurrence may indicate, instead, that new circulation patterns swept the sea floors with water largely free of dissolved minerals. In that case the fossil shells dissolved into the water and vanished. The actual population of the sea may also have diminished, due, perhaps, to some form of "pollution" introduced by nature; but most oceanographers would find it hard to believe that the gentle rain of fossil shells onto the sea floor ceased altogether.

After the *Glomar Challenger* drilled Antarctic waters in 1973, Dennis Hayes of Lamont–Doherty, L. A. Frakes of Florida State University, co-chief scientists on that leg, and their colleagues suggested that the massive delivery of ice to the southern oceans that apparently took place five million years ago may have been to blame. It could, they said, have sent frigid water coursing across the floors of the seas with sufficient intensity to erase the earlier sediment record.

A variety of explanations has been proposed for the chert that commonly overlies the gap in the fossil record. Because it is found, as a rule, only at considerable depth within the sediment; it is suspected by some that great pressure, or a long passage of time (or both), are necessary for conversion of fossil silica into this rock. The fact that, in the Caribbean as well as North Atlantic, it was laid down during the Eocene, when the Panamanian floodgate was wide open, has persuaded others that the former patterns of ocean circulation led to a bloom of silica-rich organisms such as diatoms, radiolaria, and sponges. Still a third proposal is that global volcanic eruptions of great intensity showered the sea with ash. Because such ash contains silica and phosphorus, this, it is argued, would enable organisms dependent on those substances to proliferate. Ash layers close to the depth of the chert have been found at some—but not all—drill sites.

Another episode identified in the history of the Atlantic was an extraordinary bloom of one organism, a form of plankton or drifting life known as *Braarudosphaera rosa*. During the mid-Oligocene, 30 million years ago, its shells rained on the bottom in such numbers that they now form a closely spaced succession of chalk layers over much of the Atlantic. Why this one organism should flourish at the expense of all others is unknown.

These were but some of the discoveries to come from the Deep Sea Drilling Project. Probably no other expedition of any kind had collected such a volume of specimens and new information. It rolled back the pages of history to the time when ocean basins were born, showing—what would have been thought incredible a few years ago—that none of the oceans is older than the age of dinosaurs. In the words of Aleksandr Peyve, head of the Geological Institute of the Soviet Academy of Sciences, the drilling of the *Glomar Challenger* marked the start of "a new era in geology."

When, in 1983, the Deep Sea Drilling Project was replaced by the Ocean Drilling Program, based on the drill ship *JOIDES Resolution*, 1092 holes had been drilled at 624 sites around the world in water as deep as 7000 meters (23,000 feet). The project had helped to weave the scientific world closer together. Scientific institutions in a number of countries were taking part in planning future goals for the project. One of them, the

Soviet Union, had agreed to contribute, each year, a million dollars toward its cost (or the equivalent in fuel or other services). It was a far cry from the early days of Mohole, when Cold War rivalry was uppermost.

Because of discoveries of what promise to be vast mineral and oil resources, the *Glomar Challenger*'s work has also been hailed as of major economic importance. According to Melvin N. A. Peterson of Scripps, who with William A. Nierenberg, director of that institution, had been co-leader of the project, the expedition results will provide a harvest of economic benefits "for centuries." While the drilling did not penetrate as deeply as envisioned for Mohole, it had been far more widespread and, in many respects, as productive scientifically as the ambitious goals envisioned for that ill-fated project.

Chapter 11

Antarctica — The Crucial Puzzle Piece

WHILE THE *GLOMAR CHALLENGER* WAS DRILLING THE OCEAN FLOORS OTHER VIVID evidence for drift was found among ice-shrouded, blizzard-swept mountains in the heart of Antarctica. One of the earliest proponents of the drift concept, the South African Alexander Du Toit, recognized that since Antarctica, like a central piece in a jigsaw puzzle, occupied a critical position when the southern continents were fitted back together, its little known rocks and fossils would be, at least for geologists, the ultimate test. "The role of Antarctica is a vital one," he wrote in 1937, "...the shield of East Antarctica constitutes the 'key-piece'—shaped surprisingly like Australia, only larger—around which, with wonderful correspondence in outline, the remaining 'puzzle-pieces' of Gondwana can with remarkable precision be fitted."

The fossil collections from Antarctica at the time were meager. The remains of five genera of prehistoric penguins, some apparently as tall as a small man, had been found near the Antarctic Peninsula. While this hinted that perhaps the penguins, now native to various parts of the Southern Hemisphere, might have evolved in Antarctica as that continent became colder, this did not bear directly on the drift debate. In fact some took it as negative evidence. The penguins, too clumsy on land to escape any kind of predator, evolved in Antarctica, they said, because there were no four-footed animals there.

From the start, the most dramatic fossil finds were made alongside the Beardmore Glacier—the great ice river that flows down through the Transantarctic Mountains from the South Polar ice sheet, dammed up by those mountains to depths of 2000 meters or more. The glacier, 160 kilometers (100 miles) long and many kilometers wide, was discovered early in this century by British explorers seeking to use it as a highway to the South Pole. The circumstances of those early finds are in sharp contrast

186

to the more recent discoveries in that same area. The first was made in December 1908 by a four-man party under Sir Ernest Shackleton, making its perilous way up the heavily crevassed surface of the glacier. For part of their two-week trek up the glacier they had been helped by Socks, the last of their ponies, the others having been shot and left at depots to provide meat for the return journey. "Socks, the only pony left now, is lonely," Shackleton wrote in his account. "He whinnied all night for his lost companion." But Socks, too, was doomed. On December 7 Shackleton and two others were hauling one sledge in the vanguard, while Frank Wild, the fourth member of the party, brought up the rear, leading Socks. The glacier surface was crevassed, with some of the giant ice cracks hidden under snow bridges. Suddenly the trio in front heard Wild cry for help.

"We stopped at once and rushed to his assistance," Shackleton wrote, "and saw the pony sledge with the forward end down a crevasse and Wild reaching out from the side of the gulf grasping the sledge. No sign of the pony. We soon got up to Wild, and he scrambled out of the dangerous position, but poor Socks had gone. Wild had a miraculous escape...he just felt a sort of rushing wind, the leading rope was snatched from his hand, and he put out his arms and just caught the further edge of the chasm." From then on it was man-hauling the sledges all the way, and after 10 more days the weary explorers saw a vast white expanse ahead that they hoped was the South Polar Plateau at the top of the glacier. They pitched camp and, after their evening meal, Wild climbed a nearby hill to see how much farther they had to go.

"He came down," Shackleton wrote, "with the news that the plateau is in sight at last, and that to-morrow should see us at the end of our difficulties. He also brought down with him some very interesting geological specimens, some of which certainly look like coal...Wild tells me that there are about six seams of this dark stuff, mingled with sandstone, and that the seams are from 4 in. to 7 or 8 ft. in thickness. There are vast quantities of it lying on the hill-side." If it proved to be coal, said Shackleton, "the discovery will be most interesting to the scientific world."

It was indeed. The idea that lush, coal-producing forests grew close to the South Pole seemed incredible. The specimens were a challenge to those seeking to reconstruct the Earth's history—and so were those from the same area, found with the frozen bodies of Captain Robert Falcon Scott and his companions after their ill-fated journey to the Pole in 1911–1912. Shackleton had turned back on the plateau, convinced he could not reach the Pole and return alive. Scott, in a race with Roald Amundsen, pressed on and perished on the return journey.

It was close to the spot at the head of the Beardmore Glacier where Frank Wild had found his coal that Scott's party did its most important collecting. Even at this early stage in the geologic exploration of Antarctica it was evident that a sandstone formation of extraordinary thickness and horizontal extent constitutes a large part of what are now called the Transantarctic Mountains. This great range, as its name implies, is now known to traverse the entire continent, following the Ross Sea coastline, including the continental edge of the Ross Ice Shelf, for more than 1600 kilometers (1000 miles) on the Pacific side of the continent, and then crossing to the Weddell Sea on the Atlantic side.

The British, in their exploration near McMurdo Sound and on their marches south from there to the Beardmore Glacier, had seen in the mountains that tower above the white ice sheet repeated displays of the great, flat-lying beds that they called Beacon sandstone. Scott, in his diary, remarked, as they neared the top of the glacier enroute to the Pole: "All day we have been admiring the wonderful banded structure of the rock..." But on their way down, after reaching the Pole to find that Amundsen had already been there, the situation was becoming grim. Excerpts from the diary, found with Scott's body, reflect their withering morale.

February 6—"Wilson (Edward A. Wilson, zoologist and artist) very cold, everything horrid...Food is low and weather uncertain...Evans (Petty Officer Edgar Evans) is the chief anxiety now; his cuts and wounds suppurate, his nose looks very bad, and altogether he shows considerable signs of being played out."

Two days later Scott wrote: "Had a beastly morning. Wind very strong and cold." But that afternoon, at the head of the glacier, they decided to camp by a moraine, or heap of glacier-bulldozed rock, up against Mount Buckley.

"It has been extremely interesting," says the diary. "We found ourselves under perpendicular cliffs of Beacon sandstone, weathering rapidly and carrying veritable coal seams. From the last Wilson, with his sharp eyes, has picked several plant impressions, the last a piece of coal with beautifully traced leaves in layers, also some excellently preserved impressions of thick stems, showing cellular structure." And the next day: "Wilson got great find of vegetable impression in piece of limestone. Too tired to write geological notes."

All told they collected 16 kilos (35 pounds) of rocks, hauling them on the sledge, week after week, with their dwindling supplies. Evans died on the glacier. Then Captain L. E. G. Oates, crippled by frostbite (and possibly scurvy) until he could barely walk, removed himself as a burden to the party by vanishing into a blizzard. But still the surviving three continued to haul the rocks until they could walk no more. The next summer their bodies were found in their tent, the rock collection on the sledge nearby—probably the most hard-won geologic specimens in history.

When the fossils were brought back to London, they were found, as noted earlier, to represent elements of the Gondwana flora that flourished, from India to Antarctica, in the coal-forming period 200 to 300 million years ago. The evidence that Antarctica once supported an abundant, temperate-climate vegetation, similar to that of the other lands assigned to Gondwanaland, was seized upon by the "drifters." But the skeptics argued that spores and seeds can be carried long distances across the sea by winds, currents, and driftwood. It takes only one such fortuitous event, within a time span of a million years, to transport a species from one continent to another. They pointed out that not a single bit of fossil evidence had been found that Antarctica had ever been the home of land-dwelling vertebrates (animals with backbones), apart from human explorers and their imports, such as sled dogs and ponies. The absence of any such land animals in the growing collections of Antarctic fossils was taken as strong evidence that the continent has always been isolated from its nearest neighbor, as it is today, by 1000 kilometers (600 miles) or more of stormy ocean.

Following the heroic journeys of Amundsen, Scott, and Shackleton, others made long, frigid treks into the Transantarctic Mountains, notably Laurence M. Gould, a geologist who was chief scientist on Admiral Richard E. Byrd's first expedition to

Antarctica. In 1929 a party under his leadership climbed on skis the shoulder of Mount Nansen, near where Amundsen passed on his way to the Pole and far to the east of the Beardmore. Gould reported that "not the least significant of the thrills was the realization that the flat lying rocks that cap Mt. Nansen were a great series of sandstones, containing toward their top seams of impure coaly material. This discovery definitely establishes the fact that we are here dealing with precisely the same mountain structure as those known and studied by British geologists along the western borders of the Ross Sea."

The extent of this formation was stretched even farther to the east on the next Byrd Expedition, in 1934, but in none of its explorations, nor in the examination of rock outcrops and the scattered snow-free areas elsewhere in Antarctica, were any traces of land vertebrates found. While coverage of most of the landscape by a deep ice sheet was a great handicap, those areas that were exposed were a geologist's dream, with no vegetation and virtually no soil to conceal the rock formations.

By far the most intensive geologizing began with the International Geophysical Year of 1957–1958. At its outset, a large part of the continent, including some of its most majestic mountain ranges, had never been seen. In a program that seems destined to continue for a long time to come, these areas have been photographed, mapped, and given a geologic once-over by scientists from many nations. Among those engaged in such geologic mapping was a New Zealander named Peter J. Barrett, working as a graduate student for the Institute of Polar Studies at Ohio State University. During the final days of 1967, with David Johnston of New Jersey, he was examining Graphite Peak, which protrudes above the ice behind the mountains that flank the east side of the Beardmore Glacier and thus is not far from the original British discoveries. (Figure 11.1 shows locations of fossils found in Antarctica up to the 1980s.)

The portion of Graphite Peak rising above the ice consists of a series of layers, dipping to the southeast, that is 700 meters thick, of which more than half is a triple-decker sandwich of shale and sandstone now known as the Fremouw Formation. These layers, whose contribution to the drift debate has been critical, have been traced for at least 500 kilometers (300 miles) along the axis of the Transantarctic Mountains. In a lens-shaped accumulation of pebbles near the top of the upper sandstone bluff, Barrett found what he suspected might be a fragment of bone. It was brought to the American Museum of Natural History in New York and examined by Edwin H. Colbert, emeritus curator of paleontology there and one of the world's leading authorities on the history of the amphibians and reptiles. "The specimen was too incomplete for close identification," Colbert wrote, "but there could be no doubt as to its general nature; it was a portion of the lower jaw of a labyrinthodont amphibian." The labyrinthodonts, so called because of the labyrinthine infolding of dentine in their teeth, were the first pioneers of the land. Their fish ancestors had developed lunglike organs so they could come to the surface and gulp air during dry spells, when pond water became depleted in oxygen.

In the fish that evolved into amphibians the paired ventral fins, fore and aft, turned into stubby feet (see Fig. 11.2). But why did they venture forth from the water? "Not to breathe air," said Alfred Sherwood Romer, the noted Harvard paleontologist, "for

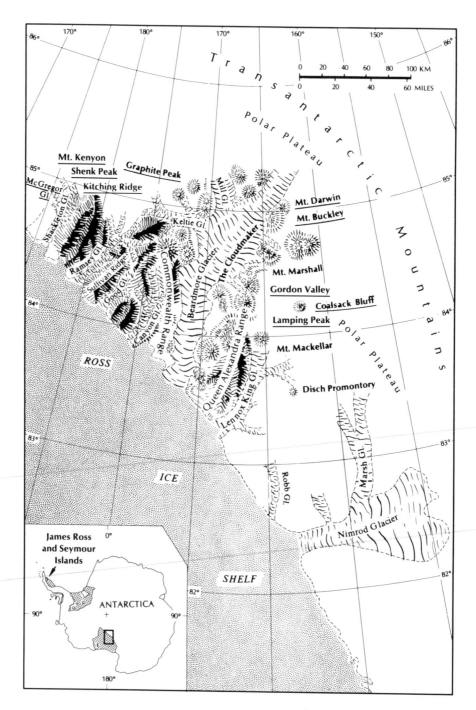

Figure 11.1 Antarctic vertebrate-fossil finds *from the time of Shackleton and Scott to the 1980s, are indicated by underlined labels. The inset shows the area covered by the detailed map.*

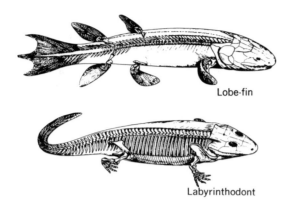

Lobe-fin

Labyrinthodont

Figure 11.2 The evolution of the Labyrinthodont from the lobefin, a fish, to an amphibian, was a subtle but critical transition.

that could be done by merely coming to the surface of the pool. Not because they were driven out in search of food—they were carnivores (meat-eaters) for whom there was little food on land. Not to escape enemies, for they were among the largest of vertebrates found in the fresh waters from which they came.''

The answer, he believes, was that the interval when the amphibians evolved in the Devonian, more than 350 million years ago, was one of droughts, when pools dried up, and the labyrinthodonts survived because they became able to crawl to other ponds when this happened. The boldness of those who thus pioneered—perhaps in one special place and at one special time—was of great importance, for these first amphibians proliferated into many species, some of which were ancestral to the reptiles and ultimately to the mammals and man.

To understand the discoveries that followed, it is helpful to keep in mind the hierarchy of the animal kingdom—the relationship of its many members that reflects their evolutionary histories. The vertebrates, including fish and amphibians, fall within one of the basic divisions, or phyla, of the animal kingdom—the chordates. Other phyla, for example, are the mollusks (clams, oysters, squids, snails, etc.), the arthropods (crustaceans, insects, spiders, scorpions, etc.), and the sponges. Each phylum is divided into orders. Those of the mammals include rodents, carnivores, and primates. The orders are split into families. The Carnivores, for example, include the family of dogs, the family of cats (from tabbies to tigers), the family of bears, and so forth. Within each family the closely related species are grouped into genera. And finally, at the bottom of the scale, are the species, sufficiently alike to interbreed.

The labyrinthodonts belonged to an order of amphibians, and what excited Colbert was his belief that, being fresh-water animals, they could not cross even a narrow body of salt water. Nor did they have the build of long-distance swimmers. He pointed out that modern amphibians cannot tolerate salt water. Perhaps some of the early ones were different; or perhaps the oceans were far less salty in those days, 200 to 350 million years ago. The remains of one variety have been found in marine sediment in Spitzbergen, but whether these animals really lived in the sea, said Colbert, "is open to question.''

Like most geologists and paleontologists he had been skeptical of the drift concept, but recent discoveries of affinities between the early reptilian fossils of South Africa and those of South America (discussed in the next chapter), had begun to shake his conviction. Nevertheless, when he received Barrett's specimen he had, until then, agreed with the majority of his colleagues that the early contiguity of Africa and South America had not been unalterably established. One "cannot completely rule out," he wrote, "the possibility of Triassic intercontinental migrations by the long way around—by a trek up through Africa and across Asia, across the Bering bridge, down through North America, across the isthmus of Panama, and into South America. Improbable, one might say, but, nonetheless, a distant possibility."

Colbert's identification of the Antarctic jaw fragment created a sensation in the geologic world. He urged on the National Science Foundation a change in the program for the 1969–1970 season of Antarctic research to place special emphasis on fossil-hunting. In particular he proposed that men trained in the techniques of hunting and collecting vertebrate fossils be included. As a result he and three other such experts joined the team led south that year by David H. Elliot of the Institute of Polar Studies at Ohio State. Elliot, a British-born geologist, had participated in a succession of scientific expeditions to the Antarctic and knew the Beardmore area well.

For about eight years, following the airlifting of a scientific station to the South Pole for the International Geophysical Year of 1957–1958 (it has been manned ever since), the United States Navy maintained a small weather station near the outlet of the Beardmore Glacier to support its annual flights to resupply the Pole Station. Longer-range transport planes and weather satellites had, in recent years, obviated the need for this Beardmore Station; but for the 1969–1970 season the Navy was asked to airlift and maintain a small research station, to be established initially on a smooth stretch of ice near a prominent and easily recognized nunatak named Coalsack Bluff. ("Nunatak" is an Eskimo term for a summit protruding above the ice.) Coalsack Bluff was 50 kilometers (30 miles) west of the Beardmore, but within helicopter range of Graphite Peak, east of the glacier, where Barrett had made his find. It seemed wisest to start the fossil hunt at the site of a known discovery. Then, some five weeks later, the camp would be moved to McGregor Glacier, a tributary of Shackleton Glacier 200 kilometers (125 miles) to the east, to enable the fossil hunters to work in that area and for other specialists to extend their studies (see Fig. 11.1).

Both sites had been visited in previous years, so the surface of the ice sheet was believed to be free of crevasses. The airlifting would be done by the Navy's lumbering C-130 (Hercules) cargo planes, equipped with skis. They were to carry in four dismantled Jamesway huts—semicylinders like the Quonset huts of World War II, but smaller and with padded (insulated) canvas roofs. Navy Seabees would assemble the huts, and the Navy was also to provide food, cooks, radiomen, mechanics, and crews to fly the three helicopters that would lift the geologists to their sites.

Such were the plans, but the polar climate that had inflicted misery, frustration—and sometimes death—on earlier explorers, with their ponies, dogs, and man-hauled sledges, was no kinder to those with helicopters, motor toboggans, and prop-jet planes. It had been hoped that exploration from the Coalsack Bluff camp could begin on November 5, but day after day bad weather at the main base on McMurdo Sound prevented planes from flying necessary equipment in from New Zealand. "Our expedition was a gamble, and a costly one at that," Colbert wrote. "We had no assurance that we would find fossils, and our chances for success seemed to diminish every day at

McMurdo, as we waited through the weeks for storms to abate. It was the stormiest Antarctic spring in years. Each day, as the winds howled past our huts, driving clouds of snow across the great ice shelf and Ross Island, on which the base is located, our long-laid plans for a concerted fossil hunt became increasingly tenuous and dislocated.''

Finally, on November 16, a team of Seabees landed at the site and erected one Jamesway hut per day for the next four days. Then, on the 22nd, the scientists arrived in the midst of a howling storm. During such stormy periods cold air pours off the Polar Plateau like water over a dam, hugging the ice surface from which it sweeps ice crystals that pellet one's face with savage force. This driving layer of "drift" may be only inches thick or deep enough to conceal an entire camp. Under such circumstances it was obviously impossible to fly in the helicopters, and without them Graphite Peak lay far beyond reach. Hence, after a "night's" sleep (so near the South Pole the sun never sets at that time of the year), the scientists began setting up their field laboratory. However, David Elliot, the party leader, and James M. Schopf, a specialist in coal and fossil plants, gazing at Coalsack Bluff, several kilometers away, got into a debate. The bluff was not one of the priority targets, but Elliot believed its horizontal bands represented the same Fremouw Formation that constituted Graphite Peak. The only way to settle the matter was to go see, so they climbed aboard motor toboggans and, together with some others, rattled over to the nunatak (Fig. 11.3).

They found that, in fact, the two or three sandstone layers, typical of the lower part of the Fremouw formation, rose as steep bluffs, being resistant to erosion, whereas the

Figure 11.3 David H. Elliot and Edwin H. Colbert from left to right on Coalsack Bluff.

shale between them had crumbled into more gentle slopes. Underneath this assemblage was an earlier formation with the coal seams from which the feature had taken its name. One proved to be eight meters thick. This older formation, dating from the Permian Period of 250 million years ago, showed the characteristic Gondwana assemblage of plants, whereas the sandstone above it, from the more recent Triassic, 200 million years ago, was largely devoid of such fossils.

During his scouting of the bluff Elliot found two fragments that looked to him like heavily weathered bits of bone. Eager for an expert verdict, he and his companions climbed back onto their toboggans and returned to the camp. Colbert took one look at the fragments and confirmed that they were bone. After lunch he and a number of the other scientists journeyed to the bluff, and the fossil experts, working their way along low cliffs on its far side, soon spotted more bones. Most were embedded in the gravel of ancient stream beds. There were no intact skeletons. Apparently these were the remains of reptiles that had lived and died along streams farther up-slope. Then, when floods occurred, their broken, dismembered skeletons were washed down and buried in the gravel beds now exposed, here and there, along the bluff.

The next day Colbert wrote to Bobb Schaeffer, chairman of the Department of Vertebrate Paleontology at the American Museum of Natural History, a letter which he signed as "A very excited old man" (he had just celebrated his 64th birthday). It said in part:

> Dear Bobb,
> We did it!
> On our first day of field work we found a cliff full of Triassic reptile bones...
> This happened yesterday.
> The bones are not articulated—it is a stream channel deposit. But they are numerous and in *good* condition. Should give a varied fauna.
> We are getting prepared for an intensive collecting program.
> The locality is Coalsack Bluff. Only a few miles from our camp. The cliff (a low series of small cliffs) faces north into the sun, and is thus relatively warm...
> We are tremendously excited.
> This really pins down continental drift, in my opinion. Antarctica had to be connected in the Trias.

Day after day the scientists worked along the face of the bluff, chipping out the bones that they found and wrapping them in toilet paper or cloth for protection until they could be more securely prepared. Normally in the field such specimens are wrapped in Japanese rice paper, covered with cloth, and then enclosed in plaster of paris for shipment back and cleaning in a museum laboratory. But the party at Coalsack Bluff had no rice paper, and water for mixing with plaster of paris would freeze before it could be used. Hence they poured molten beeswax or paraffin over the wrapped-up specimens to provide a protective coating for shipment.

In one part of the bluff there were sufficient bones to justify "quarrying" large, fossil-rich chunks that could be shipped intact for dissection back home. To excavate these they used a portable pneumatic drill, and the eternal silence of that sweeping, black-and-white landscape was broken by the rat-tat-tat normally associated with city streets. It was miserable work on the bluff. The layers of volcanic rock, polished and sculptured by the sandblasting of wind-driven ice crystals, testified to the local climate, and the faces of the geologists were similarly subjected to erosion. Sometimes the men were so absorbed in their excavations that they were unaware of the frostbite and open sores developing on their faces until the day was over and they headed back to camp.

The Antarctic, said Colbert, "does separate the men from the boys, or rather the old men from the young men." Most of the work on Coalsack Bluff was done at an elevation of about 2000 meters above sea level, and he found clambering up and down the cliffs for six hours or more at a stretch exhausting. Perched high on the cliff each collector, geologist's hammer in one hand, would try to chisel loose a specimen while grasping the fossil in the other, bare-palmed mitt lest it fly off into space when it came free. Once it had been liberated he would scribble an identification note, wrap the specimen in tissue, and push it into the specimen bag, slung over his shoulder alongside his ice ax and crampons.

As they worked over the bluff the scientists began to fill in details of the region's history—details later supplemented by exposures found elsewhere in that region. Embedded within the coal-bearing formation that lay under the Fremouw sandstones was evidence that, during the earlier Permian Period, the streams flowed southeast. With the onset of the Triassic, some 230 million years ago, the landscape was tilted and the flow was reversed to the northwest. The area became a floodplain, and it was then that successive layers of freshwater sediment were laid down to produce the sandstone and shale. Within the Fremouw sequence were fossil tree roots and, at some sites, tree trunks as much as a meter in diameter. The tree rings in these logs were sharply defined, suggesting strong seasonal variations like those in the temperate latitudes of today (although some tropical areas also have such a cyclic climate).

Overlying the Fremouw formation—at those sites where the upper part of the sequence was intact—they found an enormous deposit of volcanic debris, including a sample of congealed ash (tuff) which, from analysis of its rubidium and strontium isotopes, was found to have erupted 203 million years ago (give or take 12 million years). This predated Antarctica's breakaway from the rest of the world, but may have represented activity preliminary to that rupture. During the subsequent Jurassic Period (the Age of Dinosaurs) the entire region had been intruded and partially capped by basalts, and it is likely that this derived from the rupturing process, just as, for example, the eruption of the Palisades along the Hudson River was a by-product of the opening of the Atlantic.

Many of the vertebrate species represented in the growing collection from Coalsack Bluff were strikingly like the fossil assemblages found in the Triassic deposits of the Karroo Basin, which occupies much of the southern half of South Africa. These African fossil beds have painted an extraordinarily rich picture of African life in the Permian and Triassic periods. One primitive reptile that predominates in the Karroo deposits from this period was, to Colbert, the essential key in linking the Antarctic fauna to that of Africa. The early Triassic lakes and swamps of South Africa must have crawled with these odd-looking creatures, known as *Lystrosaurus* (see Fig. 12.4). The fossil beds where they occur are called the *Lystrosaurus* Zone, and their remains account for about 90 percent of all fossils found in that layer.

Lystrosaurus apparently lived like a hippopotamus, though the largest were no larger than sheep. The skull was almost spherical, with the eye sockets and nasal openings high on its surface and the mouth underneath. Two tusklike teeth protruded from the upper jaw, which was beaked and otherwise toothless, for *Lystrosaurus* apparently was a vegetarian. The location of eyes and nostrils near the top of its head implied a largely aquatic existence. A few bone fragments suspected of being from *Lystrosaurus* had been found in the early work at Coalsack Bluff; but, said Colbert, "field identifications can be tricky, especially when one is dealing with bones that are not highly diagnostic." Then, on December 4, while the four paleontologists were working in the field, James Jensen of Brigham Young University in Provo, Utah, found an upper jaw fragment with a tusk. Colbert immediately decided it was from a *Lystrosaurus,* an identification that he was later able to confirm by comparing its salient features with those of South African *Lystrosaurus* specimens in the collections of the American Museum of Natural History in New York and the University of California at Berkeley.

Meanwhile, Laurence Gould, chief scientist on Byrd's original Antarctic expedition and now dean of American polar scientists, had visited the camp at Coalsack Bluff with Grover Murray, president of Texas Tech University and a member of the National Science Board. They enthusiastically radioed Washington that they considered the *Lystrosaurus* discovery "not only the most important fossil ever found in Antarctica but one of the truly great fossil finds of all time."

By now the three turbojet helicopters had finally arrived. With a normal operating range of only about 200 kilometers (125 miles) they had been fitted with extra tanks for the 650-kilometer (400-mile) flight from McMurdo, and a C-130 Hercules flew with them as mother ship in case of a forced landing. The day after their arrival there was a mishap that could have been fatal. Helicopters dependent for their lift on a single rotor must have a small tail rotor to counter the twisting effect of the main one. The tail rotor pushes air sideways. Without it, with the lifting rotor spinning in one direction, the helicopter would tend to spin in the opposite direction, and there would be little lift. Hardly had one of the copters, with two scientists and a crew of three, taken off to explore farther afield when the drive shaft for the tail rotor broke. The machine crashed and was wrecked. Fortunately it had not been very high and no one was hurt, but this accident raised doubts about the other two craft, and their tail-rotor shafts were sent to New Zealand for checking, a precaution which meant that any airlifting of the geologists was further delayed.

Eventually some sites, bearing such appropriate names as Blizzard Heights and Storm Peak, were visited, and the collections made by Elliot then and three years earlier produced fossil fish from the period subsequent to that of the reptilian deposits—the Jurassic—that were believed by Bobb Schaeffer to have been freshwater species closely related to fish that then inhabited the lakes of Australia and have been found nowhere else. However, no other concentration of Triassic reptiles was located. "We were fantastically lucky in having found the Coalsack Bluff deposit," Colbert reported in his account for the National Science Foundation. "Such occurrences are characteristic of fossil vertebrates deposited in freshwater sediments: great stretches of rock exposure will be barren of fossils, or almost so, but here and there, where the conditions for burial and preservation were just right, there will be remarkable accumulations of fossils. Undoubtedly other fossil pockets comparable to Coalsack Bluff exist, but it may take a great deal of hard prospecting to find them."

He was wrong. Another extraordinary bit of luck befell the geologists when they returned the following year to work the McGregor Glacier area. The original plan to move the camp to that site after five weeks at Coalsack Bluff had been abandoned because bad weather and the various operational difficulties, particularly with the helicopters, had set the exploration behind schedule. But the rich Coalsack find that, by the time of their departure, had produced some 450 bone specimens had kept the fossil collectors so busy they did not greatly mind postponing the McGregor visit.

In looking forward to the Antarctic summer of 1970–1971 the group, again led by Elliot (with Donald A. Coates of the previous year's party as co-leader), had a special hope. In the first season of intensive fossil-hunting they had not found a single intact skeleton. They did not have a complete skull—frequently important for absolute identification—and there was little prospect of finding one at a site, like Coalsack Bluff, where dismembered skeletons had been washed into gullies before final entombment. Of particular interest, they pointed out in their report, would be discovery of "the small and very interesting mammal-like reptile *Thrinaxodon.*" This long, low carnivore, built somewhat like a weasel (Fig. 11.4), was prominent in African beds that were also rich in *Lystrosaurus,* a peculiar feature of the *Thrinaxodon* occurrences being that the skeletons tended to be found in groups.

Establishment of the McGregor Glacier camp could have become a disaster, for it was found that, despite the smooth-appearing snow surface at the planned landing strip, crevasses were hidden beneath the snow, as they were on the Beardmore Glacier where Socks, the pony, plunged to his death. Careful scouting, however, located a safe airstrip nearby, and on the first day of field activity from the new camp James W. Collinson of the Ohio State group found the almost complete imprint of a *Thrinaxodon* skeleton—the reptile they particularly hoped to find. It was an extraordinary repetition of the chance discovery of a fossil on the first field trip to Coalsack Bluff. The find was made at what is now known as Thrinaxodon Col on the lower slopes of Mount Kenyon, not far from the camp. The animal's soft tissues had decayed by the time it became buried in mud that then hardened to preserve its bony remains.

Elated by this find, the fossil-hunters, with helicopters at their disposal, visited a number of outcrops in the area. They crossed the mountains to Ramsey Glacier, flanked on the west by Sullivan Ridge, a series of snow-free summits named in recognition of this writer's peripatetic activities in Antarctica. They even flew as far afield as Graphite Peak, site of the original fossil discovery, and found there an abundance of fossils of *Lystrosaurus* and other reptiles associated with it.

It was, however, the five Triassic deposits in the McGregor Glacier area that were most fruitful. In the party was James W. Kitching of the University of Witwatersrand in

Figure 11.4 Thrinaxodon, *a weasel-like reptile from the Triassic Period.*

South Africa, perhaps the most experienced of all hunters of Permian and Triassic reptiles that occur so abundantly in Africa, and he identified the numerous finds that followed. The best preserved skeletons, sometimes fully articulated, were embedded in fine-grained sandstone or siltstone where they had been buried beneath the ancient floodplain. A number of *Thrinaxodon* skeletons were found and, as in Africa, they tended to occur in groups. These creatures belonged to a branch of the reptiles, the cynodonts, that were beginning to show some of the skeletal features that became typical of mammals.

By a strange quirk of evolutionary history, the cynodonts dominated the land in the early Triassic, whereas their competitors, the Thecodonts—also found at the Antarctic sites—were relatively obscure. Yet by the end of that period the cynodonts, although they had given birth to the first, furtive little mammals, had almost vanished. Instead those clumsy creatures of the early Triassic, the thecodonts, had evolved into dinosaurs that inherited the Earth. Some of the dinosaurs ran on two legs, leaving their giant, birdlike footprints in the mud of the Connecticut River Valley and elsewhere to awe today's visitors. Their cousins—also descended from the thecodonts—ultimately evolved into flying reptiles, birds, and crocodiles. But in the final stage of this 150-million-year drama, the big dinosaurs, or "ruling reptiles," were replaced by descendents of the little furry animals whom they had hardly noticed in their evolutionary infancy—the mammals.

What particularly impressed the fossil-hunters was that a number of the species that they found, including *Thrinaxodon,* were not merely members of the same genus as those in South Africa—were not merely cousins, so to speak—but were of the identical species. This seemed to rule out any long migration from Africa, over the Bering Land bridge (which on occasion has proved a broad avenue linking Asia and North America), then down through the Isthmus of Panama, and across a hypothetical land bridge following the curving island chains between South America and Antarctica. By 1990 fossils of early Triassic reptiles and amphibians had been found at 10 sites in the Beardmore area.

The scientists, reporting in 1972 on their discoveries, pointed out that, as first noted by George Gaylord Simpson, an authority on the evolution of higher animals, a narrow isthmus, like that of Panama, acts as a "faunal filter." Some species run the gauntlet, but others do not. Thus, although North and South America have now been linked for millions of years, if one leaves out Australia, no two regions of the world differ more radically in their fauna.

In this case the full range of animals living in South Africa during the Triassic seemed to be represented in the ice-enveloped hills of Antarctica. A broad and close connection was therefore implied, and this, the scientific team reported, "constitutes one of the strongest lines of evidence as yet adduced in support of Gondwanaland and, correlatively, of continental drift."

A major problem in seeking to assess Antarctica's role as the crucial puzzle piece is determining how it fitted into the assembled fragments of Gondwanaland. The other continents can be reassembled rather unambiguously, although India's exact position is debatable. But Antarctica, like one of those infuriating jigsaw pieces that is almost round, could have fitted into the puzzle in a number of ways.

One clue that has been used, in seeking to reassemble the southern continents, is the pattern of South Polar glaciation in the Carboniferous and Permian periods, when they presumably were all assembled (Fig. 11.5). As in the more recent ice ages, a succession of ancient ice sheets spread across the land, each melting to give way to warm and verdant periods. The ice sheets left layers of glacier-borne rock, sand, and mud (called tillites) interbedded with sediment laid down in the warmer periods. Even in India the glaciation was heavy—not localized ice cover such as one now sees in Switzerland or Alaska. These ancient tillites have been found in southern Brazil, western Argentina, South Africa, and New South Wales in Australia. Some occur even as far north as the now-tropical Congo, but they may have been left by local glaciers flowing from nearby mountains. Yet in the northern continents there is no evidence of ice ages during this period of 100 million years.

This evidence, though meager at the time, was used by Wegener in 1912, and more extensively by Du Toit in 1937, as an argument for fitting together the southern continents into Gondwanaland (Fig. 11.6). Du Toit assembled them so that areas thought to have been repeatedly glaciated during that time fitted into a single ice sheet that, it would appear, had been near the South Pole. To do so, he slipped India and Madagascar in between East Africa and northern Australia.

Since the 1950s, expeditions into the heart of Antarctica have discovered beds of tillite and rocks planed by flowing ice in Permian and possibly Carboniferous forma-

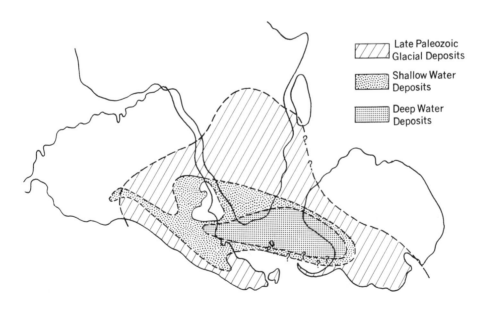

Figure 11.5 Permian Ice Sheet. *As early as the 1930s, Du Toit saw evidence in the glacial debris of Permian age found in South Africa and southern South America for a former conjunction of those two continents. More recent finds there and in Antarctica extend the area of Permian glaciation and reveal a sea enclosed by those lands. This reconstruction, by Lawrence A. Frakes and John C. Crowell, omits Australia, which presumably adjoined Antarctica to the right.*

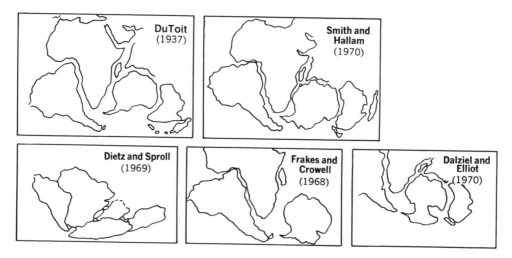

Figure 11.6 Fitting the Antarctic into Gondwanaland. For many years Du Toit's version was not questioned; however, from 1968 to 1972 no less than seven alternative orientations were proposed. Four of these alternatives are illustrated here.

tions at many sites, and this gave added strength to Du Toit's argument. In some cases the direction of ice flow could be determined, and in the Falkland Islands, rising from the continental shelf east of the southern tip of South America, elongated deposits of sand in the tillites are believed to be eskers—those peculiar "walls" that snake for hundreds of kilometers across the landscape where an ice sheet once lay. They apparently were formed by rivers that ran beneath the ice, depositing sand and other debris that, after all the ice had gone, remained as a wall-like elevation on the land. There is also evidence, in sediment deposits, that a Mediterranean-type inland sea existed in the area of Gondwanaland enclosed by South America, Africa, and Antarctica. In some places these sediments are more than two kilometers (one mile) deep, the sea having lain over features now as remote from one another as the Southern Hills region of Buenos Aires, the Cape Province of South Africa, the eastern Falkland Islands, and Antarctica south of the Weddell Sea.

It was during the Triassic Period, following the Permian, some 200 million years ago, that Antarctica became warm and vegetated enough for the reptiles and amphibians whose fossils have been found in the Beardmore area. Not until 30 million years ago did East Antarctica's present ice cover begin to evolve, according to evidence from neighboring deep-sea sediments, and glaciation of West Antarctica may have come more than 10 million years later.

Since Du Toit's day, computer techniques that match the underwater edges of continental blocks with the "smallest average misfit" have produced dramatic congruence among the various coastlines of Pangaea, as first demonstrated by Bullard and his colleagues. An American group—Robert Dietz, Walter Sproll, and John Holden—had fitted northwest Africa to eastern North America, placing Nova Scotia next to Morocco, New York opposite Spanish Sahara, and Jacksonville alongside Dakar. Sproll and Dietz also fitted southwest Africa and Antarctica together, placing the submerged rim of Antarctica, from the Weddell Sea to the Princess Martha Coast, up against the rim of Africa from Durban to the city of Mozambique. This matching of 2400 kilometers (1500 miles) showed overlaps and open gaps with a total area no larger than that of Maryland, although the rim of Antarctica, admittedly, is only imperfectly known. The two men produced a similar congruence between Australia and Antarctica, as did A. Gilbert Smith of Cambridge and Anthony Hallam of Oxford. Smith and Hallam, like Dietz and his colleagues, undertook a bold effort to reconstruct the timetable of Gondwanaland breakup, even trying to assemble the "little pieces" of the puzzle, such as Madagascar, Ceylon, and New Zealand. They pointed out, for example, that New Zealand and Australia must have been wedded during the Permian, for the fossil fauna of that period in southeast Australia resemble those of New Zealand more closely than they do other parts of Australia.

From 1968 to 1972, however, no less than seven orientations of Antarctica with respect to its neighbors were proposed and debated (Fig. 11.6), a major problem being to "dispose of" the Antarctic Peninsula and the neighboring, ice-buried archipelago of West Antarctica, which are the chief digressions from Antarctica's otherwise rounded shape. This great peninsula is far longer than Italy, with a rugged backbone so similar to the Andean system of South America that most agree it is essentially a continuation of that system. The mountains of the Antarctic Peninsula, apparently formed by subduc-

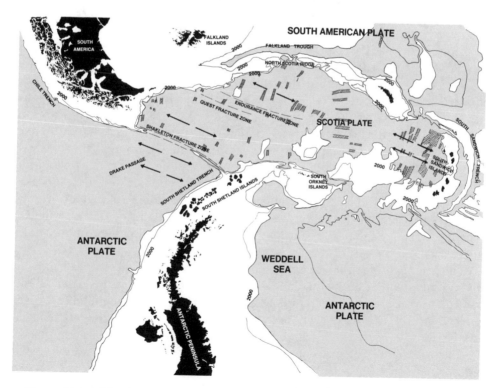

Figure 11.7 A The Scotia arc. In a more recent version, provided by Dalziel, arrows show the directions of sea floor spreading. It was published in 1985 by the British Antarctic Survey, the University of Birmingham, and the Lamont-Doherty Geological Observatory of Columbia University.

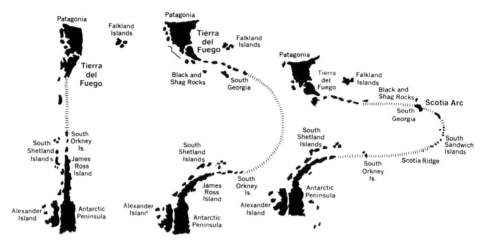

Figure 11.7 B The Scotia arc. *According to Ian W. D. Dalziel and David Elliot, the Scotia arc, between the southern extremity of South America (Cape Horn) and the northernmost extension of Antarctica (the Antarctic Peninsula), formed as South America and Antarctica were driven west leaving a section of Pacific floor behind. The arc developed where the floor of the Atlantic burrowed under this remnant of the Pacific.*

tion as it drove west, continue, intermittently, along the Pacific coast of Antarctica to the Rockefeller Mountains, within sight of the Ross Ice Shelf, making of this part of Antarctica a young geologic province very different from the great, rounded shield of ancient rock known as East Antarctica that forms the main body of the continent. The Andes of South America extend toward Antarctica as the eastward curve of Cape Horn. Like a mirror image of this pattern, the Antarctic Peninsula reaches north toward South America, then curves east, the two features being linked by a great loop of intermittent ridges and islands known as the Scotia Arc (see Fig. 11.7A). The eastern extremity of this arc, well out in the South Atlantic, consists of a curved chain of partially ice-clad, volcanic islands whose constant eruptions and earthquakes are a challenge to those seeking to untangle the changing geography of this region. The Scotia Arc looks like a fractured umbilical cord, linking two continents—Antarctica and South America—that once were one (see Fig. 11.7A). But why does it describe such an elongated loop? Why is its eastern extremity—the arc of the South Sandwich Islands—such a battleground between crustal plates, shaken by earthquakes and sending billowing clouds of volcanic ash aloft, whereas the rest of the arc is quiescent? (See Fig. 11.7B).

When, in 1819, Lieutenant Commander Zavodovski of the Russian ship *Vostok* landed on the island of the South Sandwich group that now bears his name, climbing halfway up its volcanic cone he found the ground very hot. In fact several of the islands are bare and black in a region where most land is white and ice-covered. The Zavodovski Crater is reportedly in constant eruption, emitting poisonous compounds of sulfur and hydrogen. When Captain C. A. Larsen, leader of a Norwegian whaling expedition, landed there in 1908 seeking a site for a whaling station, he was ill for some time afterward, apparently from inhaling the volcanic gases.

Some reconstructions of Antarctica's place in the puzzle have oriented Antarctica so that, had it then existed, the Antarctic Peninsula would have been superimposed on Africa, the proponents arguing that, like the Andes, the mountainous peninsula was thrown up by the westward drive of the continents during the breakup and did not exist when they were assembled as Gondwanaland. In some cases the peninsula is put west of Cape Horn; in others it is put to the east. Ian W. D. Dalziel of Lamont-Doherty and David Elliot of Ohio State have proposed that Cape Horn and the Antarctic Peninsula originally formed a straight line, but that the link between them was stretched and bent eastward into a great loop as the continents were driven west (see Fig. 11.7B). P. F. Barker and D. H. Griffiths in Britain suggested that the Andes and their continuation in West Antarctica took shape when the continents were unified and that the mountains arose as a continuous feature along the Pacific margin of Gondwanaland. Most recently a group of British and American Earth scientists have suggested a complex scenario in which West Antarctica, including the Antarctic Peninsula, consists of four crustal blocks, each with its own history.

The ultimate resolution probably will come from matching the geology of Antarctica with that of its neighbors, as one uses the color patterns on a jigsaw fragment to fit it properly into the puzzle picture. Too little is still known of the great, largely ice-covered land at the bottom of the world to do this definitively. But, with each passing year, the picture is being filled in by explorers from many nations, and eventually that "crucial puzzle piece" will fall into its proper place.

Chapter 12

Changing Geography and the Diversity of Life

NO MATTER HOW ANTARCTICA ACTUALLY FIT INTO THE FORMER ASSEMBLAGE OF continents, the fossil finds there show that, despite its present isolation and utterly inhospitable climate, it must, because of its once central position, have played an important role in the evolution of higher land animals. This discovery, and the accumulation of evidence for past changes in world geography, have forced those who seek to reconstruct the history of life on this planet to take a new look at past theories and old arguments.

Despite evidence that catastrophic impacts from space may have caused some extinctions, such as demise of the dinosaurs, little is still known regarding the conditions that brought about the great turning points in the history of life, such as the first appearance of the vertebrates, then of the amphibians, the reptiles, and finally of the birds and mammals. Information on the origin of the vertebrates is virtually nonexistent, in part because most of the early fossil record is in marine sediments and the vertebrates apparently originated in fresh-water streams. The only evidence consists of a few bony plates from fish that lived in fresh water more than 400 million years ago in what is now Colorado. Under a microscope they show structure characteristic of vertebrates. These apparently were rather pathetic bottom-dwellers who, lacking jawbones and teeth, were unable to defend themselves or forage efficiently for food. Hence for millions of years they barely managed to survive. "Mud-grubbers they were," wrote Alfred Romer of Harvard, "and mud-grubbers, seemingly, they were destined to remain." But toward the end of the Silurian Period, some 400 million years ago, they developed jaws and teeth, and thus began the rapid rise of the vertebrates. Throughout the next period—the Devonian, or Age of Fishes—the vertebrates dominated the seas of the world without challenge. They also populated the freshwater streams of the continents, and the increasing plausibility of drift has led some scientists to look more closely at the distribution of Devonian freshwater fish for possible continental connections.

Such fish traditionally have been regarded with suspicion as indicators of land links because, on a geologic time scale, they may switch rather rapidly from a marine to a lake environment—or vice versa. Nevertheless, the present barrier of the Atlantic Ocean has resulted in quite different fish populations in Europe and North America. Of 42 genera of fish found in North American lakes and streams, only 21 percent are also found in Europe. Until recently it was assumed that in the Devonian Period the situation was roughly the same; but in 1968 what Romer described as a "more intensive study" showed that, of 58 Devonian genera that lived in North America, 57 percent are also found in European "redbeds." These formations consist of the "Old Red" sandstone laid down in that period on both continents, for example in Scotland and New York State.

The drift hypothesis makes the earliest evidence for the next great step in evolution seem less incongruous—the emergence of the vertebrates onto dry land. One would expect that they picked a pleasantly warm area for this great experiment, but the oldest known amphibian remains are from Greenland—a circumstance more comprehensible if one believes, as suggested by deep-sea drilling and other evidence, that Europe, Rockall Bank, Greenland, and North America all were once joined in a single land mass that lay in a temperate or tropical latitude.

The initial discovery was made in 1932 when the remains of an amphibian—the earliest of the known labyrinthodonts—were found in a fresh-water bed in East Greenland. It was of a type, designated *Ichthyostegid*, that lived about 350 million years ago. In the epochs that followed, the descendants of these creatures, in spite of their seeming adaptation to fresh water and their clumsy build, spread to the farthest corners of the Earth (in terms of today's geography). Take, for example, the anthracosaurs, a form of labyrinthodont that lived in coal-forming swamps and thus became widely known during the early years of the Industrial Revolution. The 19th-century demand for iron ore led to heavy exploitation of blackband ironstone deposits southeast of Glasgow in Scotland, the formation being ideal because it combined iron ore with seams of coal. In 1861 the skull of a strange beast was found atop one of the coal seams, and it was sent to Thomas Henry Huxley, the biologist who became the chief champion of Darwin's theory of evolution and who had begun to catalogue and organize such finds as a means of supporting the theory.

The skull was that of an anthracosaur. Additional remains were found by coal miners and fossil-hunters in the Permian and Carboniferous deposits of England, Ireland, Bohemia, Nova Scotia, West Virginia, Ohio, and Texas, all lying close to the location of the Equator in those periods, according to reconstructions of continental drift. This would have provided a hot, damp climate suited to the vegetation of coal-forming swamps in which this creature lived.

One must, however, be mindful of a bias in such apparent distribution patterns. There has been a great deal more fossil-hunting in Europe and North America, for example, than in Africa, and in his 1970 analysis of anthracosaur discoveries A. L. Panchen of Newcastle-on-Tyne, England, pointed out that "the records of geographi-

cal distribution of fossil vertebrates, particularly rare ones, often tell one more about
the distribution of vertebrate paleontologists than that of the original animals.''

In any case, the labyrinthodonts, as they evolved into an extensive suborder,
spread even farther afield—from the edge of the Arctic Ocean in Greenland and Spitsbergen, to Madagascar, now in the tropical Indian Ocean, as well as to South Africa and
the heart of Antarctica (see Fig. 12.1).

As part of his reexamination of old evidence Romer studied the records of Permo-
Carboniferous fossils extracted from a coal mine northwest of Prague, Czechoslovakia,
and prepared a scoreboard showing the number of vertebrate genera and families

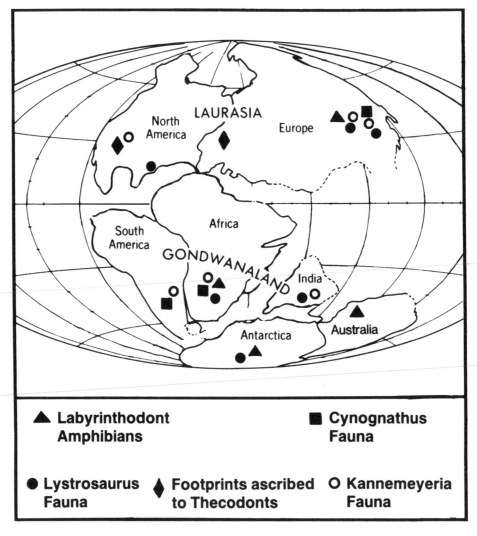

Figure 12.1 The world of the Permian and Triassic periods. *The distribution and density of
early amphibians and reptiles hints at the relative locations of the continents 200 to 270 million years
ago.*

(most of them amphibians) that were identical with those from the same period in the redbeds of Texas. He found twice as many similarities as in the present wild fauna of those regions. In fact the Texas residents of that time resembled those of Czechoslovakia as closely as they did the contemporary fauna of New Mexico, only a few hundred miles away!

Such was the rise and global spread of fresh-water amphibians—a process far more easily understood in terms of geography that welded together many of the lands now widely dispersed. But the amphibians were visitors to the land, not permanent residents. Like their fish ancestors, they laid their eggs in the water. To make the next great leap, to a fully terrestrial existence, it was necessary for evolution to develop a method for reproduction on land, and thus the reptiles and their amniote eggs came into being. Such eggs are enclosed in a protective shell and are provided with fluids for nourishment and membranes that both keep them from drying out and serve as crude lungs to extract oxygen from the air that filters through the shell.

By the Triassic Period the reptiles had spread throughout the world including, as has now been shown, Antarctica. And while the discovery of *Lystrosaurus* reptiles in Antarctica provides dramatic evidence for drift, the occurrences of this animal in other early Triassic deposits is also revealing—particularly the unearthing in western China not only of *Lystrosaurus* remains, but of other animals usually associated with it in South Africa. The Chinese animal was not the same species as that found, in identical form, in South Africa, India, and Antarctica, but it was so similar, Edwin Colbert wrote, in an analysis of such discoveries, that the faunal relationship of China to South Africa during the early Triassic "evidently was close." In his book, *Wandering Lands and Animals,* Colbert describes these far-flung occurrences of *Lystrosaurus* as "the paleontological beads from a broken string, each occurrence separated from the others by thousands of miles of deserts and grasslands, mountains, jungles, or open ocean." Yet, when *Lystrosaurus* lived, that string of beads must have been assembled so that those creatures could move freely between the places where their remains are now found.

Colbert pointed out that so far there have been no discoveries of *Lystrosaurus* and its contemporaries in South America and that even the fossil beds of India lack most of the animals associated with that beast in Antarctica and South Africa. In other words, the links, of that time, between Antarctica and Africa are more clearly demonstrated than those between India and Africa. This simply may be due to accidents of collection and preservation. However, the occurrence of this fauna in China implies that the two great supercontinents—Gondwanaland in the south and Laurasia in the north—had not yet become fully separated. Such a connection between Eurasia and Africa also existed in the previous, or Permian, period, as shown by fossils found in the Russian deposits near Perm from which that period took its name.

From a reconstruction of the breakup of the all-inclusive continent, Pangaea, Robert Dietz and his colleague, John C. Holden, have concluded that the first rifting between Gondwanaland and Laurasia began in the late Triassic and that by the end of that period they were completely separated but for a "pivot point" at what is now Gibraltar. Between the northern and southern supercontinents, to the west of Gibraltar, was an extension of the Pacific Ocean into the present relative positions of the Caribbean and Gulf of Mexico. To the east of the pivot point another extension of the

Pacific—the Tethys Sea—penetrating from the opposite direction, had opened up a precursor of the Mediterranean. The Gibraltar pivot, however, permitted some exchange of fauna.

The fossil evidence now shows that, while the two Americas during this period were relatively isolated from one another, Africa and South America were extensively contiguous. Although remains of the early, *Lystrosaurus* phase of African life in the Triassic have not been found in South America, recent excavations in strata of other periods have provided strong evidence for such a connection. Actually, one indicator has been known—but largely ignored—for a number of years. This was the puzzling geographical distribution of *Mesosaurus*, a river-dweller of the Lower Permian, 40 million years before the time of the early Triassic reptiles (Fig. 12.2). *Mesosaurus* seems to have been the earliest of the aquatic reptiles, with a slender, alligatorlike snout armed with needle teeth that it presumably used in sieving crustaceans from the brackish waters of estuaries. The remains of this creature have been found in only two places: western South Africa and almost directly across the Atlantic in southeastern Brazil. According to Romer it was almost impossible to imagine that "this rather feeble little fellow" could have breasted the waves over the 5000 kilometers (3000 miles) of ocean that now separate the two continents.

More dramatic have been the discoveries made on the Argentine side of the Andes almost at the same time that Peter Barrett picked up his first fossil fragment on Graphite Peak in Antarctica. Among the four-footed animals that inhabited South Africa following the heyday of *Lystrosaurus*, the most familiar to fossil-hunters was a carnivorous reptile known as *Cynognathus* (Fig. 12.4). Its doglike skull structure marked its advance along the line whereby this branch of reptiles finally evolved into mammals. (Figure 12.3 shows a Thecodont, representing the base of the other, ultimately dominant, line of reptiles, whose descendants became the great dinosaurs.)

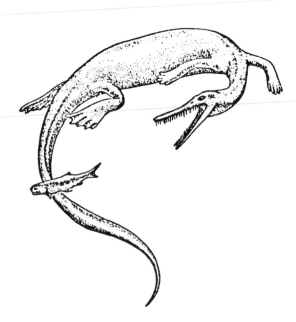

Figure 12.2 Mesosaurus, *a slender reptile, prowled the rivers of what are now South Africa and southern Brazil presumably when Africa and South America were joined during the Permian period.*

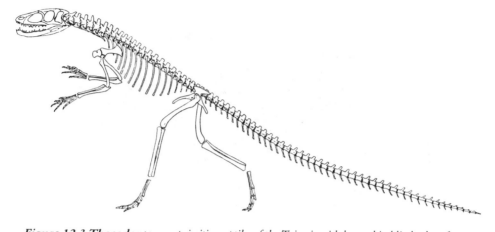

Figure 12.3 Thecodonts *were primitive reptiles of the Triassic with longer hind limbs than front limbs. They were ancestral to crocodiles, pterosaurs, birds and some dinosaurs.*

Cynognathus seems to have succeeded weasel-like *Thrinaxodon* as the roving predator of the Triassic scene. It may have been hairy, representing another step toward mammalian appearance. Its largest contemporary was a hulking monster called *Kannemeyeria*, generally related to *Lystrosaurus* and built somewhat like a rhinoceros, with a turtlelike beak. Its two tusks belied its vegetarian diet (Fig. 12.4).

When, in the 1960s, remains almost identical to those of these African beasts were found in Argentina's Mendoza Province, Romer's last vestiges of resistance to the drift theory vanished. Probably as much as any man, he had been responsible for fossil-hunting in Argentina and Brazil. Now J. F. Bonaparte, who was carrying on the work in association with Argentina's National University of Tucumán, had found unmistakable relics of these "African" animals. How could a beast as heavy and clumsy as *Kannemeyeria* have swum the Atlantic? Or how could it have migrated 32,000 kilometers (20,000 miles) overland between Africa and Argentina via the polar climates of Siberia and Alaska, as well as the equatorial tropics—through deserts and forests, over mountain ranges and other impediments? Furthermore, fossil beds with animals much like those of the *Cynognathus* deposits of Africa have been found in India and in Chinese excavations in Shanxi Province, providing further evidence of north–south connections.

In fact by 1973 it was clear that at least some of the Chinese fossil specialists were leaning toward continental drift as an explanation for these finds. Sun Ailin of the Institute of Vertebrate Paleontology and Paleoanthropology of the Academy of Sciences (Academia Sinica) in Beijing drew attention to the *Cynognathus* finds in Shanxi and the more recent discovery of fossils from the hulking *Kannemeyerias* in beds overlying those, on the north slopes of the Tianshan mountains of Xinjiang Province, that yielded *Lystrosaurus*, thus forging a link between that region of northwest China, in the heart of Asia, with South America and South Africa (see Fig. 12.1).

"Both South Africa and India are long supposed to be parts of the ancient southern continent—Gondwanaland," he wrote, and Xinjiang "was far beyond the supposed

Figure 12.4 **Kannemeyeria** *(top) was a heavy and clumsy beast resembling the rhinoceros. As opposed to smaller and more mobile reptiles like the* **Lystrosaurus** *(middle) and* **Cynognathus** *(bottom), it was difficult to argue that the migratory path of this beast could cover India, Africa, South and North America, and China as they are presently located. For skeptics like Alfred Romer, the prevalence of the* **Kannemeyeria** *was convincing evidence for the continental drift theory.* **Lystrosaurus** *remains and those of nearly identical amphibians have been found in South Africa, India, Antarctica, China, and North America. The pervasiveness of the* **Lystrosaurus** *supports the existence of the super continents, Laurasia and Gondwanaland, and their continued connection during the early Triassic period.*

northern boundary of Gondwanaland...but the vertebrate faunas show great similarities to those of the southern continents. As these vertebrates are land dwellers, it is difficult to imagine that migration could have taken place through the wide seaways.

"The recovery of the fossils of *Lystrosaurus* in Xinjiang and an analysis of the above-mentioned faunas," he continued, "seem to indicate that the existence of land connections somewhere between the two ancient land masses—Southern and Northern—at that time is highly probable."

As the continents began fragmenting, in perhaps the greatest geographical revolution of the Earth's history, a biological revolution was also under way: the emergence of the "ruling reptiles," or dinosaurs. They were to dominate the Earth longer than has any other form of land life—roughly 100 million years.

Were these two revolutions linked? Perhaps so. During Leg 19 of the Deep Sea Drilling Project it occurred to the participating scientists, notably to Thomas R. Worsley and David W. Scholl, that there might be such a connection. From evidence they found, in the expedition's cores, of far-flung volcanic activity around the rim of the Bering Sea and North Pacific during the dinosaur heyday, it could be assumed that there was extensive carbon dioxide enrichment of the Earth's atmosphere by such eruptions. This, they proposed, could have caused a "greenhouse" effect that made for a warm, cloudy climate ideally suited to huge animals that, unlike the little mammals of that time, lacked an efficient system for bodily temperature control.

Because this effect permits sunlight to pass, but inhibits escape of the resulting heat as infrared radiation, the more carbon dioxide in the air the warmer the climate (presumably) becomes. The steady rise in the burning of fossil fuels has raised carbon dioxide levels in the air, leading to widespread fears that this may lead to undesirable climate changes.

In any case, once volcanic activity associated with the rupture of Gondwanaland and Laurasia into the modern continents came to an end, the large-scale injection of carbon dioxide into the air also terminated. Furthermore, according to this hypothesis, about 160 million years ago tiny drifting sea creatures learned how to manufacture limestone shells, using carbon dioxide dissolved in the sea. At this time the evidence for a catastrophic impact at about the time of the dinosaur extinctions had not yet been discovered and the drilling scientists proposed that this depletion of carbon dioxide in the atmosphere and oceans chilled the climate enough "to absolutely raise hell with the big reptiles."

Because the great extinctions have been so perplexing, they have served as fair game for all sorts of speculation, including the idea that they were related to periods of repeated reversal of the Earth's magnetic field, or the proposal that man was responsible for the demise of the great ice-age mammals. In any case, there is widespread evidence for massive eruptions of lava along the continental margins that rifted at the breakup of Gondwanaland and Laurasia. The resulting basalts, like those that cap and intrude the Beacon sandstones of the Transantarctic Mountains, including Coalsack Bluff, are of a type known as dolerite (or diabase). There are dolerite intrusions along the entire eastern flank of the Appalachians that erupted during the late Triassic and early Jurassic, presumably in connection with the first stages in Atlantic rifting, and the eastern part of Africa as far north as Zimbabwe also shows signs of such eruptions.

As noted earlier, the Transantarctic Mountains do not lie entirely on the present rim of the Antarctic block, but some believe they mark a zone of rupture where West Antarctica—the part that is clearly a continuation of the Andes mountain system—came loose from the main Antarctic plate as the latter, during continental breakup, rotated clockwise.

Whatever validity of the proposal that extensive volcanic activity determined the destiny of the dinosaurs, there is no doubt that the mammals who succeeded them had many advantages, of which their warm bloodstream, as a device for maintaining a uniform and relatively high temperature, was but one. (Some, in fact, believe the dinosaurs, as opposed to the other reptiles, may have been warm-blooded.) A high temperature makes for efficient body chemistry. It enables birds, almost instantaneously, to produce the intense burst of energy needed for flight. It presumably allows the brain—a highly energy-dependent organ—to operate faster. Some of the larger reptiles, apparently to cope with environmental changes in temperature, developed enormous fanlike sails on their backs. These could be opened or closed for temperature control. When the animals were hot, the sail could be opened to the breeze. When they were cold, it could be turned to the Sun. But this could not match the efficiency of the mammalian temperature-control system. The mammals were also able to bring their young into the world alive and provide them with milk. But by far their most important attribute, in seeking to survive among the giant, birdlike feet of the dinosaurs, lay in their nimble wits. Their brain size grew far larger than that of other animals, and they developed an eagerness to probe, sniff, peek, and climb that, in *Homo sapiens,* can be called intellectual curiosity.

Shortly before the main continental breakup the mammals split into two basic groups. There were the placental mammals whose babies, in the womb, are equipped with an elaborate placenta, enabling them to draw oxygen and nourishment from the bloodstream of the mother until, in many species, they are sufficiently grown to walk and, to a considerable extent, fend for themselves immediately after birth. Man and the other primates are placental mammals. The other branch were the marsupials, with only a primitive placenta. Their babies, at birth, are utterly helpless. The offspring of the large type of kangaroo, for example, is only half the length of a thumb, and those of the American opossum are the size of a small bee. Such babies must be reared in the abdominal pouch, where a milk supply is provided, before they can venture forth.

The enormous diversity of the mammals seems to have been a direct result of the breakup and drifting apart of the supercontinents. One almost could argue that man, as the fruit of one line of primate evolution, might not exist had it not been for this process.

In an analysis of the role of continental drift in shaping the evolution of the mammals, Björn Kurtén of the University of Helsinki has pointed out that, during the entire Age of Reptiles, only 20 orders, or major groups, of reptiles evolved. This took some 200 million years—from the Permian through the Triassic, Jurassic, and Cretaceous, culminating in the long reign of the dinosaurs. Yet, in the 65 million years that have elapsed since then, the mammals have produced 30 orders—that is, in a third of the time they have diversified into half again as many orders. Kurtén attributed this, not to some special genius of the mammals, but to the breakup of the supercontinents into

Plate 1: **Earth.** *En route to the Moon in 1972, astronaut-geologist Harrison Schmitt looked at the Earth. Recorded by the radio monitor in Houston, he noted that he had not believed in continental drift, "but I tell you," he remarked, "when you look at the way the pieces of the northeastern portion of the African continent seem to fit together, separated by a narrow gulf, you could almost make a believer of anybody." This view of Africa, the Arabian peninsula, and the adjacent part of Asia was photographed on the Apollo 11 mission to the Moon.*

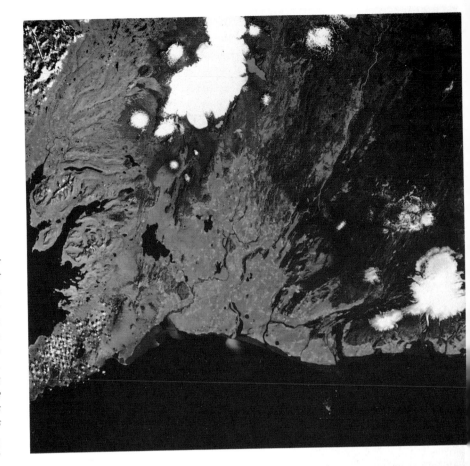

Plate 2: **Iceland**. *In this space view of southern Iceland the capital city of Reykjavik is on the north side of the Reykjanes Peninsula, dotted with clouds on the southwest. The large white area near the top is the glaciated mountain Langjökull. Next to the lake, between the city and glacier, is Thingvellir, where the world's first parliament is said to have met in A.D. 930. The glaciated region to the right is Myrdalsjökull. The dark area to its left being the Hekla volcano, traditional gate to hell.*

Plate 3: **Japan.** *This region in Japan is the product of "subduction": the descent of a vast section of oceanic floor into the mantle. The resulting volcanic activity produced Mount Fuji, whose white cone, as seen from ERTS-1, is visible in the lower part of the photograph. The outskirts of Tokyo, tinted blue, are to the right. The bodies of water in the foreground are the Suruga Bay (left) and the Sagami Sea (right).*

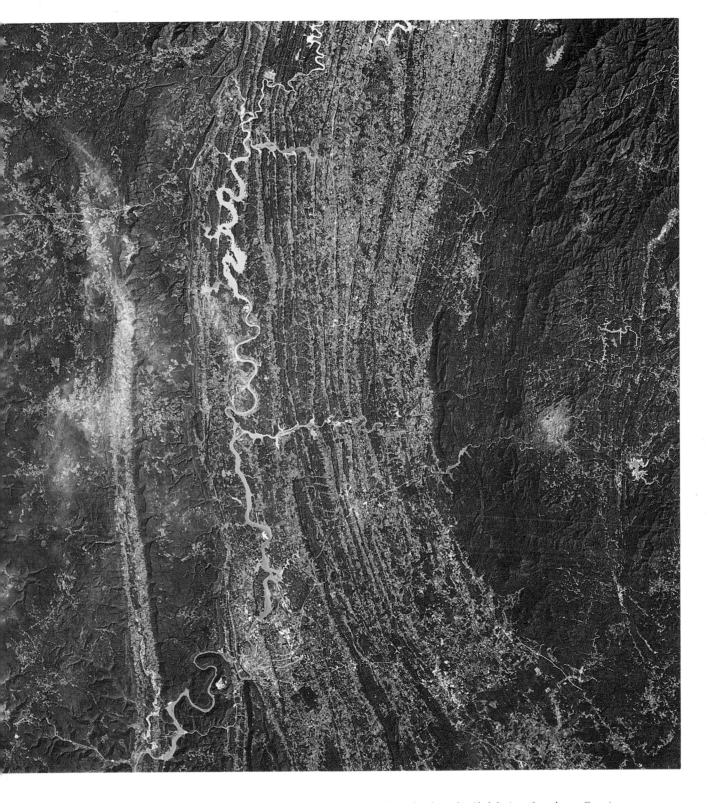

Plate 4: **Southeastern United States.** *The folding of the Appalachians shows clearly in this Skylab view of northwest Georgia, western North Carolina and southeastern Tennessee. The Tennessee River runs from northeast to southwest, the city of Chattanooga being the blue area alongside the river to the south. The pink area on the right is where vegetation was destroyed by fumes from a copper smelter in Ducktown in the southeast corner of Tennessee.*

Plates 5a and 5b: **Los Angeles.** *The susceptibility of California's largest city to earthquakes is illustrated by this ERTS image of the Los Angeles Basin and San Fernando Valley, shown above in color to discriminate surface properties and below in black and white with surface features identified. The white region is the Mojave Desert; irrigated farms are shown in red. Two great faults, the San Andreas fault and the Garlock fault, converge to the left of the desert. All of the faults shown are intermittently active. Locations of the 1971 San Fernando, 1973 Point Mugu, and 1987 Whittier Narrows earthquakes are marked. This image was made from space on October 21, 1972. Like other ERTS images, it was recorded 915 kilometers above the surface; however, freeways and major avenues in Los Angeles are still identifiable (lower right).*

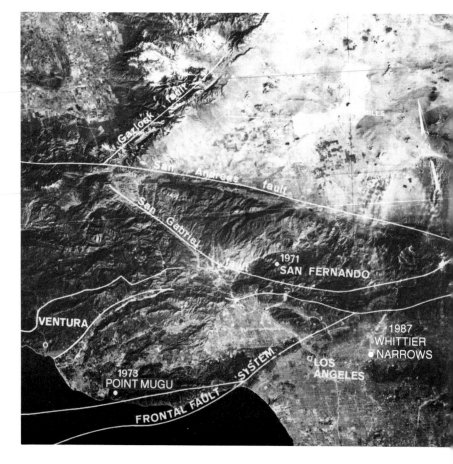

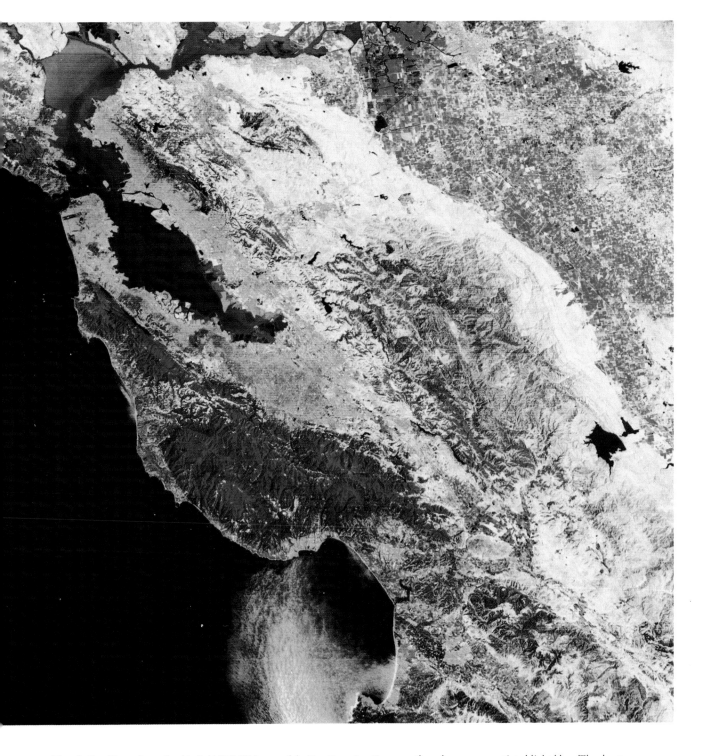

Plate 6: **San Francisco.** *In this LANDSAT image of the San Francisco Bay area the urban areas are tinted light blue. The sharp edge along the inland side of the mountains (deep red) on the San Francisco Peninsula marks the San Andreas Fault, which continues under the sea off the Golden Gate Bridge. The 1989 Loma Prieta earthquake, producing severe damage in San Francisco, occurred in that sector of the fault in the lower center of this image. Vegetation is shown in red, especially in the broad farming zone of the Joaquin Valley.*

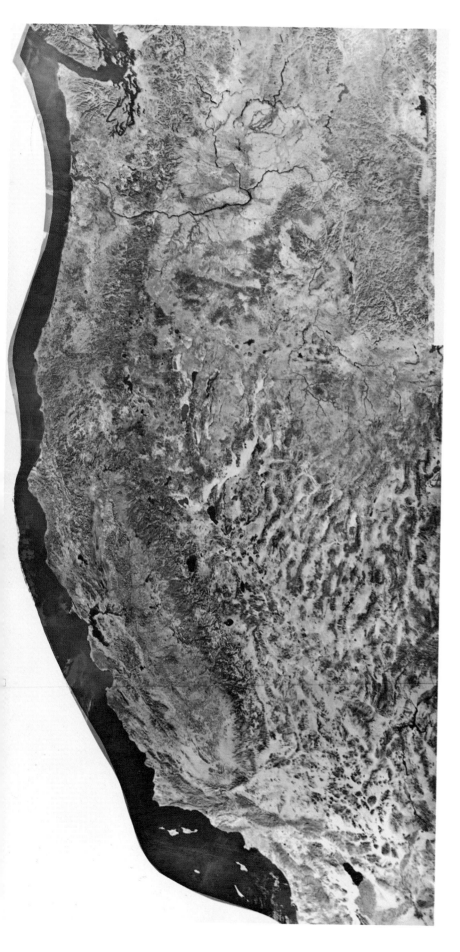

Plate 7: **Red Sea.** *(Opposite, top left) In this view from space the north end of the Red Sea splits into the Gulf of Suez, on the left, and the Gulf of Aqaba. The Gulf of Suez leads to the Suez Canal and the Mediterranean at the top. The Gulf of Aqaba ends in the great rift valley occupied by the Jordan River and Dead Sea. On the far left is the Nile Valley leading to the broad, fertile delta at its mouth. The region is clearly arid.*

Plate 8: **Himalayas.** *(Opposite, top right) The ramparts of the Himalayas can be seen in this photograph. According to the plate theory, they were thrust up when India collided with the Southern flank of Asia. Along the Himalayan front, in Nepal, lie mountains immortalized by heroic assaults, such as Everest, the world's highest, Lhotse, Kanchenjunga, and Annapurna. This image is of the southern face of the range in India, Kashmir, and Punjab. The snow line, at about 12,000 feet, is clearly visible. The dark area running through the snowy region is the Chenab River gorge which is deep enough to be below the snow line.*

Plate 9: **Makran ranges.** *(Opposite, bottom) The Arabian Sea is in the foreground of this spectacular view of Pakistan and Afghanistan. To the right is the Indian sub-continent. It moved slowly north and caused the folded sedimentary ranges of the Makran in Baluchistan to bend from east-west to north-south. According to geologists, India is like a boat whose bow is slicing through the water with the Makran and Himalayas as the "left" and "right" side, respectively, of the resultant wake.*

Plate 10: **Western United States.** *(Left) In this composite view assembled from ERTS pictures, the Cascades and Sierras form an almost continuous line parallel to the coast from Canada to Mexico. The Great Valley of California is between the Sierras and the Coast Ranges on the rim of the continent. East of the Sierras are the successive ridges of the basin-and-range province, and north of that area are the Snake River volcanics and the Columbia River basalt plateau. At the upper right, south of Flathead Lake, Montana, is the great Idaho batholith.*

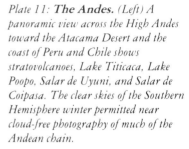

Plate 11: **The Andes.** (Left) A panoramic view across the High Andes toward the Atacama Desert and the coast of Peru and Chile shows stratovolcanoes, Lake Titicaca, Lake Poopo, Salar de Uyuni, and Salar de Coipasa. The clear skies of the Southern Hemisphere winter permitted near cloud-free photography of much of the Andean chain.

Plate 12: **The Alps.** (Lower left) In this space view of the Alps the great valley separating the European block to the north from the African block to the south cuts across the picture. The Rhone River originates within it, then makes a sharp turn north and becomes Lake Geneva. The Rhine River originates at the opposite end of the valley. The snowy peaks of the Mont Blanc Massif are on the left. The Jungfrau and its companion mountains rise to the right, north of the great valley. The Matterhorn and other peaks of the Pennine Alps are to the south, as are clouds bordering the Italian Plain.

Plate 13: **Hawaiian island chain.** (Below) Left to right, the islands are Oahu (with Pearl Harbor), Molokai, the smaller island of Lanai and Maui.

fragments each of which developed its own fauna. The reptiles, for most of their developmental period, had two homes: Gondwanaland and Laurasia. The mammals, for a considerable time, had at least eight.

There are two equal and opposite tendencies constantly at work in those processes of evolution that have made us what we are. One is "adaptive radiation" in which a few primitive species evolve in many different directions to exploit ways of life that stand open—ecological niches, as they are often called. "The world of living organisms," Kurtén has said, "is a world of specialists." Thus, from the primitive little early mammals there evolved grassland grazers, such as the horse, bison, and antelope. To forage the leafy forests there appeared deer, giraffes, and the like. To prey upon such animals there developed various carnivores from lions to wolves. To exploit the nuts, pinecones, and insects of the woods the rodents evolved, as well as small carnivores to hunt them—weasels, foxes, and such.

But the opposite evolutionary process is parallelism, where animals with completely different ancestry evolve in the same direction to fill a certain ecological niche. Such a niche, for example, is provided by the food source to be found in termites and ants. In South America, the mammalian order that includes such oddities as armadillos and sloths produced long-snouted ant bears to feast at the termite nests. In Asia and North Africa a completely different order gave birth to the pangolins, or scaly anteaters. In South Africa this niche was filled by the aardvarks and, in Australia, by the spiny anteater. The latter, and its relative the platypus, or duckbill, are relics of evolutionary transition, for, while classed as mammals, they lay eggs and have lost their teeth in favor of bills. In the case of the spiny anteater, however, this bill is elongated, like the snouts of other anteaters. Yet none of these species are blood relatives.

Because of "adaptive radiation," the primitive mammals that found themselves on each continental fragment evolved in a variety of directions to populate the land. Because of "parallelism," these lines of evolution, to some extent, ran parallel in areas remote and isolated from one another. This, Kurtén believed, led to the multiplicity of mammals.

It was during the Cretaceous Period, 70 to 140 million years ago, that the drifting apart of the continents seems to have reached its maximum. The volcanic activity associated with rifting may have begun much earlier, in the Triassic; but it was in the Cretaceous that the seas rose and flooded lower continental areas. The period derives its name, in fact, from the Latin for "chalky" because of the widespread shallow-water sediments laid down, at that time, on what are now continental areas—notably forming the famous white cliffs of Dover.

These incursions of the sea cut both Asia and South America in two and, for a time, Africa was apparently divided into three parts. In Asia a north–south waterway linking the Arctic Ocean with the Tethys Sea (now the Indian Ocean) split the continent. In South America a similar body of water filled what is now the Amazon Basin. Since the westward drive of that continent had not yet thrust up the Andes, South America then resembled two continents and North America was then simply split.

What could have caused the oceans to rise in this manner has long been a puzzle, but it now appears that this may at least in part have been the by-product of intensified sea-floor spreading. It has been proposed that, if molten rock, at that time, flowed up into the mid-ocean ridges at an accelerated rate, this would not only thrust up the ridges themselves, making the seas there shallower, but the crust of the Earth, flowing away from the ridges, would remain hot longer than it does today.

A striking feature of the ocean floors is the way in which they become progressively deeper at greater distances to either side of the mid-ocean ridges. This has been attributed to cooling and shrinking of the rigid top layer, or lithosphere, as it moves away from the ridge. But if the lithosphere remained hotter, it would stay thicker. Accordingly, during the Cretaceous, when the continents were supposedly being rapidly pushed apart and the upwelling of lava under the oceans was intense, much of the world's sea-floor would be higher than normal, displacing water that, augmented, as well, by melting of the polar ice, would overflow previous coastlines.

While not enough is known of the early history of the mammals to pinpoint where each order originated, it is clear that some 16 orders arose in the Laurasian fragments, including our ancestors the primates, whose earliest known remains are from Late Cretaceous deposits in North America. Other Northern Hemisphere orders include the Chiroptera (of which bats are the best-known members), the Insectivora (moles, hedgehogs, shrews, etc.), the Carnivora (dogs, cats, bears, etc.), the Perissodactyla (horses and other odd-toed vegetarians), the Artiodactyla (even-toed grazers and browsers such as cattle, deer, and pigs), the rodents, and the Lagomorphs (the order that includes rabbits and hares).

Many of the orders that arose in Gondwanaland are less well known or became extinct—particularly after new land links formed and they had to compete with mammals that had developed elsewhere. A number of marsupial orders evolved in Australia and South America. The most dramatic contribution of Africa were the Proboscidea— the elephants and mastodons that, when a land route opened, marched north to "conquer the world" as had the large dinosaurs before them. They even populated the continental shelf off the East Coast of the United States, when ice age seas receded and much of the shelf became fruitful pasture. Quite often their teeth are hauled up by coastal trawlers.

The role of Antarctica remains mysterious, since almost no mammalian fossils have been found there. And that of India is puzzling. Not only are the remains of early mammals scarce, but three of India's Late Cretaceous dinosaur species are strikingly like dinosaurs living then in South America. As Colbert put it, they represent "one of the enigmas in the theory of continental drift." During the 100-million-year period between 80 and 180 million years ago India, it has been widely assumed, was making the long journey that separated it, by 8300 kilometers (5200 thousand miles), from its former union with Antarctica and drove it into the southern flank of Asia. Yet the fossils are of dinosaurs similar to those living elsewhere at that time. "Obviously," said Colbert, "this subcontinental mass had connections of some sort with the rest of the world." Indeed, India's fit into the Gondwanaland puzzle, and the history of its divorce from that assemblage and its drift toward Laurasia, remain to be deciphered.

A similar puzzle, as pointed out by the British researcher, Anthony Hallam, concerns the distribution of those gargantuan creatures, the Brachiosaurs. Some of their skeletons stand 12 meters high and are 21 meters long. These long-necked plant-eaters were so ungainly that, as Hallam put it, "they would have had trouble swimming across a wide river, let alone an ocean." Yet their remains are found in Colorado, Tanzania,

Portugal, and Algeria in beds laid down when, according to most timetables, Colorado was part of Laurasia and Africa was part of Gondwanaland, the two being separated by a Tethys Sea that was open all the way through the old suture zone between the super-continents.

Another oddity is the discovery, in the Canary Islands, of the eggs of flightless birds related to the modern ostrich. The islands are separated from the African mainland by 150 kilometers (93 miles) of open water. It is assumed that these birds could not have crossed such a gap, but since the Canaries seem to be formed partly of continental rock it is suspected that they represent a fragment of Africa torn loose as the Atlantic opened. The eggs are interbedded within lava flows estimated, both from potassium–argon measurements and the timetable of magnetic reversals in the lava, to have erupted 6 to 12 million years ago. The age indicates that the islands remained attached to Africa long after the opening of the ocean began (ostriches had not evolved at the time of original rifting).

With regard to the role of drift in mammalian evolution, a feature widely discussed by paleontologists is the striking similarity between European and North American mammals as recently as the Eocene, 35 to 55 million years ago. During the early Eocene, according to Kurtén, they were "practically identical" and, in Colbert's words, "remarkably close." Traditionally, students of evolution, as Colbert points out, attributed this to links across the entire width of Asia. A land bridge across the North Atlantic seemed "so fanciful as to merit little serious consideration." Indeed, the fossil evidence shows that, by then, South America and Africa were widely separated, but the Americas apparently rotated away from Europe and Africa with pivot points, or "rotation poles," in the north so that, in the south, separation began early and moved rapidly. Furthermore the opening of the North Atlantic may have been a separate, later process than the South Atlantic opening. Hence a land bridge linking Canada, Greenland, Iceland, Rockall Bank, and Norway may have remained intact until 20 million years ago. As noted in Chapter 10, deep-sea drilling has shown that, at the start of the Eocene, 55 million years ago, Rockall Bank had undergone moderate subsidence, but paused at shallow depth throughout that period.

Meanwhile, however, the Central American "floodgate" was wide open (including a deep passage, the Bolivar Trench, across the northwest corner of South America), and hence two continents—South America and Australia—were completely isolated. It was this that determined the evolutionary history of the marsupials. This large, but generally less successful branch of the mammals seems to have originated in the Americas in the form of small creatures much like the opossums of today. As is the case with the opossum, their brains were small, and apparently they were tree-livers. For a time they did well, migrating as far as Europe. But carnivores of the placental type were too much for them and, had it not been for the opening of the Central American floodgate and the separation of Australia from Africa and Antarctica, they might have been overwhelmed. Given protection from placental predators (such as the evolving members of the cat family), the marsupials demonstrated "parallelism," or parallel evolution, to an extraordinary extent. In South America they developed their own predators—marsupial counterparts of the weasel, wolf, and cat. The Borhyaena was comparable to a puma. One form was as big as a bear and another had tusks like those of the saber-tooth tiger.

Without open water to spare them from placental predators, the North American opossums vanished during this period, and those who live in the United States today may have migrated from South America when the Panamanian floodgate was closed and the isthmus again became a land bridge. The hoofed animals, such as deer and camels, that then migrated south across that bridge were apparently too swift for South America's marsupial predators, and so the latter went hungry. The placental carnivores that came down from the north probably helped wipe out the last of these big marsupials. But one great haven remained: Australia.

Originally almost all of the native Australian mammals are believed to have been marsupials, although later a few small placental mammals, such as bats and rodents, found their way there (apart from animals brought in by man). The traditional view has been that the marsupials came down from the north, across the East Indies. But, as Wegener noted more than a half century ago, there are no marsupials in the East Indies. In fact, no fossil marsupials have been found in either Asia or Africa. They could, however, have reached Australia overland from South America—via Antarctica—before the great breakup. Soon after serving as geologist on Byrd's first expedition to Antarctica, Gould predicted that marsupial fossils would be found there, and now, at Seymour Island near the tip of the Antarctic Peninsula, he has been proved right.

In any case, according to the concept of a fracturing Gondwanaland, as stated by Romer: "Australia during the Tertiary (between 10 and 55 million years ago) would have been, so to speak, a great marsupial-filled Noah's Ark floating northeast across the Indian Ocean, finally to run aground on the East Indies."

During this long, slow voyage, the marsupial inhabitants of Australia were not idle. Like their South American cousins, they were evolving into forms to fill a great variety of ecological niches: marsupial "squirrels" (even "flying squirrels"); a counterpart of the woodchuck (the wombat); a lionlike animal that actually seems to have lived on fruit; and one form comparable to a rhinoceros, as well as such modern survivors as the Tasmanian devil (which decimates lambs and chickens), the Tasmanian wolf (with a tiger-striped back, now possibly exterminated by Tasmanian farmers), the rabbitlike bandicoot, and the cuddly-looking koala bear. The marsupials never developed grazers comparable to the hoofed placental mammals, but that role, to a great extent, has been filled by the larger species of kangaroo and wallaroo. As happened with the ice-age extinctions of large mammals in North America, many of the larger marsupials, as well as a variety of large and odd-looking grazing mammals in South America, became extinct about the time that man became a hunter, although the relationship remains to be established. It was the remains of these great mammals in what is now Argentina that amazed the youthful Darwin when he went ashore from the *Beagle* early in the 19th century. "The great size of the bones," he wrote, "...is truly wonderful." He was particularly impressed by the massive build of the extinct sloths, agreeing with the proposition that this enabled them to feed on the treetops. "The colossal breadth and weight of their hinder quarters, which can hardly be imagined without having been seen, become, on this view, of obvious service, instead of being an incumbrance: their apparent clumsiness disappears. With their great tails and their huge heels firmly fixed like a tripod on the ground, they could freely exert the full force of their most powerful arms and great claws. Strongly rooted, indeed, must that tree have been, which could have resisted such force!"

In viewing this panorama of evolutionary development, one is tempted to speculate about the possible role of Antarctica. Presumably it did not break away from the other continents until the mammals had already appeared. Might some of the most critical steps in the evolution of mammalian species have taken place on the continent that is now largely buried under thousands of meters of ice?

It is noteworthy that, long before the Shackleton and Scott expeditions discovered fossils along the Beardmore Glacier, Darwin speculated on the possible role of that region in botanical evolution. One of the mysteries that continues to baffle botanists is the origin of the flowering plants, or angiosperms, that include almost all garden varieties as well as the leafy trees and shrubs. They seemed to have burst full-blown upon the world during the Cretaceous Period, 100 million years ago, when Antarctica still had not severed all its links with the rest of the world.

In the 1840s and 1850s a succession of expeditions into the American Arctic, in search of the ill-fated expedition of Sir John Franklin, discovered fossils of warm-climate plants and trees. Such specimens were known, as well, from Spitsbergen and Siberia and, when Sir Joseph Dalton Hooker sailed to the Antarctic in the 1840s as a young surgeon on the expedition of Sir James Clark Ross, he found fossil tree trunks in the Tertiary lava beds of Kerguelen Island. The island is near enough to Antarctica to be subpolar in climate today.

In Darwin's time it was widely believed that the internal heat of the Earth was a residue of its fiery birth as a child of the Sun, radioactivity not having been discovered.

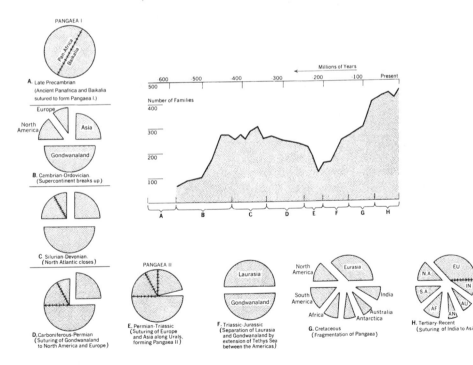

Figure 12.5 Diversity of shallow water marine life in relation to continental "marriages" and "divorces" as proposed by J. W. Valentine and E.M Moores.

Hence it was proposed by such scientists as the 18th-century French naturalist the Comte de Buffon, and a century later by Gaston de Saporta, that the polar regions were the first to become cool enough for the origin of life. As Hooker wrote to Darwin in an exchange of letters in 1881, "Buffon's and Saporta's views of life originating at a pole, because a pole must first have cooled low enough to admit of it, is perhaps more ingenious than true—but is there any reason opposed to it? If conceded, the question arises, did life originate at both Poles or only one? or if at both was it simultaneously?—but this is the deepest abyss of idle speculation."

What prompted the exchange was a request by Hooker for Darwin's advice in preparing an address that he was to give at a meeting of the British Association in York. Hooker, 23 years earlier, had been one of the first to champion Darwin's evolution theory, and now he was to present an address on the worldwide distributions of plants and their implications with regard to evolution.

"Dear Darwin," he wrote on August 4, 1881, "I am groaning over my Address for York after a fashion with which I have more than once bored you awfully." Among the subjects that he proposed to discuss, he said, was: "The establishment of the permanence since the Silurian period of the present continents and oceans" and, he asked, "Were you not the first to insist on this, or at least point this out? Do you not think that Wallace's summing up of the proof of it is good? (I know I once disputed the doctrine, or rather could not take it in—but let that pass!)"

Alfred Russel Wallace, simultaneously with Darwin, had formulated an almost identical theory of evolution. As shown by Hooker's comment, recent discoveries—particularly the strange distribution of Gondwana flora—had evoked suspicions that the world's geography has not always been as it is today.

In his reply Darwin said: "If forced to express a judgment, I should abide by the view of approximate permanence since Cambrian days" (now put at 500 million years ago, but then considered far more recent). Nevertheless, he advised Hooker in his address not to take too strong a stand on the subject: "I would speak, if I were in your place, rather cautiously..." When it came to the role of the polar regions in the evolution of plants, he bemoaned how little was known of that subject, and said: "Nothing is more extraordinary in the history of the Vegetable Kingdom, as it seems to me, than the *apparently* very sudden or abrupt development of the higher plants. I have sometimes speculated whether there did not exist somewhere during long ages an extremely isolated continent, perhaps near the South Pole." Thus, Darwin suggested, the flowering plants may have burst forth from the Antarctic after quietly having evolved there in isolation.

Probably the chief role of continental drift in shaping evolution has been in controlling the diversity of species. This effect was studied by James W. Valentine at the University of California in Davis, with emphasis on shallow-water marine species. He estimated that there are more than 10 times as many marine invertebrate species today as there were in the Paleozoic Era, some 300 to 500 million years ago. The diversity has been minimal when there was but one supercontinent. The more continental fragments there were, the greater the total length of coastline and continental shelf, and the greater the variety of environments. No such long-range comparison is possible with land plants, since the continents apparently were lifeless deserts until the "oxygen revolution" of about 400 million years ago. Only then did the oxygen content of the air become sufficient—presumably from the activity of marine plants—to provide a layer of ozone (whose molecules are formed of three oxygen atoms) that absorbed the lethal ultraviolet rays of the sun. Marine plants had been safe, as the rays do not penetrate water deeply.

From their reconstruction of the changing diversity of life in the seas, in terms of continental "marriages" and "divorces," Valentine and Eldridge M. Moores concluded that prolonged periods of stability—as opposed to episodes of rifting and rapid change—made for the greatest variety (Fig. 12.5). During such periods numerous species evolve that are suited to special environments, special food sources, or special habitats. In times of change these "ecological niches" tend to alter, producing widespread extinctions. Of more direct interest to man is the role that amalgamation and fragmentation of the land masses have played in the evolution of mammals. As noted by Kurtén, their diversity is in large measure a result of continental breakup.

During the period when an open-water moat separated the Americas, monkeylike prosimians and certain rodents (guinea pigs and porcupines) reached South America from their early home in North America. They presumably made their way—perhaps on driftwood—across the open water that otherwise isolated the two Americas from one another. Meanwhile, the prosimians had reached India and Africa, where they underwent adaptive radiation, evolving into our apelike ancestors. These early hominids, according to Kurtén's summary, then spread into Asia and Europe toward the end of the Miocene, some eight million years ago. Each of the eight or more continental fragments contributed from three to six orders of mammals, he believes, and this process of enforced diversification has not only enriched the world with an enormously varied population of furry creatures, but has perhaps stimulated the evolution of man. It is ironic that this culminating step in evolution reversed the trend. For man, by hewing down the forests, ploughing the prairies, "paving" the continents and polluting sea and air, has steadily reduced the diversity of life on the planet that he inhabits.

Chapter 13

The Afar Triangle and the Red Sea Hot Spots

WHILE MOST OF THE ACTIVITY THAT IS PUSHING OCEANS APART AND STEADILY altering the world's geography lies hidden deep beneath the sea, there are two regions, in particular, where it can be seen at first hand. One is in Iceland—a section of the Mid-Atlantic Ridge that has risen from the sea through repeated eruptions such as that which, in 1973, forced the inhabitants of Heimaey, Iceland's chief fishing center, to flee for their lives. In Iceland one can observe processes at work that are, at least in part, typical of those taking place deep under water along the line from which spreading is taking place in a fully mature ocean. The other is the Afar Triangle at the southern end of the Red Sea. Arid, incredibly hot, torn by volcanic activity, and shaken by earthquakes, it is probably as close to the proverbial concept of Hades as any place on Earth. But it is also a place where one apparently can see the birth pangs of a new ocean (Fig. 13.1).

The Red Sea, 2000 kilometers (1200 miles) long and about 240 kilometers (150 miles) wide, is the first, slitlike suggestion of such an ocean. And part of the Afar Triangle, chiefly within the boundaries of Ethiopia, is apparently a portion of the Red Sea floor uplifted so we can study it. Wegener, in his early discussion of continental drift, singled out this triangle as being of special interest. He pointed out that its low-lying terrain is composed entirely of recent lava beds and, he said, it appears to be a region of broadening in a rift that runs from East Africa north to form the Red Sea. "This idea is suggested particularly by the course of the coastlines on either side of the Red Sea," he wrote, "whose otherwise accurate parallelism is spoilt by this projection (the Afar Triangle); if one cuts this triangle out, the opposite corner of Arabia fits perfectly into the gap." This fit is dramatically evident in some of the photographs taken from Earth orbit by American astronauts. (See colored plate number 1).

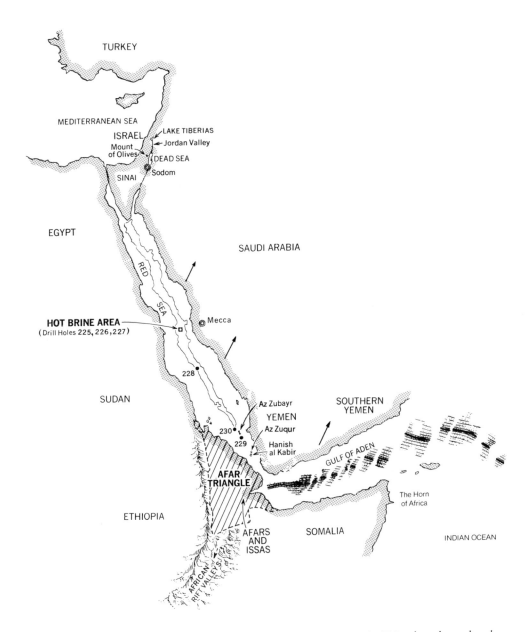

Figure 13.1 Opening of a new sea. *Wegener observed that, if the Afar Triangle at the south end of the Red Sea were disregarded as a recent addition, the two sides of that sea could be fitted together snugly. At one time they were connected; however, the sea is now widening. Evidence includes eruptions of hot, metal-laden brine in the central area of the Red Sea, opposite the sacred city of Mecca. These eruptions are very similar to the upwelling of lava in the midocean ridges. Additionally, in the Gulf of Aden, parallel magnetic bands flank a repeatedly offset rift zone testifying to Arabia's move to the north east. Arabia appears to be sliding along the Dead Sea fault past the ancient sites of Sodom and Gomorrah, opening up the Red Sea and the Gulf of Aden as it moves.*

When Harrison H. Schmitt, the Harvard-trained geologist and scientist-astronaut, looked back at the receding Earth on his way to the moon in 1972, he told the flight controller in Houston: "I didn't grow up with the idea of drifting continents and sea floor spreadings, but I tell you, when you look at the way the pieces of the northeastern portion of the African continent seem to fit together, separated by a narrow gulf, you could almost make a believer of anybody."

In 1930 the French archeologist and theologian, Pierre Teilhard de Chardin, who had explored the region, wrote: "If there is any truth at the basis of the theory of continental rupture, it is in specific and clearly defined regions, such as the south of the Red Sea, that there lies the opportunity to put it to the test and finally accept or definitely reject it."

The first hint that something extraordinary was afoot in the Red Sea had come in the early 1880s when the Russian survey ship *Vityaz* took samples of deep water along its axis. Off the sacred city of Mecca it recovered water from a depth of 600 meters that was considerably warmer than it was at shallower depth or at points farther south and nearer the Equator. Before the century was out this finding had been reinforced by Austrian and German ships that were able to take samples from even deeper water.

Sweden's *Albatross* Expedition again sampled this water in the 1940s, and a decade later the Woods Hole ketch *Atlantis* found that the deep water off Mecca was also unusually salty, even for the Red Sea. The latter, bounded on the east by Saudi Arabia and Yemen and on the west by Ethiopia, the Sudan, and Egypt (the United Arab Republic), has only one link to the world oceans (apart from the Suez Canal) and that is through the Strait of Bab el Mandeb, a narrow and relatively shallow passage at its southern end. Whereas the sea itself reaches depths of 3000 meters, this link to the Indian Ocean is only 120 meters (400 feet) deep. Since no rivers of substance flow into the Red Sea and, under a searing sun, the evaporation rate is high, it has become considerably saltier than the great oceans. Bathers in highly saline lakes, like the Great Salt Lake of Utah, know their extraordinary buoyancy. In the Red Sea this reportedly has made it difficult for submarines to dive.

The high temperature and even greater saltiness of the deep water off Mecca were at first attributed to a process whereby water in shallow areas along the coast became hot in sunlight and salty from evaporation and then, being heavier because of its salinity, slipped down to form a deep layer along the central axis of the sea. However, in 1964 the British research ship *Discovery* extracted water from a deep hole in the centerline trough of the Red Sea whose temperature was 44 degrees Centigrade (111 degrees Fahrenheit). It contained 256 parts of salt per 1000 parts of water (normal sea water contains about 35 parts per 1000, and Red Sea surface water is unusually salty at 38 parts per 1000).

Such a reservoir of hot, salty water could not have been formed by solar heating along the coast, and so this part of the Red Sea became a Mecca for research ships from a number of countries. Measurements in the ocean floor near this hot pool, now called the Discovery Deep, showed the flow of heat from the Earth's interior to be higher than at any known point in the world's oceans (79 microcalories per square centimeter per second). In 1965 the Woods Hole ship *Atlantis II* was hurrying through the Red Sea four days behind schedule on its way to the Indian Ocean, but its crew could not resist collecting at least one sample from the Discovery Deep. Its floor is 2260 meters below the surface, and lowering the sampling device was a slow business. As the scientists did so, they found to their dismay that the ship was drifting from its position over the pool; but luck was with them, for when they finally reached bottom they were over another deep reservoir of hot brine—not quite as deep as that found by the *Discovery*, but even hotter. The temperature of its bottom water layer was 56 degrees Centigrade (133 degrees Fahrenheit), and the black ooze hauled up from the basin floor looked like tar and was too hot to touch. Deeper sediment layers were extracted by coring, and, reported Egon T. Degens and David A. Ross from Woods Hole, "Marine deposits more colorful than the spectrum of blacks, blues, reds, yellows and whites...would be hard to find." The white layers were formed of shell debris from sea life or quartz washed from nearby land. Between them were brightly colored layers of metal compounds—chiefly oxides and sulfides of iron, manganese, zinc, and copper. The water salinity was 257 parts per 1000—only 13 parts per 1000 less than the saltiest water known in nature, at the bottom of the Dead Sea.

This was named the Atlantis II Deep and, when the *Chain* from Woods Hole found a third pool, in 1966, it was called the Chain Deep. All lie in a rather limited area within the trough that runs down the centerline of the Red Sea. The Atlantis II Deep is the largest, with an area of close to 80 square kilometers (30 square miles). Its water forms two distinct layers, each about 25 meters thick, with a transition zone between. The lower layer is the hottest and saltiest and is completely lacking in dissolved oxygen. Hence it is capable of dissolving large amounts of metal without the metal becoming oxidized. Samples of this water contained 5000 times more iron, 25,000 times more manganese, and 30,000 times more lead than ordinary sea water.

The upper layer does contain some oxygen and so, when water from the lower layer diffuses up into the layer above, the iron and manganese tend to oxidize, forming particles that rain onto the basin floor. As they pass through the bottom layer, they are thought to scavenge other metals—copper, lead, and zinc—from the water, and thus is the metallic ooze on the bottom formed.

Twenty months after the discovery of this hot pool a second sampling in the Atlantis II Deep produced temperature readings from the bottom layer that were 0.6 degrees Centigrade (1 degree Fahrenheit) hotter than before. Was the pool really getting hotter, or was this just a measurement error? To find out, further readings were taken by the *Chain* in 1971, and it was found that there had been an enormous invasion of hot water into the pool. The temperature of the lower layer had risen 2.7 degrees Centigrade (4.9 degrees Fahrenheit) in 52 months, and the volume of the lower layer had increased 23 percent. As reported by David Ross in *Science*, an average input of 700 gallons per second over the 52-month period must have occurred. This is roughly 200 times the flow of hot water from Old Faithful Geyser in Yellowstone National Park. During a 20-day period in March 1971, an even higher rate of temperature increase was recorded. Furthermore, the Woods Hole group reported in *Nature*, the successive samplings in 1965, 1966, and 1971 had shown a steady enrichment of the water in both dissolved iron and manganese. One sediment sample from the Red Sea deeps contained 49 percent iron oxide, and in another there was 11 percent zinc oxide and 0.01 percent silver oxide.

It was suspected that the Atlantis II Deep is the source of the brine, which then overflows into the neighboring pools. But where does it come from within the sea floor? Volcanoes erupt great volumes of water, extracted from the Earth's mantle and belched out as steam. Most of the water on the Earth's surface, including that filling the oceans, is thought to have come forth in this manner. But the relative abundances of various salts in the Red Sea brines persuaded those who analyzed them that the water originally came from the sea, worked its way down as much as one or two kilometers through sea-floor sediments or rifts in the hot rock that is forcing its way up into the central zone of the Red Sea, and then percolated back to erupt, from its own heat and pressure, into the deep pools.

The journey of this water through the sea floor may take centuries, traversing considerable distances, and, as the water flows, it leaches soluble metals from the sediment or rock (the latter presumably of basic, upper mantle composition). Probably only a small fraction of this metal-rich water erupts back into the sea. Some believe the rest forces its way into cracks in the sea-floor rock. There, as it gradually cools, veins of iron, copper, gold, silver, zinc, lead, vanadium, molybdenum, and manganese are formed. Most of the world's ore bodies may have been formed by such hot-water invasions of rock, including the Mother Lode gold of California and the silver and copper of Cerro de Pasco in Peru.

Hot springs on the continents produce miniature deposits of this sort, and, until recently, it was assumed that ore bodies were generated chiefly as a sequel to the invasion of continental formations by hot rock, or magma, from below. In this process (which clearly does account for some ore deposits), such a great body of magma, once it comes to rest, cools slowly from the outside in. As it does so, liquid and gaseous components, driven from the outer part by the hardening process, work their way inward until their pressure becomes so great that they rupture the outer rock and force their way back outward through the resulting cracks. In doing so these substances, typically in hot-water solution, enter areas that are progressively cooler, and various components precipitate out of solution and are deposited. Because the metal compounds do so at different temperatures, this process segregates them into veins of various ores.

The possibility of another hot-water process of ore formation has been suggested by the metal-rich water that comes from steam wells drilled beneath the Salton Sea of southern California. That sea, today no more than a shallow, salty lake some 30 miles long, has features in common with the Afar Triangle. The lake lies well below sea level—as do some of the salty Afar lakes—and is astride the area where the rift and fault system of the East Pacific Rise penetrates the North American continent (as the Red Sea and Gulf of Aden rift zones apparently extend under the Afar Triangle). The heat flow up through the floor of the Salton Sea is so great that wells have been drilled into its deep accumulation of sediment as a source of steam. The salinity of water from these wells is comparable to that within the Red Sea deeps, and the water contains appreciable amounts of copper, lead, zinc, and silver. It has been assumed that these metals were leached from the mineral grains of the sediment, rather than from basic rock welling up from the Earth's upper mantle, and some have suggested the Red Sea deposits might also have arisen in this manner. But the evidence from deep-sea drilling that metal-rich layers lie at the bottom of the sediments across much of the Atlantic and Pacific implies continuous extraction from the mantle beneath mid-ocean ridges.

In 1969 a three-day symposium on the Red Sea organized by the Royal Society in London brought together scientists from a number of lands who had been exploring the region, and, at the close, K. C. Dunham of London's Institute of Geological Sciences, summing up the new discoveries, said that the minerals in and under the brine pools, "though geologically small in scale are already causing a revolution in thought about ore genesis."

While geologists, hitherto, have been able only to speculate on how the so-called hydrothermal ore deposits were formed, millions of years ago, the process now seems to be taking place right in front of them—in front of their eyes, that is, if they are brave enough to venture in person down into submarine canyons where it is thought to occur. It was this prospect that persuaded American and French scientists to board deep submergence craft and descend into the canyons of the Mid-Atlantic Ridge where, they suspected, such activity might be going on. Distinctive "earthquake swarms" that occur in quick succession at or near particular spots in the Mid-Atlantic Rift Valley, with no single outstanding event, were suspected as evidence of either the geysering of metal-rich hot water or lava eruptions. The dives confirmed these suspicions and, when later made along the East Pacific Rise and other Pacific sites, brought to light a whole new world of specialized undersea life, as will be described later.

Although the Red Sea pools are limited in extent, the richness of their deposits is impressive. It was estimated in 1969 that within the Atlantis II Deep (an area roughly half that of the District of Columbia), the gold, silver, copper, lead, and zinc in the top 10 meters of sediment—the maximum depth then sampled—after recovery would be worth about $2.4 billion, based on 1968 smelter prices. A consortium of German and American firms began to explore the possibility of dredging this deposit—an enterprise that, however, would raise complex questions of international law, for the area lies almost equidistant from the Saudi Arabian and Sudanese shores of the Red Sea.

Additional brine pools have now been discovered along the axis of the Red Sea. One was sampled in 1971 a short distance northeast of the Atlantis II Deep. The following year the German research ship *Valdivia,* under contract to the Preussag Mining Company, found others, using a narrow-beam echo-sounder to detect reflections from the dense brine, and at almost the same time one was found by the *Glomar Challenger.*

Because of the surge of interest in the Red Sea brines, the ship had been assigned a series of drill sites along the centerline of that sea. Although normally the crew would not attempt to "spud in" unless there were at least 100 meters of soft sediment to stabilize the drill before it hit hard rock, in this case, because of the special importance of these sites, the rule was waived and, with great care, they nursed the bit into the bottom.

One hole was sunk into the central floor of the Atlantis Deep, but it soon met a layer of basalt that prevented further penetration. The coring tube brought up 14 meters of metal-rich sediments similar to those reported by earlier expeditions. At other sites, where greater depth was reached, what proved to be a widely occurring layer of shale rich in vanadium and molybdenum was found, plus another containing five percent zinc (by weight) and one enriched in copper. The basalt that was sampled was no richer in metals than mid-ocean basalts drilled elsewhere, and the question of precisely how the metal deposits were formed remains open. But their occurrence in some of the earliest sediment laid down after eruption of new sea floor along mid-ocean ridges is so widely observed that there seems no doubt they are a by-product of the rifting and spreading process.

Piston coring of the Red Sea has disclosed layers rich in iron and copper that, being buried under normal oceanic sediment, seem to represent earlier metal-rich eruptions during the time since the Red Sea began growing, some 20 million years ago.

Such activity may have contributed to the earliest efforts at metal-working, for the ancient Egyptians discovered copper near the shores of the Red Sea and its northwest arm, the Gulf of Suez—that is, on the Sinai Peninsula and in the eastern deserts of Egypt itself. More recently, at least seven lead–zinc deposits have been found along the Red Sea coast of Egypt, their alignment being strikingly parallel to the rift valley of that sea.

This led geologists of the Saudi Arabian government to look along the opposite side of the Red Sea, and in 1968 they found several occurrences of lead, copper, and zinc directly across from those on the Egyptian shore, as though both had been formed at the same time, along the central rift, and then carried apart by spreading of the Red Sea floor. "On the two sides of the Red Sea," said their report, "the occurrences—even the mineralized bodies themselves—are aligned...NNW–SSE, which is the direction of the great fractures which created the depression."

One of the *Glomar Challenger*'s tasks in the Red Sea was to sample a hard layer, deep beneath the soft sediments, that sharply reflected sound waves. It had been detected over much of the ocean—though not beneath the central trough. At the first drill site, 16 kilometers (10 miles) east of the Atlantis II Deep, this layer was reached beneath 177 meters of sediment. It proved to be salt, including layers of anhydrite and rock salt. Interbedded within this formation were shale layers whose fossils dated from the Miocene, between five and 20 million years ago.

It thus appears that both the Red Sea and Mediterranean repeatedly dried up during the same period, building up successive layers of salt (from the desiccation) and shale (from the periods of flooding). Like Gibraltar, the Strait of Bab el Mandeb is a narrow, potential floodgate. It is a tradition of the Somalis that their ancestors crossed the strait, from Arabia to Africa, when there was a land bridge there.

The fact that this ancient salt deposit flanks the central trough of the Red Sea has been taken as evidence that there was a prolonged intermission in the opening of that sea which finally ended with a resumption of spreading relatively recently. A drill hole on the slope of the Atlantis II Deep, within two or three kilometers (one or two miles) of the supposedly active spreading center, brought up rocks at least as old (about five million years) as Late Miocene. While this material might have slipped down the slope from a greater distance off the centerline, the research team aboard the drill ship concluded that the spreading probably did not resume until less than two million years ago, following a hiatus that may have extended back into the Early Miocene, more than 15 million years ago (the drill sites are shown in Fig. 13.1).

It is now being argued widely that the Red Sea is entirely a product of sea-floor spreading, although some still maintain that, except for the central trough, its floor is formed of continental crust. Those who believe the entire floor is oceanic say the regions flanking the central rift lie at moderate depths beneath the sea, not because they are continental, but because they are blanketed with enormously thick sediments—presumably to a large extent evaporites. The brine pools apparently occur where the uppermost of these salt layers has been exposed to sea water, perhaps by rifting along the centerline.

At the London symposium, A. B. Frazier of the Gulf Oil Company disclosed some of the carefully guarded findings of oil prospectors in the area, particularly in the Dahlak Salt Basin that lies on the Red Sea floor just north of the Afar Triangle. Several wildcat wells have been sunk more than three kilometers (1.8 miles) into this formation and the seismic work has shown parts of the Red Sea accumulation to be five kilometers (three miles) deep.

At one of the *Glomar Challenger*'s final drill sites in the Red Sea, near the southern extremity of its rift system, the cores from below 200 meters became increasingly gassy, forcing a halt in the drilling and reminding the drillers that this sea may, like its twin, the Persian Gulf, harbor important reserves of gas and oil. However, after the ship passed through the Strait of Bab el Mandeb into the Gulf of Aden, drilling there failed to disclose any such salt deposits. Apparently the gulf has never been closed and desiccated.

Since no salt layer was evident in echo-soundings of the central trough of the Red Sea, it would appear that it has not dried up in the last one or two million years. Thus, while no evidence was found for a removal (or parting) of Red Sea waters, as described in *Exodus*, it appears that man's primitive ancestors could, on occasion, have walked dryshod across that sea, then a plain of scorching salt flats.

At the London meeting it was noted by Israeli scientists that the authors of the Bible were aware of the rifting that is constantly altering the region, particularly along the great valley that, as an apparent continuation of the Red Sea trough, extends from the Gulf of Aqaba, at the top of the Red Sea, north along the Dead Sea, Jordan Valley, and Lake Tiberias (the Sea of Galilee). It has long been suspected that the east, or Jordanian, side of this valley is slipping north relative to the west, or Israeli, side; and the Israelis have sought out geologic matches, on either side of the rift, that might provide clues to the extent and rate of this slippage. They have concluded that, since the Cretaceous Period, 70 million years ago, it has amounted to 105 kilometers (70 miles) with a slippage of 40 kilometers (25 miles) in the past 10 million years.

On the side of Jerusalem facing this rift rises the Mount of Olives. According to a prophesy in the *Book of Zechariah*, "the Mount of Olives shall cleave in the midst thereof toward the east and toward the west, and there shall be a very great valley; and half of the mountain shall remove toward the north, and half of it toward the south." Indeed, from the Old Testament, it is clear that the Biblical world was all too familiar with such rifting and its accompaniments: In *Psalm 114*, "The mountains skipped like rams, and the little hills like lambs." And in *Genesis*:

Then the Lord rained upon Sodom and upon Gomorrah brimstone and fire from the Lord out of heaven;

And he overthrew those cities, and all the plain, and all the inhabitants of the cities, and that which grew upon the ground...

And he [Abraham] looked toward Sodom and Gomorrah and toward all the land of the plain, and beheld, and, lo, the smoke of the country went up as the smoke of a furnace.

It is believed that the plain where those ill-fated cities stood lay at the southern end of the Dead Sea and thus directly in the rift zone. The Biblical description could very well apply to what happened, in 1783, to communities astride another rift zone—in Iceland—as will be seen in the next chapter.

The slippage documented by the Israelis fits into the picture of plate motions in the Red Sea area deduced from other evidence. For example, a northward slippage of the Arabian plate, from the Indian Ocean to the Jordan Valley, is indicated by the transform faults that have been mapped beneath the Gulf of Aden. These faults repeatedly offset the ridge-and-rift system that passes from the Indian Ocean through the Gulf of Aden to join the Red Sea and East African rifts in the Afar Triangle (see Fig. 13.1). These offset (or "transform") faults are evident in three ways: in soundings as sea-floor features, from the pinpointing of earthquakes that occur along them from time to time, and from the magnetic patterns on the sea floor. The faults all run from southwest to northeast, implying that this is the direction in which the sea floor is spreading obliquely from the gulf centerline.

Such motion would mean the Arabian coast of the Red Sea has also slipped northeast relative to the African shore and, indeed, it is only when this motion is reversed that the two coastlines slip into a snug fit. This also brings the geology of the coasts into conformity. On the Arabian side (particularly in Yemen), as well as on the African side, there are great flood basalts that presumably erupted when the sides were joined and first began rifting.

The special importance of the Afar Triangle, recognized originally by Wegener on geographical grounds, has been greatly enhanced by the discovery that three rift systems converge there—from the Red Sea, from the Indian Ocean (via the Gulf of Aden), and from East Africa. It thus has become a critical area in seeking to learn the nature of rifting and of oceanic birth.

While much of it is low and even below sea level, there are three blocks of high ground along the seacoast, the largest being known as the Danakil Alps. Being separated from the Ethiopian block, to the west, by the Afar depression and from the Arabian block, to the east, by the southern extremity of the Red Sea, they may be slivers of continental rock left behind as Africa and Arabia drew apart. The Afar depression is bounded on the west by the Ethiopian escarpment, 600 kilometers (375 miles) long and in places more than 4000 meters high. The volcanologist Haroun Tazieff, who knows it well, has said of this mighty wall: "Here clifftops stand higher above the valley floor below them than anywhere else in the world." On the south the Triangle is bounded by an escarpment that rises as high as 1500 meters and continues east along the south shore of the Gulf of Aden to form a promontory known as the Horn of Africa.

The three rifts that converge in the Triangle are quite different from one another. That from the Indian Ocean is a continuation of the one that bisects the Carlsberg Ridge (where the magnetic evidence for sea-floor spreading was first recognized) and, with its marked offsets, is typical of the oceanic ridge-and-rift systems. The Red Sea trough lacks such dramatic offsets, although there are hints of a few small ones. There is no evidence of a ridge, perhaps because that sea is still too narrow for a ridge to be evident, particularly under the heavy blanket of sediment on both sides of the trough. A peculiar feature of this rift is that it seems to jump 200 kilometers (125 miles) to the west, at its southern end, to continue down the axis of the Afar Triangle; yet to date no evidence of a transform fault to account for this offset has been found—a subject currently much in dispute.

The African rift valleys run virtually the full length of East Africa from Lake Nyasa (Lake Malawi) north through a succession of long narrow lakes and deep cliff-walled valleys. A curving, western branch runs through Lake Tanganyika, itself more than 640 kilometers (400 miles) long. The other branch, farther east, passes through Tanzania and Kenya, flanked by such massive by-products of the rifting as the snow-capped volcanoes of Kilimanjaro and Mount Kenya and the tempestuous crater of Ngorongoro. Close by the latter, one of the best-known manifestations of the rift system is Olduvai Gorge, where Louis S. B. Leakey found some of the earliest relics of man's ancestry. Lakes along the rift in this sector tend to be saturated with sodium carbonate, and it makes bitter the waters of Lake Abbe in the Afar Triangle, directly where the three rift systems seem to converge.

To J. W. Gregory, the British explorer who described the rifts in the last century, they formed a continuous feature from the heart of Africa through the Red Sea and Jordan Valley to the Middle East. "From Lebanon, then, almost to the Cape," he said, "there runs a deep and comparatively narrow valley, margined by almost vertical sides, and occupied either by the sea, by salt steppes and old lake basins, and by a series of over twenty lakes, of which only one has an outlet to the sea. This is a condition of things absolutely unlike anything else on the surface of the Earth."

With today's knowledge of the global network of oceanic rifts the African valleys are viewed in a different perspective. They continue to be active, not only in terms of earthquakes, but also through the continuous opening of rifts and cracks parallel to the valley. In the Ethiopian sector, for example, cracks sometimes large enough to accommodate a man opened in 1956, 1966, 1969, and 1970. In the rainy season they are soon filled by collapsed soil, but during a long dry spell they may bridge over and mimic the crevasses of Antarctica as a hazard. Thus, according to a report from Haile Selassie I University in Addis Ababa, during the 1969 dry season a farmer, while tilling a field, saw his ox and plow vanish into a hidden fracture.

Although early geological exploration of the Afar, in the 1920s, was done by camelback, today the scientists use helicopters and Land Rovers. But it is still an enterprise that attracts the adventurous. Most colorful among them is the "fire-eating" volcanologist, Haroun Tazieff, who descends into craters to capture samples of the erupting gases and who, in the 1960s and 1970s, led at least five expeditions there, supported by the national research agencies of France and Italy—the Centre National de la Recherche Scientifique (with which he is associated) and the Consiglio Nationale delle Richerche (see Fig. 13.2).

Then there is Paul A. Mohr of the Smithsonian Astrophysical Observatory in Cambridge, Massachusetts, who was drawn to the area by an advertisement sent to him in jest. "I relinquished a career as an R.A.F. jet-jockey," he reports, "when my father sent me, jokingly, a copy of an ad for a teaching post in Addis Ababa." He decided to go, "knowing absolutely nothing about rift valleys—never even heard of them." (He is now an authority on that heavily rifted region.) None of the efforts of recent years, Mohr adds, would have been successful without the guidance and diplomatic support

Figure 13.2 Arrole Volcano *rises from a barren landscape typical of the Afar region.*

of Pierre Gouin, professor of geophysics at Addis Ababa, who himself did much of the early gravity and seismic surveying of the region. The cooperation of Ethiopia, of which Addis Ababa is the capital, was essential since the entire triangle is controlled by that government except for two small areas on the Gulf of Aden—the French Territory of the Afars and Issas and a fragment of the Somali Republic.

The Afar Triangle, Tazieff has written, "is one of the world's most forbidding regions. In addition to the fact that its terrain is all but impassable, the area is extremely hot; we were to find that the temperature rises to as high as 134 degrees Fahrenheit in the shade in summer and 123 degrees in winter (57 degrees and 51 degrees Centigrade). The region is inhabited only by nomadic tribes of fierce repute; the young warriors are said to mutilate male victims to offer trophies to their women, and they have been known to massacre armed parties for their weapons. Several exploring parties in the 19th century were slaughtered by the tribesmen."

More recent explorers have found the natives furtive, rather than savage. However, Tazieff, whose expeditions, he says, go unarmed, reports that in 1970–1971 he had to revise the expedition plans because of "insecurity" in the north of the country. Mohr tells of "being stroked by native Gallas who had heard of, but never seen a white man ('you must be sick,' they said, seeing the pallor of our skin)." Mohr's frights have come from causes other than native hostility, such as a Land Rover breakdown that threatened to strand him in the midst of a torrid desert. Fortunately, he reported, "from that molten sky, a black cloud gathered, headed our way and poured delicious cool rain, enough to cool the vehicle so we could touch it and tinker." Another time, after a hot day's work mapping the swarms of dikes, or wall-like intrusions of lava that are common in that region, he was imbibing "evening dew" in a corrugated-iron shack set up on the lonely desert, presumably many long miles from any surprise visitor. "I was tapped on the shoulder and turned—to meet the wicked gaze of a thirsty ostrich." It turned out, he said, to have been a local pet, "but I leapt."

The Afar landscape, some of it more than 180 meters below sea level and utterly arid, is wild scenery shaped by savage events still in progress. Its central trough, aligned parallel to that of the Red Sea, is cut by swarms of narrow rifts parallel to its axis as though the trough is being torn wider and wider. Such swarms of narrow cracks also occur in Iceland, where they are known as gja (pronounced "gyaw")—one of several respects in which the features of Afar resemble those of the island country astride the Mid-Atlantic Ridge.

Apparently, such rifting extends deep enough to permit molten basalt from below to well up into the cracks, forming the dikes of hard rock. (In some areas of the world, notably Scotland, softer rock has eroded away from the sides of such intrusions, leaving them standing in the open as walls. Hence the term "dike.") Because the rifting seems to come first and the basalt intrusion later, some, including Paul Mohr, believe the intrusions are not responsible for pushing the crustal plates away from this region. It is the other way around, they say. Something is dragging the plates aside, forming cracks into which the basalt then flows.

Along the Afar trough there are boiling springs and a chain of seven volcanoes, all of them active, either in periodic full eruptions, or in gas-breathing fumaroles, and the resulting lava has overflowed the trough. The most active crater is that of Erta Ali, which put on an impressive display in 1972, described by geologists from Haile Selassie I University. Its incandescent lava lakes, as seen from the air, glowed brightly even in full sunlight. "Fountaining activity was intense, and projections were thrown to a height of up to ten metres," they reported. "Samples of Pele's hair were collected near the central pit, and the flow was sampled."

Pele is the goddess of fire in Hawaiian legend and her "hair," formed of threads of volcanic glass, solidifies from the lava fountains of Kilauea, the crater on the flank of Mauna Loa, the great Hawaiian "shield" volcano that, despite its gently-sloping, shieldlike profile, rises almost 10,300 meters from the ocean floor—with its companion, Mauna Kea, it forms the greatest protuberance on the Earth's surface. While volcanoes of this type are typical of mid-ocean spreading zones, as in Iceland, significantly they occur also along the African rifts, including, as noted, Kilimanjaro and Mount Kenya, as well as Mount Erta Ali in Afar. While the base of the latter is 50 kilometers (30 miles) wide, its height is less than 520 meters. Circumpacific volcanoes erupt through a circular crater and, if they crack open, it is along lines radiating from that crater. But the eruptions of these volcanoes come from riftlike apertures, and those of Erta Ali are aligned with the Red Sea axis. In fact many of the past Afar eruptions have occurred along such rifts instead of from a central crater, flooding the landscape with lava in a manner also typical of Iceland. Tazieff and his colleagues have analyzed more than 100 samples of rock from volcanoes of the Erta Ali chain and found them to be predominantly basalt of a typically mid-ocean character.

At the southern end of the range is Lake Giulietti whose eastern shore is bounded by a "staircase landscape" of giant steps, ranging in height from 30 to 100 meters, that mark one side of the Afar trough. Similar step-blocks, formed apparently of large, tilted blocks leaning away from the central rift, constitute a dramatic feature of the central rift valleys of Iceland, and flank much of the Mid-Atlantic Ridge. They are particularly evident in detailed soundings of the rift valley southwest of the Azores into which American and French scientists have descended in deep submersibles. (In fact, just as the Apollo astronauts, as training for their exploration of the moon, were taken to grim, volcanic regions of the United States thought to resemble the lunar surface, so it was decided to expose the diving crews to the Afar, Hawaiian, and Icelandic lava fields before they ventured into the unknown depths of the mid-ocean canyons.)

North of the Afar volcanic range the depression is filled with evaporites and there are, in fact, several kinds of evidence that only a few tens of thousands of years ago this region was a part of the Red Sea floor. Farther south, now well above sea level, Tazieff and his co-worker, Enrico Bonatti of the University of Miami, believe they have identified a guyot—one of those submarine mountains with a flat-topped, volcanic profile first identified by Harry Hess in the Pacific—that was formed as a submarine volcano and then lifted from the sea. Others, however, have questioned this interpretation of the feature. Although the region still seems to be rising, in the long run it would appear doomed to sink again beneath the sea, as presumably happens to all such crusts generated along oceanic ridges to have their day in the sun and then submerge. "The present absence of water in the Afar triangle," reported Tazieff, "is only a temporary phase in the development of the ocean." Although some believe the Triangle may ultimately provide the key to the nature of the deep forces tearing that region apart—and hence to the basic driving force of plate movement—that key remains to be found and there is, as yet, no agreement on the relationship among the three rift systems that converge there from the Red Sea, Gulf of Aden, and East Africa.

One of the peculiarities of the northern part of the African rift, as it passes near Addis Ababa in Ethiopia, is the nature of the silica-rich rock that has erupted from the rift. Here again there is a similarity to Iceland and other outcroppings of mid-ocean rift material, but the richness of the Ethiopian rock in iron and sodium has perplexed specialists in lava chemistry. Some argued that the lava must have picked up these elements on its passage through continental crust beneath the rift, but this seems less plausible in the light of evidence that at least some of the African rift valleys represent fractures that penetrate the entire crust. Mohr has pointed out that the process that has enriched the Red Sea brines with iron could also account for the rift valley enrichment—a hint of continuity between the Red Sea trough and the African rifts.

In an effort to determine whether or not there is continental crust beneath some of the African rifts, scientists from the Universities of Birmingham, Leicester, and Lancaster in Britain have set off a series of underwater explosions with charges ranging up to 1400 kilograms (3000 pounds) on the floors of lakes Rudolph and Hannington, within the sector in Kenya known as the Gregory Rift. Such charges are fired under water to achieve a better coupling of the shock waves to the Earth's crust. The tremors were recorded at 10 sites scattered along 270 kilometers (170 miles) of the rift. While the rumbling of hot springs disrupted some of the observations, the results revealed a layer as shallow as 20 kilometers (12 miles) deep with "very high" velocities for both shear and compressional waves.

The implication was that the upper mantle material had penetrated halfway up into the crust. "A very similar structure has been found in Iceland," said the 1971 report, "so that the resemblance of the East Africa Rift system to the midocean rift seems to be more than superficial." However, the authors added cautiously, "Further work is required to determine the more subtle differences that may exist between the two structures, and perhaps to make it possible to decide whether the East African system should be regarded as the first stage in the formation of a new ocean basin, or as a spreading centre which is already reverting to a less active and anomalous state."

Critical to assessment of the Red Sea and African rifts as a replay of the Atlantic Ocean birth is the timetable of their rupture. In an analysis of the evidence, Mohr has proposed that eruptions along the incipient rifts began flooding adjacent areas of Ethiopia, Somalia, and Yemen as early as the Eocene, 50 million years ago. This continued for some 30 million years, building up the plateaus and highlands that now flank the Red Sea, Afar Triangle, and Ethiopian rift valley. Some of these basalt eruptions apparently originated as deep as 65 kilometers (40 miles) below the surface.

While the original opening of the Red Sea came earlier, the rifting within Africa—and within the central trough of the Red Sea—did not begin until the eve of the ice age epoch, or Pleistocene, a few million years ago. This timetable has been deduced by British researchers, using the decay of radioactive potassium into argon as an indicator of lava-flow ages. As noted earlier, a similar timetable has been derived from studies in the Red Sea itself. But the link between spreading of that sea and of the African rifts remains puzzling. Within the Afar Triangle, where the two systems seem to meet, the surface rifts, clearly visible in a photograph taken from Earth orbit by the Apollo 9 astronauts, suggest that the spreading pattern curves to link the Red Sea axis with that of the Gulf of Aden, rather than joining the African rifts. This is also indicated by gja-type fissures in the islands on the Red Sea centerline that hint at the birth of a new ocean ridge. Those islands abreast of the northern tip of the Triangle, Jabal at Tayr and Az Zubayr, are fractured parallel to the Red Sea axis (see Fig. 13.1), but those farther south, nearer the Gulf of Aden (Az Zuqur and Hanish al Kabir), are rifted in directions more parallel to those of the Gulf, as though the whole rifting system were turning a corner.

While the East African rifting is widely regarded as an example of the initial rupture of a continent, it has progressed very slowly (an effect that may, in fact, be typical of the early stages of such a continental rupture). It seems to have proceeded at about one tenth the rate of movement evident on the ocean floors. The African rifts, allowing for erosion, slumping and other effects, have spread no more than 15 to 25 kilometers (10 to 15 miles) in the past 15 or 20 million years. Instead of being offset by transform faults, like oceanic rifts, the African ones—notably the western branch that forms a boundary between the East African states and the Republic of the Congo—are curved like the great rift occupied by Lake Baikal in Siberia. The latter is so long and deep that it contains roughly one fifth of the world's lake water. It is thought to be only 10 million years old. The rift, or "graben," within which the Rhine River lies for a stretch of 290 kilometers (180 miles) between Basel in Switzerland and the castle-studded mountains below Mainz is estimated to be 30 million years old and is about 40 kilometers (25 miles) wide.

It may be that these continental rifts are basically different from those in the oceans. Although some, such as the ones in Ethiopia, seem clearly to mark the edges of moving crustal plates, others do not penetrate the entire crust. Their opening rate is slow, and the process seems to have been sporadic—as opposed to the steady flow of ocean floor away from the ridges. They tend to rest on top of great blisters, or humps, in the crust of the Earth, underlain by a high-temperature "rift cushion" formed, apparently, of mantle material that has been transformed to a less dense phase and, consequently, has expanded. It occasionally erupts through the blister as lava. These humps may be as much as 2000 meters high and from 200 kilometers (125 miles) wide (for the Rhine Graben) to 2000 kilometers (1250 miles) wide (in Ethiopia). This humping may crack open the crust, but, it is widely agreed, not enough to account for the rifts. However, M. J. Le Bas of the University of Leicester in Britain has proposed that such humping and rifting generated a series of weak spots along the line where Africa and South America finally broke apart, determining the S-shaped course of that rupture. It also may be that continental rifting differs from that in the oceans simply because the response of a continental crust to spreading beneath it is markedly unlike that of an oceanic crust.

An unexpected result of the 1972 mission of Mariner 9 to Mars was that its mapping photography, augmented subsequently by the far more detailed views from the Viking orbiters, disclosed an enormous rift valley 120 kilometers (75 miles) wide and some 6000 meters deep, running for 4000 kilometers (2500 miles) along the Martian equator. On Earth such a canyon would span the entire United States. The discovery has aroused hopes that, from closer study of Mars, we may learn more about the processes that have shaped our own planet.

The dating of the East African and Red Sea rifting events, coupled with recent discoveries of prehuman remains within volcanic ash deposits laid down some 2.6 million years ago, indicates that some of man's immediate ancestors witnessed—and apparently were overwhelmed by—volcanic activity typical of an ocean's birth.

In 1971 Glynn Ll. Isaac of the University of California at Berkeley, Richard E. F. Leakey of the National Museum at Nairobi, Kenya (son of the famous fossil-hunter), and Anna K. Behrensmeyer of Harvard reported finding more than 20 hominid fossils, with stone tools and hippopotamus bones, along the eastern shore of Lake Rudolf in the Gregory Rift. They were embedded in tuff, or solidified volcanic ash, whose age, from the potassium–argon method, appears to be 2.61 million years, with an error margin of 0.26 million years. They described it as possibly the oldest hominid occupation site yet found.

Were the events upon which those ancient ape-men gazed typical of an oceanic birth? In geophysical detail, perhaps not. But from what we have learned in the past few years—from deep-sea drilling, and from many other forms of oceanic exploration—it would appear that the outpourings of lava, the tremendous earthquakes, and the mountains belching steam and ash that finally buried the remains of these hominids and their tools of hewn stone, were a reasonable facsimile of the process that first began to emplace an ocean between Spanish Sahara and New York.

Chapter 14

Of Plumes, Eruptions, and Mid-Ocean Islands

IF THE AFAR REGION RESEMBLES THE CLASSIC IMAGE OF HADES, ICELAND, AS AN example of the mid-ocean rifting process, boasts what its inhabitants once regarded as the entrance to the netherworld—not without some justification. This is the volcano Hekla, that erupted as recently as 1970. And it was through another Icelandic crater, that of the Snaeffelsjökull (believed to have last erupted some 1500 years ago) that Professor Liedenbrock and his nephew descended to explore the Earth's interior in Jules Verne's *Journey to the Center of the Earth.* The professor knew the warnings of skeptics that they might encounter extremely high temperatures. But, he said, reasoning based on superficial information had in the past proved erroneous. "Neither you nor anyone else knows for certain," he said, "what goes on in the interior of the globe considering that we are familiar with scarcely a twelve-thousandth-part of its radius. My answer is that science can always be improved and that every new theory is always overthrown by a newer one. But we shall see for ourselves..." They did, ending their journey by being spewed forth from Mount Stromboli—another volcano—in the Mediterranean. The professor's fictitious comment on the fallibility of scientific theory is, of course, as valid today as it was when written by Verne more than a century ago.

In a sense Iceland is the most volcanic place on Earth—at least above water. Its eruptions are frequent and voluminous and, in fact, it is formed almost entirely of volcanic debris—chiefly great sheets of flood basalt that have welled up from long fissures in the Earth. The greatest such fissure eruption ever witnessed by modern man (and one evocative of the Biblical account of Sodom and Gomorrah) occurred in Iceland on June 8, 1783, when lava began pouring from a 24 kilometer (15 mile) rift near the Skaftarjökull. During the two years that followed this event, known also as the Laki eruption, lava spread over some 580 square kilometers (220 square miles). Volcanic ash was thrown as high as the stratosphere and spread as far as Siberia, Africa, and North America. In Scotland the ash fall was sufficient to destroy crops. In Iceland itself the ash and poisonous fumes were disastrous. The meager arable land was carpeted: 80 percent of the sheep (190,000) died, as did 75 percent of the horses (28,000), and more than half the cattle (11,500). Consequently about 10,000 people perished—20 percent of the island's population—victims of starvation, disease, or more direct effects, such as burial by lava.

Probably the event most dreaded of all by the hardy Icelanders is a *jökulhlaup*. A jökull (pronounced "yuckull" to rhyme with "turk-ool") is an icecap or ice-capped mountain, and a *jökulhlaup* is the bursting of such a cap after a volcanic eruption under it has melted much of the ice. The dammed-up water accumulates in great volume until finally the ice ruptures, and what follows is like the collapse of a dam impounding a lake filled with icebergs. Sometimes the volume of water melted before the rupture is so great that, it has been estimated, the resulting *jökulhlaup*, for a time, discharges as much as does the Amazon River.

Like the volcanoes of Afar, the Icelandic craters tend to be linear—long slits, rather than circular—and the landscape abounds with them, young and old. Most current activity is concentrated in a central rift zone that bisects the island in a northeast–southwest direction. Its southwest portion splits in two, part of it being offset to the northwest in a manner suggestive of the transform faults so typical of undersea portions of the mid-ocean ridges (see Fig. 6.6). Within the great trough of this branch of the zone, some 20 miles east of Reykjavik, are some of the most dramatic exposures of mid-ocean rifting—long tension rifts, where the crust has been stretched, producing fault scarps and down-dropped, tilted blocks, all parallel to the strike of the Mid-Atlantic Ridge in this area. Flanking the trough are flat-topped mountains formed of lava flows heaped one upon the other in a wide range of colors—black, purple, yellow, copper, and bronze. This natural amphitheater, known as Thingvellir, is said to have been the meeting place, in A.D. 930, of the world's first parliament, the Icelandic Althing. The unusual rock structures offered acoustical advantages; the site was central to the island; and, on the valley floor, there was pasture for ponies of the assembled delegates. Taking into account the known spreading rate in Iceland, it has been calculated that this valley is now several meters wider than when the Althing first met there.

The southwestern spur of the rift zone extends out to sea as the Reykjanes Peninsula, site of the seaport Keflavik. The peninsula continues under water as the Reykjanes Ridge, poking up here and there in a few volcanic islets. It was the aerial magnetic survey of this section of the Mid-Atlantic Ridge that showed it to be flanked by magnetic bands of extraordinary symmetry.

The southern spur of the rift zone is the site of Hekla, the traditional gate to hell. Where this zone extends out under the ocean there is a chain of 14 islands, large and small, following the northeast–southwest trend of the Mid-Atlantic Ridge in this region. The largest is Heimaey, eight kilometers (five miles) long, dominated by the volcanic summit of Helgafell. Geologists had determined that Helgafell had been dormant for several thousand years, and the island's seaport of Vestmannaeyjar, lying close to Iceland's rich fishing grounds, by 1973 had become the chief pillar of that country's fishing industry. The town's fleet of 80 boats hauled in one fifth of all fish caught by Icelanders, and the fish processed by the town's four freezing plants accounted for 12 percent of the country's exports.

In 1963 the crew of a fishing boat discovered that the sea had begun to boil beyond the southwest extremity of the island chain. Within hours lava and steam were erupting in the birth agonies of the island now known as Surtsey. Soon, in what could be regarded as a rerun of the birth of Iceland itself, the eruptions spread to form a second islet, Syrtlingur, and then a third one—Jölnir. After four years this activity ceased and there followed a lapse of almost six years until, on the evening of January 21, 1973, seismometers on the mainland of Iceland began to register small earthquakes whose origin seemed to lie just south of Heimaey. They were so weak that they were unobserved by residents of that island.

The following evening a new round of quakes began and at least three of them were felt by the islanders. Subsequent analysis showed that they were shallower than the first batch and directly under Heimaey. Shortly before 1 a.m. Icelandic time the next morning (January 23), a fissure began to open on the slope of Helgafell above the town. According to a report prepared a few days later by a group from the University of Iceland:

> About 5 min before the eruption started there were no signs of eruption here. Within a few minutes of the beginning of the eruption lava fountains were playing on a fissure 300 to 400 m long...It was glowing along its entire length and the height of the glowing fountains was 50 to 150 m. The direction of the fissure is north–north–east to south–south–west.
>
> As the most north–north–easterly part of the fissure runs only 200 to 300 m from the easternmost part of the town, evacuation of the inhabitants started immediately.

Firemen went through the streets rousing the population and urging them to gather on the docks. Some, however, were already up. Unnur Thorbjörnsdottir, who lived on the outskirts of town, near the volcano, had been awakened by her mother. "When I looked out the window," Mrs. Thorbjörnsdottir told Richard Eder of *The New York Times*, "the grass was burning. We got dressed and when we got outside the ground was hot as if it were on fire." As the citizens hurried down the streets, carrying hastily packed suitcases, more than 100 fishing boats, yachts, and Coast Guard vessels were mobilized in an operation reminiscent of the British evacuation of Dunkirk in World War II, and an airlift was launched from the airfields at Reykjavik and Keflavik.

Volcanic ash and fiery lava bombs began falling on the town, and the eruption column eventually climbed 9000 meters—almost into the stratosphere. Within four hours all 5300 inhabitants had been evacuated except a few hundred who remained to save as much of the town as possible. By the end of the month more than 50 homes had been destroyed, including that of Unnur Thorbjörnsdottir. They were either burned, crushed by the weight of ash on their roofs, or completely buried. The accumulation of ash in parts of the town exceeded five meters. Most of the fires were set by lava bombs crashing through windows, and the salvage crews installed corrugated metal sheets in those windows facing the eruption. This and the efforts of volunteers who shoveled ash from the heavily laden roofs saved many homes, although by May some 360 buildings had been burned or buried. One fish-freezing plant was completely destroyed and two others damaged, although it had been possible to remove hundreds of tons of frozen fish, valued at some two million dollars, thus mitigating to some extent the economic disaster that had befallen the community. Volcanic gas—chiefly carbon dioxide, but with some methane and carbon monoxide (both poisonous)—collected in low-lying areas, cellars and some buildings, causing one death, with several other people temporarily overcome.

Meanwhile the rift extended until it completely spanned one corner of the island, extending 300 meters beneath the sea to form a line of eruption three kilometers (two miles) long. Lava began pouring into the harbor in great volume, generating small, "rooster-tail" explosions in the water. The normally icy bay rose to 44 degrees Centigrade (111 degrees Fahrenheit), and it was feared the lava would cut off the harbor entrance, depriving Iceland of its chief fishing port. In a Herculean effort to avoid this the Icelanders undertook what two visitors from the United States Geological Survey, Richard S. Williams, Jr. and James G. Moore, later described as "the most ambitious program ever attempted by man to control volcanic activity and to minimize the damage caused by a volcanic eruption."

Bulldozers pushed the accumulating ash into dikes in an effort to divert the flowing lava, but the dikes were breached. The lava, creeping, grinding, rolling, and steaming, moved inexorably across the edge of the town, 10 to 20 meters high at its front, building up to 40 meters thick to the rear. Riding on top were great chunks of solidified lava carried from the erupting cone like black icebergs, some of them 200 meters wide. As the flow drove into the bay, a 30,000-volt submarine power cable and two pipes bringing fresh water from the mainland were cut.

The focus of salvage efforts was on drowning the advancing lava front with water to freeze it in place. This had been tested on a small scale during the Surtsey eruption. In March a pump ship was brought into the harbor, and high pressure pumps, as well as plastic pipe, were ordered from the United States. By late April these had been flown in by the United States Air Force. Forty-seven pumps were mounted on barges and began pumping water through a hastily improvised plumbing system that carried the water to a series of nozzles.

> The most difficult aspect of the cooling program (according to Williams and Moore) was to deliver large volumes of sea water to the surface of the flow some distance back from the flow front. The water effectively increases viscosity, producing internal dikes or ribs within the flow, causing the flow to thicken and ride up over itself.
>
> First the margin and surface of the flow is cooled with a battery of fire hoses, fed from a 5-inch pipe. Then a bulldozer track is made up onto the surface of the slowly moving (up to 1 m per hour) flow. The water produces large volumes of steam, which reduces visibility and makes road-building difficult. Then the large plastic pipes are snaked up on the flow; they will not melt as long as water is flowing in them. Small holes in the wall of the pipes help cool particularly hot spots.

For a time a plan was considered to use explosives to rupture the relatively cool crust on that part of the flow that had invaded the harbor, permitting seawater to quench the red-hot lava inside and thus check its advance. Stirling A. Colgate, president of the New Mexico Institute of Mining and Technology and an authority on the stellar explosions known as supernovae, flew to the island to help explore this possibility. However, he and Thorbjörn Sigurgeirsson of the University of Iceland found, from their calculations, that, if seawater came in contact with hot lava under such circumstances, there would be an explosive production of steam that would then rip open more of the lava, admitting more water to expand in what would quickly develop into a chain reaction. This, they feared, could propagate through the entire underwater body of lava producing an explosion equivalent to that of a hydrogen bomb (several megatons), with catastrophic consequences for the island (and, in terms of the resulting seismic sea waves, for seaports around the rim of the North Atlantic). A similar process, they concluded, could account for some previously unexplained foundry explosions, where water and liquid metal came into contact, or the gigantic explosions of island volcanoes like that of Krakatoa in 1883, whose yield has been estimated at 200 megatons.

The effort to stop the lava by means of an underwater explosion had been planned with the help of the Icelandic government, the Icelandic Coast Guard, and the United States Navy. But, as the two men reported in *Nature*,

> ...the day before the detonation we realized the awesome possibility that once mixing was initiated it might be self-sustaining in that the high pressure steam produced might cause further mixing until all the lava had exchanged its heat with the water above it. The energy released might have come to between 2 and 4 megatons. Naturally the experiment was called off.

By June the outpouring of lava had completely subsided and the harbor, to the relief of the Icelanders, was still usable.

While this eruption had basic elements in common with other such events in regions of crustal spreading—along the Mid-Atlantic Ridge, for example, or in Afar—there were differences, as well. The chemistry of the lava was not that typical of out-pourings along the ridge. While the eruption was from a rift, the latter ran almost due north, whereas most of the Icelandic rifts run southwest to northeast, conforming to the strike of the Mid-Atlantic Ridge in that sector (that is, the Reykjanes Ridge). Further-more, whereas the magnetic imprint on the sea floor implies that the regional source of sea-floor spreading is the Reykjanes Ridge, Heimaey and its sister islands are offset some 80 kilometers (50 miles) from that ridge. The group from the University of Iceland noted that over the centuries volcanic activity in that region has moved south-east in a series of leaps, from the Snaefellsjökull area (made famous by Jules Verne) to the present zone of activity that passes through Hekla and the new off-shore islands. This may mean, they said, that Iceland as a whole is moving west relative to a deep plume, or "hot spot," responsible for the eruptions.

If such eruptions have continued at a more or less uniform rate since Iceland began to be formed, then its rocks should display the same pattern of magnetic zones, symmet-rical to its centerline, as that found along the Reykjanes Ridge and other oceanic ridges. A magnetic survey reported in 1972 by J. D. A. Piper of the University of Leeds has shown that, broadly speaking, the rock in the island's central rift zone is magnetized parallel to the present field of the Earth, representing the so-called Brunhes magnetic epoch. It is flanked to either side by reversely polarized rock of the Matuyama Epoch. These zones, in turn, are flanked by rock of the Gauss and Gilbert epochs, representing still earlier periods of normal and reversed magnetization. But, here and there, lava from one epoch has invaded a neighboring zone of opposite polarity, so that the pattern is not nearly so symmetrical as that paralleling the submerged Reykjanes Ridge. It was, therefore, a stroke of luck that the first detailed magnetic survey was made over a part of the ridge with sufficient symmetry to be obvious, whereas in most other submerged areas-as well as in Iceland—the pattern has been made fuzzy or distorted by the whims of volcanic eruption. Only when the region is surveyed on a large scale do these local aberrations become less significant and allow the succession of magnetic reversals to come clearly into view.

Soon after the theory of sea-floor spreading became known, G. Bodvarsson, a scientist then with the State Electricity Authority in Reykjavik, and G. P. L. Walker of Imperial College, London, undertook a study of the ages and distribution of Icelandic volcanic intrusions to see if they supported or contradicted the concept. Everywhere they turned they seemed to find confirmation. Volcanic rock, they discovered, has been generated (and pushed aside) along a central zone some 10 kilometers (six miles) wide at a fairly uniform rate since formation of the oldest rocks on the island, 12 to 16 million years ago. The location of these older rocks—in the extreme east and west of the island—could be explained by a spreading action that has carried them steadily away from the central zone until they are now 400 kilometers (250 miles) apart.

The most significant find of the two scientists, however, was the extent of the lava injections from below that never reach the surface. These injections, which either fill cracks opened by some other process or force the cracks open, produce the vertical walls of basalt known as dikes. Where glaciers have cut fjords to below sea level the deeper dikes are exposed, and in the area of Reydarfjördur, on the east coast, Walker counted about 1000 of them along a sea level stretch of 53 kilometers (33 miles). Their cumulative thickness was three kilometers (two miles). That is, either they had forced the Earth's crust apart by that much, or some other process had done so. Furthermore, the number of dikes became greater at increasing depth, implying that at some depth the crust might be completely composed of dikes. In a 1972 analysis of the situation Piper and I. L. Gibson of the University of Leeds proposed that this region of solid dike intrusion lies only two to five kilometers (one to three miles down). Above it are great lens-shaped accumulations of lava extruded onto the surface by successive eruptions from regions of intensive activity, present or past. These lava beds are also intruded by dikes, but in lesser numbers, with only a relatively small percentage penetrating all the way to the surface.

Such a structure for the Icelandic crust, they found, was compatible with seismic evidence collected over the past decade. The total picture was one of sufficient intrusion at depth to "accommodate" an annual drift away from each side of the rift zone of one centimeter (0.4 inches), which is also the rate indicated by the magnetic timetable flanking the Reykjanes Ridge.

The Icelandic findings—the evidence for spreading in dry-land geology, rather than in submarine sampling—helped convince the last holdouts. But Sir Harold Jeffreys, widely regarded as the dean of Earth scientists, still held out. In the 1970 edition of his classic work, *The Earth, Its Origin, History and Physical Constitution*, published when he was 79 years old (the first edition came out in 1924), he took note of the new Antarctic fossil finds and the volcanic formation of Surtsey:

"A fossil dinosaur in the Antarctic and a new volcano off Iceland have also been claimed as support for drift," Sir Harold wrote. "With regard to the former, an island chain connects Antarctica with South America, and may well once have been a land connexion. For the latter, there are volcanoes on land, and whatever the explanation of volcanoes may be there is nothing surprising about the occurrence of a new one near by. These are all examples of the reckless claims that get into the newspapers."

The island chain to which Sir Harold referred is the Scotia Arc. In a sense it does, as he suggested, represent a link between the continents, but, from reconstructions of plate movements, it seems most likely that, as shown in Fig. 11.7A, it developed after Antarctica and South America separated.

One way to explore the nature of the process that seemed to be tearing Iceland apart was to measure it, and in 1967 Robert W. Decker of Dartmouth College in Hanover, New Hampshire, led an expedition there to run a series of survey lines across the rift zones to obtain precise information on what was happening. With him was Páll Einarsson of the Lamont–Doherty Observatory. Two survey lines were run. One traversed the southwestern spur of the rifted region, where the Althing first met; the other crossed the southern spur, close to Mount Hekla. Fifty-eight segments, averaging 2700 meters in length, were measured in the hope of determining the geographic distribution of the spreading process. The plan was to return in 1972—five years later—and repeat the measurements. However, in May and June of 1970 Hekla erupted, spreading an estimated 200 million cubic meters of lava over the terrain. Since the volcano was close to the survey profile, Decker and his colleagues (now including Paul Mohr of the Smithsonian, the authority on Afar rifting) decided to make their comparison measurements there forthwith.

This time they used one of the most sophisticated and precise distance-measuring devices known—a geodimeter in which the light beam generated by a helium–neon laser acts like a radar beam, only with greater accuracy. The results suggested a widening of the zone by six or seven centimeters (two or three inches)—a distance which seemed considerably greater than the error margins of this equipment. Part of the survey line, it was found, had been compressed and part of it stretched. From this it was possible to identify, along that line, where the spreading originated, and it lay 25 kilometers (15 miles) northeast of Mount Hekla. The zone from which spreading had occurred was apparently not in direct line with the volcano's fissures, whose extension passed 10 kilometers (six miles) to the west. Early analysis of earthquakes associated with the eruption, however, showed some near the deduced spreading area.

The trio of scientists hoped from their measurements to throw some light on the nature of plate movement, and they considered three possibilities:

1. The plates are being *pushed* apart by the injection of lava into the rift zones.

2. The plates are being *pulled* apart by the weight of their descending edges—the aprons of crustal material that have become so cool and dense that they are sinking back into the mantle below the ocean trenches.

3. The plates are being *dragged* apart on the backs of currents flowing in the upper mantle.

The three scientists decided that at least the upper part of the crust, in Iceland, was being pulled apart, allowing keystonelike blocks to subside in the rift zones, and to them this looked more like a dragging apart of the plates than a pushing apart by lava injection. However, they pointed out that local sinking of blocks has been seen where lava is forcing its way into the cracks, as in Hawaii, or where the surface has collapsed after an eruption has emptied an underground lava reservoir. Since this last process occurs after an eruption, rather than at its start, its occurrence should be identifiable, in Iceland, if closely spaced measurements of land deformation were made before, during, and after an eruption. Hence, the three reported, "More detailed knowledge of the sequence of ongoing horizontal and vertical movements of the ground surface over the Icelandic rift zones should provide some definite conclusions on the 'push' versus 'pull' or 'drag' mechanisms." And, they added, "future measurements of both horizontal and vertical ground-surface deformation on a more regional scale across active rift zones will provide important insight into the driving mechanisms of the separating plates."

The nature of the process that drives the plates continues to remain elusive. When a number of those exploring this problem assembled in Flagstaff, Arizona, for a symposium on the subject in 1970 they were reminded that the evidence, in this regard, is very meager. The plates of the Earth's crust can be likened to ice floes that are being pushed hither and yon by currents beneath them—a situation in which it is difficult to determine the water movements responsible for such motions.

By then a variety of proposals had been advanced. Bryan Isacks and Peter Molnar of the Lamont–Doherty group argued that it was the weight of the cold, dense apron on the most ancient edge of a plate that pulled it down into the mantle along a sea-floor trench, much as one side of a towel, when it becomes soaked, drags the rest of the towel under water.

Other proposals, some of them harking back to the earliest days of drift theory, involved forces related to the Earth's spin, tidal tugging by the moon, and "downhill" slippage by plates because processes within the mantle have given the Earth an asymmetrical shape. As pointed out by William M. Kaula of the University of California at Los Angeles, an authority on the gravity field of the Earth (whose irregularities reveal the lumpiness of the Earth's interior), precise mapping of that field, using radar altimeters to record the orbital wobbles of Earth satellites, should show the real extent of such global hills or other signs of up- and down-currents within the Earth. Since the flight path of an Earth satellite is controlled by the local gravity field, deviations of that path from symmetry are a sensitive indicator of deviations from symmetry inside the Earth.

A proposal in 1971 that, in 1991, continued to generate wide interest was that of W. Jason Morgan of Princeton, who suggested that some 20 "plumes" of hot, plastic rock are rising from deep in the mantle beneath key volcanic areas around the world. Each plume is, perhaps, 100 kilometers (60 miles) in diameter and, according to the hypothesis, they constitute a series of pipelines that, each year, bring up 600 cubic kilometers (150 cubic miles) of deep-mantle material, transporting upwards, as well, all of the heat generated within the depths. At the top of the mantle these plumes spread out like the clouds at the top of a thunderhead, and it is this spreading motion, immediately under the crust, that is dragging the crustal plates apart.

This was a very different concept from that of shallow "conveyor-belt" cells of convection limited to the upper mantle, and it avoided the strange geometry called for by the latter hypothesis. It had been hard to explain why rising lava should push up into the zigzagging rift zone of the Mid-Atlantic Ridge if that were the process driving the plates apart. If the plates were being dragged apart by plume action, then those rifts were caused, not by the injection of material into them, but by activity elsewhere. Lava simply welled up along the ridges to fill the opening cracks.

In a sense the plume concept was conceived as early as the 1950s by Tom Gold, that prolific generator of new—often controversial—ideas, who was co-inventor of the "steady state" concept of the universe. He suggested that channels of molten material might rise through an otherwise solid mantle and that this should be verifiable since such up-currents, being liquid, would absorb certain earthquake waves. Then, in 1963, Tuzo Wilson, as noted earlier, proposed that the Hawaiian Islands and other island chains whose components become progressively older from one end to the other were formed by a steady upwelling of molten material beneath a moving crustal plate. The islands of Hawaii, he pointed out, are one of seven parallel chains of islands and sea mounts in the Pacific whose production he associated with the general southeast-to-northwest movement of the Pacific Plate over what might be called "island-making machines" or lava upwellings from beneath the crust. Wilson assumed that, if a great turnover or convective flow in the mantle was responsible for the drift of the Pacific floor, the fountain of upwelling material responsible for the Hawaiian chain might be in motion, too. He proposed that it was, in fact, moving, but at a slower rate, being rooted in a deep, slow-moving part of the convection system, or "cell."

Not only did this process account for the island chains of the Pacific—a one-way spreading pattern—Wilson said, but also for the two chains of submerged islands and ridges that extend away from the Mid-Atlantic Ridge near the eruptive island of Tristan da Cunha in the South Atlantic (Fig. 14.1). One, the Walvis Ridge, reaches toward Southwest Africa. The other, the Rio Grande Ridge, extends toward Rio de Janiero. Both have been formed by the same upwelling of lava that most recently has produced Tristan da Cunha (whose inhabitants had to be evacuated in 1961 because of an eruption), some of the material being carried away by the east-moving plate and some by the west-moving plate. The resulting ridges, or chains of submerged mountains, are oblique to the Mid-Atlantic Ridge, according to the hypothesis, either because the plates, in addition to moving away from it, have a northerly component of motion, or because the plume is moving south.

Morgan's plume hypothesis differs basically from that set forth by Wilson eight years earlier in that it sees the rising and spreading material as the driving force responsible for the entire process of plate movement and continental drift. It is also due largely to the contributions of Morgan and Dan McKenzie of Cambridge University that the concept of the Earth as being paved with a spherical mosaic of moving plates has come to be known as "plate tectonics"—a term that today symbolizes the revolution in our understanding of the Earth.

From the geometry of these plates and their motions Morgan and McKenzie deduced that they must meet in a variety of "triple junctions" whose evolution has determined the configurations of the sea floor and nearby land in many areas. The basic motions of the plates (six of which are of vast extent), they pointed out, tend to remain constant over great periods of time, both as to speed and direction, but a triple junction where they meet may be intrinsically mobile. In other words, the manner in which

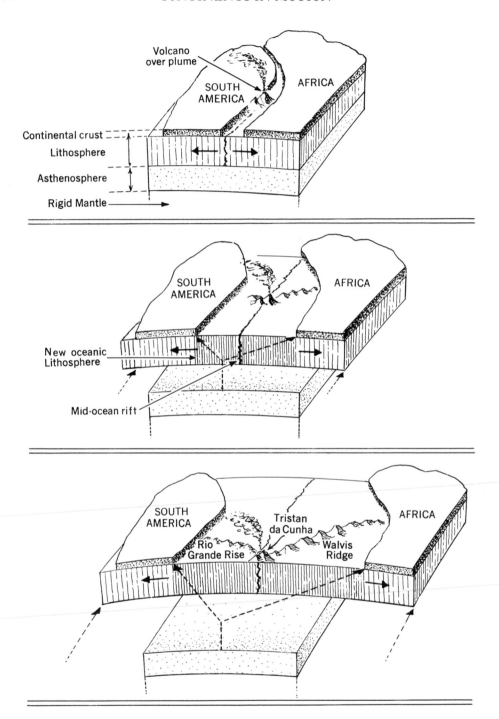

Figure 14.1 Tristan da Cunha Plume. *Two forms of motion may have generated diverging South Atlantic ridges. In the top picture, a rising plume separates the African and South American plates. Volcanoes erupt over the plume. As the plates slide north relative to the plume (middle), the old volcanic cones move and new volcanos arise in their place. In the South Atlantic today (bottom), the Walvis Ridge and the Rio Grande Rise, both radiating from the Tristan da Cunha Plume, mark the splitting apart of South America and Africa.*

three adjoining plates spread from ridges and sink into trenches can bring about migrations of their junction and, it would appear, past migrations of such points can explain a variety of features of the Earth's surface, including a puzzling "elbow" in the magnetic patterns of the Pacific floor, and, as will be seen later, the development of the western United States. Central Japan constitutes a triple junction in which three "subduction" zones (where oceanic plates are descending into the mantle) converge: one extends northeast from the junction as the Japan Trench; one runs south toward the Marianas Islands along the Bonin Island Arc, and the third extends southwest to Okinawa as the Ryukyu Island Arc. Morgan and McKenzie proposed that the depth distribution of earthquakes in this area has been determined by migration of this junction.

Of special interest would be a junction from which three ridges, or spreading zones, radiate, since this should occur above a rising plume. Morgan and McKenzie cited a location west of the Galapagos Islands from which three plates are radiating: the Pacific Plate, riding northwest from the East Pacific Rise, the Cocos Plate moving northeast toward Central America from the Galapagos Rift Zone and the Nazca Plate being carried from the East Pacific Rise toward South America. The proposition that subduction of the Cocos Plate under Central America had lifted that isthmus out of the water was reinforced by the drilling results of the *Glomar Challenger.*

To explore the triple junction hypothesis an expedition, staffed by Princeton and Navy scientists, sailed to the area aboard the Navy research ship *USNS DeSteiguer.* They

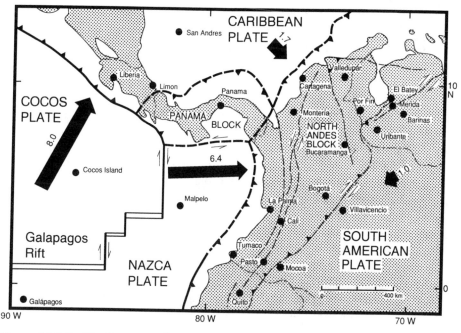

Figure 14.2 Pacific Ocean sea-floor motion. *The bold arrows represent the direction and intensity of present-day plate motion. The numbers indicate the average slip rates in centimeters per year during the last five to ten million years.*

found the Galapagos Rise, or rift zone, cut by a series of deep valleys parallel to its axis, with a central rift deeper than any yet discovered on the mid-ocean ridges—deeper, in fact, than the Grand Canyon. They also detected the magnetic patterns typical of a source of spreading. The evidence pointed to motion of all three plates away from their triple junction (as anticipated), the component of motion away from the Galapagos Rise being very slow compared to that from the East Pacific Rise—a pattern suggested, as well, by the bottom topography. Ridges from which the floor is spreading rapidly—such as the East Pacific Rise—develop broad, relatively smooth contours, whereas slow-spreading ridges become extremely rugged, with long, faulted blocks that tilt away from the centerline. This rugged pattern has proved typical not only of the Galapagos Rise but, as already noted, of Afar and the northern Atlantic Ridge (both underwater and, where exposed, in Iceland).

If, as Morgan presumed, some of the plumes are semipermanent features, embedded deep within the mantle, their surface manifestations could be used to retrace plate movements. He proposed that several oceanic ridges along which there is no earthquake activity, such as the great Ninetyeast Ridge in the Indian Ocean, represent lines inscribed on the moving crust by the deep-buried plumes. The Ninetyeast Ridge runs due north and south along Longitude 90° East.

His proposal received support from drilling conducted by the *Glomar Challenger* on two of its legs across the Indian Ocean in 1972. The ship drilled five holes along the 4000-kilometer (2500-mile) length of the ridge. In cores brought up from depths as great as 2000 meters were the shells of fossil clams, oysters, and other shallow-water organisms overlying lignite (low-grade coal) formed in swamps when parts of the ridge were at the surface. At the north end, which, according to Morgan's hypothesis, was the first part of the Indian Ocean Plate to drift over the plume, the core barrel brought up sediments laid down in shallow water some 75 million years ago on top of newly erupted lava. This material was much older and more deeply submerged than that at sites farther south along the ridge. At the southern end it was only 15 million years old (Fig. 14.3).

"Thus we can visualize the Ridge," said the scientific team aboard the drill ship, "as having once been a chain of islands, perhaps rather like the Hawaiian Islands, with active volcanoes at the southern end and a chain of low, submerged, older, extinct volcanoes extending northward away from the volcanic center." The volcanic activity that produced the Ninetyeast Ridge, segment by segment, therefore seems to have left us a timetable of the northward drive of the plate that finally brought India hard up against Asia. In 1988, ages were obtained by the Ocean Drilling Project for the bedrock in three more drill holes along the ridge. Then, in 1989, British scientists offered evidence for a plume, now under the submarine plateau from which Kerguelen and Heard Islands protrude in the southern Indian Ocean. It was this plume, they said, that broke up the supercontinent Gondwanaland, pushing apart Antarctica, Australia, India, and Africa to form the Indian Ocean. The effect of this plume, they reported, was reflected in chemistry of the rocks hauled up from the Indian Ocean floor. The nearer they were to the hypothetical plume, the more they were chemically contaminated by the plume. They also pointed out that volcanic rocks on all those continents, including

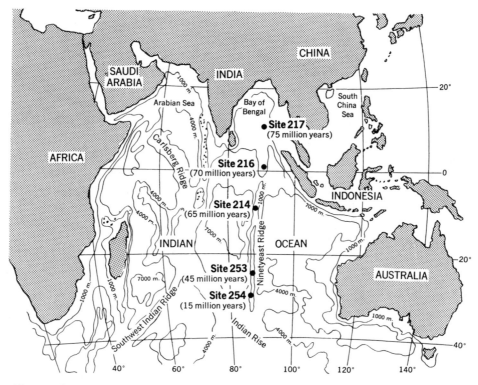

Figure 14.3 Ninetyeast Ridge *may be the product of a stationary plume beneath the north-moving Indian plate. Drilling cores indicate that the oldest sediments are found farthest to the north. Newer sediments are found closer to the presumed plume as found by the* Glomar Challenger *in 1972. More recent data confirm this picture.*

the ones in the Prince Charles Mountains of Antarctica, had been erupted as the plume began to tear the land apart, 100 to 130 million years ago.

Such evidence for a plume under Iceland had been reported earlier by J. G. Schilling of the University of Rhode Island, based on the analysis of 4000 kilos (9000 pounds) of lava specimens dredged from the full length of the central axis of the Reykjanes Ridge and collected ashore in the rifting zone of Iceland. The sampling was done from the research vessel *Trident* in 1971, and the analyses showed a systematic change in chemistry of the lava at increasing distances from Iceland—that is, from the hypothetical plume. Schilling's deduction was that plume material, with a composition possibly characteristic of the deep mantle, was spreading from the plume through the asthenosphere (beneath the rigid plates), progressively diluting the asthenosphere material that wells up along the ridge to form new sea floor as the plates move apart. A similar effect has been reported by Schilling for a hypothetical plume under the Afar Triangle. The basalts erupted along the Red Sea trough and the Gulf of Aden at increasing distances from the Triangle show a systematic change in their inclusions of those elements known as rare earths.

Another proposed product of plume activity was a chain of submarine volcanoes, known as the New England Sea Mounts (or the Kelvin Sea Mounts), that extends for

1000 kilometers (600 miles) toward the Mid-Atlantic Ridge from the American continental margin off Cape Cod. Most of the 27 known peaks of this chain have been discovered since 1950 and, by 1989, eight of them had been dated according to Robert A. Duncan of Oregon State University. Two had been penetrated by deep-sea drilling and samples from the rest were obtained by the deep-diving submarine *Alvin* or by dredging. The dates, Duncan said, clearly show the timetable expected if they were produced by a plume under the west-moving Atlantic floor. The oldest, at the west end, is 103 million years old and the age of the youngest, at the east end, is 82 million years. While this chain may extend inland as far as Ottawa, accounting for volcanic features there, in Montreal and the White Mountains, the dates are more ambiguous, Duncan added. Some, skeptical of the plume theory, have proposed that such features lie along old transform faults, such as those that join offset segments of the oceanic ridges. In that case earthquakes like those which shook Boston in 1755, dislodged a great turbidity current from the Grand Banks in 1929, and brought havoc to Charleston, South Carolina, in 1886 could represent residual activity along such faults.

The validity of plumes as reference points in determining absolute movements of the plates depends on their remaining in fixed positions—an assumption that Xavier Le Pichon, a leading French investigator of plate movements, Tanya Atwater and Peter Molnar while at the Massachusetts Institute of Technology, as well as others, have questioned. Like many in the field, Le Pichon was a convert. In the mid-1960s, as a young, bright-eyed scientist working with the Ewing brothers, Maurice and John, at the Lamont observatory, he published with them an article on the mid-ocean ridges stating, in its abstract, that for the past few million years "there is no reason to believe that significant continental drift or expansion occurred." Yet he was soon to become one of the best-known proponents of plate movement.

Since promulgation of the plume theory several scientists have found what they consider supporting evidence for the idea including the Kerguelen Island plume as origin of the Indian Ocean. Iceland, with its enormous outflow of lava, is an obvious candidate as the surface expression of a giant plume that has been pushing apart the Eurasian and North American plates. It has been estimated that in recent centuries one third of the lava erupted from all of the world's above-water volcanoes has spread across that island. When Tuzo Wilson presented his early version of the plume concept, in 1965, he proposed that the submarine ridges linking Iceland to Greenland in the northwest and Scotland to the southeast were formed by lava from the Iceland plume that was deposited on the sea floor as it moved in opposite directions away from the Mid-Atlantic (see Fig. 10.4).

Peter Vogt of the United States Naval Oceanographic Office has proposed that there are "successful and unsuccessful" hot spots. "The Icelandic center," he wrote, "was strong enough to fracture the Laurasian plate 60 my (million years) ago, while Hawaii is still struggling unsuccessfully to break the Pacific plate."

Adherents of the plume hypothesis have identified several other features that they believe were produced by movement of the crust over a plume. One is a line of ancient volcanoes from the Eifel Mountains, near Germany's border with Belgium, past the famous Drachenfels, or Dragon Rock, with its ruined 12th-century castle overlooking the Rhine near Bonn, and along the Czech-East German border as far as Wroclaw

(Breslau) in Poland. The ages increase more or less progressively from the Eifel (where the eruptions took place some 10,000 years ago) to the Polish sites, where they have been dated as 29 to 34 million years old, the implication being that the lava rose from a plume that remained stationary in the mantle as the European plate was carried east from the Mid-Atlantic. It has been proposed that the volcanic features leading eastward to Yellowstone Park are a plume product, as were the great outpourings of lava that blanketed Washington and Oregon 15 million years ago, or those that left a series of granite outcrops from Nigeria, in the bight of Africa, to the Sahara. The latter, it is said, become increasingly old from south to north, as though formed during the northward drift of Africa over a plume.

Such a proliferation of hypothetical plumes has evoked some skepticism. The earlier explanation of such features was that the crust of the Earth ruptured progressively along a line, producing a succession of erupted features, and a number of geologists still find this a more plausible explanation. Yet those who question the concept of plumes as enduring features with roots extending almost to the liquid core are forced to concede the existence of "hot spots" producing such islands as Iceland and Hawaii. And the recognition that they rise beneath moving plates of the Earth's rigid lithosphere has cast in a new light those islands that heretofore seemed random specks, sprinkled across the oceans. Take, for example, St. Helena, where Napoleon spent his final, sad years. It lies on the eastern flank of the Mid-Atlantic Ridge, between Brazil and Angola, and turns out to be a fossil counterpart of Iceland. The two shieldlike volcanoes of which it is formed—now deeply eroded—erupted through the mid-Pliocene, a few million years ago, and then became dormant; but they are cut by swarms of dikes oriented north–south and thus parallel to the ridge, as in Iceland.

Ascension Island, farther up the ridge and 145 kilometers (90 miles) west of its centerline, proves to be a great volcanic cone formed within the past 1.5 million years—presumably the time it took for the volcano to develop and be carried away from the active part of the ridge. Bouvet Island, landmark of early Antarctic explorers, sits almost directly on the ridge crest, where the latter turns east to pass between Antarctica and Africa. Its age is apparently less than 700,000 years, and fumaroles indicate that it is not entirely dead volcanically.

Thus, in terms of age and structure, each dot of an island in the oceans of the world, far from being a random occurrence, manifests an epoch in the growth of that ocean. More recently, the possibility of such broad plumes has been deduced from the speeds with which earthquake waves traverse various parts of the Earth's interior, but the matter is far from settled. Even more puzzling is the manner in which plate movements have helped raise mountains far from any coastline.

Morgan's plume hypothesis has been elaborated by his Princeton colleague, Kenneth S. Deffeyes, into a model involving circulation of the entire mantle, from top to bottom. Material that has risen through each plume spreads as much as a fifth of the Earth's circumference before going down a trench into the mantle again. The ocean floors are formed of the uppermost part of this flow that has hardened into a rigid layer—the lithosphere—(recalling the argument of Hess and Dietz that the ocean floors are really the top of the mantle). As the floor marches away from the spreading centers, the oceanic lithosphere gets cooler, thicker, and heavier until finally it slips down a trench into the asthenosphere. As it descends, the plate disperses and becomes part of the upper mantle.

Meanwhile, the hottest mantle material, near the Earth's molten core, flows away to feed the plumes. Because of this continuous removal of material from the base of the mantle, the entire mantle (except the plumes) very slowly sinks. Thus the circulation is complete: Hot material rises through the plumes. As it nears the surface and is less compressed, part of it melts and erupts as basalt. The heavier residue spreads beneath the lithosphere (here and there carrying continents on its back), cools and descends along trenches to disperse through the asthenosphere and sink through the mantle to be reheated and recycled.

A factor that, in this model, helps drive the circulation is the change of material, while still deep in the rising plume, to a less dense phase. A phase change is where crystal structure is altered but not composition, as when graphite (a moderately dense phase of carbon) changes to diamond (a superdense phase of the same element). In 1971 Gerald Schubert of the University of California at Los Angeles and D. L. Turcotte of Cornell University proposed that phase changes may play a key role in the plate-driving process. If, because of the very high temperature in the rising material, the change to a less dense (and therefore lighter) phase occurs relatively early—hundreds of kilometers below the surface—the plume must move upward with renewed vigor. If, when the material goes down into a trench, the opposite effect occurs, increasing the density at a shallow depth, this must help pull the plate down.

As evidence for a slow cycle of circulation through the entire mantle, Deffeyes cited studies done by Virginia M. Oversby of Australia and Paul W. Gast, who then were both at the Lamont Observatory, of lead in lava freshly erupted onto oceanic islands. From the abundances of certain forms, or isotopes, of lead generated by a long sequence of radioactive decays it appeared possible to determine how much time had elapsed since this material, cut off from outside influences, began its long residence in the mantle.

Gast and Oversby looked at lead in the volcanics from six Pacific islands (from Japan and Iwo Jima to Easter Island) and from islands on or near the Mid-Atlantic Ridge, such as Faial (in the Azores), Tristan da Cunha, Saint Helena, and Trinidade (rising from a chain of sea mounts extending from Brazil toward the ridge and not to be confused with Trinidad in the West Indies). The two scientists concluded that some sort of transformation had occurred about 1.4 billion years ago, "rejuvenating" this deep-seated material. Deffeyes proposed that this was when slow circulation of the mantle had last brought it near the surface. The fact that the evidence had been found on islands in two widely separated oceans, including some thought to stand directly over rising plumes, was taken as confirmation of a globally uniform circulation of mantle material.

Chapter 15

Wherefore, Then, Were the Mountains Thrust Up?

OF THE GRAND FEATURES OF THE EARTH'S SURFACE—THE SWEEP OF THE ROLLING ocean, the expanse of the sky-domed plains—there is no sector that commands such awe as the mountains. To our pride they stand as a perpetual challenge. To our spirits they are the loftiest monuments on this planet accessible to man. Wind and rain, the chemistry of time, and the cracking action of frost all work on the mountains, as they work on lower lands. Were it not for some rival process, the world, long since, would have been worn down until it was largely (if not entirely) submerged beneath the seas. The earliest inquirers into nature marveled at the mountains and pondered how they came to be. It was obvious that, while some, like the Appalachians, are old and drastically eroded, others are relatively young and even still growing. As recently as the 1960s Walter M. Elsasser, the noted physicist and co-inventor of the dynamo theory of the Earth's magnetism, wrote: "The problem of the formation of the continents and mountains is one of the few great unsolved questions of physical science."

In about 1450 the Swiss naturalist Felix Hemerli recognized that some of the fossils embedded in the Alpine rocks, far from the sea and thousands of feet above sea level, were of creatures that had lived in an ocean. Likewise, the first geologists to penetrate the Himalayas—the highest mountains of all—found deep-sea fossils of the relatively recent Triassic period. Since then it has become evident that virtually all the great mountain ranges contain large amounts of ocean floor material, even though some are hundreds or thousands of miles from the sea.

"Mountains, then, stand on the sites of old seaways," says a geology textbook of the 1950s. "Why have the sea floors of earlier times become the lofty highlands of today?" Despite new findings, says the book, "This riddle is still unanswered..."

No less puzzling to scientists of the late 18th and early 19th centuries were the observations first made by Hans Conrad Escher von der Linth and his son Arnold. This was a time in the Alps when the revolutionary armies of France, and those opposing them, marched back and forth across the passes in pursuit of one another. It was a time, too, when those of the intelligentsia who were not of a scientific bent had little love for the Alps. Beauty, to them, lay in symmetry and order, and the mountains were chaotic. It is said that Madame de Staël, the novelist and agitator against Napoleon's autocracy, ordered the blinds of her carriage pulled when she ventured into the Alps from her chateau near Geneva because she found the mountains unbearably hideous and distressing.

Hans Escher and his son Arnold felt otherwise. In the Glarus region east of Lake Lucerne they sketched the remarkable folding and contortion of rock layers that, it was reasonable to assume, had once lain flat. Such evidence of compression by forces of a magnitude difficult to comprehend can be seen in cliffs and highway cuts throughout the world. But what particularly struck the Eschers was the evidence, in rock exposures that did not seem to have been crushed or turned topsy-turvy, that a sheet of relatively young limestone lay under a dark sandstone (known today as the Verrucano graywacke) that was clearly much older. It was as though a fossil bed with human remains had been found lying relatively flat under one full of dinosaur bones.

Since then it has been found that such inversions of the layering sequence occur in various parts of the Alps (see Fig. 15.1, for example). In some places the underlying formation is a continuation of the one above it that was flipped over in a fold. Its internal layering is, therefore, in reverse order. If the layers of the upper formation are numbered 1, 2, 3 (in order of increasing age from the top down), then the layers of the lower formation are in the sequence 3, 2, 1. This can be demonstrated by folding a stack of blankets back on itself. More often, formations have over-ridden themselves so that the sequence 1, 2, 3 repeats itself again, lower down. This can be seen, for example, by the motorist traveling southeast from the Lake of Zurich to the lake known as Walensee. It was here, between the lakes, that Hans Escher won for his family the hereditary title of "von der Linth" ("of the Linth"), after his drainage of the Linth River plain liberated its 16,000 residents from the scourges of malaria (it is said to have been the only such title ever granted by Switzerland). Escher sketched the mountains on the north side of Walensee, and today the motorist traveling the main highway on the south side of the lake can see the repetitive sequence of layers in the opposing mountain wall.

When the younger Escher described such formations, he said they covered the region like great *"nappes"* (from the French for "tablecloth")—a term that has become part of the geologic vocabulary. Of the underlying limestones he said: "One must be inclined to consider their present cover by older rocks as a consequence of a colossal overthrust or a folding of the strata." Nevertheless, he added, this interpretation "presents very great difficulties as well."

What he saw before him was evidence that some great horizontal movement from the south had thrust one sheet of sea-floor limestone across a large part of Switzerland

Figure 15.1 Overthrust of sediments. *Normally, in relatively flat-lying formations, such as this one seen across Lake Walensee in Switzerland, the layers become increasingly ancient from top to bottom. Here, layers, all typical of sea floor sediments, are numbered in terms of increasing age. As can be seen, one entire sequence of sea floor sediments has been thrust over another in what is known as a "nappe."*

and then thrust another layer of older rock on top of it. More recent surveys have shown that, in the eastern Alps, some of these overthrusts extend for 100 kilometers (60 miles) or more. The Glarus nappe is thought to have been six kilometers (3.7 miles) thick, before being eroded, and to have pushed 35 kilometers (22 miles) across the landscape. But the younger Escher hesitated to promulgate such a thesis: "No one would believe me, they would put me into an asylum!"

When, at the turn of the 20th century, the Simplon Tunnel was cut 19 kilometers (12 miles) through the spine of the southern Alps, the concept became more plausible. In the tunnel, which in places is under 2000 meters of rock, it was possible to identify successive nappes rising from their roots in northern Italy and heaped, one upon the other, to form the Alps farther north. The Simplon cross section of Alpine geology has now been supplemented by numerous mine shafts and tunnels hewn from the rock to carry highways, railroads, and mountain torrents. Rivers have been diverted through the hearts of mountains to concentrate water in high-elevation reservoirs above a few hydroelectric power plants.

The Dent Blanche Nappe, named for the great mountain that stands near the Matterhorn on the Swiss–Italian border, has been traced as far south as the Mediterranean coast at Savona, west of Genoa. The upper part is a sequence of sedimentary rocks, known as the Arolla series, that accumulated to a thickness of 3000 meters on an ancient sea floor. It is underlain by another great sedimentary layer cake, the Valpelline series. Both have largely been eroded away, but some remnants, like the towering spire of the Matterhorn, which is entirely a remnant of the Dent Blanche Nappe, soar in unparalleled splendor. The Matterhorn's summit pyramid is formed of the Arolla series, but most of the mountain is of the Valpelline formation. The whole mountain—indeed the whole Dent Blanche nappe, where it survives, rests on sea-floor rocks, some of which were slippery or malleable enough to allow an overlying layer several kilometers thick to slide over it. In laboratory tests it has been found that, if water in the pores of limestone or other such rock is under very high pressure—comparable to that exerted by a great pile of rock on top of it—the porous rock shears readily to permit slippage. Such high pore pressure may exist, for example, in rock originally laid down beneath a very deep sea floor. As will be seen later, discovery of the importance of pore pressure in permitting fracture and slippage has important implications in explaining—and possibly controlling—earthquakes.

Signs of great overthrusts were not limited to the Alps. They were also to be seen in the mountains of Norway, Sweden, and the Scottish Highlands on the European side of the Atlantic, and in the Appalachians on the American side. Not only was the cause of such thrusts a puzzle, but so were the enormous thicknesses of sediments represented in their rocks. As early as 1859 James Hall, in the United States, found that, while sediments laid down on states west of the Appalachians, such as Illinois and Iowa, during the 300 million years of the Paleozoic Era were relatively shallow, the accumulations of the same era that were folded to become the Appalachians were extremely thick. Where the folds had been worn down until the oldest strata were exposed, it was evident that, in places, the original accumulation exceeded 11 kilometers (seven miles). Furthermore, the embedded fossils showed that, from top to bottom, this had occurred largely in a shallow shelf sea, as though the sea floor had subsided at the same rate as the sediment accumulation, always keeping the uppermost layer near the water surface.

Hall recognized, from what had been written of other mountains, that such deep sediments were typical of rocks forming many of the world's great ranges, and he proposed that the subsidence of the sea floor had been due simply to the weight of the accumulating sediments. His contemporary, James Dwight Dana, however, saw a flaw in this argument, for the sediment was too lightweight to displace its own volume of the Earth's dense interior, just as an iceberg cannot sink entirely beneath water. He proposed, therefore, that these great accumulations took place in troughs formed gradually from wrinkling of the Earth's crust as the planet's hot interior cooled and shrank. Such wrinkling, he said, thrust up mountains parallel to the troughs, and erosion of the mountains helped contribute to the vast accumulations of sediment in the troughs themselves. Dana called these sediment-filled dips in the crust "geosynclinal," and the term "geosyncline" is still used by geologists, although now they have differing views as to the nature of such features and how they came about.

Even though the concept of crustal shrinkage persisted well into this century, Clarence Dutton of the United States Geological Survey pointed out that it was hard to see how it could account for thrusting that was uniform in direction over very large areas—for example, the entire Appalachian system or much of the Alps. As an apple dries, its skin wrinkles rather uniformly all over; there are no great overthrusts.

When Wegener's theory was published, proponents of overthrust saw that here, at last, was an explanation. Among those who seized upon it was Emile Argand, who, early in the century, had used the new information from the Simplon Tunnel in building a comprehensive picture of the nappes—one that today has come to be generally accepted. He proposed that Africa plowed north, pushing the floor of the Tethys Sea before it and thrusting it, in slabs that heaped one upon the other, over Europe. Then, in the early Tertiary, some 60 million years ago, Africa itself pushed over the rim of Europe, with the climax of Alpine deformation occurring in the Oligocene, about 30 million years ago.

This idea was picked up by Rudolf Staub, who calculated that the resulting compression must have been on a vast scale. "If we smooth out only the Alpine folds and sheets over the transverse section between the Black Forest (in Germany) and Africa," he wrote in 1924, "then in relation to the present-day distances of about 1800 km (1100 miles) the original distance separating the two must have been about 3000 to 3500 km (1800 to 2000 miles) which means an alpine compression (in the wider sense of the word Alpine) of around 1500 km (900 miles)." As the extent of the nappe layering became known, Staub revised upward his estimate of the crustal compression. The nappes, he said, are probably stacked 12-fold, and if they were spread out to their original positions, far to the south, they would span a distance 10 to 12 times greater than today.

His view was that, in the early breakup of Pangaea, Europe and Africa drew apart to form the Tethys Sea, but then, he said, "the mighty colossus (Africa) finally restrained little Europe in the Middle Tertiary, extruded the floor of the former ocean, which then lay between them, as a vast mountain range over Europe and thrust it further north."

To doubters of continental drift all of this was preposterous. The Alps, to them, were the product of direct uplift and local folding. When Sir Edward Bailey, in the 1930s, argued for great overthrusts, he was ridiculed. "I am treated as a geopoet and geomystic, which troubles me little," he wrote; "I am also treated as a buffoon, which is more disagreeable." Indeed, who could believe that such monuments as the Matterhorn could be products primarily of horizontal, rather than vertical, movements?

Just as the drifters and their opponents became entrenched in two embattled camps, so did the proponents and opponents of overthrust. As Rudolf Trümpy of the Federal Institute of Technology in Zürich put it, "ultranappism" (which assigned no role whatsoever to uplift) resulted in the backlash of "antinappism," and the latter, he wrote in 1960, "is only now losing some of its exaggerated violence."

In any case, both schools were agreed on one thing: the Alps, like the Appalachians and other similar mountain systems, were derived from a "geosyncline" that accumulated sediments of great diversity and extraordinary depth. The protagonists were divided only on where it lay—whether it was roughly in the present location of the Alps or far to the south. It was evident that the sea floor had lain close to one or more land areas, for some of the Alpine rocks contained material that clearly had been washed off high ground. Indeed, as in all geosynclines, the only way to account for the vast accumulation was a combination of normal oceanic sedimentation and material from heavy erosion, as from newly arisen mountains.

It is only with the growing acceptance of continental drift, in its modified form of plate tectonics, that the "nappists" have begun to win over their skeptical colleagues, some offering a compromise in that the descent of one plate beneath another may also have caused considerable uplift.

A new and exceedingly important element has been introduced into the debate on the origin of the Alps and other mountains. This is the discovery of the true nature of a perplexing sequence of rocks that occur in elongated zones within some mountain systems. The sequence is known as the ophiolite suite. It puzzled geologists because the lowest units of the suite (which in its entirety may be several kilometers thick) consist of very "basic" rocks—the dense kind that typically erupt from deep in the Earth. It was assumed that they had invaded the sedimentary formations where they were found, at the very high temperature typical of molten intrusions. Yet often the rocks around them showed no evidence of having been baked by such an invasion. Moreover, the time when these basic rocks were last molten was found, from radiation measurements, to have been much more ancient than the ages of the adjacent rocks.

As more was learned about structures beneath the sea floor, from seismic probing, bottom sampling, and drilling, some began to suspect that the ophiolite suite represented an entire, top-to-bottom cross section of the oceanic crust—from the sediments, down through the lavas, past the "Moho," and into the upper mantle to the base of the rigid plate or lithosphere. If so, the ophiolite suite would at last reveal the nature of the region, known as Layer Three, identified deep beneath the sea floor by seismic soundings. At the base of Layer Three was the Moho and above it was Layer Two, about 1.5 kilometers (one mile) thick and thought to be formed largely of pillow lavas—so called because, having erupted under water, they form pillowlike lumps. Layer One, in this terminology, was the carpet of overlying sediment.

Because this whole oceanic sequence typically lay beneath several kilometers of water, with its deeper parts well beyond the reach of the *Glomar Challenger* drill, convincing confirmation of the hypothesis was difficult. Some geologists said the lower part of the ophiolite sequence was clearly mantle material, comparing it to specimens from oceanic islands like Saint Paul Rocks, in the Mid-Atlantic, that they believed were also from the mantle. But the skeptics said true mantle rock was so heavy that it would rarely, if ever, be pushed up within reach of a geologist's hammer. Uncertainty as to whether or not a true specimen of mantle material had ever been obtained had, in fact, been the rationale of the Mohole Project.

In 1971 Fred Vine, who had solved the riddle of the magnetic patterns on the sea floor, and Eldridge M. Moores, of the University of California at Davis, proposed that an ophiolite suite on the island of Cyprus might be a section of sea floor that originated on a mid-ocean ridge in the now-vanished Tethys Sea. They pointed out that great numbers of basalt dikes had intruded the island, representing a 100-kilometer (62-mile) expansion of the crust in that area.

The following year Vine and co-workers from the universities of Leicester, Birmingham, and East Anglia decided to use explosions to test the similarity of the Cyprus formation to that underlying the deep sea. "For," they wrote, "if it can be established that this hypothesis is correct, the study as a whole will greatly increase our knowledge and understanding of the composition and structure of the deep-sea floor. As well as defining its economic potential it might resolve such fundamental problems as the composition of the seismic 'Layer 3' and the nature of the oceanic upper mantle. Clearly it could, potentially, yield more information than the expensive and ill-fated American 'Mohole' project."

Vine and his co-workers set off 29 small explosions in drill holes that were filled with water to plug the hole and direct more energy into the deep rock layers. The response of those layers proved, in fact, to be like that of the sea floor (allowing for the absence of a heavy overburden of water). Then, in 1973, I. G. Gass and J. D. Smewing of Britain's Open University painted a picture of Cyprus geology strikingly reminiscent of Iceland. Where the base of the layer of pillow lavas—which they took to be the counterpart of the Second Layer—was exposed, dikes formed 60 percent of the material. And, within what was assumed to be the Third Layer, intruded dikes represented 90 to 100 percent of the rock. In other words, as in Iceland, the deepest region was formed entirely of dikes, like a pack of cards on edge.

When the Cyprus findings were reported in *Nature*, the journal pointed out editorially that as early as the 1950s—before Hess proposed the sea-floor spreading concept—"remarkably prescient" scientists in both Cyprus and Iceland had recognized that in each area the dikes manifested great tension, or pulling apart of the crust. With the information newly in hand, said the editorial, "the Cyprus rocks can, in conjunction with the volcanic successions developed in Iceland, be employed to give what is probably the most complete model deduced from exposed rock of an ocean ridge in cross-section."

The sequence on Cyprus thus appeared to bear out the Vine-Moores hypothesis and represented an ophiolite suite the deepest layers of which formed on an active ridge in the Tethys Sea. As this part of the ancient sea floor moved away from the ridge, it gradually became capped with sediment, then was thrust up and eroded into the island of today. Some believe the islands of Juan Fernandez, where Robinson Crusoe met his man Friday, were similarly formed, having been carried eastward, toward Chile, by sea-floor spreading from the East Pacific Rise. The ophiolites, however, typically have been thrust up into mountain systems where two continents have collided, crushing a section of sea floor between them. Normally, where such an intervening sea floor is being squeezed between converging continental plates, the oceanic plate goes down a trench into the mantle, but some top-to-bottom cake slices of the oceanic plate are thrown against the far side of the trench, or are thrust up in the final stages of collision, becoming embedded in all kinds of scrapings and slumpings incorporated into the resulting mountains. In this way mantle-type rocks that erupted along a ridge long before, become emplaced, cold, among rocks that are much younger.

The recognition that ophiolites mark suture zones between continental plates has led to a worldwide search for such rocks as clues to ancient plate boundaries and plate movements. It has been found that they occur in well-defined bands, as, for example, in the Himalayas or along the axis of the Urals. Another runs through Yugoslavia, Albania, and Greece, paralleling their coasts, but well inland (see the third map, Fig. 15.7). One of the best-known ophiolite exposures lies at the base of the Matterhorn, visible from the Gornergrat railway station above Zermatt. It marks the suture between the Pennine Alps (which, with Italy, may have been part of the African Plate left behind when Africa retreated after its last great thrust at Europe) and, to the north, the Limestone Alps, comprising what was once the southern rim of Europe.

Among the clues used recently to decipher the radical changes that have occurred in the geography of this region are the scars left by turbidity currents that once swept across the vanished sea. Just as such currents today periodically spread great deposits of material across the deep basins of the Atlantic, so did similar "submarine avalanches" in oceans and seas that vanished long ago. The sand and gravel layers laid down by turbidity currents in the Tethys Sea turned into sandstone that was then thrust far inland across Europe. Yet, remarkable as it may seem, that sandstone, known by its Swiss name, "flysch" (pronounced to rhyme with "fish"), displays scars that still show the directions in which currents traversed various parts of that ancient sea (Fig. 15.2).

Flysch deposits are found as far inland as Poland, along the northern slopes of the Carpathians—themselves formed in the same manner as the Alps. It is evident that the flysch was laid down in a deep oceanic basin, not in shallow coastal water. Each episode of deposition, as noted earlier, left a characteristic sequence in which the largest particles—the first to drop out of the current—formed the bottom of the layer and fine-grained sand formed the top, with an "icing" of clay laid down by the gentle rain of very fine particles and fossil shells that fell to the bottom as part of normal sedimentation. It was on this "icing" that the next flood of debris was laid. Eventually these successive layers, after deep burial and prolonged time, turned into sandstone like that now visible in the Alps as flysch.

The clues to the direction of flow, in the vanished sea, are various. In some places within the sandstone there are charcoal smears along the line of movement. Sand grains or little grooves may be similarly aligned. But these clues are ambiguous, since the flow could be in either of two opposite directions. However, the so-called flute casts—little gouges in the former sea bed that are narrow at one end and broad at the other—reveal the actual direction. By the late 1950s such analysis of current flow in former oceans had been developed to a rather precise art, and one of its practitioners was Kenneth Hsü, who later joined the team aboard the *Glomar Challenger* when it ventured into the Mediterranean.

He surveyed flow directions recorded in flysch from many parts of Switzerland in an effort to reconstruct the pattern of turbidity currents, and hence the geography, of

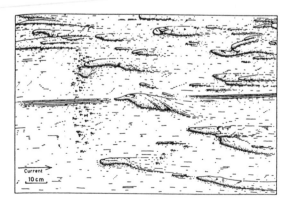

Figure 15.2 Miocene turbidity currents left casts of grooves and flutings in sandstone of the Apennines. These casts indicate that the current flowed from left to right.

the sea from which this material was derived. In each outcrop of flysch he made some 30 observations, plotting the directions as one does winds recorded over a period of time (known as a wind rose), to display the average direction of movement. In one line of outcrops along the northern front of the Alps east of Lake Geneva (the Gurnigel Flysch), the flow had been consistently from the north or northwest. Yet in another line, farther east (the Schlieren Flysch), the currents had come from the southwest (Fig. 15.3).

Others had found additional clues. Some flysch sandstones contain volcanic debris, including the andesite typical of eruptions where a plate is descending into the mantle. The volcanic chips are of a sharp-edged variety that could not have been washed long distances. Careful analysis of the rock grains also revealed bits of granite, schist, gneiss, limestone, and other rocks that must have been eroded from a large land area. Plant remains showed that this land had been heavily vegetated. And hence there developed the concept of a long-lost volcanic island, perhaps comparable in size to Crete, whose erosion had contributed to the flysch. It came to be known as Habkern Island (although some believe it was a chain of islands—an island arc), whose volcanoes were fed by material extracted from a descending oceanic plate, and which then was somehow plowed under during the continent–continent collision that formed the Alps (Fig. 15.4).

In 1971 Hsü, on the basis of his measurements of ancient current directions in the flysch, and Seymour O. Schlanger of the University of California at Riverside, pro-

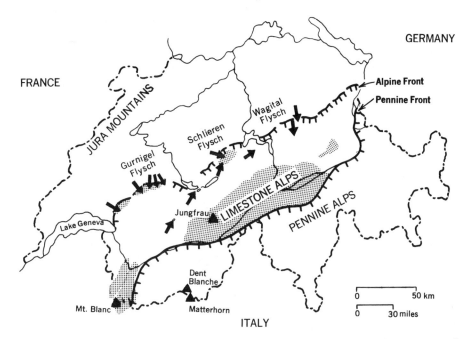

Figure 15.3 Directions of the turbidity currents *that scoured the sea floor of various parts of an ancient sea were determined from "flysch," or sandstone of deep-water origin in the Swiss Alps. Flow directions in different Alpine formations are shown by arrows.*

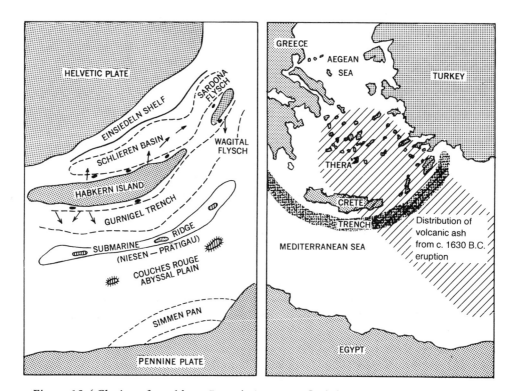

Figure 15.4 Closing of an old sea. *By analyzing various flysch deposits (see Fig. 15.3) and determining the flow directions of turbidity currents in the sea that was crushed when the Alps were formed, Kenneth Hsü and Seymour O. Schlanger deduced that these currents flowed away from two sides of either Habkern Island, a now vanished volcanic island, or an arc of islands in that sea. As shown in the map to the left, the sea and its islands lay between two converging continental plates: to the north, the Helvetic Plate which was part of Europe and to the south, the Pennine Plate. The two scientists believe that Crete, the island arc of which it is a part, and the trench alongside it, represent a modern counterpart of the earlier situation. In the modern case, the African plate is pushing the Mediterranean floor under the rim of the European Plate. The catastrophic eruption of Thera in about 1630 B.C., widely believed to be the basis of the Atlantis legend, was presumably a by-product of this converging of continental plates.*

posed that the two kinds of flysch—the Gurnigel Flysch with its traces of flow from the north and the Schlieren Flysch with its record of flow from the southwest, were formed on opposite sides of the island arc that included Habkern Island. The Gurnigel Flysch, laid down on the southern side, contained shale layers with the remains of foraminifera and other microscopic fossils constituting a complex like that found, for example, at a depth of three to six kilometers (two to four miles) in the Peru–Chile Trench. It was laid down, Hsü and Schlanger proposed, in a trough formed where a north-driving "Pennine Plate" was descending under the "Helvetic Plate" and its island arc.

The other formation—the Schlieren Flysch—was laid down on the north side of the island arc (between it and Europe) in an elongated basin much like the Aegean Basin of today. In fact the two scientists proposed that the situation predating the Alps has its modern counterpart in the eastern Mediterranean, with the island arc that includes Crete and Rhodes, the trench to the south of it, and the deepening zone of earthquakes that bear witness to descent of the African plate as it pushes north under those islands (see Fig. 15.4).

One of the most striking features of the Alps and Carpathians is their formation in great arcs, both of which bulge toward the north. If the mountains were produced by the descent of a north-moving plate, they should be curved in the opposite direction, like the Pacific arcs. Such curvature is typical where plates that fitted the spherical surface of the Earth begin descending into that sphere. The front along which a rigid section of such a plate begins its descent forms an arc bulging in a direction opposite to that of descent.

The explanation for the "wrong way" Alpine curvature proposed by Hsü and Schlanger was that in the late Eocene and early Oligocene, some 40 million years ago, there was a "flip" in the direction of plate descent (Fig. 15.5). Whereas the north-moving Pennine Plate had been burrowing under the Helvetic Plate, the latter now began to descend under the Pennine Plate. Such a flip of the direction in which a plate is descending had been proposed two years earlier by Dan McKenzie of Cambridge as an explanation of what happens when the trench in front of an island arc "tries to consume a continent."

The process can be illustrated in terms of the present situation in the eastern Mediterranean. The line of islands that includes Crete and Rhodes rides the rim of a plate formed, in that area, of thin, dense oceanic crust. It is being underthrust by a similar apron of crust riding north as an oceanic extension of the African Plate. Libya and Egypt seem ultimately doomed to collide with Greece and Turkey. Before that

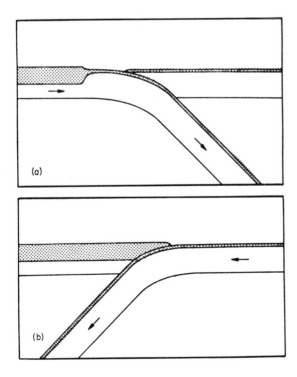

Figure 15.5 "Flipping." With these diagrams, Dan McKenzie illustrated how a "flip" may occur when one plate is descending beneath another. In the upper diagram the oceanic portion of a partially continental plate is moving toward the right and descending beneath another oceanic plate. In the lower diagram, the continental part of the left-hand plate has reached the "battleline" between the two plates, and, since it is too light in weight to descend, there is a "flip" and the opposing plate begins to descend beneath the lighter plate.

occurs, the north-moving front of Libya and Egypt will reach the trench south of the island arc and the descending plate on which those African countries are riding will start to drag them down into the Earth's interior, via the trench. But being formed of lighter continental rocks, Libya and Egypt will be too buoyant to descend thus. There will be a brief impasse, and then a flip will occur. The dense oceanic plate to the north, with the island arc on its rim, will begin descending under the African continental plate.

This, according to the hypothesis of Hsü and Schlanger, was what happened earlier to reshape the curvature of the Alps. There was a flip, and the Helvetic Plate, carrying Habkern Island on its back, began descending under the Pennine Plate. Thus the vestiges of the plants, animals, and volcanoes that once thrived in that "lost world" have been carried into the depths beyond our ken. According to the Hsü–Schlanger reconstruction, the death throes of this "lost world," as it descended ever more deeply, generated the intrusions of granite that now stand clear against the sky as the lofty summits of the Bergell in that skier's paradise of Switzerland's southern Engadine. Geologists are puzzled, however, for elsewhere in the world, where island arcs have been thrust against a continent, at least part of their typically volcanic ruins have remained, bearing witness to their passing. In the Alps, apparently, this is not the case (apart from the volcanic debris in the flysch).

The most striking relict of the battle between the Pennine and Helvetic Plates—that is, between Europe and Africa—is the great trough that separates the Pennine Alps along the Swiss–Italian border from the Limestone Alps of central Switzerland. It is within that trough that both the Rhine and Rhone Rivers, flowing out of it in opposite directions, have their birth. To the south of it the mountains, from the Matterhorn to the Grand Combin and Saint Bernard Pass, are formed of sedimentary rock that originated along the northern fringe of Africa and was thrust over the basement rocks of Europe as the Dent Blanche Nappe and other nappes of the Pennines. To the north the limestone nappes, derived from the southern flank of Europe, have been worn away in places to expose the uplifted granites of what was once Europe's rim.

These granites now challenge the alpinist as the summits of the Jungfrau, Mönch, and Eiger, in the Limestone Alps, and the highest summits of the Mont Blanc Massif. This is the erosion-resistant material that forms the soaring pinnacles, or aiguilles, around Mont Blanc, as well as Mont Blanc itself and the massive blocks of the Grandes Jorasses, Tour Noir, and Dolent. These, it seems—like the Jungfrau group—were once covered by the original surface rocks plus a succession of nappes which have been eroded from the summits and only show at lower levels. (A view of this region from space appears as colored plate number 12.)

But how did this thrusting come about? If Africa had simply pushed north, driving the foundations of a great sea down under Europe and nappes of sedimentary rock over that continent, one would expect a certain degree of fit between the opposing coasts of Africa and Europe and a zone of former volcanic activity above the line where the oceanic plate descended. Instead, the southern flank of Europe is marked by three great peninsulas—Spain, Italy, and Greece—reaching toward Africa, and, except in Italy, there is almost no evidence for the volcanic activity typical of such situations, especially of the type known as andesite.

It turns out, from drilling by the *Glomar Challenger* in the western Mediterranean, that the basins of that sea are relatively young, and it is also evident from the Alpine rocks that the ocean basins from which they were bulldozed were a great deal younger than, say, the ocean floor now folded into the Appalachians. With the history of the expansion of the Atlantic Ocean relatively well documented in the magnetic timetable inscribed on its floor, it has been possible to determine the movements of Europe and Africa, relative to one another, generated by that expansion. It was these movements that, in turn, helped shape the western Mediterranean, the Alps, and other features of southern Europe and northern Africa.

The picture that has emerged is one of small plates of the Earth's crust, or "microcontinents," such as Spain, Italy, Greece, and the clusters of Mediterranean islands, caught between two giant land masses that move first one way and then another, now pulling apart to form ocean basins that fill with sediment, then coming together to roll the microcontinents between them, like ball bearings, and crush the sediments into mountains (Fig. 15.6).

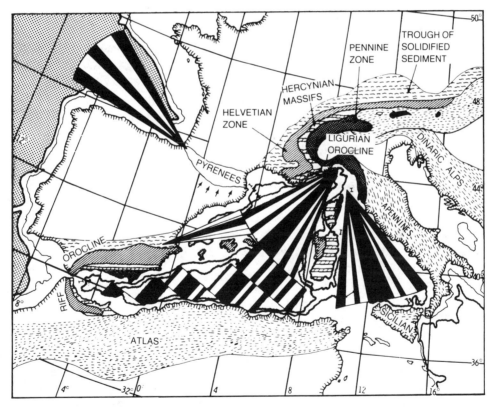

Figure 15.6 Motion of the Iberian peninsula *away from France and the Western Mediterranean is shown here schematically as it was envisioned by Peter Vogt and his colleagues at the Naval Oceanographic Office in Washington. They determined the peninsula's movement by analyzing the magnetic patterns under the Mediterranean Sea and the Bay of Biscay, both of which opened as the peninsula moved.*

Trying to decipher this history is like looking at a photograph taken just as a football referee has ended a play with his whistle and trying to reconstruct the play. Those who have sought to do so, like Hsü, Dan McKenzie and A. Gilbert Smith of Britain, Xavier Le Pichon of France, and others, have exploited the clues available: the history of relative movements by Europe and Africa, inscribed on the Atlantic floor, the directions of old sea-floor currents (enshrined in flysch), and the sutures between old microcontinents (marked by ophiolites).

Since one band of ophiolites runs close to the eastern borders of Yugoslavia and Greece, most of these two countries, plus Albania, may have been part of an Adriatic plate that perhaps also included eastern Italy. Additional ophiolites along the northern Apennines, that form the spine of the Italian peninsula, suggest the welding together there of two ancient plates.

At the end of the Triassic Period, about 200 million years ago, dinosaurs could walk from southern Europe to Africa across a broad land connection, and there apparently was still considerable similarity between the reptiles of China and those of Antarctica. The Tethys Sea extended from the Pacific into what is now the Eastern Mediterranean; but the microcontinents that now constitute Spain, Italy, Yugoslavia, Greece, Sardinia, Corsica, Sicily, and the Balearic Islands formed a single land bridge between Europe and Africa in what is now the western Mediterranean.

Then, in the Mid-Jurassic, 165 million years ago, the Central Atlantic began to open, pushing Africa and North America apart. Europe remained wedded to North America, and so Africa began moving east, relative to Europe. As a result, the land bridge broke up into a mosaic of plates that were dragged east and rotated counterclockwise. The rotation and transverse slippage of Spain began to open the Bay of Biscay, as well as a trough where the Pyrenees now stand, which became filled with the sediments later crumpled into those mountains. The Adriatic microcontinent was dragged by Africa until it collided with what are now the southern Balkans. The intervening sea floor was destroyed or crushed to form the ophiolite belt of Yugoslavia and Greece.

Meanwhile, these movements had created deep basins in what became the western Mediterranean, and these filled with the sediments later to be rolled up, not only into the Alps but into other mountain systems generated by these movements, such as the Apennines and the Betic ranges that flank the Mediterranean in southeast Spain.

Then, from Middle Cretaceous to Late Eocene (80 to 40 million years ago), the relative motion between Europe and Africa was reversed because now the North Atlantic was spreading, driving Europe east more rapidly than Africa. This pushed the Adriatic Plate west, to collide with a fragment that included Corsica and Sardinia, producing the Appennines and their ophiolites. Spain was pushed back to form the Pyrenees and then, as the motion of Africa became more northerly, a small plate, comprising Sicily and the Italian boot, was welded to Italy along a line in Calabria marked by ophiolites. Finally the Adriatic Plate and its neighbor to the west were driven north against the underside of Europe to produce the Alps.

Hsü, in offering this reconstruction (Fig. 15.7), argued that Italy and the Balkan part of the Adriatic Plate are so intimately welded to Africa across a shallow part of the Mid-Mediterranean south of Sicily that they should be considered, structurally, to be part of Africa rather than Europe. However, in closing his presentation, he said that it

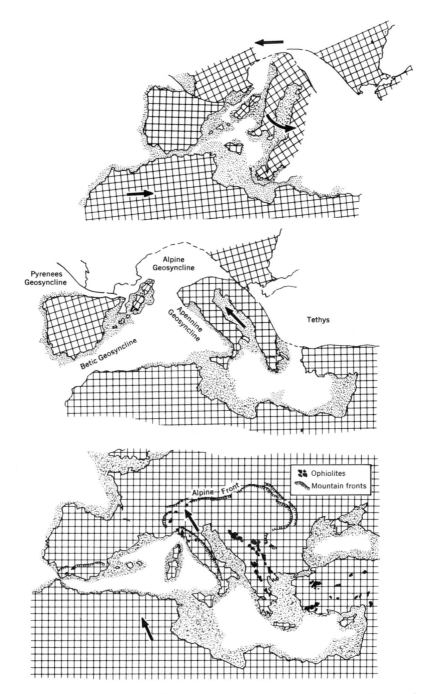

Figure 15.7 Development of the Mediterranean and the Alps *as envisioned by Kenneth Hsü. During the Middle Jurassic period, 165 million years ago, the separation of Africa and North America pushed Africa east, relative to Europe, and intervening plates were rotated counterclockwise (top). Then, between eighty and forty million years ago, the separation of North America and Europe reversed relative motion between Europe and the plates to the south (middle). Finally, Africa pushed north to complete formation of the Alps (bottom).*

was not offered as a "final solution" but rather to stimulate the collection of further data "to check, improve, or disprove this working hypothesis."

Today the pattern of earthquakes in the Mediterranean area shows that the African Plate may still be underthrusting Eurasia, although some have questioned whether it is Africa or Eurasia that is in motion. The Eurasian Plate is also being carried away from the Mid-Atlantic, and this may be responsible for the Asian counterpart of California's San Andreas Fault. This is the Anatolian Fault that spans the full east–west length of Turkey and, since it entered a new period of activity in 1938, has produced at least 13 major earthquakes, taking thousands of lives. As on the San Andreas, the motion, as one looks across at the far side of the fault, is always to the right. This could be because Eurasia is moving east or, as McKenzie has proposed, because a small plate south of the fault is moving rapidly west.

An unexplained discovery regarding these movements is the orientation of the pyramids at Giza, in particular the Great Pyramid of Cheops, which, although built 4500 years ago, is still said to be the most massive structure ever erected by man. The precision of its layout implies a high degree of surveying ability. Each side is 230 meters long, within a margin of 20 centimeters (eight inches). The sides meet at almost perfect right angles; but, while they are oriented roughly in a north–south–east–west alignment, the pyramid is noticeably slewed in a counterclockwise manner. The possibility has been explored that this could be because Africa has rotated in the 45 centuries since the pyramid was built—but the opening of the Red Sea implies a rotation in the opposite direction.

A variety of timetables of plate motion have been advanced to explain the origin of the Alps, only a few of which have been outlined here. The final decipherment, if it ever comes, will be a tour de force by someone who, with intellectual capabilities somewhat akin to those of a great chess master, will be able to digest and integrate into a coherent theory the whole corpus of accumulating evidence—from deep-sea drilling, from Alpine geology and tunneling, and from the developing concept of plate tectonics.

Chapter 16

The Appalachians — Born of a Vanishing Ocean

ANYONE WHO FLIES OVER THE SERIES OF PARALLEL RIDGES AND VALLEYS THAT RUN from Central Pennsylvania south across the Maryland panhandle and the Virginias can look down on one of the most perfectly ordered series of fold mountains in the world. They are like huge waves that were frozen in place as they rolled across the terrain. Many are so clean and straight it is hard to believe they are not artificial, the work of some mammoth earth-moving machine. This is the ridge-and-valley province of the Appalachian system which extends, in its various manifestations, from Newfoundland to Alabama, a distance of about 3000 kilometers (2000 miles). Compared to the Alps even the higher ranges, such as the Great Smokies on the Tennessee–North Carolina border, are puny. But there is reason to believe some of the Appalachians were once on a scale with the loftiest and most rugged mountains of today.

Those seeking the origin of the Appalachian system recognized early that it was largely formed of rocks that had once lain as an enormously thick accumulation of sea-floor sediments. It was from this that there developed the belief that other mountains, as well as the Appalachians, had arisen from the crumpling of a sediment-filled trough, or geosyncline. In some cases the original sediment accumulation must have constituted tens of thousands of cubic kilometers. It was recognized that this material falls into two distinct categories: One consists of sedimentary rock formed on an underwater platform that, in the manner typical of a geosyncline, sank (most of the time) no faster than the rate of sediment accumulation. Hence fossils, even several kilometers down in the sequence of layers, are from shallow water varieties. (Such a formation came to be known as a miogeosyncline.)

For many years the other type of accumulation was perplexing. It contained ash and lava, indicating volcanic activity in its vicinity. But it was not until the 1940s that Marshall Kay of Columbia University began to persuade the geologic community that this material, even where lying well inland, had been laid down in deep water. This was indicated by the nature of the fossils found in such a formation (which came to be called a eugeosyncline). Since there was evidence that one or more chains of volcanic islands—perhaps analogous to the island arcs of the Western Pacific—followed the coast (which then lay considerably farther inland than the present one) from Newfoundland to Georgia, Kay suggested that the deep-water sediments accumulated in troughs that lay between those islands and a shallow, sediment-covered platform close to the coast. Some driving force then thrust the deep-water material up over the shallow platform. Subsequently, those shallow-water sediments were themselves pushed far inland over the continental basement.

In the following decade Charles L. Drake, working on his doctoral thesis at Columbia, realized that Kay's reconstruction of the coastline from which the Appalachians had been built had much in common with the East Coast today (apart from the hypothetical volcanic island chain). The shallow-water platform was comparable to the continental shelf, bounded on the seaward side by a slope that at first descended steeply and then more gently to the abyssal ocean floor. Seismic soundings, by this time, had shown not only the extent of sediment accumulations on the shelf, but the existence of a great wedge of deep-sea sediment heaped against the outer slope of that shelf. Sampling of the latter sediments showed they were derived, in part, from turbidity currents and closely resembled the old deep-sea material found in mountains to the west (the eugeosyncline).

But if the Appalachians were derived from such a coastal sequence, what extraordinary process could have pushed all that material up over the continent? By this time the use of radioactive "stopwatches" to determine the ages of various formations in those mountains had nailed down the timetable of their building in a manner that helped clear the way for a plausible explanation.

It was known that the mountains were formed in three primary episodes, and their dating, by radioactive methods, at first deepened the mystery by showing how well-defined and widely separated were these chapters in Appalachian history. When the first major episode of Appalachian mountain-building began some 450 million years ago, the eastern part of North America bore the ruins of a former upheaval that had occurred about 600 million years earlier. The handsome but heavily eroded mountains of the Adirondacks are the most dramatic remnants of that so-called Grenville episode, and they have survived, apparently, because they were raised by later activity. Erosion of the Grenville summits built up a many-kilometer accumulation of sediment in Tennessee and North Carolina, later to be thrust up as the Great Smoky Mountains.

The first important episode of Appalachian upheaval, between 450 and 420 million years ago, is known as the Taconic for the Taconic Mountains that remain as one of its modest mementos (Fig. 16.1). The Taconics, along the New York–Massachusetts border, but extending south into Connecticut and north into Vermont, are actually formed from slabs of material that slid westward, off newly uplifted mountains where there now remain only the Berkshire Highlands. Thus, part of eastern New York was

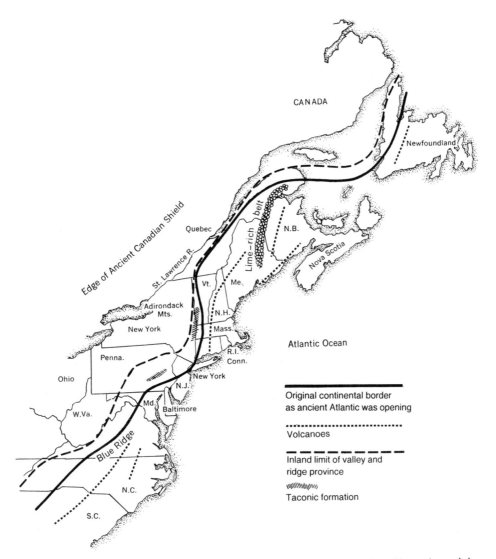

Figure 16.1 Development of the Eastern Margin of North America. *Mountains and the remnants of volcanic arcs testify to the collision and separation of the North American and Euro-African plates.*

originally in Massachusetts. Such sliding, chiefly on the inland side of a mountain system, is well known to Alpine geologists, who call the resulting slabs "klippen." The Taconic slabs slid into an immense, shallow sea that extended west from the region between Vermont and Alabama, where it was rather deep, as far as Minnesota and Texas, separated from the ocean by the newly raised mountains. This flooding of the eastern part of the continent apparently occurred partly because the land was subsiding near the new mountains. The chain of mountains formed by the Taconic upheaval continues north as the Green Mountains of Vermont and, in the other direction, passes through the Hudson Highlands (including Bear Mountain) and across northern New

Jersey and the Delaware Water Gap to become the Blue Ridge range that runs south through Pennsylvania and Virginia.

While this range was being formed, there also arose a chain of volcanic islands whose remains now lie along the east side of the Connecticut River Valley from Long Island Sound to Maine. At that time the rim of the continent presumably lay to the west, along the Green Mountain and Berkshire foothills. Almost all that is left of these islands is a series of some 20 domes of gneiss, or granitelike rock, from the Lyme Dome, at the mouth of the Connecticut, up that river to one southwest of the Rangely Lakes in Maine. The domes, as a rule, are elongated along the axis of this chain, and extensive volcanic deposits around them testify to the labors of their ancient volcanoes.

Since another chain of volcanic islands apparently lay farther out to sea, there came to be two parallel island chains off the rim of the continent: one nearer what was then the rim and the other farther to seaward. Between them there developed a tranquil sea, isolated from heavy continental sediment, accumulating alternate layers of clay and lime-mud whose residue, in the form of "ribbon limestone," has made the soil of northern Maine and nearby New Brunswick ideal for potato crops. Farther south there was also uplift at this time, as recorded by sediments that flowed northwest into Virginia and Tennessee, but the effects were largely wiped out by subsequent upheavals. Also, during this period, there were volcanic eruptions in this southern area so cataclysmic that they spread their ash as far north as Ontario.

The next, and apparently most radical episode of mountain-building, at least in the northern Appalachians, was the Acadian. It was preceded by some 200 million years in which sediment in the developing geosynclines accumulated to depths as great as 20 kilometers (13 miles). Then these layers, from Long Island Sound north to central Maine and Newfoundland, were thrust westward, burying the former islands along the Connecticut Valley under a massive succession of nappes.

In 1971 Richard S. Naylor of the Massachusetts Institute of Technology pointed to evidence that this heaping of nappes reached an extraordinary height. He focused his attention on the granites quarried in Vermont, such as those found near Barre. Some 20 intrusions, or plutons, of this "Barre granite" lie in Vermont, ranging in diameter from half a kilometer to 10 kilometers. This stone is widely used for monuments and other structures because of its clean, bright, beautifully uniform texture.

These granite bodies are enclosed in envelopes of clay-derived rock containing relatively rare aluminum silicates (notably kyanite) formed only, it was believed, under pressures such as those exerted by an overburden many kilometers thick. Assuming, as Naylor did, that the pressure was caused by nappes stacked to a height of 12 to 15 kilometers (seven to nine miles), this would make the resulting mountains far higher than Everest (which rises a little more than 8.8 kilometers). Actually, the weight of these nappes must have caused the crust to subside until erosion wore the mountains away and permitted the surface to rebound, but in any case the northern Appalachians at that time, some 380 million years ago, must have been a magnificent spectacle.

The Barre granites apparently intruded the nappes after the latter had completed their sliding, as the culmination of this mountain-building process—one perhaps similar to that which intruded the granites that now stand naked as the summits of the Bergell region in the Alps. What particularly impressed Naylor, however, was confirmation in the Barre granites that the Acadian phase of this history occurred within a rather short timespan. It had been recognized, he said, as an abrupt event in northern Maine (where granite intrusions have left such features as Mount Katahdin, the state's highest), but this now seemed applicable to much, if not all, of this phase of Appalachian mountain-building. The whole process, thrusting and heaping the nappes into a stack many kilometers high, must have occurred within a period of 30 million years. "The actual span," he wrote, "was probably considerably shorter."

Until the development of plate tectonics it was hard to explain what process would suddenly thrust and crumple the landscape in this manner. The erosion of these great mountains spread sediment to the west into New York State and Pennsylvania, producing the heavy deposit that was later uplifted to form the Catskills.

There followed a warm, humid time when great swamps, filled with lush vegetation, developed into what became the Appalachian coal fields. In fact, American geologic terminology refers to these coal-forming periods as the Mississippian and Pennsylvanian. Then, between 230 and 260 million years ago, came the final phase of Appalachian construction, known as the Alleghenian. It was this episode that crumpled the landscape from the Poconos of Pennsylvania through the Allegheny Mountains of Pennsylvania and West Virginia to the Cumberlands of Tennessee, forming a succession of folds and also, apparently, initiating the uplift of the Catskills. While the ridges formed from Pennsylvania to Virginia as a consequence of the folding are strikingly uniform, they are not the original folds. The latter were largely leveled by erosion, which then went to work on what remained. Limestone layers in the folds were quickly eaten away, forming the Great Valley, west of the Blue Ridge, and many parallel valleys. The more durable rocks resisted and now stand as ridges, their layers sometimes protruding vertically from the hillsides.

There were two diametrically opposed schools of thought regarding the crumpling process. John Rodgers of Yale referred to the protagonists as "thin-skinned geologists" and "thick-skinned geologists." The former believed that only the sedimentary rocks, lying several kilometers thick over the basement rock, were folded by a thrust from the side, the basement remaining flat as before, like the floor under a crumpled carpet. This is believed to have occurred when uplifting in the central Alps caused sedimentary rocks to slide north, forming the folds of the Jura Mountains, north of Lake Geneva, even though the underlying basement remained flat. The "thick-skinned" geologists believed that, under the Appalachian folds, slabs of the basement rock overthrust one another, as though, using the earlier analogy, the floor boards beneath a crumpled carpet had also been forced to overlap. The debate was partially resolved by COCORP, the Consortium for Continental Reflection Profiling, whose lines of trucks, thumping in coordinated rhythms, were able to obtain seismic echoes from deep in the crust. By the 1990's Jack Oliver of Cornell University, COCORP leader, said the technique had revealed far more over-thrusting on both sides of North America than had been suspected, but showed folding limited to the sedimentary layers, strongly supporting the "thin-skinned" school.

The first real break in the century-old search for an explanation of how the Appalachians came to be occurred in 1966, soon after Tuzo Wilson and a limited number of his fellow earth scientists became convinced that, indeed, "all the world's a set of slabs." Wilson, it will be recalled, suggested that there was an ancient Atlantic Ocean whose opposing shores, during the Early Paleozoic Era (some 450 million years ago), came together in a series of monumental collisions. He found evidence for this in the puzzling distribution of fossil animals laid down in coastal sediments during that time. These fossils fall into two clearly defined faunal realms: one typical of Western Europe and northwest Africa at that time, which we will call the "European" fauna, and one characteristic of North America, which we will designate the "American" fauna (to geologists they are known, respectively, as the "Atlantic" and "Pacific" faunas). Each of these faunal realms is strikingly uniform and clearly distinct from the other, suggesting that they were widely separated, as though by an ocean. For more than a century, Wilson pointed out, scientists had noted that fauna of the "European" type are found in certain areas along the rim of North America and that "American" fossils lie in certain European rocks (Fig. 16.2).

The latter occur along the coast of Norway, in northern Scotland and northwest Ireland, yet in nearby England and Wales the rocks of that period contain "European" fossils. On the American side, southeastern Newfoundland, facing the open Atlantic, is in this sense "European," whereas on the inland side of that island, across the great fault system that splits it in two, the fossils are of the American type. Similarly Nova Scotia and southern New Brunswick are European, but the rest of New Brunswick displays American fossils. The line of demarcation cuts across southern Maine, a coastal patch of New Hampshire, and then runs down the middle of Massachusetts and through eastern Connecticut along a succession of fault lines to Long Island Sound. Seemingly out of sheer perverseness geologists have named the chief fault of this series for a small body of water in Massachusetts called Lake Chargoggagoggmanchaugga-goggchaubunagungamaugg. It is also known as the Lake Char Fault, since on maps the full name spans a large part of Massachusetts. The older geologic structures on either side of this fault seem unrelated.

Wilson's explanation was that an ancient ocean separated these faunal realms until the North Atlantic closed to form a single supercontinent, Laurasia. Then, when the land broke apart again, giving birth to the modern Atlantic, the line of cleavage was not identical to the earlier line of suture. Part of "America" stuck to Europe and part of "Europe" and "Africa" stuck to America. Thus Boston was once part of Northwest Africa, whereas northern Scotland, much of Ireland, and coastal Norway were welded to Greenland and what is now the Canadian Arctic.

In other words, a full-sized ocean once lay between the lands now occupied by Boston and Pittsfield, and that same ocean separated London and Inverness or Oslo and Bergen. Even though geologists were on the threshold of radical changes in their concepts, Wilson knew his proposition would be hard to swallow. "It may seem strange," he wrote, "to propose that a former position of the Atlantic Ocean lies through New England and that its full significance has not been realized, but it must be remembered that throughout the area outcrops are poor and that most of the mapping is old. Surface mapping reveals few faults, but new tunnels have shown that faults abound under the drift-filled valleys and that some of these are major."

The Appalachian mountain-building occurred, Wilson proposed, when the Paleozoic Atlantic closed. In the final stages of its closure, he said, the continents on either

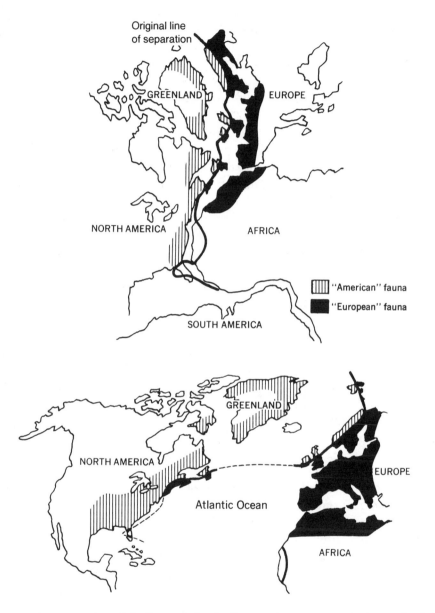

Figure 16.2 Formation of the Proto-Atlantic. *For more than a century, scientists wondered why the remains of shallow water marine animals typical of Europe during the early Paleozoic Era, some 450 million years ago, are found in the sedimentary rock of eastern New England, and why coastal fauna typical of America are found in parts of Norway, Scotland, and Ireland. Tuzo Wilson proposed that these marine organisms lived on opposite sides of an earlier ocean that later closed, welding the continents together (top). These continents subsequently broke apart to form the modern Atlantic, but the line of cleavage did not exactly correspond to the earlier line of suture. Later fossil finds in the coastal Carolinas and Florida show that they had once been part of what is now Africa.*

side "would have touched, first at one promontory and then at another. It can be expected that high mountains would have been formed locally and that they would have produced alluvial fans (of sediment) on both continents. As the ocean diminished the climate would have become increasingly arid." This led to desertlike conditions that, he said, produced the Old Red Sandstone found so extensively on both sides of the Atlantic. Then, as the ocean began opening again in the Jurassic, or age of dinosaurs, regions near it such as Scotland became increasingly maritime in climate, as shown, Wilson said, by the fossil record.

Whether the Old Red Sandstones were really laid down in an arid climate is controversial. Those of a contrary view cite evidence that these areas were not only tropical but at least intermittently moist in annual rainy seasons, as are parts of the Sudan. Such rains can leach the soil until it becomes rich in iron oxides. Then, in the hot, dry season, the soil is baked to a reddish, bricklike material known as laterite. Catastrophic conversion of farmland to laterite in this manner has occurred in Brazil and elsewhere when all protective vegetation was removed. The Old Red Sandstones derive their color, at least in part, from iron oxides, and some seem to have formed in much the same way as laterite.

Today, Wilson pointed out, the Atlantic is a slowly expanding ocean, whereas the Pacific, with garlands of island arcs along its western and northwestern perimeter, is clearly being squeezed smaller. Since island arcs are typical of such a closing ocean, he said, one would expect them to have formed when the Paleozoic Atlantic began to close. This, then, would account for the vestiges of such arcs and their prolonged volcanic eruptions found in the Appalachians.

The first attempt at a detailed, play-by-play reconstruction of how the Appalachians—and their apparent continuation in northern Europe, the Caledonians—might have been formed by the opening, closing, and reopening of an ancient ocean was carried out, appropriately, by two geologists from opposite sides of that hypothetical ocean. They were John M. Bird, then at the State University of New York at Albany, an authority on the American mountains, and John F. Dewey of Cambridge University. Dewey knew the Scottish and Norwegian part of the story and already had made a stab at a partial reconstruction.

As a starting point they drew on a concept of Robert S. Dietz (who had helped develop the sea-floor spreading concept and had given it its name) and his colleague, John C. Holden regarding the manner in which offshore sediments accumulate. The two had gone a step further than Drake's work of a few years earlier in showing to what extent the present situation off the East Coast of North America matches the hypothetical structure of the continental rim that was bulldozed to form the Appalachians (Fig. 16.3).

The key element in their proposition was the role played by the enormous accumulation of sediment on the deep-sea floor just seaward of the continental shelf. This forms a wedgelike heap leaning against the slope of the shelf and, they reasoned, becomes so heavy that it slowly forces the crust of the Earth along the rim of the shelf to subside. As a result, while sedimentation piles up enough material to maintain a relatively uniform water depth throughout the shelf, the accumulation on the outer part of the shelf is greater, since that sector is sinking.

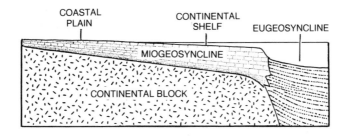

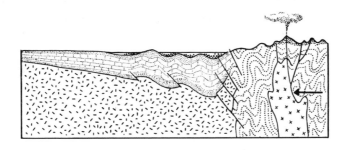

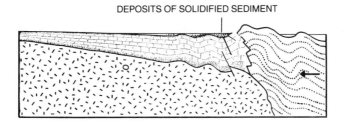

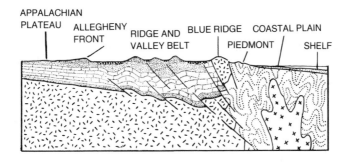

Figure 16.3 Growth of the Appalachians. *Robert S. Dietz proposed that the Appalachians were formed when sediment laid down on the continental shelf of an expanding ocean, a miogeosyncline, and sediment laid down beyond the rim of that shelf, a eugeosyncline, were crushed together by the contracting ocean. Arrows indicate the direction of pressure on the eugeosyncline.*

This increase in the thickness of sediment layers on the shelf at greater distances offshore had long puzzled geologists. One would expect the reverse situation, with more accumulation near the shoreline from which much of the sediment was derived. But a series of wells sunk to bedrock across the coastal plain in the Virginia–Carolina area, including one quite far out on the continental shelf at Cape Hatteras, had confirmed this thickening of the layers to seaward. What impressed Dietz and Holden was that this same effect is seen in sedimentary rocks that now form the Appalachians. Their layers become increasingly thick to the east—toward the earlier ocean that, Wilson said, once lay there. Dietz and Holden facetiously labeled their explanation for the greater sediment thickness out near the rim of the shelf as that "certain sinking feeling." It seems typical of any coastline facing an expanding ocean.

In this respect, South America today displays, in a relatively straightforward manner, the two types of coastline generated by continental drift: Its west coast, with continuous mountain ranges, volcanoes, earthquakes, and an offshore trench, is typical of a sea floor descending under a continental plate; its east coast, riding placidly westward as an integral part of the plate forming the Southwest Atlantic, is accumulating undisturbed sediment on its submarine shelf and on the fringing deep-sea floor, the continental rim slowly sinking under the weight of this accumulation. The subsidence can also probably be attributed to slow cooling and shrinking of the upper mantle that supports the continental rim. This material was hottest when it rose toward the surface along the mid-ocean ridge and has cooled progressively as sea-floor spreading carried it away from that zone. The East Coast of the United States from Maine to Virginia, as reported in 1972 by the National Geodetic Survey in Washington, is sinking at a rate of from half a meter to a full meter per century, although this is partly attributable to sea level rise. In the short span of 11 years, from 1959 to 1970, parts of the Gulf Coast in Texas and Louisiana sank a half meter. By the 1990's concern over sea-level rise had been greatly enhanced by the threat of global warming and extensive melting of polar ice.

Bird and Dewey recognized in the Dietz–Holden proposal a key element in explaining how the Appalachians could have been formed by the opening and closing of a Paleozoic Atlantic (Fig. 16.4). The traditional view that the building materials of such a mountain system—the geosyncline (more precisely, miogeosyncline and eugeosyncline)—had to form in troughs where the mountains now stand was no longer necessary. This material, in an expanding ocean, could have accumulated offshore—even far out to sea—much like the sediments now being laid down along the eastern margins of North and South America.

The first step in development of the Appalachians they saw as the opening of the Paleozoic Atlantic and the initiation of heavy sediment accumulation on the continental shelf and beyond its rim. The initial mountain-building episode—the Taconic—then began, in the Bird–Dewey reconstruction, when, some 450 million years ago, the by then fully grown ocean started to contract. As it did so, the sea floor began descending under the rim of North America, which at that time had been eroded into a coastal plain much like that of the Gulf Coast (although possibly with an offshore chain of extinct volcanoes). This coastal plain had subsided below sea level (possibly because of

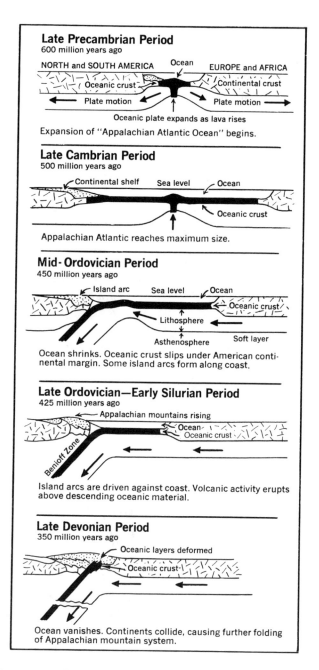

Figure 16.4 Formation of the Appalachian Mountains *as envisioned by John M. Bird and John F. Dewey.*

upper mantle cooling during expansion of the Paleozoic Atlantic) and a shallow sea reached as far inland as Minnesota. As the sea floor began to descend under the submerged rim of the continent, a trench was formed, new volcanic islands were born, and mountains arose along the continental edge, much in the manner of the Andes.

The next chapter in this history—the Acadian—began with the first close approach of the converging continental margins. The offshore sediments were thrown over the island arcs and coastal mountains in great nappes. As the sea floor was driven down under the continent, the island arcs themselves may have been shoved westward, their granite roots surviving as the domes that lie along the east side of the Connecticut Valley. Similar processes were probably taking place on the opposite side of the closing ocean, and the outer line of volcanic islands, in Kay's double rank of island arcs, may have formed along the opposing coast and been swept into the convergence zone (see Fig. 16.5).

"How far offshore these islands lay is quite uncertain," says John Rodgers of Yale, since the intervening ocean floor vanished into the mantle. The overlying sedimentary material "was folded like an accordion and converted into the present metamorphic core of the northern Appalachians..." There is, in fact, disagreement as to the width of the Paleozoic Atlantic. Some reckon it in thousands of kilometers, others in hundreds.

The big squeeze, when it finally closed, seems to have been initially in New England and then farther south. In Newfoundland and the Maritime Provinces of Canada the ancient volcanic debris along the coast is separated from the folded basement rock of the earlier continent by about 300 kilometers (200 miles). Yet in the central Appalachians the separation is only 40 kilometers (25 miles), and in Baltimore

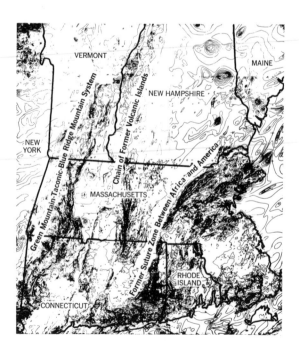

*Figure 16.5 **Magnetic survey of New England** plotted by Isadore Zeitz and his colleagues at the United States Geological Survey. The map clearly shows the suture zone between the former Euro-African and American plates, as well as other relics of the region's history.*

the two zones seem to come together. Rodgers believes this sector may be only one quarter to one fifth its original width.

It was during this period that parts of the Euro-African rim became welded to North America, leaving behind fringe zones from Canada down through New England and again in the eastern Carolinas and most of Florida. Since the Euro-African side of the ocean had a similar history, it is possible that some of the volcanic ruins now embedded in Norway may have originated as islands that formed along the American side of the ocean and then, after the collision, became incorporated into Scandinavia.

At this time, according to present reconstructions of plate motions, the Appalachian–Caledonian mountain system lay parallel to, and not far from, the Equator (as shown in the map for the Permian Period, Fig. 5.8). Although the ocean was greatly diminished, and its northern part completely snuffed out, the climate was moist enough to produce great coal-forming swamps on both sides of the suture line—from Spitsbergen through Nova Scotia to Mississippi on the American side and through Britain, Belgium, and the Ruhr region of Germany on the European side. Thus, some 300 million years ago, the coal was formed that made possible the Industrial Revolution of the 19th century.

Then, about 230 million years ago, the final onslaught began. Until then only what is now the northern part of the coastlines had collided, with ocean remaining to the south (in terms of today's "north" and "south"). Bird believes that continued closure along what is now the southern Connecticut coast and across Westchester County, north of New York City, pushed the lower part inland as the opposing continental mass came into contact with the American side. The rugged landscape behind such Connecticut coastal towns as Cos Cob may be a relict of fracturing along this hypothetical "transform fault." The pressure farther "south," in terms of present geography, produced the Alleghenies and the ridge-and-valley province of Pennsylvania and the Virginias. If those features look as though formed by some gigantic earth-moving machine, it is because that, in a sense, is what happened, the "machine" being the bulge of West Africa that, in the final contraction of the Paleozoic Atlantic, drove against the American plate off Cape Hatteras. While the folding is awesome to the motorist or air traveler, its full scope can be seen only in the images transmitted from earth satellites (as shown in plate number 4).

It appears that what was once a single mountain system has now been broken into fragments as widely scattered as the ranges of Spitsbergen, Norway, Scotland, and the Appalachians, stretching from Newfoundland to Alabama and perhaps beyond. It was a noble range, probably more than 6400 kilometers (4000 miles) long, and testifying to its continuity is the consistency of the sedimentary sequence along most of its length.

Among the by-products of the sea-floor bulldozing that produced these mountains was a succession of marble deposits formed by the sweeping up, heating, and kneading (under great pressure) of limestones and other rocks laid down in the fringing seas by the steady rain of tiny shells. The resulting marbles are now found from the quarries of western Vermont and Massachusetts, down through the Inwood marble of New York (which at White Plains reaches a thickness of 600 meters) to the Cockeysville marble of Maryland, the Murphy marble of North Carolina and Georgia, and the Sylacauga marble of Alabama.

The events that shaped the American side of the ocean had their counterparts in Europe and, to some extent, Africa. Patrick M. Hurley of the Massachusetts Institute of Technology, whose group has reconstructed geologic timetables by extensive age determinations of rock formations, has pointed out that the Grenville phase of American mountain-building also shaped the Channel Islands, off the French coast, as well as parts of Normandy, Scotland, and Norway. The Taconic and Acadian episodes in America were matched by the Caledonian mountain-building that molded much of the landscape in Britain and Norway. Finally the Alleghenian upheaval, chiefly manifest in the southern Appalachians, coincided with similar activity in southern Europe and North Africa, known there as the Hercynian.

Just as the collisions of opposing coastlines left their mark on the continents, so did the rupture that finally split them apart to produce the Atlantic Ocean of today. As in Africa, several rift valleys seem to have been split apart by this process, one of which finally tore open to produce a new ocean. Others survive as the Newark Trough, extending south from Haverstraw and the Jersey Meadows, east of the Hudson, to Virginia, or as the lower Connecticut Valley, from the Vermont–New Hampshire border to Long Island Sound. There are even deeply-buried traces as far south as Florida.

Like the African rifts, these troughs apparently held a succession of elongated lakes and swamps whose mud hardened into reddish rock that has colored the landscape ever since. In the Connecticut Valley it is evident in highway cuts along the full length of Interstate 91, from the coast to Vermont, and there are sites where the rock still carries the three-toed footprints of dinosaurs who once prowled the valley. In the Newark Trough this material has tinted the soil from New Jersey down into Virginia.

Again as it did in Africa, the rifting opened cracks through which lava erupted to produce a system of basalt dikes, sills, and lava flows on both sides of the developing Atlantic. Some of these great basalt beds were tilted, their exposed edges forming great cliffs. The most dramatic are the Palisades along the west bank of the Hudson River, but other examples include the battlements overlooking Paterson, New Jersey, and New Haven, Connecticut. An almost continuous series of bluffs and mountains extends from Lake Saltonstall, east of New Haven, north to the Holyoke Range, topped by Mount Tom in Massachusetts. Travelers up Interstate 91 climb over the shoulder of this feature about midway between New Haven and Hartford (Fig. 16.6).

Across the juvenile Atlantic similar features were forming in Africa and, eventually, in Europe, as the rifting migrated north. In the final rifting that separated Greenland, Iceland, Rockall Bank, and Britain, basalts that erupted along the coasts of Scotland and Ireland cooled and shrank to form the famous hexagonal columns of Fingal's Cave, lapped by the sea in the Hebrides, and the Giant's Causeway on the Irish side of the water.

A striking feature of the dikes on both sides of the ocean, when they are plotted on a global map with the continents reassembled, is that they seem to radiate from the area embracing the Bahamas and the submarine plateau off Georgia. Does this mean that a plume, rising beneath that area, was responsible for pushing the continents apart and giving birth to the modern Atlantic? Paul R. May of the Inexco Oil Company has proposed that this alignment of dikes could represent "a stress field imposed on the

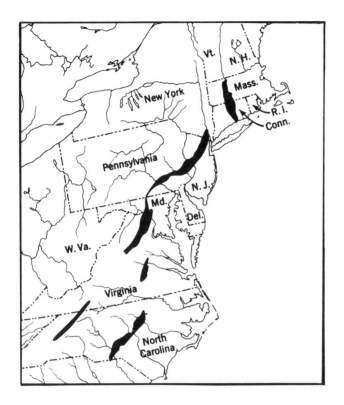

Figure 16.6 Newark Series of Triassic intrusions from the Holyoke Range in Massachusetts to the Hudson River Palisades and south to the Carolinas fill rifts produced when the Euro-African plate pulled away from the American plate.

continental crust by movements in the upper mantle at the onset of North Atlantic seafloor spreading."

If, as Bird, Dewey, and others assumed, an early version of the North Atlantic was wide open during the Ordovician Period, 450 million years ago, where was Africa? In 1962, French geologists from the Institut Français du Pétrole found extensive signs of glaciation in the Sahara and in 1970 the Algerians organized an international expedition to confirm that Africa may have been near the South Pole! The French, searching for oil accumulations in the Ordovician sandstones of the Sahara, had reported widespread glacial scars where such rocks were exposed on the surface. Yet the idea of a great polar ice sheet in the region that today is the world's hottest, with temperatures as high as 58 degrees Centigrade (137 degrees Fahrenheit), seemed incredible to some, and it even was suggested that the scars might have been produced by landslides when there was more slope to the terrain than today. Hence, in January 1970, the Algerian Petroleum Institute organized an expedition to explore the evidence. With attention to political neutrality it invited scientists from 10 countries in both East and West—a mixed bag of specialists in various fields. They crisscrossed the desert by plane, by Land Rover, and, to some extent, on foot, their search centering on an extremely rugged massif that rises from the central Sahara, known as the Hoggar. It is ringed by oases with names, like Tamanrasset, which evoke memories of the French Foreign Legion and frontier battles. On its northern flank is the broad Tassili Plateau, famous for its stone-age cave paintings.

The expedition members began to recognize a variety of clues. As reported afterward by Rhodes Fairbridge of Columbia University, one of the participants, they found, for example, "magnificent outwash plains"—great aprons of sand, gravel, and other material typically spread by water flowing from the frontal cliffs of an ice sheet when it begins melting. The entire South Shore of Long Island, including part of New York City, is such a plain, laid down by ice sheets whose farthest advance never progressed beyond the midline of that island. The outwash plains in the Sahara were cut by channels carved by the meltwater of an ancient ice sheet that, according to Fairbridge, extended "over thousands of square kilometers." The expedition found ice-etched boulders whose composition indicated they had been transported long distances from their native formations and scrape-markings on the sediment of a sort typically made by icebergs dragging their bellies across a shallow coastal shelf. The weight of the land-based ice may have depressed this part of the continent below sea level, and the seas themselves may have been higher. The continental ice sheet would then have pushed out over the water like the great 300-meter-thick ice shelves of Antarctica. The explorers found a remarkable sequence of giant ripples in the sandstone, spaced three meters from crest to crest and extending for several hundred kilometers. Such ripples are formed on the floors of ocean channels by very powerful tidal currents. But, Fairbridge asked, might not the sandstone ripples of the Sahara "represent the decanting of millions of tons of meltwater from the margin of the Ordovician glaciated region during the melt period"?

He and his companions saw what seemed to be the vestiges of eskers—those snaking ridges of rocky debris left on the landscape where a river once flowed beneath the ice. In aerial photographs they had noticed circular rock features from 100 to 1000 meters wide that some had proposed were meteorite craters in the desert; but from on-site examination Fairbridge concluded they were volcanoes that had erupted under the ice sheet, much like the process that in Iceland today produces a *jökulhlaup* or disastrous release of meltwater from beneath the ice cap of a volcano.

"Then," wrote Fairbridge, "the great day came. In the eastern part of the Hoggar, in the valley of the Wadi Taffassasset, we came across mile upon mile of superb parallel striated pavement. It was running from south–southeast to north–northwest, just like some giant bulldozer scraping. We traced the striated pavements on the ground for miles. Later, from the air, we followed them across hundreds of miles of territory, and others have confirmed their continuation right across northwest Africa. We had proof positive!" Such grooves are found only where a great ice sheet has passed by, dragging rocks that carve the bedrock like some huge machine tool. They can be seen in Antarctica, or in those areas of Eurasia and North America where ice sheets once existed. They are clearly evident, for example, in New York's Central Park, in ledges between the zoo and skating rink in the park's southeast corner.

"So we were convinced," Fairbridge said. "And our good hosts broke out the iced champagne, there, in the middle of the desert, and we toasted the explorers."

It appears that the Ordovician ice sheet covered all of northwest Africa, from Libya and Chad in the east to the Atlantic seaboard, at least as far south as Sierra Leone (Fig. 16.7). While continental drift may not be entirely responsible for such ice ages, its role has been critical. When a continent lies at one of the poles, as does Antarctica

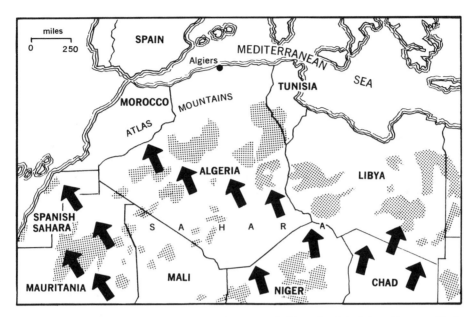

Figure 16.7 Saharan glaciation. *Scars and grooves etched in rocks of the Sahara Desert testify that a great ice sheet once flowed from there to the sea. The arrows show the directions of flow indicated by these scars. The shaded areas are ergs, or desert areas of shifting sands.*

today, a great ice sheet forms. When the pole is within an open ocean whose circulation continually transports warm water toward that area, an ice age in that hemisphere is unlikely. Today the flow of warm water into the Arctic Ocean is geographically constrained and so ice ages have periodically occurred. The reason for their cyclic behavior at intervals of about 100,000 years during the past one or two million years has been widely attributed to a combination of cycles affecting the tilt of the Earth's axis relative to its orbit around the Sun and changes in that orbit, determining its distance to the Sun at critical seasons.

The idea of an ice age in the Sahara might seem more absurd were there not other hints that the South Pole was near Morocco during the Ordovician. For example, magnetic analyses of Ordovician rock samples from Africa, South America, and Australia point to a magnetic pole in that vicinity. Marine fossils of that period found in northern Europe are warm water varieties, and coral was flourishing along the coasts of Russia and Siberia, but in southern Europe the fossils testify to cold waters, suggesting that a polar region lay to the south. While part of Africa was glaciated, as well, in the Permian period, 200 million years after the Ordovician ice age, this occurred in South Africa, lapping over into South America, Australia, and Antarctica, at a time when the general northward drift of the continents had moved the geographic location of the South Pole a large fraction of the way toward its present position (Fig. 11.5).

Evidence has also been found that, during the Proterozoic Era, some 600 million years ago, all lands that combined into the continents of today lay near the Equator. Yet all, at one time or another during that period, were glaciated. In 1987 an international project was initiated, with two geologists, one Soviet and one American, as leaders, to seek a solution to this paradox. It has even been proposed that the Earth, at that time, was encircled by a ring of icy particles, like Saturn, shadowing the Equator. As of this writing no obvious explanation is at hand.

Perhaps the most important discovery, from the attempts to reconstruct the growth of the Appalachians, is not how they came to be, but how the continent grew in the process. While some land was stolen from Africa and Europe at the final breakup, more was added by the sweeping up of island arcs and sea-floor sediment onto the eastern seaboard of North America. It seems possible that this process, repeated many times over the vast span of the Earth's history, has done much—perhaps a great deal—to build the continental plates on which we live.

Chapter 17

The Search for Universality

CAN ONE BELIEVE THAT 7000 KILOMETERS (4300 MILES) OF OCEAN FLOORS HAVE slipped under the western rim of North America? Or that an equal amount has vanished beneath the Pacific coast of Asia? Or that an ocean possibly as large as the Atlantic once lay between Russia and Siberia? Such concepts tax one's credulity. Yet they have emerged from efforts to apply the new theory of crustal plates to all mountain systems of the world, including those that now lie far inland, such as the Urals and Rockies.

These bold attempts at universality were made by a number of geologists and Earth scientists. Bird and Dewey, who tried specifically to explain the Appalachians in this way, also explored applicability of the theory to other regions and ranges. Dan McKenzie and Jason Morgan tried to explain the role played by various types of encounters between moving plates. Warren Hamilton of the United States Geological Survey applied the theory to the Andes and Urals and, with a number of others, tackled the complex history written in the mountains and valleys of the western United States.

At first it all seemed beautifully simple. Mountains like the Alps and Himalayas derived from continent–continent collisions. The parallel ranks of mountains, or cordilleras, flanking the western shores of the Americas arose along the rim of a continent that was overriding an oceanic plate. When an oceanic plate dives under another oceanic plate, an island arc is formed, such as the Aleutians. But, as so often happens when phenomena that look simple from a distance are examined in detail, these situations have proved complex. While there has been an outpouring of proposals in such journals as *Nature* and *Science* seeking to apply plate tectonics to every part of the Earth's surface, there also have been published reminders that universal application of the theory still raises serious problems—problems which many Earth scientists believe eventually will be solved.

The most recent continent–continent collisions were those between India and Asia and between Europe and the plates (African and Mediterranean) lying to the south. In both regions the pressure is still on, pushing the Himalayas and Alps still higher. From deep-sea drilling in the Indian Ocean it appears that India first began nudging the Asian plate some 50 million years ago, but that the growth of the Himalayas has been far more recent—chiefly during the past 10 million years. While erosion could not keep pace with the rise of the Himalayas, great gorges have been cut and vast volumes of sediment carried to the lowlands and oceans. The Ganges and Brahmaputra Rivers have carpeted the Bay of Bengal with sediment that, in places, is 16 kilometers (10 miles) thick—the greatest such modern accumulation anywhere. A lesser carpet has been laid down on the other side of the Indian subcontinent by the Indus River. Much of the alluvium of China's rice-growing region consists of soil eroded from the new mountains and spread by floods of the Yangtze and Yellow Rivers, preparing the way for the rise of one of the earliest and greatest of civilizations.

The Himalayan Plateau now lies more than 4000 meters higher than the region to the south. "Looking north at the Himalayan rampart from the resort city of Darjeeling," Dietz has written, "one sees the only scarp in the world as grand as the continental slopes, which surround all continents." The reason is straightforward enough, he said. One is, in fact, looking at a continental slope like those which elsewhere lie buried under the sea. "The Indian continental plate," he said, "has underridden the Asian plate, producing a double continent 70 kilometers (44 miles) thick." The actual picture, however, must be more complex, for the continental plates, as rigid sections of the lithosphere riding the plastic asthenosphere or soft upper mantle beneath them, are probably more than 100 kilometers (60 miles) thick, and, if one piled atop another, the resulting mountains would be awesome indeed.

What some believe occurs, in a continent–continent collision, is "interfingering." The colliding upper layers of relatively light rock invade one another, like the fingers of two joining hands, piling up, as well, and crumpling to form a broad mountain zone. Deep down, however, the heavy, underlying rock of one contending plate "surrenders" and is forced down into the soft asthenosphere, perhaps (at least in part) to soften and disperse. While this process may produce an unusually broad mountain belt, like the Himalayas and the Tibetan Plateau behind them, such a collision may eventually stall and force a basic change in global plate movements. By the 1980s the Himalayas, Tibetan Plateau, and Western China had been intensely studied, leading to the hypothesis (elaborated in the last chapter) that not one but several successive collisions had occurred as India sailed toward Asia. (A view of the serried ranks of the Himalayas, as photographed by astronauts in Earth orbit, appears as colored plate number 8.)

The idea that an ocean once lay between Russia and Siberia has arisen from efforts to explain the origin of the Urals that now separate those regions, forming a natural boundary between Europe and Asia. Soon after Wegener presented his arguments for continental drift, Emile Argand and others who saw an explanation for the Alps in the crushing of the Tethys Sea, proposed that the Urals might also have been formed in this manner. The possibility that Russia and Siberia were once widely separated was later reinforced by the analysis of magnetic "compasses" frozen into rocks formed in those two regions between 400 and 700 million years ago. Only if they were far apart could they have pointed to the same magnetic pole. As progressively younger rocks were studied, their magnetic orientation came more and more into conformity, as though the lands were slowly drifting together. According to the Soviet scientist P. N. Kropotkin, these findings "show that the Siberian and Russian tectonic platforms approached each other during the Paleozoic by some 3000 kilometers (2000 miles). In this interval of time, the formation of the folded system of the Urals, Western Siberia and Kazakhstan, which are distributed between these two platforms, proceeded."

Warren Hamilton of the United States Geological Survey has proposed a detailed scenario for this process that includes the subsequent opening of the Arctic Ocean and a counterclockwise rotation of northern Alaska away from the archipelago of the Canadian Arctic: During the early Paleozoic, an ocean that then lay between Russia and Siberia began to shrink, and that section of its floor forming an integral part of the European plate descended under an island arc an undetermined distance from the Russian shore. As this plate-consuming process continued, the island arc finally came up against the continental part of the European plate along the line where the Urals now rise. Since the continent would not go down under the island arc, the direction of descent reversed itself in the same kind of "flip" postulated for the Alps (see Fig. 15.5). The sea floor beyond the arc—that is, on the Siberian side—began to descend under that arc and under the European plate up against which the arc had been pushed. Meanwhile, a similar process was occurring along the rim of the Siberian subcontinent, and when the two plates collided, at the end of the Paleozoic some 200 million years ago, much of the island arc material was crunched into the newly forming mountain belt. The result of this long and complex history of ocean-floor swallowing was a series of mountains, fragments of which survive on the Taymyr Peninsula (facing the Arctic Ocean), in Novaya Zemlya (the long, curving Arctic island where the Russians conduct nuclear tests), down through the Urals (where one finds the ophiolites typical of such suture zones) to mountains along the southern margin of the old Siberian plate, in Kazakhstan. Hamilton proposed that this mountain system may even have extended into northern Greenland and the islands of the Canadian Arctic before the newly mated land mass of Eurasia pulled away from North America, opening the Arctic Ocean and swinging Alaska away from the Canadian Islands.

Assuming that the total surface area of the Earth has remained the same and that the percentage of that surface occupied by continents has grown only at a slow rate, then, when an ocean like the Atlantic closes, another ocean must open to keep the "budget" of continental and oceanic areas in balance. Aleksandr V. Peyve, director of the Geological Institute of the Soviet Academy of Sciences and one of the first members of the Soviet scientific establishment to accept at least some elements of the plate theory, has proposed that, as the ocean between Russia and Siberia (which he thinks was once comparable to the Atlantic) shrank, the Tethys Sea grew. And, as Africa moved north and the Tethys shrank, the Atlantic and Indian Oceans grew. On a recent visit to this country Peyve likened the origin of the Urals to that of the Appalachians.

While, as in the photograph taken at the end of a football scrimmage, it is difficult, where continents have collided, to reconstruct the steps leading up to that situation, it is evident that some of those steps are taking place within the oceans of today, notably in the formation of island arcs such as the Aleutians and coastal mountains like the Andes.

Study of these processes has thrown light on how a variety of mountain systems came about.

As noted earlier, the line along which an oceanic plate begins its descent is marked by a trough in the sea floor, and the path of the plate down into the Earth's interior is a sloping penetration under the continent or island arc delineated by earthquakes of ever-increasing depth. As the descending plate penetrates deeper, it reaches levels that are increasingly compressed and hot. At a depth of about 80 kilometers (60 miles) some components of the plate begin to melt and, being lighter in weight than the surrounding rock, force their way upward to erupt as volcanoes. Along the length of the Japanese arc this zone forms a clearly defined "front" marked by the first rank of volcanoes. The front typically lies between 100 and 250 kilometers (60 and 160 miles) from the trench, depending on the steepness of the plate's descent—that is, how far from the trench it reaches depths of 80 kilometers or more.

As the plate penetrates deeper, components richer in potassium tend—at least in some areas—to be included in the erupted lava. Therefore, in an island arc the potassium content of the volcanic rock tends to become greater at increasing distances from the trench (Fig. 17.1). This is more than a scientific curiosity. The potassium levels in lava across the volcanic zone generated by descent of a long-vanished oceanic plate have been used by some geologists as the one-way sign on a city street, indicating, in a mountain range like the Urals or the Sierra Nevada, the direction in which the plate made its descent when those mountains were being formed. This tells a great deal about why the mountains grew where they did and has been used both by William R. Dickinson of Stanford University in California, working with Trevor Hatherton of New Zealand, in the Sierras, and by Hamilton in his analysis of the Urals.

Another critical process that occurs where a plate descends is the manufacture of great masses of molten granite that work their way up until they cool and solidify into agglomerations known as batholiths. These may remain buried as reservoirs of hot rock or be exposed by erosion. Some of the great batholiths are tens of thousands of square kilometers in extent.

Such granite intrusions, as previously described, occur in the Alps, along the Connecticut Valley, and as major features in western North America (Fig. 17.2). Although granite is the chief building material of the continents, its origin has long been debated without resolution. Now, while the plate theory has not provided all the answers, it has explained the setting in which batholiths occur: This is where material melted out of a descending oceanic plate has worked its way up through a great thickness of rock. Such batholiths do not appear near the surface in most island arcs, but are widely viewed as being the "roots" of the great volcanic intrusions that have occurred along continental rims. In fact there is a growing suspicion that such activity is a means whereby continents have grown, enabling them to increase their area from successive episodes of drift and collision.

Another category of "continental" rocks, the chaotic mixtures of flysch, lava, and other forms previously hard to explain, has now become comprehensible in terms of what happens in and near the trenches. Where an oceanic plate descends at a trench into the mantle, a certain amount of material riding its back is scraped off onto the far side of the trench. Top-to-bottom chunks of sea floor (ophiolites) are also broken off, and all

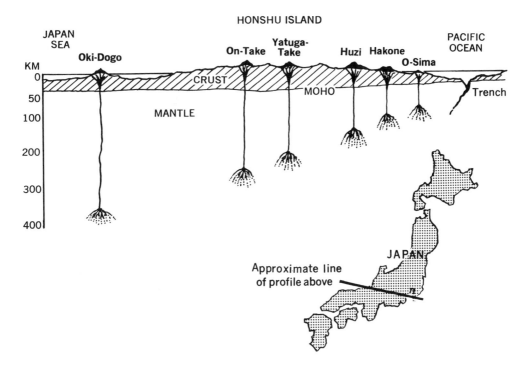

Figure 17.1 Formation of Japanese volcanos. *The composition of lava erupting in Japan changes with its distance from the Pacific Ocean. In the 1950s, Hisashi Kuno proposed that this systematic change was related to the depth at which the lava had originated—the greater the potassium content, the greater the original depth. William R. Dickinson and others later showed that this change is characteristic of other areas where one plate of the earth's rigid surface is descending beneath another. This diagram is adapted from one published by Kuno in 1959 in the* Bulletin Volcanologique.

this material is mixed with sediment and volcanic debris from the islands or the coast beyond the trench, much of it delivered by turbidity currents. The result, after an ocean floor thousands of kilometers wide has "gone down the drain," is a hodgepodge accumulation, on the inland side of the trench, that may become incorporated into the mountains thrown up by a continent–continent collision or other plate interaction. Examination of a pile of snow and dirt swept up by a bulldozer may disclose a chaotic mixture of snow, grass, dirt, gravel, and clay. If one had no idea that a bulldozer had been at work, it would be a very puzzling formation. Similarly the chaotic coastal accumulations are hard to understand until one thinks of a whole oceanic plate being swallowed down a trench. Mélanges of this type are found in the Apennines, forming the spine of northern Italy, in the Coast Ranges of California, in Japan, New Zealand, and other locations around the Pacific.

A number of questions center on the island arcs that fringe ocean basins, chiefly in the Western Pacific, and on the marginal seas that lie behind them, such as the Japan Sea and Philippine Sea. The youngest part of the sea floor behind these arcs—that is, between them and the mainland—is that which lies closest to the arc, the floor becoming progressively older (and displaying decreasing heat flow) toward the continent, as

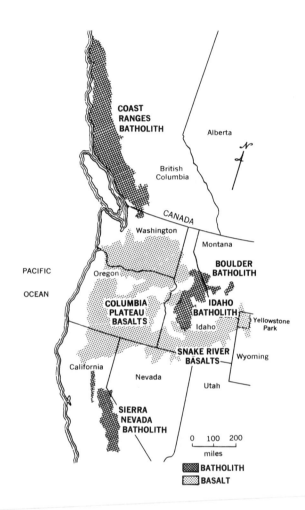

Figure 17.2 Batholiths of Western North America are believed to be the roots of great volcanic intrusions. The basalts, on the other hand, were formed by great outpourings of lava.

though the island arc were slowly migrating seaward, leaving new, shallow sea floor behind it. In 1989 cores extracted from the floor of the Japan Sea by the Ocean Drilling Project indicated that the sea began forming when Japan broke away from the mainland 20 million years ago.

One wonders why some fronts along which an oceanic plate is descending manifest island arcs whereas others, like the west coast of South America, do not. Why do the arcs, at least currently, tend to occur in the western parts of the oceans? In 1972 Tuzo Wilson and Kevin Burke proposed that where a continental plate is actively advancing, overriding an oceanic plate, one finds a coastal trench and cordillera, as in South America. But where it is the underthrusting sea floor that is the "aggressor," moving relative to the deep interior, an island arc forms. This, they said, could explain why the Southeast Asian region is surrounded on three sides by island arcs: the Burmese–Indonesian Arc, the Philippine Arc, and the Taiwan–Luzon Arc, all seemingly formed by the descent of plates under a relatively stationary block. (Figure 17.3 shows a volcano in Bali, evidence of the descending plate there.)

Figure 17.3 Volcano formation on the island of Bali in Indonesia seen in a side-scan radar image. It is almost perfectly conical.

Some have suggested also that the occurrence of island arcs on the western sides of oceans is related to a tendency of the plates to drift westward as a consequence of the Earth's spin to the east. Or, it has been proposed, such arcs simply form in the early stages of convergence between continental and oceanic plates and are later swept up against the continent, as hypothesized for the Appalachians and Urals.

Ever since the concept of sea-floor spreading was born, the Southern Andes have been regarded as a classic example of what happens when a continent overrides an ocean floor. There is a trench parallel to the Chilean coast, an earthquake zone that slopes down under the continent from below the trench at an angle of about 25 degrees and volcanic mountains that begin to rise some 300 kilometers (200 miles) from the trench line, where the descending plate has reached sufficient depth to begin producing lava. But a comprehensive and generally accepted explanation of how the Andean ranges developed as a consequence of plate motions has not yet come forth.

In the Central Andes (see color plate 11), notably in Peru and Bolivia, two ranges parallel the coast. The most ancient is the inland, or Eastern Cordillera, formed in the Paleozoic partly from ocean sediment that seems to have been thrust eastward, away from the sea. Nearer the coast is the Western Cordillera that is younger (Mesozoic and Tertiary). An accumulation of sediment between them that, in places, may exceed 15 kilometers (nine miles) in depth constitutes the Altiplano, where the Incas once ruled. The Western Cordillera consists of two belts, the inland one being of folded sedimentary rocks of shallow-water origin (a miogeosyncline). The other belt, closer to the sea, is formed from trench or other deep-water material (a eugeosyncline).

As elsewhere above a descending plate, there is volcanic activity along much of the Andean system. In 1986, for example, the volcano Nevado del Ruiz in Colombia sent a deluge of soggy ash over the town of Armero, burying some 21,000 people. Cotopaxi, rising 5943 meters (19,498 feet) above sea level in Ecuador, is the tallest active volcano in the world. But among the Andean puzzles is the manner in which volcanic activity there has migrated eastward, away from the coast. If descent of an oceanic plate were plastering new material onto the continent, causing it to grow westward, one would expect the trench, the volcanoes, and other manifestations of the process to move westward as well. Instead volcanic activity has moved in the opposite direction. David E. James of the Carnegie Institution of Washington, much of whose career has been devoted to Andean studies, believed this eastward movement may have occurred because prolonged descent of the Pacific slab (the Nazca Plate) under the continent has cooled the upper mantle there. Such cooling would mean that the descending plate would have to penetrate deeper—and farther inland—than before to reach high enough temperatures to induce volcanic activity.

It has been calculated that, because of the relative westward movement of South America and the eastward motion of the Nazca Plate off the coast of Chile, that oceanic plate may be descending under the continent at the extraordinary rate of one meter (or more) per decade. Presumably thousands of kilometers of sea floor have disappeared under the continent during the present phase of plate movements. Not only should this have added to the Pacific side of the continent, but that is where one should find the youngest rocks. Yet in southern Peru, relatively near the coast and some 300 kilometers (200 miles) west of what is thought to have been the earlier continental margin, there are rocks of continental type at least 400 million years old. James believes these may be remnants of a miniature continent or peninsula that lay off shore and was swept up against the coast.

Another problem, noted by skeptics early in the debate on sea-floor spreading, is that sediment layers on the floor of the offshore trench show, in many cases, no sign of the deformation one would expect if the ocean side of the trench were descending so rapidly. The sediment layers within reach of a sub-bottom sounder seem to lie quite flat. Part of the explanation may be that the typical sounding device is unable to "see" the fine-structured evidence of deformation. It has also become evident that near active trenches, earthquakes and turbidity currents lay down layers of sediment in such rapid succession that they mask the effects of subduction (sea-floor swallowing). In the Middle America Trench, for example, dating of the layers by analysis of their radioactive carbon has shown sedimentation rates as high as one meter per thousand years, and drill holes near the Aleutians have disclosed ice age rates as high as 2400 meters per million years. Another proposal is that earthquakes flatten the soggy layers by shaking them.

While the mountains of western North America are presumably the children of processes much like those that produced the Andes, there are obvious differences, particularly in the United States. From Cape Horn, at the southern tip of South America, to Alaska the belt of mountains and past volcanic activity facing the Pacific forms a roughly uniform zone 500 kilometers (300 miles) wide. But in the United States this zone triples in width, and the belt displays oddities such as the "basin-and-range province" that lies between the Colorado Plateau and the Sierra Nevada, consisting of a succession of more than a dozen north–south ranges, separated from one another by broad troughs.

Attempts to learn how these mountains came to be have been greatly aided by new discoveries and new techniques, notably the use of radioactive components of rocks to determine when they were formed. The earliest well-documented upheaval on the West Coast occurred during the Jurassic Period, heyday of the dinosaurs, when granitic batholiths began forming (probably underneath volcanoes) along the axis of the Sierra Nevada. Dickinson at Stanford and Warren Hamilton have concluded that this material was derived from an oceanic plate descending quite steeply (at about 50 degrees). This is the descent slope typical of an island arc, and Hamilton has suggested that the northern Sierras and the nearby Klamath Mountains include remnants of such an arc that was swept against the West Coast.

Beginning in this period, some 170 million years ago, great slabs slid or were pushed eastward across Montana and Wyoming. Chief Mountain in Glacier National Park is an eroded remnant of such a slab, the Lewis Overthrust, that probably moved as much as 50 kilometers (30 miles) east. It is hardly a match for the Matterhorn and other remnants of the great Alpine nappes, and some believe the moving force may have been downhill sliding, as in the Taconics, rather than a shove from the west.

As in the Andes, the zone of volcanic activity also migrated eastward. Before 80 million years ago it was largely confined to California and western Nevada. From 70 to 65 million years ago it was active in north-central Nevada and east to the southern Rocky Mountain area. By this time a very fundamental change had occurred beneath western North America, and the mountain states, particularly Colorado, had begun to rise in what has come to be known as the Laramide Revolution. The rocks formed before this, in the Rocky Mountain area, were mostly marine shales. Now continental sandstones began to be laid down.

A later phase, which still puzzles many of those seeking to explain this history, was the uplift of the Colorado Plateau—the region surrounding the "Four Corners" where Colorado, Utah, Arizona, and New Mexico meet. Like the floor of some huge elevator, this plateau was lifted vertically almost two kilometers (more than a mile) with little or no disturbance of its flat-lying sedimentary rocks, as is now dramatically evident where those layers have been exposed. As it rose, water cut into the layers to produce some of the most magnificent natural wonders of today's world: the Grand Canyon of the Colorado, as well as Glen and Zion canyons and countless buttes and mesas.

Meanwhile the central Rockies were growing. In the region now occupied by the lofty San Juan Mountains of southwest Colorado there were great eruptions that, by the Oligocene, some 35 million years ago, had blanketed much of the area with andesite— the rock typically erupted along the Andes and in similar zones above a descending plate. Then, about 30 million years ago, a major transformation in the nature of this volcanic activity was initiated, beginning in such southern areas as West Texas and moving northwest. The erupted lava changed to the kind of basalt one finds in regions where a continental crust has been torn apart, like the African rift valleys and along the shores of the Red Sea.

Within a few million years this change had affected the entire region south of a line from central Colorado to southern California. By 10 million years ago this line had moved north until it extended from southwest Idaho through central Nevada and central California. Prior to this, eruptions of the earlier type had flooded a vast region of northwest Wyoming and southwest Montana with andesite whose composition was strikingly like that produced by volcanoes in Kamchatka and Indonesia (which, in both areas, rise above descending oceanic plates). The resulting Absaroka volcanic field in and near Yellowstone now embraces 24,000 square kilometers (9000 square miles) but probably was twice as large originally. Geologists have found 13 of the vent complexes that fed its great volcanoes. While these relics conjure up visions of cataclysmic activity, only a few volcanoes may have been in eruption at any one time, as is currently the situation in the Philippines or Indonesia.

When the Earth's crust began to be pulled apart in this region and composition of the eruptions became basaltic, thousands of north- and northwest-trending rifts opened along the border between Idaho, Washington, and Oregon, flooding the landscape with lava. The Columbia Plateau basalts, blanketing much of the Northwest, must have resembled one of the lunar "seas" (see Fig. 17.2). Following these eruptions the Snake River basalts began to form in a zone that migrated east across southern Idaho, approaching the most recent area of eruption in the Yellowstone region, where deep, hot intrusions manifest themselves in geysers and pools that spit hot mud.

The tension that opened rifts and produced this flooding also stretched the basin-and-range province farther south so that today it is estimated to be twice as wide as it was to begin with. The successive north–south ranges and basins are a product of the faulting and eruptions that accompanied this rifting.

The final, and in some respects most remarkable phase in Rocky Mountain development was the most recent—a lifting of the entire west–central United States to form a ramp running uphill from Kansas to Yellowstone. Some 335,000 square kilometers (130,000 square miles) were raised about two kilometers (6500 feet), suggesting an enormous intrusion of material beneath the crust in this region and lifting what had been modest ranges to form the Rockies of today. In addition to the regional uplift, individual blocks have been thrust up to an even greater extent. This is seen vividly in the Flatiron Mountains of the Front Range in Colorado, where slabs of originally flat-lying sedimentary rock have been lifted into the sky to form a dramatic backdrop for the city of Boulder. Some of the blocks, such as those constituting the Grand Tetons and Centennials, apparently began their rise quite recently (within the past two million years) and are still growing considerably faster than the continuing regional uplift of a few millimeters (a fraction of an inch) per year.

The growth of the central Rockies proceeded slowly enough so that more than two dozen rivers were able to maintain their routes even though hard rock was rising all about them. They eroded their canyons faster than the rate of uplift (although other rivers, unable to do so, were completely rerouted). The result, familiar to transcontinental travelers by car or train, is a series of magnificent gorges such as the Royal Gorge of the Arkansas River and the Black Canyon of the Gunnison, both in Colorado.

Meanwhile, the western rim of the continent gradually assumed its present form. Development of the Sierras took many millions of years, the most recent phase lifting and tilting the entire range to produce a steep wall 700 kilometers (440 miles) long on the east-facing side. Volcanic activity along the Sierras gradually died out, first in the south and progressively later in the north (a process that can now be related to the history of plate movements in the area). The Cascades, from northern California to Washington, have erupted within historic times, including the great 1980 eruption of Mount St. Helens, in which a searingly hot jet blew off the north side of the moutain, killed three score people, and devastated 400 square kilometers of forest (Fig. 17.4). The mountain, whose beautiful conical symmetry was destroyed, is still restless. The Cascades form a strikingly straight line from Mount Garibaldi, north of Vancouver, British Columbia, past Mounts Rainier, Adams, and Hood to Lassen Peak in California. They recall such other circumpacific volcanoes as Fujiyama in Japan and Shishaldin in

Figure 17.4 Mount Saint Helens. *Eruption of May 18th 1980.*

Alaska, likewise formed above a descending plate. They are produced where the small Juan de Fuca Plate burrows under Oregon and Washington, but there is no parallel trench in the sea floor because sediment from the vast discharge of the Columbia River fills it as fast as it grows.

The formation of the Coast Ranges, from California to Oregon and Washington, seems clearly to have resulted from battles between the North American Plate and oceanic plates to the west. A characteristic component of those mountains is the Franciscan Mélange, a chaotic mixture of deep-ocean materials and turbidity current debris that apparently formed on the landward slope of a coastal trench. In some sectors there are other relics of past warfare between continental and oceanic plates. Thus in the Red Mountain area of the Diablo–Mount Hamilton Range behind San Francisco Bay there are ophiolites—sea-floor fragments that may mark where sea floor from the west plunged into the Earth's depths, then were driven onto the continent. Inland from these coastal mountains lies the Great Valley of California, floored with sandstone, shale, and siltstone that apparently accumulated in relatively shallow water, such as that on a continental shelf. Thus some regard the coastal zone as the typical outer edge of a continental plate thrust up from the sea, the Franciscan Mélange being the deep-water part (eugeosyncline) and the Great Valley deposits being from the shelf (the miogeosyncline).

It is the earlier history that is most perplexing, for to the east, reaching deep within the present continent, there seems to be a far more ancient repetition of the coastal sequence. In its prospectus for the International Geodynamics Project of the late 1970s, the committee of the National Academy of Sciences responsible for American participation in that program cited evidence that the mountains farther inland originated in much the same way as the Coast Ranges:

> Very similar but older rocks are present in the Sierra Nevada to the east, in the Klamath Mountains in northern California, and in a broad arcuate belt extending through the Blue Mountains of eastern Oregon to the Idaho batholith and again westward along the international border to the vicinity of Vancouver Island and then northward in British Columbia to the Alaskan ranges. This belt of great mountains contains ancient deep sea sediments, volcanics, and ultrabasic rocks—suggestive of modern island arcs and resembling the rocks of the Coast Ranges of California, Oregon and Washington—that extend backward in time to early Paleozoic. These tantalizing glimpses of ancient episodes of tectonic movements are sufficient only to suggest critical areas for further geological and geophysical investigation.

Farther east, extending inland to the region presently occupied by the Rocky Mountains, the rocks, many of them now uptilted and thrust high into the sky, tell a tale of long sediment accumulation on a shallow, flat-lying sea floor. These sediment layers thicken westward toward the now distant ocean, just as the Appalachian sediments thicken eastward, toward the now distant Atlantic. Were these layers, then, laid down on a coastal shelf as a typical miogeosyncline? Might the West Coast once have lain through Idaho, Utah, and Arizona?

> Plate tectonic theory (said the Geodynamics prospectus) would suggest that the miogeosyncline represents a continental shelf and slope through much of Paleozoic and Mesozoic time and that the eugeosynclinal rocks were deposited in a deep sea environment to the west; the two would subsequently be brought together along a boundary between converging plates. The hypothesis is attractive and perhaps applicable, but entirely speculative at present.

In any case it is evident that the factors shaping the West have fundamentally changed with time. The nature of volcanic activity has changed. The situation along the coast has shifted from one like that off South America, with a trench and coastal ranges, to one in which the scrapings of a trench have been heaped up to broaden the continent. It is a situation, too, where the deep-sea floor slopes gently away from the coastal zone, rather than the other way around, as in the Atlantic, where the bottom slopes up toward the mid-ocean ridge on both sides of the ocean.

Decipherment of the reasons for these changes has been aided greatly by the timetables of previous plate movements inscribed in magnetic band patterns on the sea floor. It is evident that the Northeast Pacific has been formed by spreading from a ridge whose northern part has vanished almost entirely under the western rim of North America. In the Atlantic, with its ridge on the centerline of that ocean, the chronology of magnetic reversals can be followed, starting at the ridge and reading symmetrically away from it toward both east and west.

Off the Pacific coast of North America, however, this is largely not so. Almost the entire eastern half of the pattern is missing—and so is much of the ridge from which the western half was generated. Since the chronology of broad and narrow zones (representing long and short periods of magnetic polarity) can be identified in terms of the known history of the Earth's magnetic field, it is possible to ascribe an age to each magnetically defined band of the sea floor. From this it can be seen that the sea floor along the central California coast was exuded from the vanished ridge 29 million years ago and that the ocean bed closest to Southern California was formed along the ridge some 16 million years ago.

Early in development of the plate theory it was proposed that the missing ridge—a northern continuation of the East Pacific Rise—now lies under the basin-and-range province between the Sierras and the Rockies, where the mountains are still undergoing periodic uplifts from basins marked by earthquakes and hot springs. It was argued that material welling up along this buried ridge had spread eastward to lift the Rockies, the Rocky Mountain Plateau, and the great ramp that slopes up from Kansas to the mountain states. The increasing depth of the Pacific, westward from the continent, would be the counterpart of the slope of sea floor away from a mid-ocean ridge, only in this case the ridge would be hidden under North America.

While this explanation of the sea floor in terms of its derivation from a vanished ridge is now generally accepted, another concept as to what happened to that ridge has come forth. It derives from the paper of Dan McKenzie and Jason Morgan on the geometry of relative plate motions, particularly with regard to "triple junctions." They pointed out that the point where three plates meet tends to migrate, depending on the relative motions of those plates toward or away from one another. One of their examples was a junction that currently lies off Cape Mendocino, California.

The three plates that meet there today are the North American Plate (comprising the northwest Atlantic and all of North America, except the southern rim of California and Baja California), the Pacific Plate (comprising the main body of the Pacific Ocean), and the small plate nestled up against the coast off Oregon, Washington, and British Columbia known as the Juan de Fuca Plate. The last named has been generated by the formation of new sea floor from a series of short ridge segments including the Juan de

Fuca Ridge, off Oregon and northern California, whose flanking magnetic bands were first recognized by Vine and Tuzo Wilson as evidence of sea-floor spreading.

McKenzie and Morgan used the magnetic timetable of plate motions that had been charted in this region to reconstruct their history—as though they were running a moving picture backwards. This showed that the Juan de Fuca Plate was once a great oceanic plate that has almost entirely vanished under the West Coast. This plate is often called the Farallon Plate for the islets of that name off San Francisco, although the islets are actually chunks of the American mainland pulled out to sea by the northwest drift of the Pacific Plate.

Some 30 million years ago, according to the McKenzie–Morgan reconstruction, the relative westward movement of North America brought the American Plate into contact with the ridge separating the Pacific and Farallon plates (the ridge whose southern part, as noted, survives in the East Pacific Rise). This encounter separated the Farallon Plate into two fragments that shrank southeast and northwest, respectively, as the Pacific Plate continued to meet more and more of the West Coast. Today the remnant of the northwest fragment, the Juan de Fuca Plate, lies northwest of Cape Mendocino and what remains of the Farallon Plate, in the opposite direction, consists of a small fragment, the Rivera Plate, off central Mexico, and the Cocos Plate, still farther down the coast.

A new situation developed as the Pacific Plate came up against the western edge of the American Plate. Whereas the doomed Farallon Plate had been moving northeast, fighting a frontal battle with the American Plate (which the Farallon Plate was "losing" in that it was descending under the continent), the Pacific Plate was moving northwest. So, instead of descending under the continental edge, it slid along it, giving birth to the San Andreas Fault (Fig. 17.5). It is chiefly this slippage along the San Andreas (which skirts San Francisco to seaward) that has given California its reputation for earthquakes. (Portions of the San Andreas Fault in the vicinity of San Francisco Bay and Los Angeles, as seen from space, are shown in colored plates 5a, 5b, and 6.)

This shift in plate motions now appears to have been responsible for many of the more recent events that have shaped western North America. In particular it has helped account for the revolutionary changes that began, 30 million years ago, when the Pacific Plate first nudged the coast and the zone of contact gradually spread, bringing progressively to a halt the penetration of the Farallon Plate beneath the continent—the process that had been responsible for volcanic activity and mountain-building in the West (Fig. 17.6).

One of the first to recognize this shifting of gears and explore its implications was an enterprising young graduate student at the Scripps Institution of Oceanography named Tanya Atwater. She proposed that activity along the San Andreas Fault might be only part of the story. The dragging of the Pacific Plate against North America, she suggested, has distorted the entire region from the Rockies westward, obliquely distending the basin-and-range region by a considerable amount. If the dragging effect has been in operation since the plates first came in contact, displacements well in excess of 1000 kilometers (600 miles) could have occurred, she said. In 1963 Donald U. Wise had published a paper entitled "An Outrageous Hypothesis for the Tectonic Pattern of

Figure 17.5 Carrizo Plain area in Southern California, showing how right-lateral motions on the San Andreas fault have displaced rivers crossing the fault.

the North American Cordillera," and she suggested, in effect, that it might not be so outrageous after all.

Wise had elaborated an earlier suggestion by Carey, the Tasmanian proponent of Earth expansion as an explanation for drift, and the idea has been developed by others. If, it was pointed out, California were shoved a few hundred kilometers southeast, the Sierras would line up with their counterparts in British Columbia, whereas now they are offset to the west. It was therefore proposed that California has slipped northwest a few hundred kilometers since the Sierras were formed. Although it is definitely known that the coastal rim of the state from San Francisco south has been slipping in this manner along the San Andreas Fault, according to this hypothesis there has been additional motion of the entire coast region, with consequent distortion of the basin-and-range zone. The proposition is supported by indications that the Pacific Plate is moving five centimeters (two inches) a year, relative to the North American Plate, but that only

70 percent of this motion is manifest along the San Andreas Fault. That parts of the American landscape have traveled extraordinary distances is also hinted by the geology of Southeastern Alaska, where ancient continental rocks (as old as Ordovician) lie to seaward of younger oceanic formations. These continental rocks seem unrelated to regional geology and, it is suggested in the Geodynamics prospectus, "perhaps have been transported long distances (thousands of miles)."

Not only did Tanya Atwater find "enticing" the idea that slippage of the West Coast occurs along a "wide, soft boundary," but she realized that the postulated disappearance of the ridge, upon its arrival at the trench, was compatible with a concept of sea-floor spreading that was gaining popularity. This was that ridges are simply cracks between plates that are being pulled apart, rather than zones in which currents rising from deep in the mantle push the plates away from one another, as originally proposed. If a ridge manifested a deeply rooted driving force it would not, it seemed, disappear so readily.

The gradual transformation of the forces shaping the Western States from those associated with descent of a plate to drag and stretching now offered an explanation for the change in volcanic activity that swept the region from south to north, beginning some 30 million years ago. This was recognized by three scientists from the Denver center of the United States Geological Survey—Peter W. Lipman, Harold J. Prostka, and Robert L. Christiansen. It was initiated, they proposed, when the Pacific Plate first touched the western rim of North America—probably in southern California—bringing to an end, in that sector, penetration of the Farallon Plate under the continent. This terminated the eruption of andesite lava generated by descent of the plate there, and, as the Pacific Plate came up against more and more of the continent, the effect migrated away from the area of original contact to the northwest (and probably southeast into Mexico). As pressure from the Farallon Plate was relieved and the coast, instead, was subject to drag by the northwest-moving Pacific Plate, new forces came into play. The continental crust stretched, releasing basalt eruptions that, too, migrated northwest, producing such features as the basin-and-range province and the Columbia Plateau.

Following the efforts of Tanya Atwater, as well as McKenzie and Morgan, to read the magnetic messages off the West Coast of North America, Roger L. Larson and Walter C. Pitman III of the Lamont–Doherty Geological Observatory and Clement G. Chase of the University of Minnesota found what they believed to be an even earlier record of Pacific history. In magnetic maps of the western Pacific they identified an identical "message," or sequence of narrow and broad magnetic anomalies, imprinted in three different parts of that ocean. The "message" apparently represented a magnetic timetable so ancient that it constituted a "floating" chronology beyond the known history of magnetic field reversals.

One of these sequences lay parallel to the Kurile Arc, northeast of Japan, and seemingly had been generated by spreading from a ridge long since vanished into the Kurile Trench. Another formed a series of east–west magnetic zones near the Phoenix Islands of the South Pacific. It was apparently produced by a ridge to the south. The third lay west of the Hawaiian chain, running northwest to southeast, and seemed to represent sea floor that had marched across much of the Pacific from the ridge system that now has partially vanished along the coast of North America.

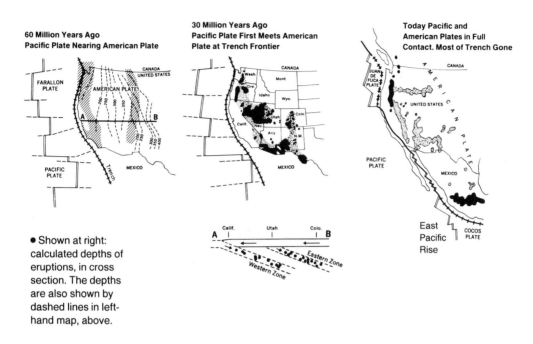

60 Million Years Ago
Pacific Plate Nearing American Plate

30 Million Years Ago
Pacific Plate First Meets American
Plate at Trench Frontier

Today Pacific and
American Plates in Full
Contact. Most of Trench Gone

● Shown at right: calculated depths of eruptions, in cross section. The depths are also shown by dashed lines in left-hand map, above.

East Pacific Rise

Figure 17.6 Formation of the Western United States. From the history of plate movements inscribed in magnetic patterns of the Pacific floor, the geologists of the United States Geological Survey in Denver have reconstructed the sequence of events that shaped the western United States. During the earlier phase (left), the Farallon Plate descended via a trench along the coast, causing great eruptions of lava. From their analysis of the lava's potassium content, the geologists concluded that the lava originated at increasing depths from California to Utah, thus confirming that the lava came from a descending plate. Farther east, however, the pattern repeats itself, as though the material derived from another descending plate. By the time the Pacific Plate first touched the American Plate (center), the eruptions had emplaced great volumes of rock (andesite) typical of that from a descending plate. The situation then changed (right). The Pacific Plate, moving northwest, dragged and stretched the western region, producing basalt eruptions. Black denotes areas where evidence of eruptions survives. Areas of probable volcanism are stippled.

All of these patterns recorded the same timetable of magnetic reversals, but their age was uncertain until they were correlated with the published results of an Atlantic study conducted by Peter R. Vogt and his colleagues using data collected by the Navy research ship *USNS George D. Keathley* in 1967 and 1968. The ship had steamed a series of 39 east–west lines, spaced about 30 kilometers (20 miles) apart and covering a region of 1.5 million square kilometers (600 thousand square miles) between Bermuda and the American continental shelf. The chief discovery was a well-defined sequence of north–south magnetic zones with a total width of 300 kilometers (200 miles) which, with the aid of dates obtained from three of the *Glomar Challenger* deep holes, the Navy scientists assigned to magnetic epochs from 115 to 155 million years ago in the Jurassic–Cretaceous periods. They called this, for their ship, the Keathley sequence.

Larson and his two co-workers compared the pattern of reversals in the Keathley sequence with the one inscribed on three separate parts of the Pacific floor and found they matched, thus linking the ancient, floating chronology to the established one. A hole sunk by the *Glomar Challenger* within the Phoenix Island series, providing an age for one of the magnetic bands, seemed to confirm the ages deduced from this link and, using this greatly extended timetable, the three scientists ran the "moving picture" of Pacific history back to its earliest stages.

They concluded that the Pacific floor had been manufactured along at least five ridges, meeting in two triple junctions. All but one of these—the East Pacific Rise and its apparent continuation off Oregon, Washington, and British Columbia—have long since vanished into the peripheral trenches of that great ocean. If one calculates the amount of sea floor that must have gone down those trenches to account for all this motion, it comes to a minimum of 7000 kilometers (4300 miles) descended under North America, 7000 or 8000 kilometers vanished under Asia, and 5000 gone down under South America, west Antarctica, or both. Probing of the sea floor in the Western Pacific has shown that, during the Cretaceous, volcanism on a stupendous scale flooded a submarine region as large as Europe with basalts. They may be as much as seven kilometers thick, dwarfing the great "seas" of basalt on land, such as that formerly erupted onto the American Northwest. As reported by the Scripps Institution of Oceanography, samples from more than 100 sea mounts have shown that, during the early Cretaceous, they rose above sea level. Presumably many more have descended into the western Pacific trenches.

One of the implications of the magnetic timetable deduced from the Keathley sequence was a prolonged period of unbroken, normal magnetic polarity between 85 and 115 million years ago. It was a time, according to their analysis, also marked by a worldwide increase in the rates of sea-floor spreading. Since the magnetic timetables imprinted on all the oceans seem similar, if periodic changes in spreading rates have occurred, they presumably were global and could be verified only by sampling and dating some of the anomalies. Deep-sea drilling has done this at only a few of the very old sites and so the inferred spurt remains to be confirmed, but its proposed occurrence was during the Cretaceous Period, when the seas rose and flooded much of the land. This flooding, in addition to melting of polar ice, may have been caused when intensified sea-floor spreading made for shallower ocean basins.

Larson also attributed a number of batholith intrusions around the Pacific to this hypothetical spurt and to the consequent intensification of ocean-floor subduction under the continents and island arcs. In any case, it is generally agreed that the batholiths that constitute much of the exposed granite in western North America were derived from a descending oceanic plate. Virtually the entire coastal zone of British Columbia is a batholith formed by a succession of intrusions 60 or 70 million years ago. Another fills much of Idaho, and nearby is the Boulder batholith whose intrusion into western Montana is apparently associated with the ore deposit at Butte that has provided the United States with much of its copper (see Fig. 17.2).

Whereas the rocks in the Coast Ranges and Sierras can clearly be related to the descent of an oceanic plate under the continent there, the Rocky Mountains are a different matter. They lie far inland, remote from the coast where the oceanic plate presumably was decending. While the lava that erupted 50 million years ago in the Absaroka volcanic field of Montana and Wyoming is similar, chemically, to that produced in Indonesia, the Indonesian eruptions occur 300 kilometers (200 miles) from where the plate responsible for those eruptions begins its descent into the Earth. The Absaroka Field is four times farther from the present continental rim. The volcanics of the San Juan Mountains in Colorado are even more distant from the sea. Furthermore, except in their northern part, the Rockies were built primarily from vertical uplift, rather than by crumpling from the direction of the sea.

An ambitious effort to explain this in terms of the plate theory had been made by the three scientists from the Geological Survey, Lipman, Prostka, and Christiansen, using new information on the ages and chemistry of eruptions that helped form the entire region from the Pacific Ocean to the eastern limits of the Rocky Mountains. Their conclusions were reported to a meeting of the Royal Society of London in November 1970.

The volcanic activity that shaped much of the West began, they believe, between 70 and 80 million years ago when North America commenced overriding the Pacific floor (that is, the Farallon Plate) "especially rapidly." (This was when the opening of the North Atlantic went into full swing.) Their explanation for the resulting occurrence of volcanic activity as far inland as West Texas was derived from the analysis of potassium abundances in the rocks erupted during that period. In the West, as elsewhere, the Denver group assumed, the amount of potassium relative to silicate in the rock was an indication of the depth of the descending slab of sea floor from which the molten rock was derived.

On this basis they deduced that the slab from which volcanic material in the Coast Ranges had come was 100 kilometers (60 miles) down. Farther inland, under the Sierras, the slab was 200 kilometers below the surface and still farther east, under the Nevada–Utah border, it was even deeper—300 kilometers (200 miles), conforming to the typical pattern of a plate that reaches greater and greater depths as it penetrates inland. But then something unexpected was found. Rocks to the east of this region seemingly had been derived from a relatively shallow plate (although its inferred depth of close to 200 kilometers was probably within the soft asthenosphere). From there on east to the Front Range of the Rockies the depths again increased in a systematic manner (Fig. 17.6).

The implication was that two down-sloping plates were responsible for the eruptions—and hence for the Rocky Mountains themselves. But how could the subduction process generate two such plates? One proposed explanation is that, where an overriding plate, like North America, is in rapid motion relative to the Earth's deep interior, whereas the subducting plate is not, the descending slab, as it penetrates to depths of 700 kilometers (430 miles) or more, "stubs its toe"—that is, becomes embedded in "stationary" mantle material under the soft asthenosphere. The overriding plate finally breaks off the embedded part of the slab, and the oceanic plate is forced down a new route.

In this way early subduction of the Farallon Plate along the West Coast could have been broken off beneath the North American Plate, which continued to press westward until the broken slab lay below the hinterland, its shallowest edge along the front of what is now Wasatch Range, facing the Great Salt Lake Desert of Utah and, farther south, along the western edge of the Colorado Plateau. Today, apparently because of the tensions that produced the basin-and-range region, the Wasatch Range is still restless, a band of freshly exposed rock along its base indicating that, not long before Brigham Young led his Mormon band into the valley, the range leaped up about 10 meters in one or more great earthquakes. Farther north, where this front extends into Montana, the Hebgen Lake earthquake of 1959 produced scarps as much as six meters high.

Where an oceanic plate begins its descent, as under a coastal trench, heat flow up through the Earth tends to be reduced, since this is the coolest part of a crustal plate. It has been noted that zones of low heat flow exist in each of the areas where these two slabs of the Farallon Plate would be at their shallowest depth. The volcanic activity and eventual uplift of the Rockies then would be explained by the gradual heating and expansion of the slab beneath that region in response to the high temperature of the surrounding upper mantle material.

The slope of the two slabs, as reconstructed from the potassium data, was relatively shallow—some 15 degrees to 20 degrees. Under island arcs the slope tends to be steeper (45 degrees or more) but, as noted earlier, under South America it is 15 to 30 degrees, and Lipman, Prostka, and Christiansen pointed out that if, as proposed by Tanya Atwater and others, the basin-and-range province has been stretched 100 kilometers beyond its original width, this would make the slope of the slab beneath it look less steep than it really was (the "corrected" slope, they said, would be 20 degrees to 25 degrees). Because the overridden slab had sunk into the soft asthenosphere and hence was "decoupled" from the rigid plate, or lithosphere, moving westward above it, the mountains generated from this deep slab would show little or no evidence of horizontal thrusting—as is the case.

The forces that stretched the basin-and-range region are still at work, not only in activity along the San Andreas Fault, but farther east—for example, along a less prominent fault known as Walker Lane, because it lies in part along the Walker River (Fig. 17.7). This feature extends northwest from the vicinity of Las Vegas and has been a source of earthquakes and rifting in which, as with the San Andreas, the west side shifts northwest, relative to the east side. Not far from this feature have been quakes that would have been devastating (and better known) had they occurred in densely inhabited areas. One was the 1932 Cedar Mountain quake, and two more occurred in 1954 (the Dixie Valley and Fairview Peak quakes). All three of them were in western Nevada.

All of this activity, to a greater or lesser degree, relates to the northwest–southeast stretching and wrenching of the region. Even the small-scale local faulting that has followed underground nuclear testing in Nevada has resulted in such motion. It was similar movement that tore Baja California loose from the Mexican coast some four million years ago and has transferred it 260 kilometers (160 miles) northwest to its present position.

The shifting of gears that began 30 million years ago, as the Farallon Plate vanished along the coast of Southern California, caused the Los Angeles basin to subside, and, less than 20 million years ago, a prolonged series of basalt eruptions began build-

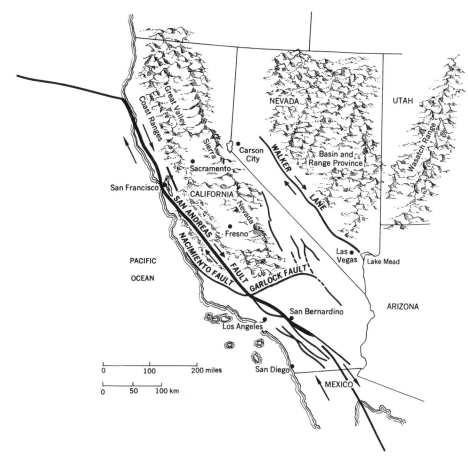

Figure 17.7 Typical fault motions in the Western United States

ing the nearby Channel Islands. As this activity moved up the coast, it produced the Berkeley Hills, overlooking San Francisco Bay, between five and 15 million years ago, as well as the Sonoma volcanic field, whose residual heat is responsible for the commercial production of electric power from natural steam.

Today the earthquakes that drew the attention of Vine and Wilson to the Juan de Fuca Ridge as a possible source of sea-floor spreading show residual activity there, but, where the plate generated by that ridge—the Juan de Fuca Plate—descends under the coast from Northern California to British Columbia, the coastal trench is filled with sediment, and there are few quakes along the descending slope of this plate beneath the coast. It is these symptoms that have been taken to indicate that the plate is stalled and that volcanic activity in the Cascades attributable to its descent is likely to fade, but this view has been shaken by the 1980 eruption of Mount Saint Helens and new fears for a great earthquake near Seattle.

Are we to believe the proposition of the Denver group that the mountain belts of the west have been generated over two oceanic slabs, one of them as far inland as the Rockies? Is it more plausible to believe that these features arose because an oceanic plate thrust under the continent at a very shallow angle, thus reaching far from the sea? Or did the original western rim of the continent lie somewhere between the Rockies and the Sierras? The diversity of these ideas reminds us that efforts to understand the world's mountains in terms of the plate theory are in their infancy and many revisions are certainly in store for the history set forth in the preceding pages. But the scope of the processes that have shaped the landscape of the continents and islands on which we live is so grand that even this preliminary view is awesome. And, for Americans, it may seem particularly strange that, of all the ranges in the world, none seems more difficult to explain than the Rockies that stand, fresh, young, and still growing in their own back yard.

Chapter 18

Earthquakes: Prediction or Prevention

THROUGHOUT HUMAN HISTORY A GREAT EARTHQUAKE HAS BEEN THE MOST terrifying and unpredictable occurrence in nature. Even primitive man could recognize the telltale signs of an approaching storm, but to have the very Earth thrown into upheaval has been without compare for inspiring fear. It was a terror well known to those living astride the Dead Sea–Jordan Valley fault in Biblical times, and, to the author of the Psalms, nothing could be a more complete demonstration of faith than to remain steadfast during such a catastrophe. "Therefore will we not fear," he wrote, "though the earth be removed, and though the mountains be carried into the midst of the sea; though the waters thereof roar and be troubled, though the mountains shake with the swelling thereof."

Because earthquakes seemed to occur with no warning whatsoever they often were viewed as the act of an angered deity. In some cases their toll has been horrendous. In 1556 an earthquake of extraordinary intensity shook down walls and roofs throughout the Chinese province of Shanxi. An inventory of those killed listed some 830,000 by name. How many more died is unknown. On All Saints Day in 1755 the populace of Lisbon was at worship when a quake brought the cathedral and churches down on their heads. An estimated 60,000 died, including those killed by the consequent seismic sea waves. This prompted Jean Jacques Rousseau to comment that such disasters were the price mankind paid for being "civilized." If we still lived out of doors, he said, we would not have to fear earthquakes—a point of view then satirized by Voltaire in his famous account of the Lisbon quake in *Candide*.

In spite of Rousseau, cities continued to grow, and by 1923 the population of the Tokyo–Yokohama area was some two million. By official count 99,331 of them died in an earthquake on September 1 of that year. Of these, 38,000 had taken refuge in an open area of Tokyo only to perish in a fire storm that anticipated the effects of fire bombings and atomic bombs in World War II.

In 1949 the 12,000 residents of Khait, a town at the foot of a long, wild valley in the Pamir Mountains south of Tashkent in Soviet Central Asia, felt an earthquake and noticed, far up the steep-walled valley, a great scar indicating that there had been a landslide. Quakes are so common in this region that the townspeople paid it little heed, unaware that the landslide had formed a dam across the valley, behind which a great mass of water was accumulating. A day or two after the quake, the dam broke, and the town, in minutes, was buried under more than 30 meters of mud and rock, sweeping over it at 100 meters per second. Few, if any, escaped. In 1970 a somewhat similar sequence of events wiped out the resort town of Yungay in "the Switzerland of Peru." The quake took more than 70,000 lives, almost half of them in Yungay, making this the greatest disaster in the history of the Western Hemisphere.

Californians have been told so many times that a great earthquake there is inevitable that they are inured to such warnings. Estimates of the damage to be expected run to billions of dollars. It has been suggested that a million might perish, and a television documentary on San Francisco shown nationally in 1973 was entitled, *The City That Waits to Die.* (Figure 18.1 shows some of the faults that cross the region.) A more

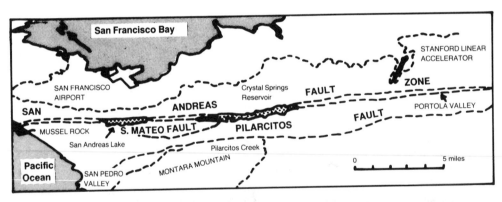

Figure 18.1 San Andreas Fault system shown in a side-scanning radar image of the topography south of San Francisco. Evident, as identified in the key map, are the San Andreas Fault System, San Francisco International Airport, and the two-mile-long Stanford Linear Accelerator. The city of San Francisco lies to the left, skirted by the main fault.

conservative study for the United States Office of Emergency Preparedness in 1972 proposed that under the worst credible combination of circumstances, a quake as severe as that which killed 452 San Franciscans in 1906 and occurring at the peak rush hour of 4:30 p.m., would take some 10,360 lives and leave 40,360 sufficiently injured to require hospitalization. Whereas the 1906 quake occurred early in the morning, when most San Franciscans were abed, one at 4:30 p.m. would find the streets and freeways filled with commuters and shoppers subject to falling bridges and masonry. (The estimates omitted any casualties that might result from broken dams in the Bay Area.) Since residents of that region now number in the millions, these figures would represent a relatively small percentage of the population. The estimates assumed that the building codes in force since most of the structures were built, including all the high-rise buildings, would reduce the per capita casualties below those of 1906. When a magnitude 7.1 earthquake struck south of San Francisco on October 17, 1989, only 63 died, partly because the quake was not close to the city and partly because of earthquake-resistant construction. The loss of life occurred chiefly in the collapse of a two-tiered freeway that had been built on soggy ground in Oakland (Fig. 18.2).

By contrast, in cities such as those of the Near East with poorly mortared masonry lacking metal reinforcement, the loss of life can be terrifying. In 1968, for example, 1200 of the 1700 residents of Dasht-e-Bayaz in Iran were killed, and, in a 1960 earthquake in Agadir, Morocco, an estimated 12,000 were killed and 12,000 injured out of a population of 33,000. When, on December 7, 1988, a severe, but not great, earthquake struck Soviet Armenia, poorly reinforced housing in several cities collapsed, killing 25,000 people.

Casualty estimates for San Francisco in the Office of Emergency Preparedness study may be overly optimistic. A 1973 report by the National Academy of Sciences cited an estimate of 50 billion dollars in damage from a major quake in either San Francisco or Los Angeles. But even if the lesser casualty figures are accurate, the popu-

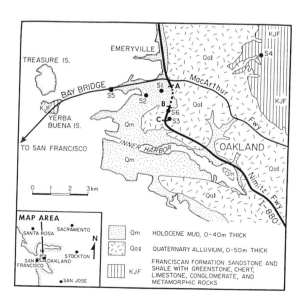

Figure 18.2 Freeway collapse. *Map of Oakland showing surficial geology. Stations S1–S6 are sites occupied during the study of aftershocks of the October 17, 1989 Loma Prieta earthquake (M$_s$ = 7.1) in the vicinity of the two-tiered section of the Nimitz Freeway, known as the Cypress structure (AC). AB indicates the freeway segment that collapsed, having been built on muddy terrain. BC indicates the segment that was damaged but did not collapse.*

lation would endure an experience familiar to Asians and Europeans, who have been subjected to mass bombing, but new to Americans.

The possibility that these great disturbances of the Earth can be predicted and, at least to some extent, controlled, has been greatly enhanced by emergence of the plate theory which, in many cases, explains why these events take place where they do. Beneath the moving plates is a vast storehouse of energy that, in ways still uncertain, drives the plates against and away from one another, manifesting itself either in the form of universal, gentle heat flow up through the plates, so subtle it is beyond human perception, or in catastrophic releases, such as quakes and volcanic eruptions that typically (but not always) occur along plate margins. Also of major importance have been recent laboratory experiments that have revealed the altered conditions of crustal rocks that seem to set the stage for a quake.

Until now the prediction of earthquakes has been the purview of astrologers, mystics, seers, and charlatans. Some thought they saw a relationship between such occurrences and various arrangements, relative to the Earth, of the Sun, Moon, and planets. All of these propositions collapsed when long-term records were analyzed. Playing with cycles—for example, linking events on Earth with the 11-year sunspot cycle—is a dangerous game, but one explanation for what seems a certain cyclic pattern in earthquake occurrences has evoked serious scientific interest.

It concerns the Chandler wobble—that 14-month component in the smallscale, roughly circular wanderings of the Earth's spin axis. As noted earlier, the elasticity of the Earth's interior is such that this wobble gradually should be damped out. Yet in some way it is periodically reactivated and, in 1967, two scientists at the University of Western Ontario, Lalantendu Mansinha and Douglas Smylie, proposed that this occurs when big earthquakes give the planet a jolt. Then Charles A. Whitten, chief geodesist of the National Ocean Survey in the United States, found a correlation between the rate at which the spin axis was changing as a consequence of the wobble and the total worldwide release of earthquake energy. This seemed to hold true, he said, for the whole of this century up to 1970. Perhaps, he suggested, the more rapid migration of the spin axis, causing slight departures from the stable distribution of mass within the Earth, helps trigger the release of cumulative strain along fault lines. When big quakes result (as well as many little ones) this could help keep the wobble alive—an idea that, typical of the whole history of attempts to understand the Chandler wobble, evoked great interest, as well as doubtful comment.

A discovery of greater practical value would be one that made it possible to predict with reasonable accuracy when and where a quake will occur, and a variety of premonitory effects—some of them rather dubious—have been reported. Take, for example, the great Alaskan quake of 1964, when a section of the Pacific Plate 800 kilometers (500 miles) wide thrust some 20 meters under Alaska at a 9-degree angle of descent. A little more than an hour beforehand there was a local disturbance of the Earth's magnetism. But other quakes have occurred without such a "warning" and magnetic disturbances are, in any case, common in Alaska. Another "advance symptom" that has been reported is lightning, attributed (on the basis of laboratory experiments) to the development of electric charges in the quartz of deep rock when subjected to deformation.

Efforts to find local signs of an impending quake have been pursued with special vigor in Japan, the Soviet Union, and, more recently, the United States. The Japanese have sought to measure every variable that might change as an earthquake becomes imminent. Such variables include ground tilt, crustal strain, and water levels in deep wells, as well as the imperceptible anticipatory tremors known as foreshocks. Changes in tilt were used to predict swarms of moderate earthquakes that shook the town of Matsushiro for prolonged periods from 1965 to 1967, forcing some of the villagers to sleep in nearby fields rather than in their creaking and groaning homes.

Near the city of Niigata on the west coast of Honshu—across the island from the side where the Pacific Plate begins its descent—a gas field has been under exploitation and, to watch for any subsidence of the terrain as a consequence of gas removal, detailed leveling surveys have been carried out routinely for many years. Some vertical movements had been recorded, but in 1958 a rapid rise began which, within a year, amounted to almost five centimeters (two inches). The uplift then tapered off, and, five years later, a major quake struck the city, centered in the area that had risen the most.

The Japanese then undertook an intensive program of periodic surveys in other regions to see if this was a universal and reliable warning sign. It seemed, they reported, that the extent of the area uplifted was an indicator of the magnitude of the quake to follow. For Niigata, they said, this would have provided several years' warning had the signs been read properly.

At a 1972 symposium on earthquake prediction, held at the National Academy of Sciences in Washington, Cecil B. (Barry) Raleigh, then at the National Center for Earthquake Research in Menlo Park, California, told of a more dramatic uplift in Japan, some two centuries ago. When residents of a coastal village saw their harbor being drained of water, they resisted the temptation to run out and collect the fish that were flopping on the mud flats, knowing that such retreat of the sea was the typical prelude to a "tidal wave," or tsunami. Instead, they fled to the hills. Nine hours later an earthquake threw much of the village down in ruins, and the villagers hurried back to salvage their effects, only to be inundated by a tsunami that then did sweep in upon them. The draining of the bay had been caused by uplift, not by subsidence of the sea, and the dreaded tsunami did not occur until after the earthquake to which this uplift was preliminary.

In the Soviet Union deformations of the Earth similar to those observed in Japan have been detected before quakes. Other clues have been found as well. In 1967, as part of an effort to develop a prediction capability that could prevent a repetition of the disaster that buried Khait in 1949, the Russians began measuring electrical conductivity of the crustal rock along a six-kilometer (four-mile) line spanning that rugged, quake-prone region of Tadzhikistan known as the Garm District. A marked rise in conductivity was recorded before each quake that occurred within 10 kilometers (six miles) of the line. Such an increase in conductivity has now been used successfully to predict a quake in the Kamchatka area of the USSR, where the Pacific Plate is descending under Asia.

At Tashkent, which lies north of the mountainous Garm District, in Uzbekistan, water from deep wells showed an increase in radon content before two quakes. Radon is a gas produced by the radioactive decay of radium. Its half-life is 3.835 days (that is, half of a given amount of radon decays within less than four days). Its increase in the water implied that some process deep in the ground was releasing this short-lived gas into the well at an accelerated rate—perhaps, the Russians thought, the squeezing of the rock in the final stages of strain accumulation.

The finding that evoked the most excitement was reported by Soviet scientists, apparently beginning as early as 1962, but because of language and other barriers, it did not become known to Western scientists until an international conference in Moscow in 1971. As part of the effort to detect some sort of advance warning before quakes in the Garm District, the Soviet scientists had set up a network of 10 seismic stations so positioned that they could monitor seismic waves traversing that region. There was enough generation of minor tremors to make such monitoring possible, and it was hoped that these waves would reveal premonitory modification of the crustal rock.

In studying the records of these stations for the 1950s and 1960s the Russians found that before each of three earthquakes (although these events varied greatly in intensity) there was a marked change in the velocities of waves through the area of the impending quake. It was not clear at the time what, exactly, was the nature of this change, but it manifested itself in a closing of the gap between the first arrivals of pressure waves and shear waves from a quake center (Fig. 18.3).

A pressure wave (such as a sound wave) is one in which particles move forward and backward along the wave path. A shear wave is one in which the particles move sideways along the path. Pressure waves travel considerably faster and thus get to an observing station first. What the Russians found was that this lead time of the pressure waves decreased weeks or months before a quake, then gradually returned to normal. When the travel times reached normal again, the quake occurred. The pattern of all three events was the same, except that the more prolonged the digression from normal travel speeds, the more severe the ensuing quake. For the smallest of the three quakes this advance period of abnormal wave behavior was 34 days. For the moderate-sized quake it was 45 days, and for the strongest it was three months. The implication was that, if this effect was monitored, it could be used to predict magnitude and time of occurrence.

The fact that the lead time of the pressure waves became abnormally small could be explained in three ways. It could mean that those waves slowed down, that the shear waves speeded up, or that both changes occurred. The Soviet data could not discriminate between these possibilities. The change, in any case, was too great to be accidental, amounting in all three cases to about six percent in relative velocities. Thus, while the effect for a big quake was more prolonged, it did not represent a greater change in velocities than occurred before a small quake.

When Carl Kisslinger described these findings to the 1972 symposium on earthquake prediction at the National Academy of Sciences, he said: "If this behavior is verified for future observations and for other seismic zones, it is a powerful, if poorly understood, forecasting tool." It was noted by participants in the symposium that the region of the Soviet observations was one, according to plate theory, of continent–continent collision, and there was no assurance that the same warning "signal" would

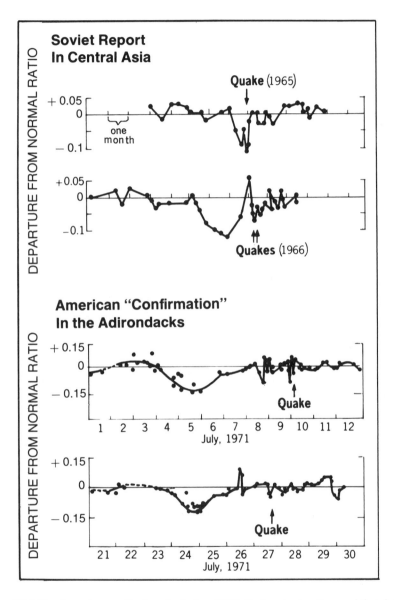

Figure 18.3 Earthquake prediction. *In the late 1960s Soviet scientists discovered that, before an earthquake in the quake-prone Garm district of Central Asia, there was a change in the ratio between the travel velocities of pressure waves and shear waves. Normally pressure wave velocities are much higher than those of shear waves. Just prior to the earthquake, however, the ratio between the velocities declined, then began increasing. When the ratio reached its normal level, the quake occurred. This same effect was then found in the Adirondacks, prior to much smaller quakes, and was present before the severe San Fernando quake of 1971. When the departure from normal was prolonged, as in the 1966 Soviet quake, the subsequent quake was severe, but this effect has not proved universal.*

be generated where an oceanic plate was descending under an island arc, as is the case in Japan, or where it was scraping the edge of the continent, as in California.

Among those inspired by the Soviet report to assess its applicability elsewhere was a group from the Lamont–Doherty Geological Observatory that included Lynn Sykes, Marc L. Sbar, and two graduate students, Yash P. Aggarwal and John Armbruster. On May 23, 1971, two earthquakes were recorded near Blue Mountain Lake, in the Adirondacks, and, while they were small, they were followed for several months by thousands of microearthquakes that could be recorded by instruments of high sensitivity. Within a few days of the original quakes the Lamont group had set up 10 portable seismographs in the area and were rewarded with three more quakes, comparable to the original ones, on June 20, July 10, and July 27. Before each of them, they said, the records of microearthquakes showed the effect reported by the Russians. Conforming to the Soviet pattern, the duration of these premonitory changes was directly related to the magnitude of the quake that followed (which, in these cases, was so small the warning change spanned only a few days).

Even the most severe of the quakes for which these warnings had been identified, in the Soviet Union and the United States, was moderate compared to those that produce extensive damage and loss of life. However, four months after the Lamont group submitted their report, another team under Don L. Anderson at the Seismological Laboratory of the California Institute of Technology reported finding the same effect prior to the earthquake of February 9, 1971, that took 59 lives and did an estimated $500 million damage in and near San Fernando, a northern suburb of Los Angeles. Among those killed were 46 in a Veterans Administration hospital building that collapsed. Two died under a fallen freeway bridge, others were crushed in their homes, and two being kept alive by iron lungs in a local hospital died when the power failed. Four of the area hospitals, although desperately needed in this emergency, were knocked out of commission.

The Cal Tech Study focused on 19 earthquakes from 1961 to 1971 all of whose tremors passed through or near the San Fernando area and were recorded by the Cal Tech seismic stations at Pasadena and Riverside. The telltale change in Earth tremor travel velocities began three and a half years before the event—a prolonged period that was thought to be an indicator of the havoc to come, conforming to the previously observed relationship between length of the effect and magnitude of the ensuing quake.

Following the same pattern, the premonitory effect for a truly great earthquake, like that which hit Alaska in 1964 (and was hundreds of times stronger than the San Fernando quake), should become evident 40 years or more prior to the event; but the supersensitive seismic records needed to identify it do not go back far enough to span such a period, and those few areas where such information is available for a lesser interval often do not include metropolitan regions where major quakes are most feared.

The Cal Tech study indicated that the effect came about because rock in the region where an earthquake was to occur became a poor transmitter of pressure waves. Thus it was the slowing of such waves—not the speeding up of shear waves—that was responsible. Since pressure waves can be generated efficiently by explosions, whereas shear waves cannot, this meant that where swarms of natural quakes are not available for monitoring purposes, artificial waves could be generated, or those from mine explosions used. Despite early excitement regarding this effect, however, it was not widely observed during the 1980s and may not be as common as hoped.

Another line of attack on earthquake prediction came from laboratory experiments as well as field observations that came about quite by accident in Colorado. The latter demonstrated, once again, that something unpleasant can turn out to have beneficial side effects. In 1961 the United States Army, to get rid of waste water contaminated with nerve gas or other lethal material, drilled a well 3.6 kilometers (2.2 miles) deep at its Rocky Mountain Arsenal east of Denver. After penetrating impermeable rock the drillers bored through 20 meters of gneiss that was vertically fractured, and it was assumed that any water injected into this fractured zone could never contaminate ground water at higher levels. In 1962, therefore, the Army began periodically pumping waste water down the well.

Three years later David M. Evans, a consulting geologist in the Denver area, pointed out that, since the pumping began, seismographs at the University of Colorado, the Colorado School of Mines, and nearby Regis College, had recorded 710 small earthquakes in the vicinity of the well. He cited a proposal by M. King Hubbert of Stanford University and William W. Rubey of the University of California at Los Angeles that, if water pressure within the pores of deep rock were increased, this might make the rock more easily fractured and thus facilitate strain release. The quakes were occurring in a zone a kilometer deeper than the well, but the fractured region beneath the well bottom presumably allowed the water to work its way down.

John H. Healy and his colleagues at the United States Geological Survey office in Denver then began studying the quakes, most of which were too weak to be noticed by local residents. The tremors originated within a narrow zone 10 kilometers (six miles) long that ran across the arsenal and almost directly beneath the well, and, it was found, they occurred primarily during periods of heavy water injection. Although half a million tons of fluid had gone down the well, its volume did not seem enough to cause the quakes by gross distension of the deep region, particularly since the water was supposedly dispersing in the fractured gneiss beneath the well. Rather it appeared that the water somehow was permitting slippage along an old fault that previously had been "locked." Meanwhile, some of the tremors had become moderately severe, and Denver residents were complaining of cracked plaster and other damage. Lawsuits were instituted, and, in September 1965, the Army cut off further injections (leaving itself with a controversial disposal problem). (See Fig. 18.4.)

The quakes began to slacken (although by 1968 some 1600 had been recorded), and it was proposed that, if water were withdrawn from the well, this might stop them altogether and also might demonstrate, for the first time, a capability of turning quakes off as well as on. However, so little was known of the processes taking place several kilometers down where the quakes were occurring that it was feared any tampering with the water might produce a severe quake instead of preventing further tremors.

Then an opportunity to test the idea unexpectedly presented itself. In western Colorado the Chevron Oil Company for seven years had been pumping water 2000 meters down a series of "dry" wells around the perimeter of the Rangely field to force

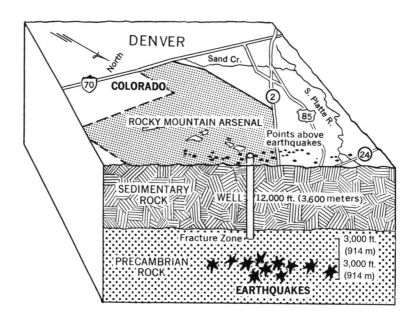

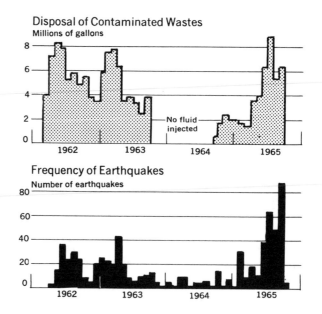

Figure 18.4 Coincidence of earthquakes and waste disposal

oil up within reach of otherwise dry wells in the center of the field. Water pressure in deep rock below the field had thereby been increased, it was estimated, by as much as 60 percent, and small quakes were occurring at a rate of 15 to 20 a week along a fault traversing the field. Many of them originated well below the maximum depth of the wells themselves. The oil company was persuaded not only to stop pumping water in but, during a four-month period, to withdraw water from four wells near the center of quake activity. There was an immediate drop in such activity, and occurrences near the withdrawal wells virtually stopped.

This encouraged hope that along faults where quakes originate at relatively shallow depth, such as the San Andreas, water injections and withdrawals could be used as a control device. The naturally occurring quakes of the San Andreas are much deeper than those generated by water pumping in Colorado. However, few, if any, are below 10 kilometers (six miles), a depth which is shallow as natural quakes go. One problem, it was pointed out, is to make sure that the consequences are really controlled. King Hubbert, one of those who proposed that increased water pressure makes the rock fracture more readily, warned that, where enormous strain has developed in a fault, releasing it could be disastrous. He likened the situation to a giant mousetrap set so that a slight disturbance can release it. Water should be injected only when stress in a fault is minimal, he counseled, saying, "If you're going to mess with a mousetrap, mess with it when it's not set."

The laboratory experiments that led ultimately to an explanation both for the advance warning and the special role of water began in the 1960s, using devices that could simulate conditions deep in the Earth. The first surprising discovery was made by various researchers including a group at the Massachusetts Institute of Technology consisting of William F. Brace, B. W. Paulding, Jr., and Christopher H. Scholz. These three found that when crystalline rock like granite is subjected, in a pressurized environment, to stress at from one third to two thirds its breaking point, it begins to swell. The extent of this swelling is independent of the pressure, at least within the range of pressures then available in the laboratory. But, most important, as the rock swells it becomes more difficult to break. Brace and his colleagues proposed that this is because in such rock, whose tiny pores are filled with pressurized water, swelling of the rock increases the pore sizes and, perhaps, opens up new pores and microscopic cracks devoid of water. With the pores no longer saturated with pressurized water, it was deduced, the rock no longer broke as easily.

Then an Israeli geophysicist, Amos Nur, and Gene Simmons at MIT found that rock with dry pores, as opposed to that which is saturated, transmits pressure waves more slowly. They showed this to be true of a variety of local rocks including some, like Chelmsford granite, whose porosity is no more than one percent. Nur suggested that the Soviet observations could be explained if the rock first swelled under the stress preliminary to an earthquake, then slowly saturated with water until the pores were filled and it became relatively breakable. Scholz (by then at Lamont) and James H. Whitcomb of the Cal Tech group elaborated this concept. They noted that crustal rocks beneath shallow depths typically are water saturated, but, they reasoned, when such rock is dilated by stress, cracks and pores are opened that, to begin with, must be dry.

This hardens the rock, preventing the immediate occurrence of a quake, and there is a sharp drop in the velocity of pressure waves passing through the area. Then water slowly diffuses into the cavities, and, as the rock becomes increasingly saturated, the wave velocity climbs back to normal. When saturation is complete, the quake occurs. Furthermore, the greater the area under stress, the longer it takes for this return to saturation—and the larger the resulting earthquake.

Both the Lamont and Cal Tech groups pointed out that this explanation for the prelude to an earthquake could also account for other recently reported long-term forms of warning. Dilation of a great volume of crustal rock could explain the lifting of the landscape reported by the Japanese as having occurred five years before the quake at Niigata. The five-year period would be the time required for the dilated rock again to become saturated.

The increased electrical conductivity of the crust as the time of a quake approaches, reported by the Russians, would arise from the final filling of pores with water. The more water in the rock, the better a conductor it becomes—a phenomenon which could also account for some of the reported magnetic precursors. The sharp increase in the radon content of water from deep wells, recorded several years before the Tashkent earthquake of 1966, would arise from rock fracturing, accompanied by flushing out of the radon by water movement. Being short lived, radon can reach the surface only if it does so relatively soon after its production by the decay of radium.

Although these discoveries have, for the first time, made earthquake prediction a realistic possibility, the change in pressure wave velocity, which seemed the most powerful tool in that respect, has been seen primarily in regions where earthquakes occurred at shallow depths (less than 16 kilometers or 10 miles) and involved one form of slippage: underthrusting. While the San Fernando quake was clearly a by-product of the northwest movement of the Pacific Plate, relative to North America, it did not involve side slippage, as occurs along the San Andreas Fault. The Los Angeles slab pushed north under the San Gabriel Mountains, lifting them as much as a meter. In 1974 MIT scientists reported evidence that the pressure wave change had occurred, from 1964 to 1968, before five out of six large Japanese earthquakes involving side slippage. Nevertheless, hopes for finding such changes prior to great earthquakes had faded by the 1980s. The effect seemed by no means universal.

The most successful prediction has been by the Chinese in 1975, before a severe quake hit Haicheng in northeast China. When a combination of premonitory signs first led to the evacuation of homes, a month or two earlier, it proved to be a false alarm. Then, early in 1975, the symptoms became even more alarming and with great difficulty the population was again persuaded to flee into the snowy open. The quake struck and caused extensive damage. Unfortunately, when a far larger quake occurred the next year, directly under the industrial city of Tangshan, 240,000 died and more than 700,000 were injured. Predicting quakes in China is complicated by the occurrence of many faults, rather than a single dominant one, like the San Andreas Fault of California. It is under compression from several directions, including the westward motion of the Pacific floor and the northward drive of India.

Probably the most successful way to predict quakes, in general terms, has been the "gap" method, identifying sectors of a mobile fault long overdue for a quake. The rupture zone of the great 1964 Alaskan quake had lain quiet since 1900. The last great earthquake in southern California took place more than a half century ago, and the same goes for San Francisco. It has been suggested that, in the San Francisco area, movement along the fault may be four meters "in arrears." By 1973 more than 15 such "seismic gaps" had been identified around the margins of the Pacific—in Alaska, the Aleutians, the Kurile–Kamchatka sector, Japan, Mexico, Central and South America—as well as the Caribbean. One of them was wiped off the list of these "most expected" sites when a major quake occurred near Sitka, Alaska, in 1972. In the 1980s more gaps were identified in Mexico and in the Shumagin and Yakataga sectors of the Aleutian–Alaska front.

Those sectors of the San Andreas Fault nearest Los Angeles and San Francisco both seem to be "locked" and accumulating strain, whereas slippage occurs regularly and in small, usually harmless, increments in the intervening part of the fault that passes near Hollister. It is there that a winery of the Almaden vineyards has become famous, for in 1939 it was inadvertently built astride the fault (where an earlier structure, for reasons then unknown, had collapsed). The building, slowly being torn apart by motion along the fault, is cluttered with scientific instruments as well as hogsheads of wine.

Whereas seismologists have been able to identify prime danger zones for quakes in the West, leading to the enforcement of building codes that require quake-resistant construction, in the East there is no such clear-cut pattern, and residents of that region have tended to regard themselves as immune from major quakes. Such events are, in fact, so rare that their cause is little understood and their prediction, in terms of what is now known, virtually hopeless. On February 5, 1663, an earthquake apparently of major scope struck the region around Three Rivers in Quebec. Lurid, and not necessarily reliable, accounts tell of mountains thrown down and forested slopes sliding into the St. Lawrence River. Pewter was shaken from shelves as far away as the colonial seaports on Massachusetts Bay. In 1755, only 18 days after the quake that devastated Lisbon, residents of Boston were awakened during the night by a sound like approaching thunder. Some, on running to the windows, saw trees along the streets waving back and forth in a wild manner. Chimneys and walls fell in nearby towns, but no deaths were reported.

The most devastating earthquake since settlement of the East Coast occurred in Charleston, South Carolina, in 1886. Much of the city was destroyed, almost no house escaping damage. At least 60 died. However, what was probably the most powerful quake ever recorded in North America—little known because few witnessed it—occurred, not in California nor along any known plate margin, but in the heart of the continent, centered at New Madrid on the Mississippi in southern Missouri.

Shortly after 2 a.m. on the morning of December 15, 1811, according to a summary published many years later by the United States Coast and Geodetic Survey:

...the inhabitants of the region were suddenly awakened by the groaning, creaking and cracking of the timbers of the houses and cabins in which they were sleeping; by the rattling of furniture thrown down; and by the crash of falling chimneys. In fear and trembling they hurriedly groped their way from their houses to escape the falling debris, and remained shivering in the winter air until morning, the repeated shocks at intervals during the night keeping them from returning to their weakened and tottering dwellings.

Daylight brought little improvement to their situation, for early in the morning another shock, preceded by a low rumbling and fully as severe as the first, was experienced. The ground rose and fell as Earth waves, like the long, low swell of the sea, passed across the surface, tilting the trees until their branches interlocked, and opened the soil in deep cracks as the surface was bent. Landslides swept down the steeper bluffs and hillsides; considerable areas were uplifted, and still larger areas sunk and became covered with water emerging from below through fissures or little "craterlets," or accumulating from the obstruction of the surface drainage. On the Mississippi, great waves were created which overwhelmed many boats and washed others high upon the shore, the returning current breaking off thousands of trees and carrying them out into the river.

Unfortunately, the plate theory has provided no clear explanation for such occurrences. There has continued to be a moderately active fault zone along this sector of the Mississippi and along the Saint Lawrence Valley, the alignment of the two zones suggesting a connection. The damaging earthquakes of 1929 and 1944 that struck, respectively, Attica and Massena, New York, lie along this line. Likewise small to moderate quakes occur rather often along the coastal zone east of the Appalachians. One of these was the so-called Amityville earthquake of 1884, which shook the entire New York City area and was felt as far away as Burlington, Vermont and Baltimore, Maryland. Another, in 1893, was centered beneath the area from 10th to 18th streets in New York, creating consternation in Manhattan pool halls, and early in 1973 a small quake shook the Philadelphia area. Yet, strange to say, quakes seem rarely, if ever, to occur in the Appalachian zone itself. (Figure 18.5 shows sites of recent earthquakes in the United States.)

While major quakes in the East are rare, the damage to be expected when they do occur is greater than in the West. The constant warring and rifting of lithospheric plates has so fractured the crust in the West that earthquakes tend to lose their force within a relatively short distance. They can do much damage locally, but not over a wide area. Carl Kisslinger of the University of Colorado has pointed out that the Earth's crust in the East is relatively intact and serves as an efficient propagator of earthquake waves. Hence, he has written: "Although the number of large earthquakes east of the Rocky Mountains is much smaller than to the west, the much larger areas of high intensity for a given magnitude in the East makes the long-term risk, in terms of potential damage to property and loss of lives, roughly as great as in the West."

When the possibility of prediction first began to seem real, at the 1972 symposium on the subject in Washington, some of the participants expressed concern about what will happen when forecasts are made. The social, economic, and political problems will be formidable: How to avoid panic, how to cope with traffic jams, looting, or economic chaos when a great city is told that a quake is likely at a certain (perhaps not very precise) time. An ill-prepared prediction could do more harm than the quake itself—particularly if the latter were not as severe as expected. The Lamont group ended its 1973 review of the situation by stating: "Studies on how best to use earthquake prediction as a social tool should also begin at once, since the day when earthquake prediction is a practical reality may not be so far away."

Furthermore, predictions by themselves would not be enough. A month before *Nature* published the report of the Lamont group on its Adirondack observations, an earthquake leveled much of Managua—the second time in a half century that the Nicaraguan city had largely been destroyed. In both cases damage was particularly

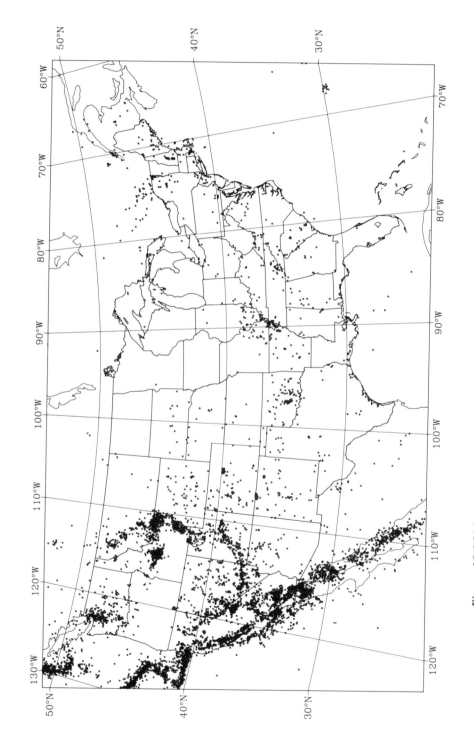

Figure 18.5 U.S. Seismicity, 1960–1988 compiled by the U.S.G.S. The concentration of quakes along the western plate boundary is evident as well as along the Cascades and other mountain ranges.

heavy because buildings had not been built properly for quake resistance. In its commentary on the Lamont report *Nature* said: "There is little to be gained by predicting earthquakes...if afterwards there is no city to which the temporarily evacuated inhabitants may return."

The ideal solution, of course, would be prevention, but the idea of putting reins on the Earth to control such activity seemed, until recently, quite preposterous. The Colorado well experiments have now shown that, at least in suitable situations, this should be possible. The development of a theory to explain the controlling role of water in such circumstances makes for a more educated attack on the problem. The most widely discussed scheme would be to sink a series of deep wells along a fault like the San Andreas for periodic strain relief. Water would be withdrawn from wells at the two ends of a sector chosen for relief (to lock the fault firmly at those points). Then water would be pumped into wells within that sector until something gave way. Such experiments obviously should not be carried out near inhabited areas until they have been thoroughly tested elsewhere. The method also may be limited to one type of earthquake zone, such as that in California where quakes are shallow and slippage is sideways, as opposed to areas of deep underthrusting. Even that, however, would be a boon, for some of the most devastating quakes elsewhere in the world occur along similar faults, as in Turkey and Iran.

The immediate problem, however, is to determine how universally applicable the newly found prediction effects may be. In an editorial on the discoveries Philip H. Abelson, the editor of *Science* and a geophysicist who was also president of the Carnegie Institution of Washington, wrote:

> If we wish to understand and be able to predict the rare, large earthquakes, we should be seeking premonitory signals everywhere that earthquakes have been known to occur. We should invest in new ideas, development of new instrumentation, and in the establishment of observing networks. Other countries should be encouraged to do likewise, and we should assist them whenever feasible.
>
> The task of minimizing earthquake disasters is a large one (he continued), and may require decades to complete, but what are decades in a span of millions of years?

Chapter 19

Can the Genie Within Be Tamed?

IF CONTENTION BETWEEN CRUSTAL PLATES OF THE EARTH ON THE ONE HAND produces havoc through quakes and volcanic eruptions, can we on the other capture, for our own benefit, some of the vast energy involved in these processes? The truth of the matter is that in some parts of the world this has been done for generations. Since 1904 the Italians (astride where the African Plate has pushed under the European Plate) have been generating electric power from the steam jets, or *soffioni*, that erupt from the ground in the Larderello region, where the provinces of Pisa, Siena, Leghorn, and Grosetto come together. The heat has been attributed to a great batholith of hot granite two to six kilometers (one to four miles) below the surface that, because of the low conductivity of the overlying material, is cooling off very slowly. Production of the power plants there in 1986 was 348 megawatts (one megawatt being a million watts).

In Iceland, on the Mid-Atlantic Ridge, an unconventional public utility known as the Hitaveita Reykjavikur, or Reykjavik Municipal District Heating Service, sells metered hot water derived from the scalding lava rock that underlies that region. Iceland's most famous gusher of steam and hot water is the Geysir (from which the English word for geyser is derived), which lies not far from the capital city.

In 1969 some 90 percent of the homes in Reykjavik were heated by geothermal water. It warmed more than 95,000 square meters of hothouses where, within sight of snow, one can see bananas and grapes ripening, and some farmers' wives bake bread in ground that is exceptionally hot. At least 80 swimming pools are geothermally heated, and as of 1969, 40 percent of the country's 200,000 inhabitants lived in homes serviced by one of the five regional geothermal water systems. Heating in this way is, for Icelanders, about 57 percent cheaper than heating with oil shipped to that remote land, and from 1961 to 1969 the use of geothermal energy for space heating and industrial purposes doubled, saving the country oil imports of 210,000 tons. A not inconsequential advantage of such heating is that no fire hazard is involved.

Iceland's chief industrial use of geothermal steam has been at a Lake Myvatn plant where 42,000 tons of diatomite, mined from that lake, have been processed each year for export. Diatomite, or diatomaceous Earth, is formed of diatom shells—a million of them per thimbleful—and is widely employed for filters, high temperature insulation, paint additives, and many other uses. Its presence in a great deposit on the floor of Lake Myvatn testifies to millions of years when that section of the Mid-Atlantic Ridge was submerged and subject to the gentle rain of diatom shells that is part of the life cycle of the sea. To furnish steam for this plant wells have been sunk to 1200 meters. The steam is used to dry the diatomite and also to turn the turbines of a pilot power plant generating three megawatts.

By 1987, 18 countries, including China, Japan, Kenya, New Zealand, the Philippines, the Soviet Union, and the United States, were using geothermal energy for commercial generation of electric power. As the plate theory has developed, it has become evident that the crust of the Earth is gridded with a network of superheated zones, like some giant, global toaster with an odd pattern of glowing electric wires. Not only do these zones produce volcanic eruptions and earthquakes, but they are the source of hot springs, occasional steam jets, and, far more widely, great intrusions of hot rock that lie at depths shallow enough to be within reach of drilling.

During the years preceding 1984 the use of geothermal power increased at 17 percent a year, its expansion stimulated by the energy crisis of that period. As oil prices plummeted, the expansion leveled off, but a new "crisis" in 1990 promises a revival. The most efficient method uses dry steam to turn the turbines. This is applied at the Geysers, 90 miles north of San Francisco, where the Pacific Gas and Electric Company derives 1800 megawatts from a complex of wells—about 38 percent of total world geothermal production (Fig. 19.1). At most sites, however, liquid water droplets are mixed with the steam as it comes out of the ground and the mixture must be "flashed" into dry steam by reducing its pressure. This is done at the 13 plants of Cerro Prieto, south of the Imperial Valley and Mexico's chief source of geothermal power. In 1989 the Cerro Prieto plants were generating 620 megawatts, the second highest geothermal output on the continent. Plants on the California side of the border, in the Imperial Valley, including the Salton Sea, had a combined output of 265 megawatts. For comparison, the giant Ravenswood fossil fuel plant in New York City generates 1742 megawatts.

Because of popular opposition to atomic power plants (no one wants one in his backyard), PG&E produces more power from geothermal steam than from atomic reactors, although its chief power source is still fossil fuel. The cleanest form is, of course, hydroelectric, but there is little opportunity for expansion of that source, and California is exceptionally fortunate in the large dams that stand near by. Fossil fuel plants, while smoky, do not evoke as emotional an opposition as do atomic plants.

Geothermal plants are not necessarily pollution free, although they probably can be made so. The first thing that strikes the visitor to Big Sulfur Canyon in the Geysers steam field is the powerful smell. It has been estimated that the amount of sulfur released at the Geysers (in the form of hydrogen sulfide, the gas that smells like rotten

Figure 19.1 California geothermal plant. *Wells at the Geysers geothermal field in California emit steam. How much heat within the earth can we tap for power?*

eggs) is equivalent to that coming from a power plant producing the same amount of electricity from low-sulfur fuel. Mexico's plant at Cerro Prieto, where a black volcanic cone rises south of the California border, may, according to some estimates, release as much sulfur as a plant burning high-sulfur fuel.

There are other problems. The steam that roars from the 2100-meter-deep wells at the Geysers sometimes carries sand and grit which must be removed in centrifugal separators before the steam enters a turbine. Steam or hot water from such wells typically contains other contaminants which may become corrosive on contact with air. While 75 percent of the water at the Geysers is discharged into the air as vapor from cooling towers, the residue contains boron and ammonia, and, while the amounts of those substances are small, the water cannot be dumped into local streams and must be directed down injection wells drilled for that purpose. In the Salton Sea area of southern California the salt content of geothermal water may reach 20 percent (compared to 3.3 percent in sea water), although a short distance to the south, at Cerro Prieto, it drops to two percent.

The exploitation of geothermal power is a technological challenge. The steam, where available, is relatively cool and at low pressure: seven times sea-level atmospheric pressure at 205 degrees Centigrade (400 degrees Fahrenheit) at the Geysers, compared to 200 times atmospheric pressure and 550 degrees Centigrade in a typical fossil fuel plant. Consequently, turbines in use at the Geysers, especially designed for low-pressure steam, are about one half as efficient in terms of electric production per unit of steam energy as those in conventional power plants. However, the added turbine cost per kilowatt of achieved energy is more than compensated by there being no need for boilers and water treatment facilities.

Since "dry" steam, with no water admixture, is rare, flashing must be used in most locations, including Wairakei in New Zealand, Otake in Japan, Pauzhetka in Kamchatka, and El Salvador. The flashing method also can be used to obtain fresh water from the saline water in some geothermal areas. Hot water is passed into a low-pressure chamber where part of it flashes to steam which, while too weak for driving a turbine, can be condensed into clean fresh water. The remaining water is then passed to a chamber at even lower pressure, and more steam is flashed from the water. Thus, in several stages, considerable amounts of water can be extracted for use in irrigation. Most of the irrigation for Southern California's Imperial Valley, however, is diverted from the Colorado River, just before it enters Mexico, via the All American Canal, making the valley in winter a truck gardener's paradise that produces a large percentage of the carrots and lettuce eaten by Americans during those months. Leaching of irrigation waters as they drain back into the Colorado has made the river twice as salty as allowable for drinking.

Under a 1944 treaty the United States agreed to allow at least a small part of Colorado River water to flow into Mexico. The salinity of this water was not specified, but in 1974 Congress passed the Colorado River Basin Salinity Control Act whereby the water flowing into Mexico would not be more than 115 parts per billion (ppb) saltier than when it reached the Imperial Dam, north of the border. To achieve this and not allow water diverted to the Imperial Valley to become too salty for irrigation, the Bureau of Reclamation is building, at Yuma, Arizona, the world's largest desalination plant using the reverse osmosis method. It is due to begin operation in 1991, producing 67,000 acre feet per year with a salt content limited to 295 parts per million (ppm) and at a cost of about one dollar per thousand gallons. The highly saline residue, at 9800 ppm, will flow down the Wellton–Mohawk bypass drain that is being extended through Mexico almost to the Gulf of California. Also, to increase the river's flow, the Coachella Drainage Canal from the Imperial Valley has been lined, to reduce seepage, and additional wells have been dug in southern Arizona.

Another concern has been that geothermal and irrigation water extracted from the Imperial Valley might cause it to subside, despite efforts to counter this by injecting saline water back into the Earth. The current threat to the beautiful Renaissance city of Venice is an example of what happens if too much water is pumped from a sedimentary basin. Industrial plants around the rim of the Venetian lagoon have drawn so much water from beneath its floor that Venice has been sinking at an alarming rate, its palaces and the famous Piazza San Marco now subject to flooding during every storm surge of the sea. Similarly, Long Beach, California, subsided until parts of the city were below sea level because so much oil was withdrawn from below. Now, however, oil companies under such circumstances pump water into an oil field to replace the oil that is being withdrawn. If water is imported, perhaps from the Gulf of Mexico, to replenish the deep hot water beneath the Imperial Valley, the region may provide geothermal power almost indefinitely.

One proposal is to dredge a ship canal to Yuma, Arizona, where the Colorado River enters Mexico. Yuma was a seaport, via the lower Colorado, before the river was dammed there to divert water through the All American Canal to the Imperial Valley. If a ship canal were dredged as far upstream as Yuma Mesa and a Mexican–American seaport were built near San Luis, as has been proposed, ocean water could be channeled from there for injection down wells around the perimeter of the Imperial Valley, both on the American and Mexican sides of the border. (Some, however, have questioned whether such massive injection would be feasible.) The new port would enable ships to haul out potassium chloride, sodium chloride, and calcium chloride derived from the geothermal water. Commercial use of such compounds (for example, in snow removal) depends on cheap bulk transport to potential markets.

Another concern is the possibility that tampering with underground water reservoirs in an area of many active faults, like the Imperial Valley, might produce earthquakes. As noted earlier, it is the injection of water, rather than its removal, that seems to generate quakes, but the effects of any change of the water regime in such an area will have to be watched closely.

It is also possible that extracting hot water in earthquake-prone geothermal areas may reduce the danger of quakes. Laboratory studies by Brace and his colleagues at MIT have indicated that faults are most apt to creep in response to strain when the rock is hot. At lower temperatures, the rock is more likely to resist slippage until great strain has accumulated. Hence, in a comprehensive study of the geothermal potential of the Imperial Valley, a group from the University of California at Riverside has proposed that, in fault zones, withdrawal of hot brines from great depth may heat rocks nearer the surface, encouraging them to creep rather than store up strain for a destructive earthquake.

One way to produce electricity from water not hot enough for efficient, direct production of steam is to use the water to heat a substance whose boiling point is very low, such as isobutane. The latter vaporizes at a temperature well below the boiling point of water. Its vapor is denser than steam, making it an efficient driver of turbines. However, it is flammable and must be used with special precautions, including a well-sealed circulation system. (Other fluids, such as freon, that are not flammable, can be used, but they are more costly.)

In a 1986 survey, Ronald DiPippo of Southeastern Massachusetts University found that there were such plants at 13 sites in the United States (in some cases with multiple small plants). There were three sites in California, five in Nevada, two each in Oregon and Utah, and one in Idaho. In Kamchatka, where 63 active volcanoes form a chain over the descending Pacific Plate, the Russians have built a pilot plant whose turbines are driven by freon vaporized by the heat of geothermal water. The plant is at Paratunka, a suburb of Petropavlovsk-Kamchatsky, and its modest 800-kilowatt output supplies local apartment houses. Hot water from the ground is used, as well, to heat a greenhouse producing 2000 tons of vegetables a year.

There is also the possibility, yet to be demonstrated, of exploiting the heat of hot rock intrusions that are dry—that is, to which ground water at present has no access. From the growing knowledge of plate boundaries and subsurface structures it is evident that such intrusions are more widespread than had been realized.

Actually, geothermal heat exists beneath every populated place in the world, but in most areas it is probably too deep for practical exploitation, at least in terms of present technology. To extract energy from the shallower deposits of hot rock, a way must be found to deliver water to them and then recover it as steam or hot water. Such deposits are dry because they are sealed off from surface water, and they tend to be great masses of granite with few cracks that permit the entry and circulation of a fluid. It will, therefore, be necessary to drill into such rock and fracture it in some way, such as by the injection of water at a pressure of about 470 atmospheres—a technique used in oil extraction and known as hydrofracturing. Explosions at the bottom of such a hole also have been proposed, and in 1971 the Atomic Energy Commission and the American Oil Shale Corporation studied the possibility of using a nuclear explosion for this purpose as part of the so-called Plowshare Program ("They shall beat their swords into plowshares...") for converting nuclear explosions to peaceful purposes. However, such explosions are costly; they would require safe handling of the radioactive steam or hot water that would then be produced; and the AEC was encountering much public opposition to the Plowshare projects in general.

Once the hot rock reservoir has been fractured enough to permit water circulation, another hole would be drilled so that cold water could be injected via one hole and hot water could escape via the other. The hot water would vaporize isobutane, which would drive turbines, and the now cool water would be returned down the first hole, eliminating any pollution problems (except the almost negligible one—in this case—of heat released to the environment).

A group at the Los Alamos National Laboratory in New Mexico has been exploring this technique and trying to develop an economical method of deep drilling. They have produced a small drill "bit" that melts its way down through volcanic rock, instead of boring a hole, in the hope of developing what they call a "Subterrene" that can sink wells into rock so hot that, at such temperatures, ordinary drill pipe gets almost as soft as spaghetti. They hoped rock melted by the drill would form a glassy wall inside the bore hole. For geothermal test drilling a site has been found nearby at Jemez Mountain, where a hot rock residue of volcanic activity at 300 degrees Centigrade (570 degrees Fahrenheit) lies 2300 meters below the surface.

It has been proposed that the amount of "dry" geothermal energy available in this form is 10 times the combined steam and hot water reserves, but the fracturing technique remains to be demonstrated; to date it has been used primarily in sedimentary, oil-bearing rock—not in granitic batholiths. The Los Alamos group hopes that, once they have cracked the granite at the bottom of a drill hole, the fractured area will enlarge itself as heat is withdrawn. When the granite cools, they reason, it should contract and fracture further, opening up new areas of hot rock to the circulating water. In 1989 Sandia National Laboratories, to test its geothermal potential, began drilling a six-kilometer hole into the Long Valley caldera near Mammoth Springs, California. It was said to be the deepest penetration, so far, of a potentially active volcanic feature.

A study of the Imperial Valley, described by Robert W. Rex and his colleagues from the University of California at Riverside in 1972, found that the valley is a sedi-ment-filled trough six to seven kilometers (roughly four miles) deep, containing 10,000 cubic kilometers (2400 cubic miles) of geothermal brines—"a potential re-source of remarkable dimensions." The reservoir is about equally divided between the Mexican and American parts of the valley. The water temperature is higher than 250 degrees Centigrade (500 degrees Fahrenheit) at more than 130 times atmospheric pressure. When this reservoir is pierced, pressure in water entering the well drops and the water starts boiling, as it does in a geyser. The resulting mixture of steam and water shoots to the surface at close to the speed of sound. In fact, at some operating geother-mal fields the roar, when a well is vented to "clear its throat," constitutes a form of "noise pollution" comparable to the sound of a jet taking off.

The great accumulation of sediment in the Imperial Valley is in large measure a consequence of the uplift of the Colorado and Kaibab plateaus and the subsequent cutting of the Grand Canyon by the Colorado River. The resulting sand, gravel, silt, and clay were deposited in the valley as the river periodically shifted its channels. The trough, without these sediments, would look in cross section much like the deep, southern end of the Gulf of California, of which it is clearly an extension. Its walls are formed of outward tilted, subsided blocks with a relatively flat basement floor.

This is the pattern found in oceanic rift and fault valleys throughout the world. However, the geometry of the ridge-and-fault system here is remarkable in that the ridge segments are extremely short and the "transform" faults linking these offset segments are extremely long (the longest being the San Andreas Fault itself, extending almost the entire length of California from a series of short rifting segments in the Gulf of California, and possibly some beneath the Imperial Valley, to a short ridge beneath the Pacific off Cape Mendocino). These bits of ridge represent a highly fragmented extension of the East Pacific Rise up through the Gulf of California (see the right-hand map, Fig. 17.6).

The ridge sections have been identified as localized spreading centers through analysis of earthquake records. For example, in 1969 more than 70 shocks with Richter magnitudes between 4.0 and 5.5 (the San Fernando quake of 1971 was rated at 6.6) were recorded in a period of six hours near Consag Rock, a young volcanic outcrop on the centerline of the Gulf of California at its upper end. The directions of movement in these quakes were typical of spreading from a mid-ocean ridge.

Apparently, some four million years ago the northwest motion of the Pacific Plate, relative to the American Plate, which is dragging with it the continental rim from San Francisco to Mexico, tore Baja California loose from the mainland and since then has carried it 260 kilometers (160 miles) up the coast, while the oblique element of spreading in the Gulf of California pulled the peninsula away from the coast. Some believe the entire slice as far north as San Francisco eventually will be split off from the continent, and the region from Baja California north through San Diego, Los Angeles, Santa Barbara, Monterey, and Santa Cruz will become an island.

There is ample evidence for crustal movement of this type in the Imperial Valley. The United States Coast and Geodetic Survey conducted a series of triangulations in the area, beginning in 1931, showing the motion to be similar to that all along the San Andreas Fault (where the far side always moves to the right). Such activity resulted in a major earthquake (Richter magnitude 7.1) in 1940, and where the Imperial Fault, a continuation of the San Andreas system into Mexico, crosses the border, it was found that the west side of the fault had moved three meters north relative to the east side. (Neither Mexico nor the United States has complained—in fact no one is sure which country gained and which lost territory, although the border, had it been marked by a fence, would have been displaced that much.)

From temperature measurements taken in wells as deep as 2.8 kilometers (1.7 miles) in the so-called Buttes area at the south end of the Salton Sea, Rex and his colleagues estimated that, near the bottom of the sediment, six kilometers (four miles) down, it must be hot enough to melt granite. The Buttes are a series of volcanic domes that apparently erupted underwater from 16,000 to 55,000 years ago, when the Salton Sea was deeper than it is today. Fragments of rock brought to the surface by these ancient eruptions are similar to those erupted along the crest of the East Pacific Rise. But also common are bits of granitic rock, which seems to form, typically, on a continental rim, rather than on a ridge. Despite the high salinity and dissolved metal content of the water from wells near the Salton Sea, it would appear from more than 10 geothermal wells drilled elsewhere that most of the hot water beneath the Imperial Valley is less salty than sea water. And, according to Rex and his co-workers, "abundant iron and manganese chlorides, plus small amounts of copper, silver, zinc and lead" in the Salton Sea brines "may have an economic potential themselves."

Rex has estimated the electric generating capacity of the Imperial Valley brines at from 20 to 30 thousand megawatts, or five to 10 percent of current American power production. The investment needed to achieve this generating capacity, in his view, would be about five billion dollars, and the fresh water that could be derived from the plants would constitute a flow comparable to half that of the lower Colorado River. No wonder replacement ocean water would be needed!

His estimate of the power potential of this reservoir has been used in a study of California's long-term power needs, conducted by the Rand Corporation in 1972 for the California State Assembly, with support from the National Science Foundation.

The authors urged various measures to slow the soaring demand for electric energy and then said:

> Slowed growth of demand, combined with new technological developments such as use of geothermal resources, could be extremely effective in resolving California's electricity quandary. If, for example, slowed growth in electricity demand were supplemented by rapid development of the geothermal resources of the Imperial Valley to their full 30,000-MW potential, then *no new fossil-fueled or nuclear power plants would be needed in California after 1985.* (Italics theirs.)

Not all potential reservoirs of geothermal energy lie along plate boundaries. As some of the deeper oil wells along the Gulf Coast of Texas, far from such a boundary, became depleted, they began to produce superheated steam and water. Well head pressures ran as high as 500 atmospheres or more. A well in Matagorda County, 5859 meters deep, tapped water at 273 degrees Centigrade (523 degrees Fahrenheit).

It appears that there is a huge reservoir of very hot water deep in the Gulf Coast sediments. It has been studied extensively by Paul H. Jones of the United States Geological Survey office in Baton Rouge, Louisiana. The sediment accumulation there is of extraordinary depth, due in part to slow subsidence of the bedrock in the manner typical of a geosyncline and in part to events inland: first, uplift of the Rockies that led to heavy erosion and sediment transport down the Rio Grande, Missouri, and Mississippi river systems; then, the ice ages, whose melting periods spread vast amounts of sediment over the Gulf. The deposits are well known, for some 300,000 wells have been sunk by oil prospectors between the Rio Grande and Alabama. The deepest accumulations—in both cases more than 15 kilometers (almost 10 miles)—are off the south Texas coast and along the Louisiana shore. Most of the material was laid down in the last 26 million years.

Beneath most of this accumulation is a thick salt deposit dating from the Late Triassic to Mid-Jurassic, when reopening of the Atlantic was at an early stage and the Gulf was cut off from the world oceans, apparently drying up periodically in the manner that later took place in the Mediterranean. It is this salt that has pushed its way up through the sediment in great fingerlike "diapirs," three to eight kilometers (two to five miles) in diameter, whose crests are known as salt domes (Fig. 9.4). More than 400 of them have been penetrated by drillers.

Salt is an efficient conductor of heat, and at the depths where the main salt layer now lies the Earth is very hot. Hence, Jones has written, "the immense volume intruded into the overlying sediments from depths of 15 km. or more, must have a profound effect upon the geothermal regime of the basin...The salt diapirs actually resemble a nest of heating rods thrust upwards into the basin deposits."

The sedimentary layers thus being cooked are predominantly shale, and the process extracts much of the water chemically bound into the rock, yielding water that represents from 10 to 15 percent of the bulk volume of the shale. This combines with water already there to form a high-pressure reservoir of very hot water. A sheetlike layer of waterlogged sediment over the entire Gulf basin serves as an efficient insulator, impeding upward movement of this heat. The water is salty at shallow depths, but below about three kilometers (two miles) it becomes progressively fresher. Unlike the water in most geothermal reservoirs, it is not replenished by rainwater (or snowmelt) that works its way down to the hot region. It is "manufactured" locally.

For this reason, like the oil being extracted from that region, it is a depletable resource, but Jones believed the extent of this hot water is vast. He did not envision its exploitation in the immediate future for, as he said, "the power generation technology is based on oil and gas, which provide a relatively inexpensive source of energy." However, he pointed out that there may be other such accumulations elsewhere in the world where oil and gas are not locally available. Since the water comes out of the ground so hot that it readily turns to steam and is thus "self-distilling," it could be used for irrigation—a consideration of little value along the Gulf Coast with its moist climate, but of potential importance in some underdeveloped lands.

In the early 1950s the United Nations began exploring the potential of geothermal energy as a power source for nations just embarking on the road to industrialization. A glance at the geography of plate margins (as shown by the map of global earthquake activity in Fig. 6.7) shows that many of them pass through or alongside developing countries. The African rift valleys were an obvious area for exploration, and, under United Nations auspices, aerial surveys of infrared emissions from the land have been carried out in Ethiopia and Kenya. Just as infrared "light" generated by the heat of an electric iron can be used to take photographs on infrared-sensitive film, so the infrared "glow" of the landscape can serve as an indicator of variations in ground temperature over large regions. According to Joseph Barnea who, as director of the Resources and Transport Division of the United Nations, was responsible for that organization's role in this work, it has been estimated from the aerial scanning "that a part of the Afar region in Ethiopia may have an exploitable geothermal potential sufficient to meet the present need for electric power for the whole of Africa." He added that there are "other areas in Ethiopia that are believed to have a geothermal potential of similar magnitude." If, as Jason Morgan and others believe, the Afar Triangle stands over a rising mantle plume, the region's potential in this respect should be no surprise. Haroun Tazieff the volcanologist, with his on-the-ground knowledge of the hidden heat resources of the area, has proposed that power-demanding industries, such as aluminum plants, be established on the nearby Red Sea coast.

In opening an international seminar on geothermal energy at the United Nations in January 1973, Barnea noted that, a few years earlier, his proposal that the rift areas should harbor vast geothermal resources was not popular. The idea of extracting energy from within the Earth was still unorthodox and, he said, "I personally recall the criticism to which I was exposed." However, he continued, "I am glad to report that practically all of the countries in the African Rift Valley have now requested United Nations assistance in geothermal exploration."

The Larderello company, veteran exploiter of the Italian steam fields, has helped survey potential areas in Uganda and at the El Tatio geysers near Antofagasta, Chile, where United Nations advisors have taken part in discussions of a possible plant. The island nations around the Pacific "ring of fire" have many potential sites, and Barnea estimated that two thirds of Turkey should harbor underground energy resources (although excessive carbonate in the water is an impediment). United Nations technical assistance has been provided to two Central American countries—El Salvador and Nicaragua—where an earthquake-generating oceanic plate descends beneath the volcanic isthmus linking the Americas, and Barnea has proposed that, since the region seems to have much more potential energy than it could use, the surplus power could be exported to the United States via long-distance transmission lines. One must, of course, be wary of optimistic statements regarding a technology that is still in its infancy. But if only a substantial fraction of these predictions prove valid the implications for a power-hungry world could be enormous.

The United Nations group has also pressed for greater direct use of hot geothermal water for house heating and for manufacturing processes, like those in the pulp and paper industry, that require only moderately high temperatures. For as long as historical records reach into the past hot springs have been exploited for recreation, for health spas, and, to some extent, for home heating. The Romans established their baths around the entire rim of the Mediterranean and even as far afield as Bath, England. Early in their history the Japanese began using the hot springs that abound in that land. By the last century Europe was dotted with spas and many sprang up in the United States.

Today Iceland is not the only part of the world that heats homes with geothermal water. In southwest Idaho the Boise Hot and Cold Water Company has been selling its "product" since 1890. In France oil prospecting has disclosed a reservoir of water at 70 degrees Centigrade (158 degrees Fahrenheit) beneath the Paris basin. The water, withdrawn from a well 1800 meters deep, is used to provide 3000 apartments at Melun with hot water and heat. Being saline it is then returned down another well. About two fifths of the annual needs of the apartments are provided, but the reservoir is said to be large enough to provide this service to one million apartments for 1300 years.

Two of the United Nations group, John Banwell and Tsvi Meidav, reported to the 1971 annual meeting of the American Association for the Advancement of Science that the Pannonian Basin in Hungary contains such a vast amount of hot water that its energy equivalent equals roughly half that of the entire world's petroleum reserves, as known at the end of 1963. The field, like that under the Paris basin, was discovered while drilling for oil and by mid-1969 was being used to heat 480,000 square meters of public buildings and housing. While the water is not corrosive, it is so laden with calcium carbonate that it must be run through settling tanks before use, and the upper parts of the 80-odd wells must be reamed periodically to clear them of deposits that accumulate there as the water flashes into steam and sheds its dissolved solids.

The Russians have reported that the Soviet Union's geothermal potential of economic significance (used, to date, chiefly for space heating) probably exceeds the USSR's combined resources of petroleum, coal, and lignite. Soviet authorities have said that, in 1970 alone, 15 million tons of fuel were saved through the use of energy from underground. Since 1916 Japan has heated greenhouses in this way, at one place growing tropical vegetation, alligators, and crocodiles as a tourist attraction. With its heavy dependence on imported fuel, Japan has also shown much interest in geothermal power generation, and the Otake plant, producing 15 megawatts, was but a first step in that direction. By 1986 there were 10 geothermal plants in Japan.

As in other refrigerating systems, energy (in the form of geothermal heat) can be used for cooling. Not only do the Russians make such refrigerating units, but the resort hotel at Rotorua, the famous New Zealand thermal area, is air conditioned in this manner (using a lithium bromide absorption system).

In their plea for wider use of geothermal water, Banwell and Meidav conceded that its energy, unlike that of oil, is not readily transportable and in many cases the temperatures are not high enough to be a likely source of electric power. But they urged that such water be used more extensively to heat buildings. "Why," they asked, "should a rapidly depleting fuel such as petroleum be used for heating boilers in those large regions of the Earth where the more abundantly available geothermal energy can do the same at a comparable or lower cost, and without the attendant air pollution associated with fossil fuel burning?" They appealed for energy management on a national level to achieve a more rational policy in this regard.

The realization that fossil fuel resources are limited and the growing public irritation with pollution from the combustion of such fuel have resulted in agitation for a major American effort in the geothermal area. In September 1972, a conference on the subject was held in Seattle at the initiative of Walter J. Hickel and the University of Alaska, with funding supplied by a program of the National Science Foundation designed to give support to "research applied to national needs." Hickel had been governor of Alaska (a state with its own geothermal potential, particularly where the Pacific Plate descends beneath the volcanic zone of the Alaska Peninsula), and he was later Secretary of the Interior. The resulting report recommended a 10-year research effort, at a cost of $684.7 million, to explore geothermal resources and develop new drilling methods and utilization technology. It set as realistic goals a generating capacity by 1985 of 132,000 megawatts, and by the year 2000 of 395,000 megawatts. (The 1986 survey found geothermal plants in the United States producing 2006 megawatts.) By contrast, said the report, the entire power system of New England now produces 15,000 megawatts. The projected output for 1985, according to the analysis, would constitute a saving of almost 30 percent of oil imports anticipated to meet the power requirements of that year. "Geothermal energy, therefore, can have a major impact on U.S. national self-reliance," it added.

Rex, who directed the Imperial Valley study, argued that national planning has ignored geothermal energy out of ignorance. Lack of data, he wrote, "was construed to indicate an absence of the resource, even though the United States has an enormous extent of recent volcanic rocks and an abundance of dormant volcanoes." Heat flow observations in the region from eastern Oregon down through eastern California and Nevada, he said, suggest that "tens of thousands of square miles of the basin and range area have geothermal potential." He cited as other such areas not only Hawaii, Alaska, and the Gulf Coast, but also the Appalachian region, whose ancient folds still harbor reservoirs of heat whose manifestations bubble to the surface from the relatively tepid waters of Sand Spring, two miles south of Williamstown, Massachusetts, to Hot Wells, Texas. It is no accident that towns named Hot Springs are to be found in three of the Appalachian states—Virginia, North Carolina, and Arkansas. Rex cited, as particularly promising, areas in the Virginias, western Maryland, and the Ozark and Oachita mountains of the midcontinent. Much of the new knowledge concerning thermal areas he attributed to the inspiration of Francis Birch, a Harvard professor who had championed the importance of heat flow measurements and sent his students (including Rex himself) into the field to use this tool as a way of exploring processes at work beneath the surface.

In a sense geothermal power has its lobby. In a number of states where atomic plants have been projected and environmental impact statements have been drafted by the Atomic Energy Commission, as required by law, the statements have been challenged on the grounds that alternative, allegedly pollution-free sources of power—including geothermal energy—had been ignored. This has occurred, for example, in Colorado, Washington, and Kansas.

Since this buried energy is constantly replenished by the upwelling of molten rock from the depths, it is in a sense inexhaustible. However, the volume of water in a reservoir at any one time is limited, and if it is withdrawn more rapidly than replenished and heated, the resource is depleted. The steam wells at the Geysers start out producing seven to 15 megawatts of thermal energy but slowly decline to levels of one to three megawatts, requiring the drilling of new wells. To justify investment in a new geothermal source its potential and lifetime must be assured. "A utility, by nature, has to be conservative," according to Carel Otte, who headed the geothermal division of the Union Oil Company of California, leader of the consortium that drilled the Geysers wells. "It can't commit funds unless there is a guaranteed return. That's the nature of the utility business." Heretofore, he added, "all the wells in a field had to be drilled and tested before commitments could be made for generating facilities, and equipment ordered. That meant a five-year delay between drilling the wells and installing the generators." In the oil and gas industry the behavior of gas reservoirs is well enough known so that only a few test wells are needed to persuade management to go ahead with a big investment. While steam is also a gas, it was not certain that the same rules could be applied to a steam reservoir. Was it possible, for example, that when a whole cluster of steam wells were opened at once, as at the start-up of a power plant, the pressure would drop radically as it does when all the taps are opened in a house with limited water supply?

Consequently, in January 1967, 21 steam wells at the Geysers were opened full blast for a three-week period, venting into the sky, with a horrendous roar, energy that could have been used to make electricity (and profits). "That was wasteful," Otte said. "It's like determining the reserves of a gas field by opening up all the wells and letting them blow. It won't happen again." In other words, he believes geothermal power has come of age.

A major factor, he says, has been the opening of federally owned lands to geothermal prospecting. In Nevada, a likely area for such work, 85 percent of the land is under federal ownership, and this is true of about 60 percent of California land. The Geothermal Steam Act of 1970, enacted in response to the growing power demands and public concern over environmental threats from ever bigger fossil-fuel and nuclear plants, authorized the leasing of federal lands for the development of such resources. It provided that the Secretary of the Interior could designate a "Known Geothermal Resources Area," or KGRA, within which sites could be leased through competitive bidding (Fig. 19.2). Such areas are defined as those "in which the geology, nearby discoveries, competitive interests, or other indicia would, in the opinion of the Secretary, engender a belief in men who are experienced in the subject matter that the prospects for extraction of geothermal steam or associated geothermal resources are good enough to warrant expenditures of money for that purpose."

Pursuant to this, the Secretary of the Interior had the Geological Survey identify such areas, and in 1971 a map was published showing a number of them, plus many hot spring sites in the Western states. Also identified by the Survey were areas "valuable prospectively" in which the outlook was not as certain as in a KGRA. From the map it is evident that the distribution of these sites is related to the tectonics of the Western

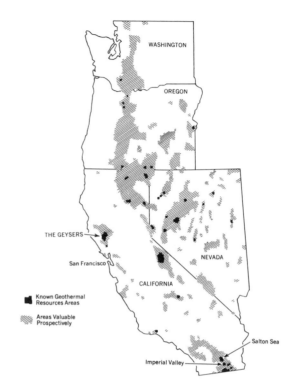

Figure 19.2 Geothermal resource areas

states—to the forces that have shaped the Rockies, the basin-and-range province farther west, and the coastal zone.

While Rex and other specialists believe the Geysers steam field is probably the world's largest, with a sustained production capacity perhaps as great as 10,000 megawatts, the actual capacity and lifetime of such fields remains uncertain. At the 1973 United Nations meeting Banwell tackled this problem, focusing on the two fields that have been exploited the longest—the hot water or wet steam field at Wairakei in New Zealand and the Larderello steam field in Italy. Early in exploitation of the New Zealand field, from 1957 to 1964, there was a rapid drop in pressure of about 20 atmospheres, and the ground subsided as much as 25 centimeters (10 inches) a year in some areas. There was an almost complete halt in geyser and hot spring activity in the valley, long a tourist attraction, but "spectacular increases" in the number and size of steam-venting fumaroles on the opposite side of the production field. After that, however, the situation stabilized and natural replenishment seemed adequate to meet the withdrawals.

At Larderello there was also a drop in pressure early in exploitation of the field. This then leveled off; but it has been found that, if the number of wells within a given area surpasses a certain density, there is no net increase in productivity, apparently because of limits to permeability of the formation from which the steam is derived. Banwell said that in such fields, to sustain heavy production far into the future, it will probably be necessary to recharge the underground reservoir and avoid overexploitation through rational management.

The experience at Wairakei should be kept in mind by anyone wishing to exploit the Yellowstone geothermal area with its 200 geysers and 3000 hot springs. It could be that to do so would "turn off" Old Faithful and many others of that region's wonders. However, since they are within a National Park, they are presumably immune from such economic utilization.

In any case, estimates for the lifetime of a typical geothermal field vary from dozens of years to thousands, with the large figure seeming more likely if exploitation is controlled. As Banwell pointed out to the United Nations meeting, there is no argument as to the extent of the stored energy within the Earth. According to Banwell and Meidav, in the upper 7.5 kilometers (4.7 miles) this energy, per square kilometer, is equivalent to that in 21 million tons of oil! In other words, this energy reserve exceeds by a vast margin the most optimistic estimates of the world's reserves of fossil fuels.

The debate concerns the percentage of this geothermal energy that can be tapped in economic fashion. Barnea, in his welcoming talk to the 1973 meeting at United Nations headquarters, said of the growing search for such buried heat resources: "...it is likely that geothermal energy indications will prove more frequent than petroleum indications and that in fifty years, geothermal energy will be recognized as an energy resource of even greater significance than petroleum."

Hickel, in his introduction to the 1972 Seattle conference report, likened present knowledge of such resources to the time when oil seeps on the surface were the only clues to petroleum reservoirs. Their counterparts, for geothermal energy, are the geysers and hot springs, he said.

While such comments may be tinged with enthusiasm, entry into the field of big businesses like Union Oil suggests that within a generation an appreciable part of the world's power may be derived from such stored energy. When, as Otte put it, companies like his oil firm and Pacific Gas and Electric put their resources on the line, "suddenly everybody sits up and takes notice."

Drilling for steam or hot water has a certain appeal to a petroleum industry, which faces ultimate depletion of the reservoirs of oil and gas that have been their mainstay. The technologies have much in common, although steam wells tend to be larger and high temperatures must be dealt with.

Many of those studying the world's future energy needs doubt that geothermal energy will replace other sources, but they hope it will make an important contribution. Prognostication is difficult because of uncertainty not only as to accessibility of the resources, but also regarding the technology of generating power from hot water and wet steam. Yet, if man is able to extract only the tiniest fraction of energy that is pushing the surface plates of the Earth hither and yon, such power, in terms of his relatively puny needs, will be substantial. And it should endure so long as this planet remains habitable.

Chapter 20

Earth's History and Man's Destiny

THE WORLD IS AS WE FIND IT, WITH DRY LAND, OCEANS, ATMOSPHERE, ORE AND fuel deposits—in fact, with man and all his works—because of the forces within the planet that keep its surface plates in motion. In its infancy the Earth was a ball of bare, sun-scorched rock. The atmosphere and the water vapor that fell as rain to erode the land and fill the seas came, primarily, from volcanoes generated by the planet's internal ferment.

The world today probably would be submerged beneath an endless sea, were it not for the internal churning that rejuvenates the land. Like the life resident upon it, the world is constantly renewed. The land forms—mountains, rivers, lakes—are born of this constant change. They pass through youth, middle age, senescence, and death, making way for new forms. The Yangtze, Amazon, and Mississippi, as they exist today, have been shaped by the uplifts of the Tibetan Plateau, the Andes, and Rockies, although they may have had ancestral roots in earlier river systems. As it turns out, even the ocean basins are mortal—far more so than the continents.

The processes that renew the Earth also generate its deposits of metallic ore and create the conditions for oil accumulation. It was these resources that made possible the rise of civilization from the bronze age, to the iron age, to the Industrial Revolution with its dependence on fossil fuels and minerals (and even its reliance on the precious metals used to symbolize wealth). And the ultimate destiny of our civilization depends to a considerable degree on understanding how and where these deposits were formed.

A major point at issue, discussed further in the last chapter, is to what extent the reworking of the crust and remolding of the landscape by plate movement typical of the

338

past 200 million years has characterized the Earth's long history (the past four billion years or more). Attention has focused on the slabs of very old rock that form the nuclei of continents. They are known as cratons, and some continents are built around several of them. In fact, eight have been identified within Africa, all roughly two to three billion years old, separated by much younger belts, 600 million years old or less.

The Canadian Shield is a typical craton. That portion of it uncovered by younger rocks lies in the region between the Great Lakes and Hudson Bay and extends down into the central United States (the oldest rocks found on the continent being gneisses of the Minnesota River Valley formed 3.8 billion years ago). The rocks of cratons, while they give a superficial appearance of uniformity, like the largely flat, granitic rocks and greenstones of the Canadian Shield, on closer examination show that in their youth they were radically reworked. It is evident that they, too, were formed by a long cycle of rifting and rewelding. The history of the world's continents seems to have been one of repeated divorce and remarriage from a very early time.

Rocks typical of those formed along island arcs are sometimes found as elongated zones within the cratons. Although such bands are more than three billion years old, it is still possible from the character of the rocks (such as, at least in some cases, the variation in potassium content) to guess the ancient direction of underthrusting. Such a "chemical polarity" apparently exists, for example, in the greenstone belt of southern Ontario, and, since the geometry of ore deposits typically is related to the direction of underthrusting (and the resulting eruptions), such indicators can ultimately prove of major importance.

That forces within the Earth once tried to tear North America apart is suggested by a zone of intensified gravity and highly magnetic rock that extends from Lake Superior southwest to Abilene, Kansas. In the northern part dense, lavalike rock that erupted between 1.14 and 1.12 billion years ago lies exposed. In the southern sector it is buried under more than 1000 meters of surface deposits, although it has been sampled by well drilling. Because of this gravity "high," a pendulum clock that runs at a certain rate in Chicago will run a tiny bit faster in Minneapolis, which is astride the feature. After magnetic surveys conducted in the 1960s by planes flying back and forth along lines one or two miles apart, the striking nature of this feature became evident, and in 1973 it was proposed that this might have been an ancient rifting similar to that which more recently produced the rift valleys of Africa. Or it may mark the edge of an ancient continental plate. "Is there significance," asked a report by the National Academy of Sciences, "to the coincidence of the base metal districts within echelon offsets of the midcontinent gravity high?" To the north of that high, in Minnesota, lie the iron deposits of the Mesabi Range that helped make the United States an industrial giant. From 1892 to 1950 it contributed a billion and a half tons of high grade ore. In fact, with its associated ranges, the Mesabi has produced 65 percent of all iron ever mined in the United States (see Fig. 20.1).

Was this the product of plate interactions more than a billion years ago? And, if so, do further iron deposits lie buried under the heavy sediment deposits farther south, helping to account for the complex magnetic contours of that region revealed by the aerial magnetic surveys? If an ocean that once divided North America was snuffed out along this feature, one would expect to find those seafloor chunks known as ophiolites, but none have been identified.

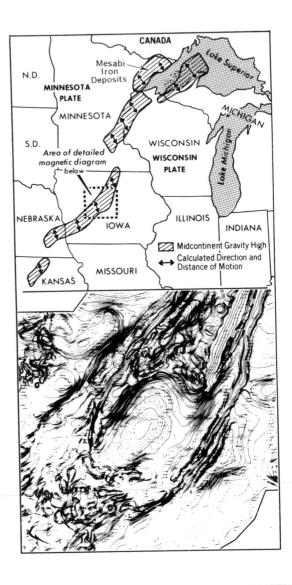

Figure 20.1 A belt of abnormally high gravity in the Midwest extends from Lake Superior to Kansas. Scientists postulate that this belt was formed more than a billion years ago by rifting and volcanic eruptions caused by two sections of the North American Plate splitting apart. The directions of the plates' movement are indicated by arrows on the top map. As with midocean ridges, these rifts are offset from each other. Within this high gravity zone, before it was exploited, existed one of the richest iron deposits in the world in the Mesabi Range. More iron deposits may be deeply buried under sediment in the southern sector. The lower diagram is a detailed magnetic map of a small section of this zone, outlined by dotted line in map above, that illustrates the complex magnetic features throughout the rift region.

Elsewhere, however, as in Africa, they have provided clues to continent–continent collisions. Such an ophiolite zone running inland from Accra, on the coast of Ghana, separates the West African craton from the Congo craton. It has been proposed that Himalayan-type mountain ranges once lay where African cratons collided, but now have been eroded almost entirely away.

This was suggested when some 65 scientists met in March of 1972 at the Airlie Conference Center in Virginia to assess the implications of plate theory with respect to the origin of continents. The chief division was between the "big bang" school, which believes the continents were largely formed in a single period, some three billion or more years ago, when a "froth" of light rock rose to the surface from the hot, churning interior, and the "steady state" school (now in the ascendancy), which sees the continents gradually increasing in area as a consequence of plate motions, associated sedimentation, and volcanic activity. (The terms "big bang" and "steady state" were stolen from the cosmologists and, as used here, of course have nothing to do with the origin of the universe.)

Another area of confrontation was between those who believe the distribution of oceans and continents has remained relatively stable and those who say oceans have been opening and closing throughout continental history. The former maintained that, prior to the splitup of Pangaea, there was rifting of the continent but that this produced only giant ditches and volcanic activity, rather than new oceans. The eruptions and sediment accumulations in such troughs, they felt, could account for the formations in and between cratons so typical of those associated with contemporary island arcs.

Proponents of the view that the Earth's geography has been in flux from the start cited the current northward drift of Australia as an example of how island arcs could become embedded in the heart of a continent. If the drift continues, they pointed out, Australia will sweep up the islands of Indonesia before colliding with Asia to form a new supercontinent. In that case, they concluded, according to the summary report of the conference, "the intervening zone will contain a compressed jumble of older island-arc and microcontinental remnants such that the over-all result might not be unlike the Churchill Province (of Canada), or parts of Africa."

Warren Carey, the articulate proponent of Earth-expansion, was at the meeting to champion his thesis as a better explanation for crustal behavior than the plate theory. The report, identifying him merely as "a stimulating and provocative skydiver from Tasmania," said he had shown "that a great many of the structural-tectonics features of the Earth can be elegantly modeled without recourse to a plate-tectonics approach."

One of the arguments used by Carey and other proponents of expansion, such as Laszlo Egyed of Hungary, was the evidence of a progressive emergence of the continents from beneath the sea over the past 500 million years or more. Calculating the areas of land above water in each period of geologic time from the distribution of marine fossils reveals up-and-down fluctuations but, on a broader scale, indicates a progressive lowering of relative sea level by some 300 meters. Since volcanic activity constantly adds to the Earth's water budget, one might expect the opposite trend. The explanation, said Carey, was that the ocean basins were expanding, causing sea levels to drop.

A year earlier, however, Anthony Hallam at Oxford had suggested an alternate explanation, namely, what has been described as "underplating" of the continents with new material. It is widely assumed that, as part of the plate-driving process, lighter components of the otherwise heavy rock of the mantle are being extracted and added to the continents either in volcanic eruptions that heap new material on top of the land or through "underplating" of the continents. If, as seems to be the case, sea level relative to the continents has subsided 300 meters in the past 500 million years, this could be explained by the addition of only about one millimeter (one 25th of an inch) of material to the underside of the continents every thousand years, causing them to rise progressively higher. With such a "plausible" explanation available, Hallam said, recourse to lowering sea levels as argument for an expanding Earth "cannot reasonably be sustained." He proposed, as others have, that the ancient cratons, in their youth, were relatively thin—perhaps only 12 to 15 kilometers (seven to nine miles) thick—and have been getting thicker ever since. It has been suggested, for example, that the internal temperature of the Earth initially was so high that it impeded formation of a thick and entirely rigid crust.

Hallam also proposed that past fluctuations in sea level were controlled chiefly by variations in the rates of sea floor spreading. An increase in spreading would mean a swelling of the ridges and a consequent rise in sea levels. Like others, he saw as the most likely explanation of worldwide flooding in the Cretaceous Period the development of more extensive oceanic ridges. Such an extension of the ridges would have occurred as significant dispersal of the continents got under way.

In seeking out the old, "original" fragments of continental material—the cratons—geologists have found themselves up against a time barrier. Whereas some of the rocks brought back by astronauts from the Moon have been found to be well over four billion years old, and while there is considerable evidence that all bodies of the solar system, including the Earth, Moon, and meteorites, formed about 4.65 billion years ago, very few rocks have been found on Earth much more than three billion years old and they have largely been of volcanic "greenstone" type.

A notable exception was the 1971 discovery, by a group from Oxford University, of granitic gneisses in West Greenland slightly more than 3.7 billion years old. Some now believe extensive granitic cratons formed that early, with remnants surviving in Greenland, Scotland, and the Soviet Union, but that they were largely buried under volcanic material by some sort of catastrophic development 700 million years later— that is, about three billion years ago.

The history of the Earth seems to have differed fundamentally from that of the Moon in that the older parts of the lunar crust formed at least 500 million years earlier than any surviving parts of the Earth's surface—even the most ancient Greenland rocks. The explanation could be that, with a hotter youth, the Earth was slower to form a solid crust. Or it may be that processes like those typical of plate tectonics (which do not seem to have been very active on the Moon or Mars) have recycled crustal material so thoroughly that none of the early rock remains.

It is too early to predict the role of the plate theory in leading to new oil and ore deposits, but certainly, for the first time, that theory has begun to place such searches within the framework of a relatively comprehensive picture of the Earth. For example, it now appears that the formation of oil and gas reservoirs has been determined to a considerable extent by the history of plate movements over the past 200 million years. While the precise manner in which oil forms within the Earth remains uncertain, the predominant belief is that it is of biological origin. Over hundreds of millions of years the plant life on Earth, through the chemical magic of photosynthesis, has captured the energy of sunlight to produce organic material. The plants are eaten by animals who manufacture other organic compounds. It has been estimated that, were it not for this activity of the plants and animals, which removes carbon dioxide from the air, the atmosphere of the Earth would be like that of Venus—almost pure carbon dioxide— and the Earth's surface, consequently, would be intolerably hot. Instead, the carbon thus robbed from the air is now largely locked up in the world's sedimentary rocks, young and old. Philip H. Abelson, former head of the Carnegie Institution of Washington and an authority on rock chemistry, told a 1972 meeting on the implications of the plate theory for the petroleum industry that the amount of organic material synthesized by all living things since life began was probably equal to the entire weight of the Earth. Most of this production, of course, involved recycling old organic material—dog-eat-dog, so to speak.

Of the organic residue in the world's rocks only a very small fraction has been subjected to the treatment needed to break it down into oil. And of this only a small fraction, originally in the form of widely dispersed, microscopic droplets, was able to percolate through the rock and accumulate in reservoirs suitable for recovery by energy-hungry man (see Fig. 20.2). In a 1973 article in *Nature*, D. H. Tarling of the University of Newcastle-on-Tyne, assessing the role of plate movements in creating oil accumulations, argued that oil has originated chiefly from the breakdown of animal proteins. Vegetable matter, according to this hypothesis, tends to produce natural gas. Among the requirements for gas and oil production, he said, are rapid deposition of sediment, as off a river mouth, so that the remains of plants and animals become buried before they are eaten or consumed by decay bacteria. Under ideal conditions oxygen content of the bottom water should be very low, further impeding decay. Heating—as in regions of moderately high heat flow upward through the Earth—speeds the conversion of organic residues into oil, and Tarling pointed out that the highly productive gas fields of northwest Siberia occur over what seems a continental extension of the active ridge that traverses the Arctic Ocean and presumably generates high heat flow. Similarly the productive oil fields of the Los Angeles Basin lie where a former ridge has given way to a series of transform faults. Very intense heat flow, as over a "hot spot" or "plume," will break down the hydrocarbons that constitute petroleum into their simplest form—chiefly methane gas—and, if prolonged, may disperse an accumulation entirely.

More than 90 percent of known oil and gas reservoirs are associated with deep salt (evaporite) deposits, and one of the situations likely to form such salt accumulations, according to Tarling, is where continents have begun drifting apart, as occurred early in the history of the present Atlantic. The initial result would be a series of shallow basins, cut off from water replenishment from the world oceans. As the continents drew apart, the floors of these basins would slowly subside, permitting successive, shallow-water salt layers to be laid down to produce the deposits—some of them many kilometers deep—that push up as salt domes through the later depositions of sediment. An overburden at least 600 meters thick is needed to produce such upward surges of the salt.

Because of the critical role of deltas in producing oil accumulations, the identification of ancient deltas is of great importance. Yet, Tarling pointed out, this is complicated by the gross changes in geography that have occurred. For example, it appears that, when Africa and South America were joined and the westward drift of South America had not yet thrust up the Andes, an ancestor of the Amazon flowed from Africa across South America into the Pacific Ocean—that is, in a direction opposite to that of the river's present flow. In the past few years, along the inland side of the Andes, extensive oil deposits have been found whose origin, Tarling proposed, may have been related to this earlier river's outflow.

"The formation of mountain systems, along subduction zones or by continental collisions," he wrote, "...drastically changes the drainage pattern so that, for example,

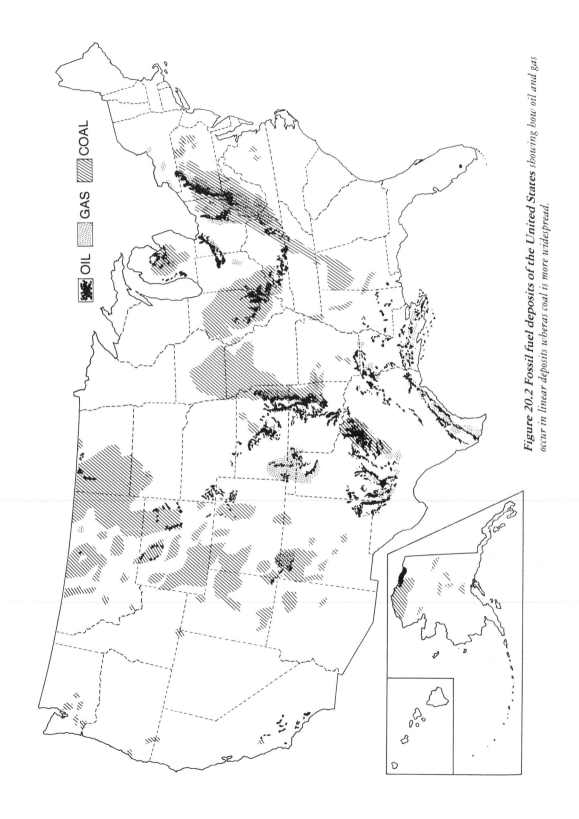

Figure 20.2 Fossil fuel deposits of the United States *showing how oil and gas occur in linear deposits whereas coal is more widespread.*

dispersed oil fields of the Andes may once have formed part of the proto-Amazon delta when the Amazon drained from Africa.''

While the prerequisites for oil accumulation may occur in either a hot or cold environment, Tarling pointed out that the great salt deposits, so often a prerequisite for oil migration and accumulation, seem always to form in hot, arid climates. Since continental drift, over the past 400 million years, has, at some time or other, placed most of the northern lands near the Equator, they and their fringing seas, he said, are more likely repositories of oil than the southern lands. An exception would be the great accumulations of gas recently discovered beneath the sea near Tasmania and New Zealand. The latter reservoir is said to be one of the largest known. While the Tasmanian and New Zealand fields now lie far apart, if and when those lands and Antarctica were assembled into Gondwanaland (as shown in one of the maps, Fig. 11.6), those two fields and the area off Antarctica where gas was found in the *Glomar Challenger* cores may have been close together.

Early in the Deep Sea Drilling Project it was recognized, from reports by various oceanographic expeditions, that structures shaped like salt domes have thrust up through the sea floor sediments on both sides of the Atlantic. While the *Glomar Challenger* drilled one such dome and found it to be a volcanic intrusion, other findings indicate widespread occurrence of buried salt beds, laid down when the Atlantic was young and narrow and then overlain by vast accumulations of sediment such as that, rich in organic matter, discharged by the great rivers of Africa and South America. Heating of the region, as fullscale spreading of the Atlantic got under way, could have stimulated the conversion into oil of organic material in the accumulating sediments.

When the Atlantic was still a small sea, between 205 and 160 million years ago, according to Peter Rona of the Atlantic Oceanographic and Meteorological Laboratories of the National Oceanic and Atmospheric Administration in Miami, that incipient ocean had features in common with the Red Sea which, during the past few million years, has accumulated salt beds that, in places, are 5000 meters thick. The Atlantic salt deposits, he proposed, have been split and now lie on opposite sides of the fully grown Atlantic, and there are indications of oil reservoirs in both regions. The former belief that salt domes are limited to shallow ocean depths has been annulled by drilling into the deepest parts of the Gulf of Mexico and the Mediterranean. In a 1973 article for *Scientific American,* based on a study he did for the United Nations, Rona wrote:

> It is reasonable to expect that petroleum accumulations will extend seaward under the continental shelf, the continental slope, and the continental rise to water depths of about 18,000 feet (5500 meters) along large portions of both the eastern and western margins of the North and South Atlantic.

One of the programs of the International Decade of Ocean Exploration was a four-year project, initiated in 1971, to study continental margins of the Eastern Atlantic. The *Atlantis II* of Woods Hole, in 1972–1973, recorded 95,800 kilometers (60,000 miles) of sea floor profile off West Africa and along most of this distance profiles of gravity and magnetism were also made. Three regions of deep sediment and salt dome structures were identified: one from Angola up the coast to Gabon, one off Nigeria, and the third off Morocco. In some cases oil is already being extracted from similar formations along the nearby coast, and, although offshore exploitation did not begin until the late 1960s, it now accounts for about 30 percent of West African production. (Domes off Angola are shown in Fig. 9.4.)

While such features exist on opposing sides of the ocean, just matching those on one side with regions along the opposite coast is not enough to identify promising drill sites. The detailed local structure, be it a salt dome, fold, or fault, is critical—a fact grimly evident to prospectors for Shell Oil on Sable Island off Nova Scotia. At the Princeton conference on plate tectonics and oil accumulations it was noted that Shell had sunk 24 wells, all of them dry, around a site where Mobil then struck oil. The Princeton meeting brought together a number of leading specialists in oil prospecting, as well as such developers of the plate theory as Jason Morgan, to consider the proposition that, were it not for the openings and closings of ocean basins and the folding and faulting of sediments associated with those movements, the world's accumulations of oil would be meager indeed.

H. Douglas Klemme of Weeks Natural Resources, Inc., assigned a special role to movement of the Earth's surface plates over "hot spots" and "plumes," such as those envisioned by Morgan, as well as to the regions of high heat flow on the continental sides of island arcs. A certain amount of heat, delivered over a suitable length of time, he said, is needed to break down the lipids, proteins, and hydrocarbons within sedimentary rocks to produce oil. But excessive heat carries the breakdown process too far, producing gas. When the temperature reaches 107 degrees Centigrade (250 degrees Fahrenheit), Klemme said, the oil begins converting into methane and other gases, and above 350 degrees Centigrade (660 degrees Fahrenheit) virtually all of it becomes gas. This means that any hope of oil prospectors for new reservoirs of oil at great depth are illusory. Since heat increases at greater depths within the earth, little oil can be expected much below 4000 meters, and even below 3000 meters the proportion of oil production, relative to less valuable gas, begins to decline.

The plumes of hot rock, rising beneath drifting continents, Klemme told the conference, act like Bunsen burners—the gas flames used for heating retorts in the laboratory. And the sedimentary basins that have been carried over these plumes, as the continents drifted, "were like tank ships" laden with oil. This process, he said, can cook the sediment just the right length of time to extract oil from the shale or other highly organic rock. It can also extract residual water, making the rock sufficiently porous for the oil to migrate and accumulate.

But where the "flame" burns too hot or too long, or where the heat is trapped by a heavy overburden, the oil may be converted to gas that can be used only if within pipeline reach of consumers, or by costly liquefaction. James D. Lowell of the Esso Production Research Company told the meeting that a test well near the south end of the Red Sea had produced only gas, presumably because local heat flow was high and was sealed in by the thick layer of salt on top of it. The rock below the salt beds of this area, according to Klemme, is so hot that it softened the drill bits.

The view that former continental margins, as well as present ones, are likely sites for oil accumulation is borne out by the discoveries of great reservoirs on both sides of the Urals, as well as on the inland side of the Andes. Insofar as the Red Sea and Gulf of Aden seem to represent the earliest stages of continental margin development and the oil-forming processes associated therewith, a number of geologists are looking at the region very closely. This area includes the Afar Triangle which, as previously noted, was until recently one of the least-known land areas on Earth. In a typical study, David J. J. Kinsman of Princeton, to learn more about the genesis of a continental shelf, has examined what seems to be an incipient shelf that is now dry land along the north coast of the Somali Republic, flanking the Gulf of Aden. If that gulf and the Red Sea expand into an ocean, the mantle beneath their shores, as it gets farther and farther from the central ridge, will cool and subside. What is now coastal land therefore will sink to become continental shelf and what are now bluffs that face the newly forming ocean basin will become the underwater continental slope, descending from the shelf to the deep-sea floor. Taking advantage of the new perspective on the world provided by the plate theory, Kinsman could drive his vehicle over the counterpart of a continental shelf and broaden understanding of what, the world over, is one of the most economically important of geologic provinces.

If, from the new theory, we understand better why oil and gas are to be found in some places and not in others, this is even truer for ore deposits. Most of the continental veins of metallic ore upon which our civilization has been built were deposited from hot-water solutions. In many cases the metal was combined with sulfur, and notable among such sulfide deposits is the copper on the island of Cyprus. The demonstration by Fred Vine and others that the ophiolite suite on Cyprus—the Troodos Massif cutting across that island—was a cross section of sea floor generated on a mid-ocean ridge has been of major importance in relating ore genesis to the new theory.

As early as 3000 B.C. copper was being mined on Cyprus, and a long succession of nations struggled for control of the island's rich deposits—Egyptians, Assyrians, Phoenicians, Greeks, Persians, and Romans. It was, in fact, Rome's chief source of supply, and the Romans called the metal Cyprium (later corrupted to "cuprium" and, in English, to "copper"). The Cyprus deposits, which occur in the "pillow lava" sector of the ophiolite suite, are also rich in other metals. The Mavrovouni ore body, with an estimated weight of 15 million tons, contains 43 percent iron, 48 percent sulfur, four percent copper, and small amounts of zinc, gold, and silver. The mines on Cyprus are still profitable, and, in 1970, the island exported 30 million dollars worth of mineral products.

The idea that such deposits are formed by activity along mid-ocean ridges and then are transported away from them on the moving plates has been reinforced by deep-sea drilling. Metal-enriched layers were encountered at the base of the sediments at Atlantic, Pacific, and Indian Ocean sites. At one, 650 kilometers (350 miles) southeast of New York, a small vein of pure copper was brought up in a core. On the crest of the Ninetyeast Ridge in the Indian Ocean the *Glomar Challenger* drill penetrated copper within basalt under the sediment, showing that hydrothermal deposition had occurred in the bedrock as well as the sediment.

One of the uncertainties, with regard to the deposition of various metals under the hot pools of the Red Sea, has been whether they were extracted and concentrated from the sediment, or whether they had been derived from the upwelling mantle rock underneath, adding to the world's gross inventory of accessible metals. The question is still being debated, although several researchers, from a variety of clues, have concluded that the metal injected into sediment along the East Pacific Rise has been extracted by hot water circulating to depths of several kilometers in the basalt of that ridge.

In any case, most of the ore bodies exploited to date have been continental—chiefly in mountain ranges that arose over a descending oceanic plate. Not only do the mountains lie parallel to these zones of subduction, but so, too, do many of the metal-rich areas. This is true, for example, of the ranges that extend, parallel to the coast, from Alaska, down the West Coast of North America to Mexico, Central America, and western South America. Furthermore, there is a suggestion that, just as the potassium content of the erupted rock, at least in some regions, tends to increase inland, where the rising lava has come from deeper and deeper parts of the descending slab, so does the nature of the ore vary. Nearest the coast—and hence in formations derived from shallowest depth—one finds iron. A little farther inland, typically, are copper and gold deposits, as in the Foothills Copper Belt and Mother Lode of California and the Comstock Lode of Nevada. Still farther inland, according to this rather oversimplified sequence, come the silver, lead, and zinc deposits, like those in the hinterland of British Columbia and in northern Idaho. The same trends are evident in the Andes of Peru, Bolivia, Chile, and Argentina. Richard H. Sillitoe of the Institute of Geological Investigations in Santiago, Chile, has proposed that the manner in which these ores have been extracted from the Earth's interior has something in common with the process of potassium extraction.

There are, of course, ore deposits that are remote from any coast and difficult to relate to any obvious plate movements. This, for example, is true of the Rocky Mountain zone (Fig. 20.3). Frederick J. Sawkins of the University of Minnesota has suggested that the pattern of ore deposits in that region may indicate that the Pacific slab descended at a very shallow angle; or the pattern may mean—as the Geological Survey group in Denver proposes—that one section of slab broke off and was overrun by the west-moving American Plate. It is noteworthy that some of the continent's richest copper deposits, from Butte, Montana, through the mining area near Salt Lake City, to the extensive deposits in Arizona, lie along the low-potassium zone that is assumed by the Denver geologists to be over the shallowest rim of the ancient slab.

East of that line the sequence, to some extent, repeats the coastal pattern, going to the gold of Cripple Creek, the lead and silver of Leadville, and the fabulous molybdenum deposit at Climax (all in Colorado).

Sawkins sees the descent of an oceanic plate under Indonesia and Southeast Asia as responsible for the tin and tungsten of those regions. The same is thought to be true regarding Japan's copper—one of that country's few major mineral resources. Microscopic water inclusions in the Japanese ores suggest their formation by underwater eruptions, and strikingly similar, but very much older, ores are found in Canada, far from any present plate boundary. In Manitoba the belt of copper–nickel sulfide ores from Thompson Lake to Moak Lake has been related to the former descent of a plate beneath an ancient island arc in that area. Other inland deposits flank the Urals and can likewise be attributed to the descent of a vanished ocean floor. Virtually all of the known porphyry copper deposits are spatially related to three well-established subduction zones: the region from the eastern Mediterranean to Pakistan, the circumpacific belt, and the Antilles.

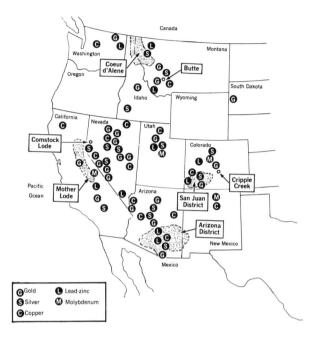

Figure 20.3 Metal deposits of the Western United States, charted from data assembled by James A. Nobel, reflect the developmental history of that region.

The process responsible for emplacing ore deposits on continental rims, as seen by Sillitoe, is an extremely prolonged one—a thought that should give pause to those who regard our metal reserves as almost unlimited. The metals of the cordilleras, or parallel ranges, from Chile northward, in his view, were extracted from the mantle along the East Pacific Rise and were slowly carried toward the South American coast by spreading from that ridge. Then, as the ocean floor and part of its sediment load descended beneath the continent, this metal, here and there, became part of the rising lava.

If thousands of kilometers of oceanic crust descended in this manner, that would account for the long duration of the activity that produced these collections of ore bodies. The tin of Bolivia was emplaced from the late Triassic, 200 million years ago, until perhaps only a few hundred thousand years ago. The emplacement of copper in Chile extended from the Jurassic to the Pliocene (170 to five million years ago).

These attempts to relate the world's metal resources to plate theory are very new, having been initiated only in the early 1970s, and, as Sawkins put it, "a consensus has yet to be reached among economic geologists" regarding cordilleran ore deposition. But an understanding of these processes is of the utmost importance to those seeking to stretch the world's dwindling metal supplies. Without knowing how ore bodies are formed it is difficult to seek out those that lie buried with only meager surface clues to their presence.

In 1973 the United States Geological Survey issued a voluminous assessment of American and world ore deposits. "The real extent of our dependence on mineral resources," it said, "places in jeopardy not merely affluence, but world civilization." At present rates of consumption all of the world's known copper resources will have been exhausted within a half-century, it said. The use of zinc has increased so rapidly that as much of the metal was used from 1950 to 1970 as in the entire history of mankind up until that time.

American resources have been severely depleted. The United States imports a third of its iron, 87 percent of its aluminum, and virtually all of its chrome and manganese (the last American manganese mine closed in 1970). The metal is essential in steel production. Low-grade ores for various metals are available, but their exploitation would be costly, requiring profligate use of energy, and could lay waste much of the landscape.

In the first seven months of 1973, the United States imported 35 percent of its gas and oil, and this percentage was rising steadily. According to the Commission of the European Communities, the energy requirements of the world as a whole will increase 150 percent by the end of this century. Vincent E. McKelvey, director of the Geological Survey, said in presenting the report: "If resource adequacy cannot be assured into the far-distant future, a major reorientation of our philosophy, goals and way of life will be necessary." In other words, it will mean a return to the simpler rural life of a century or two ago—changes that some (unmindful of hardships they never knew) might welcome. As H. E. Goeller of Oak Ridge National Laboratory in Tennessee has pointed out, strict recycling procedures can do much to defer such a return to the ways of the past. Also many of the roles now played by copper and perhaps other metals can be fulfilled by substitutes.

The depletion of American resources has had a serious effect on the country's import-export balance, draining off dollars and degrading the value of the dollar abroad. This balance of payments deficit began growing in 1964, and by June 1989, the margin of imports over exports was 8.2 billion dollars.

Some of the metals in short supply on land can be found on the sea floor in relative abundance. High hopes were expressed for the manganese nodules that abound on wide areas of the sea bed, and a craft named the *Deepsea Miner* attempted "vacuuming" the Atlantic floor to test the marginal economics of this resource. In 1972 the American research ship *Discoverer* dredged from the median rift of the Mid-Atlantic Ridge a slab of rock 4.3 centimeters (1.7 inches) thick formed of 40 percent manganese. Presumably this was from an extensive deposit laid down by mineral-laden hot water flowing through rock beneath the rift valley. The location was roughly midway between Florida and Africa and implied the contemporary generation there of deposits like those lifted from the sea and being mined on Cyprus. But how are those beneath the deep sea to be found and exploited? Describing in *Scientific American* the dredge-haul of manganese by the *Discoverer* and other recent finds, Peter Rona said:

> Consider how much we would know about the mineral deposits of the continents if our sampling procedure were limited to flying in a balloon at an altitude of up to six miles and suspending a bucket at the end of a cable to scrape up loose rocks from the surface of the land. What are the chances that we would find the major known ore bodies, which generally underlie areas of less than a square mile?... Averaged over the world's oceans, the distribution of ocean-floor rocks that have been sampled to date is only about three dredge hauls per million square kilometers!

Discovering and extracting mineral deposits on the deep sea floor would be costly and challenging. While the technology for such operations is being developed, it is aimed chiefly at the extraction of oil and gas and so far has been confined to the continental shelves. Such operations have become routine in the Gulf of Mexico and are spreading across the North Sea. Already one fifth of world oil production comes from the sea floor. However, with regard to the mining of metals it probably will be a long time before continental deposits are sufficiently depleted to make sea floor extraction profitable.

In Antarctica, for example, there are mineral deposits, not only beneath its massive ice sheet, but along its coast. In 1957, when this writer was hiking across the snow-free terrain of Clark Peninsula in Wilkes Land, I noticed, in the gneiss that formed one of the ridges of that area, a vein of black rock with a metallic sheen. In joints of the rock were green stains, presumably of malachite (a copper carbonate). Specimens of the dark rock were brought back to Brian Mason of the American Museum of Natural History in New York and proved to be tephroite containing more than 65 percent manganese oxide. This mineral, Mason noted, typically is associated with ore deposits of importance, such as the zinc at Franklin, New Jersey, the lead–zinc ores of Broken Hill, Australia, and the iron–manganese of Langban, Sweden. Presumably plate interactions have enriched Antarctica with ore bodies, as they have elsewhere; but, as with the seabed, difficulty of access (through the fringing belt of pack ice) and uncertainty as to sovereignty may defer exploitation for some time. Regarding the seabed, in 1969 the United Nations General Assembly passed a resolution stating that, until an international regime responsible for the ocean floor had been formed, "States and persons, physical and juridical, are bound to refrain from all activities of exploitation of the resources of the area of the seabed and ocean floor, and the subsoil thereof, beyond the limits of national jurisdiction."

Of less importance to our well-being than ore deposits, but of great interest to gem hunters, has been the application of plate theory to the hunt for precious stones, notably diamonds. It has long been known that diamonds originate in unique structures known as pipes, and in recent years an explanation for them has developed that casts into the shade some of the most dramatic ideas of the believers in catastrophism.

Early in the study of diamond as the hardest substance found in nature it was recognized that such a highly compressed form of carbon must be formed under conditions such as those very deep in the Earth. Carbon atoms can be arranged in a rather open structure which we know as carbon. They can be compressed into a tighter configuration called graphite. But, if sufficiently squeezed (or pounded or coaxed), they form a very tightly knit crystal known as diamond.

The depths within the Earth at which the requisite pressures occur are greater than 145 kilometers (90 miles). Hence, it was assumed that diamond pipes represented eruptions from such depths. However, a most remarkable feature of these deposits was that, far from resembling ordinary erupted rock, they were formed of countless fragments, often cemented together, but strikingly rounded. Some were as large as stream boulders and some smaller than beach pebbles. Furthermore they represented a wide variety of minerals instead of the uniform material one would expect from a deep eruption. Because such mixtures are characteristic of the famous diamond pipes at Kimberley, in South Africa, the formation is known as kimberlite. It characterizes diamond pipes in many parts of the world, from Siberia to Arkansas. In some cases the occurrences are elongated as "dikes," rather than round, and they are typically a few hundred yards in diameter. Not all of them have produced diamonds—for example, those in southeast Utah and northeast Arizona.

The only important diamond pipe in the United States is the Crater of Diamonds Mine at Murfreesboro, Arkansas. In 1924 it produced Uncle Sam, the largest diamond ever found on this continent, weighing 40.23 carats. However, at last report the mine was no longer economically productive and was serving as a tourist attraction.

Large diamonds also have been found in gravel and other glacial till spread by the ice sheets of the last ice age across Wisconsin, Michigan, Indiana, and Ohio (see Fig. 20.4). These include Theresa, weighing 21 carats, and Eagle, with a weight of 15 carats, both found in Wisconsin. Eagle was stolen from the American Museum of Natural History in New York, in 1965, by a team of acrobatic thieves that included Jack Roland Murphy, better known as Murph the Surf. They later pleaded guilty to the theft, but Eagle was never recovered.

Somewhere to the north of these bulldozed finds is the diamond pipe—or several of them—from which they were swept by the flowing ice. Apparently the source is buried under surface soil, and a favorite summer pastime of geology students, for many years, has been to scour the region south of Hudson Bay, swatting black flies and mosquitos in a search for this hidden treasure trove.

Clues to the manner in which diamonds have reached the surface have come from the development of new analytic techniques, and an attempt has been made to relate their eruption to plate movements. One line of attack focuses on a special form of wine-

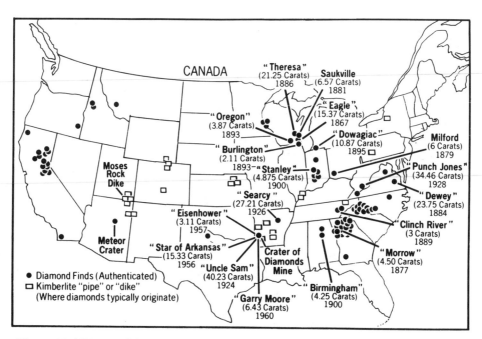

Figure 20.4 Diamond finds and related structures. In addition to locally occurring diamonds, some were carried by ice from a Canadian source into the Northern United States.

red garnet, rich in chromium and magnesium, that typically is found in the diamond pipes and other kimberlite formations. The device used for analysis of these garnets is an automated electron microprobe which fires a very narrow beam of electrons at a speck of target material. The resulting x-ray emissions indicate its composition. In this way it has been possible to determine relative abundances in the garnet of two forms of a substance known as pyroxene—a critical finding, for the ratio of these abundances indicates the temperature and pressure of the environment within which the garnet was formed. Since these factors increase progressively with depth, the analyses have made it possible to estimate how far down in the Earth various kimberlite garnets were formed.

In delivering his presidential address to the Geochemical Society in 1972, Francis R. Boyd of the Carnegie Institution of Washington reported that some garnets from the South African pipes had formed 145 kilometers (90 miles) down at a temperature of 940 degrees centigrade (1750 degrees Fahrenheit). Other samples proved to have crystallized 200 kilometers (125 miles) below the surface at 1360 degrees centigrade (2500 degrees Fahrenheit) or more.

As the nature of ophiolites, representing top-to-bottom samples of the oceanic lithosphere, became evident, it was also recognized that the mixture of rounded rock types in the pipes represented specimens collected from an entire cross section of the Earth to a depth of 200 kilometers—all the way down to the relatively soft asthenosphere.

Thomas R. McGetchin of the Massachusetts Institute of Technology, and others, pulled together these clues to propose a rather sensational scenario for the formation of diamond pipes and other such eruptive features, known collectively as diatremes. The eroded nature of the fragments is apparently the result of their extremely violent and rapid journey up from great depths in the Earth. Even the diamonds themselves, when found, are rounded, despite their extreme hardness. The eruption is driven by an accumulation of substances—chiefly water and carbon dioxide—within the extremely hot, high-pressure environment of the upper mantle. These gases, in liquid form, begin forcing a passage up through the lithosphere, moving slowly at first. But as they near the surface they turn to vapor, because of the reduced pressure, and begin to drive the eruption at ever-increasing velocity until it breaks through the surface with a roar like that of a thousand supersonic jet engines.

McGetchin and his MIT colleague, G. Wayne Ullrich, viewed such an occurrence as having "much in common with flow through a rocket engine, where the rocket is stuffed in the ground upside down." But the gas comes out cold. The chilling effect of a rapidly expanding gas is familiar to anyone who has used a carbon dioxide fire extinguisher. As the gas emerges, it freezes, becoming "dry ice." The same apparently applies to diatreme eruptions. Geologists who had found unaltered coal around diatremes had wondered how material erupting from the red-hot mantle could penetrate such formations without baking them.

When a diatreme, or diamond pipe, erupts, fragments swept up along the entire route apparently are thrown high into the air, falling back to form a crater somewhat like those resulting from high-velocity gas eruptions like one that occurred at Nilahue, Chile, in 1956. But the great eruptions from 200 kilometers down are so rare that none probably ever has been witnessed by man—to date. And so much time has passed since the last such event that its crater has long since eroded away, leaving only the throat of the pipe, clogged with debris that was not blown clear before exhaustion of the erupting gas. Formation of the South African pipes has been put at 120 million years ago and those of Utah and Arizona at 30 million years ago—times when the African and, later, the North American plates apparently were in rapid motion.

A notable feature of garnets from the bottom 40 kilometers (25 miles) of the African pipes, Boyd told the Geochemical Society, is that they are almost all severely sheared, whereas those from shallower depths are not. The explanation may be, he said, that the shearing occurred in this bottom layer of the rigid African Plate as it began to move relatively fast over the asthenosphere, generating stresses that sheared the garnets and heat that helped promote the diamond eruptions.

The idea that compressed water is the prime driving force of these events has received support from the experiments of Charles E. Melton and A. A. Giardini of the University of Georgia, who have been crushing diamonds to see what materials might be hidden within them. They have found that they are, in fact, "crystal bottles" containing small amounts of water, alcohol, and other substances derived from that region of the upper mantle where they were formed. The researchers crush the samples in a high vacuum so that any gases that escape from within the crystal could be identified by an instrument known as a mass spectrometer. In a Brazilian diamond they found that gas constituted one part per 100,000 of the diamond itself. This gaseous component, in turn, consisted of 85 percent water, 0.1 percent ethyl alcohol, and small fractions of carbon dioxide and other gases. A Kimberley diamond also contained traces of butene, a hydrocarbon with three carbon and six hydrogen atoms.

The finding that the predominant gas in these diamonds is water not only supports the view that expanding water vapor is responsible for the eruptions, but falls in with the hypothesis that the presence of free water in the asthenosphere (as opposed to water in chemical combination) reduces melting temperatures and makes that region soft enough to lubricate plate motion.

The new discoveries concerning diamonds, in particular the role of certain types of garnet as "tags" for diamond pipe deposits, have enabled the Russians to retrace the route whereby flowing ice of the last ice age spread material from an unknown diamond source in northern Siberia. Using these garnets like the bread crumbs dropped by Hansel and Gretel to retrace their route (but with greater success, since the crumbs were eaten by birds), the Russians have followed this garnet trail 300 kilometers (200 miles) to find a major diamond deposit. Perhaps some day it will be possible, in this manner, to seek out the hidden diamond pipes north of the Great Lakes.

It is ironic that those who perpetrated the great diamond hoax of the 1870s used garnets of this special sort to "seed" the hilltop which, they claimed, was a fabulously rich diamond deposit. At the time even the most sophisticated geologist was unaware of the relationship of these chromium- and magnesium-rich garnets to diamond occurrences. Lowell S. Hilpert of the United States Geological Survey in Salt Lake City has found the hilltop, near the point where Colorado, Utah, and Wyoming meet, that was seeded with diamonds and other bright stones as part of the hoax. Careful scouring of the site a century later has produced some of the stones (including a few tiny diamonds). Hilpert has found that the garnets, from their composition, apparently came from the deep-rooted diatremes of northeast Arizona. The charlatans responsible for the hoax may have thought they were rubies or believed they could be passed off as such. In any case it appears that the hoaxers obtained the "rubies" in the West, but that they got their raw diamonds in London. (While excitement over the "discovery" was running high in New York, the city editor of *The Times* in London heard from diamond dealers there that some Americans had purchased uncut diamonds under suspicious circumstances, suggesting a possible relationship to the "new find." When this report reached New York it did not create much of a stir. By then, it seems, the Rothchilds and Tiffanys had invested in the bogus scheme, adding weight to its respectability. Not until Clarence King, a government geologist, and two companions on the 40th Parallel Survey found the site and recognized the implausible scattering of gemstones there, was the fraud recognized.)

The discovery that diatremes contain lumps of rock collected along the entire, 200-kilometer journey of an eruption from the base of the lithosphere is of great interest to geologists. As McGetchin has put it, these formations are "natural Moholes." And microprobe analysis makes it possible to take a bucketful of diatreme specimens and determine the depths from which many of them originated. In other words, said Boyd in his address, the technique "takes a miscellaneous collection of mantle rocks and puts them back the way they were."

The new understanding of the Earth is bound to have a profound impact on the search for raw materials. As Tuzo Wilson put it in his ringing 1967 address on the revolution in geology, "the Earth and its economic deposits are part of a system—an Earth system—all parts of which react on other parts. New methods for the first time give us the opportunity to explore the whole system—its interior and its ocean floors as well as its land surface...What an exciting challenge this is! What a chance for great discoveries!"

It is true, of course, that the world's resources, including those still hidden, are limited. We may learn to mine some of the metals that have been emplaced in sea-floor rock, as witnessed by the deposits on Cyprus, but that lie deep under water. We may dig deeper or use new energy sources (such as atomic fusion) to process low-grade ores. But for the really long haul, as (it is to be hoped) our civilization gets older and wiser, we should be able to reduce our demands for fresh resources to the slow rate at which the sea-floor spreading process extracts them from the mantle and injects them into accessible regions. It will be an important part of man's reconciliation with the limits of the planet on which we live.

Chapter 21

"But Still it Moves!"

ON JUNE 22, 1633, AN ELDERLY MAN SANK STIFFLY TO HIS KNEES IN THE GREAT HALL of the Convent of Santa Maria Sopra Minerva in Rome. Before him were the cardinals of the Inquisition. Through a door to one side, it is said, he could see the torturer standing ready. He began reading from the document that had been handed to him:

> I, Galileo, son of the late Vincenzio Galilei of Florence, my age being seventy years, having been called personally to judgment and kneeling before your Eminences, Most Reverend Cardinals, general Inquisitors against heretical depravity...do swear that I have always believed, do now believe, and with God's aid shall believe hereafter all that is taught and preached by the Holy Catholic and Apostolic Church. But because, after I had received a precept which was lawfully given to me that I must wholly forsake the false opinion that the sun is the center of the world and moves not, and that the Earth is not the center of the world and moves...I wrote and published a book in which the said condemned doctrine was treated, and gave very effective reasons in favor of it without suggesting any solution, I am by this Holy Office judged vehemently suspect of heresy...

It is said that, his recantation complete, Galileo, as he struggled back onto his feet, muttered *Eppur si muove*!—"but still it moves." Some believe his famous (and perhaps apocryphal) remark was made later, when he was safe from retribution, but for generations it has brought comfort to those who believed that scientific truth cannot long be downed by dogma.

The "inquisition" with which Alfred Wegener was confronted was far more gentle. Those who questioned his concept of continents in motion had legitimate reasons for doubt—reasons that only now seem to have been laid to rest. What has emerged is a view of the Earth as revolutionary—as remote from what one would believe from casual observation—as the Copernican concept of the solar system that brought Galileo to his knees before the Inquisition and then radically altered man's concept of nature. It is utterly natural to suppose that the stars and other celestial bodies move and that the Earth stands still. We feel no motion, but we see the stars rise in the east and set in the west. It is equally reasonable to suppose that the continents remain steadfast.

To have discovered, instead, that the Earth is a dynamic, ever-changing thing, to understand as never before its surface features and internal ferment, is an experience not likely to be repeated often in human history. We are confronted with the almost incredible fact that, although our ancestors evolved in the sea, the oceans of today are all younger than the earliest land mammals. The geography of the Earth is constantly in flux. In terms of the puny span of a human life the movements are not great. During one lifetime the Atlantic becomes wider by the height of a man. If Columbus repeated his great voyage today, it would be longer by the width of a football field. If those who sailed from Greece across the Mediterranean to Alexandria in the time of Socrates did so now, the trip would be shorter by the length of a football field. The lesson of the drift theory—so seemingly absurd, in its genesis—should not be forgotten by those of future generations who seek the truth about nature. Beware dogma! Thomas Henry Huxley, that champion of Darwin's radical proposition of more than a century ago, put it well:

> The improver of natural knowledge [he said] absolutely refuses to acknowledge authority, as such. For him, scepticism is the highest of duties: blind faith the one unpardonable sin. And it cannot be otherwise, for every great advance in natural knowledge has involved the absolute rejection of authority, the cherishing of the keenest scepticism, the annihilation of the spirit of blind faith; and the most ardent votary of science holds his firmest convictions, not because the men he most venerates hold them; not because their verity is testified by portents and wonders; but because his experience teaches him that whenever he chooses to bring these convictions into contact with their primary source, Nature—whenever he thinks fit to test them by appealing to experiment and to observation—Nature will confirm them. The man of science has learned to believe in justification, not by faith, but by verification.

So was it with the early champions of drift, even though formidable and, at times, seemingly irrefutable arguments were raised against them. Today the field is theirs. The battle is won—though not over. The debate as to the nature of the deep process responsible for drift continues. We have read some of the messages on the ocean floor and in the continental rocks, but if they hint at the nature of the process immediately below the lithosphere—in the "soft" asthenosphere—that is pushing the plates together in some regions and apart in others, we have been unable to decipher that information with any assurance. We are hampered by the fact that only the most recent epoch of Earth history is inscribed on the sea floor and, on the continents, the record of ancient battles between plates is complex and confusing. As Patrick Hurley of MIT has put it, "geologists have a new game of chess to play, using a spherical board and strange new rules." It is a game likely to continue for generations to come.

One continues to hear highly diverse explanations for the deep force that is driving the plates. For example, the old idea that the Earth's spin plays a role has been revived. It has been proposed that the basic drift is westward because of the eastward rotation of the Earth. While the European side of the Atlantic might seem to be moving eastward, away from the Mid-Atlantic Ridge, according to this hypothesis it is really the ridge which is moving westward, leaving new sea floor in its wake. America and the Western Atlantic, then, would be moving west even faster so that both plates are spreading away from the ridge.

The obvious force to cause such westerly motion is gravitational tugging at the Earth by the Sun and Moon—the same force that causes tides. However, Leon Knopoff and A. Leeds of the University of California at Los Angeles have suggested that another as yet unidentified force may be responsible. For years it has been widely believed that tidal tugging by the Sun and Moon has been chiefly responsible for the steady, though very slight, slowing of the Earth's easterly spin, causing the day to lengthen a fraction of a second per century. In 1922 Sir Arthur Eddington proposed that such tugging may drag at the entire lithosphere of the Earth, causing it to drift more or less westward. Because of varying degrees of resistance to this motion beneath the lithosphere, he said, the motion would not be entirely to the west and the latitude of surface features would thus change, perhaps explaining the ice ages. Walter Munk and Gordon Mac-Donald, in their book-length analysis of the Earth's rotation, concluded that such tidal dragging is far from adequate to account for the observed slowing of so massive an object as the Earth. They suggest that some other force must be responsible.

In their 1972 paper, Knopoff and Leeds proposed that this unknown force might be dragging the plates westward. To test this idea they analyzed the motions of the 10 major plates and found an overall tendency toward westerly drift. The plates studied included the six originally identified by Le Pichon: the American Plate (comprising the Western Atlantic and the Americas), the Eurasian Plate (formed of Eurasia and the Northeastern Atlantic), the Pacific Plate, the African Plate (including the Southeastern Atlantic and Western Indian Ocean), the Indian Plate (including the rest of the Indian Ocean and Australia), and the Antarctic Plate. The additional four plates were a separate Chinese Plate, a Philippine Plate, and the Cocos and Nazca Plates west of Central and South America, respectively. It was assumed by Knopoff and Leeds that the Antarctic Plate, being almost centered on the Earth's axis (at the South Pole), would not respond to spin effects and would remain stationary. In fact, alone among the plates, Antarctica is completely encircled by spreading ridges, making it the pariah among plates—the one from which all others are fleeing. Hence it was used as a reference for determining motions of the other plates. More than half the net westerly drift deduced in this manner is contributed by the motion of the vast Pacific Plate. After publication of the paper Knopoff explained that he considers some form of convective flow within the Earth to be the most likely driving force, but that the possibility of another factor remains open. Others have questioned any link with the Earth's spin on the grounds that in the past the movements of continents were northerly or southerly and so seemingly unrelated to spin. However, the extent of axis stability in the past remains uncertain.

A more popular school considers the plume theory the most likely explanation—the idea that some 20 plumes of hot, molten rock are rising from deep in the mantle to spread under the plates at strategic points (such as beneath Iceland), forcing them apart. The theory, as developed by Jason Morgan and others, provides that most mantle heat is carried to the surface via plumes. The heat flowing up through the sea floor is derived from spreading plume material, and the energy carried up from the depths also manifests itself in volcanoes, earthquakes, and the processes that keep the Earth "alive" by ever building new mountains and raising new plateaus to defeat the leveling efforts of erosion.

To adherents of the plume concept there is a striking analogy between the newly recognized dynamics of the Earth's interior and more familiar processes within the atmosphere and oceans. All three of the grand divisions of our world—its gaseous envelope, its watery blanket, and its not so solid interior—have features in common. We are familiar with the mobility of the atmosphere—great sheets of cloud are carried over our heads by air currents—and a key role in this circulation pattern is convection: the rising of moist, energy-laden air in the cumulus cloud systems of the tropics. Much of the energy responsible for weather phenomena the world over is carried aloft in this manner, and a somewhat similar process is envisioned for the hypothetical plumes of the mantle.

In the sea, surface currents, such as the Gulf Stream, have been known and exploited by mariners for centuries, but in recent years a more comprehensive picture has been obtained of the total circulation of the oceans, including vertical motions such as those in the Norwegian and Weddell Seas of the Arctic and Antarctic.

While the depth and universality of the plumes continues to be debated, there is no doubt that, as Tuzo Wilson predicted, the Hawaiian chain has been produced as the Pacific Plate drifted over a plume or "hot spot." It is also apparent that the motion of that plate changed some 43 million years ago, having produced an earlier chain, the Emperor Sea Mounts, that extends the Hawaiian chain in a northerly direction for a combined distance of 5900 kilometers to where it first began to form in the northwest corner of the Pacific 70 million years ago (as shown in Fig. 10.5). The change may have been related to the impact of India against Asia, changing global plate movements at that time. The same sequence of increasing ages from southeast to northwest has been found in other Pacific island chains, such as that of the Marquesas. However, despite the current popularity of the plume concept and its many elaborations, skeptics point out that no evidence for the existence of such plumes has been found as convincing, for example, as that for sea-floor spreading (notably in the magnetic "messages" that flank the ridges).

There have been several attempts to develop "budgets" of plate gains and losses. Assuming that the Earth is not expanding and the continents are not growing at more than a very slow rate, the total area of the planet's surface covered by ocean floor must remain fairly constant. For every square kilometer of new sea floor generated along ridges, one square kilometer must go down into trenches elsewhere on the surface. According to the calculations of Kenneth Deffeyes at Princeton, 2.6 square kilometers of oceanic plate are created and destroyed each year. He suspected that a plate, during its long, slow journey across the "soft" asthenosphere, increases in thickness as the top of the asthenosphere cools and becomes part of the rigid plate. In this way the plate, before it plunges back into the mantle, would grow to a thickness of about 100 kilometers and the annual "throughput" of the plate-making process would be 260 cubic kilometers. All told, he said, the plumes carry up more than twice that much material, but not all of it gets close enough to the surface to become part of the rigid lithosphere.

Furthermore, if the plates that cover this planet like a great, spherical mosaic are growing along ridges on one margin and vanishing down trenches along another, it should be possible, by determining the relative movements of all the plates, to construct a global budget of speeds and directions, as well as gains and losses. An ambitious effort at such budget-balancing was made by Xavier Le Pichon and his colleagues at

France's Brittany Oceanological Center near Brest. They compiled available data from throughout the world on such clues as the magnetic "timetables" imprinted on the sea floors (which indicate both direction and speed of sea-floor spreading, but whose mapping is still fragmentary), on lines of earthquake activity (marking plate boundaries), and on evidence accumulated by geologists on all the continents. However, in their book, *Plate Tectonics,* published in 1973, Le Pichon, Jean Francheteau, and Jean Bonnin cast a somewhat jaundiced eye on efforts to explain mountain systems and other continental features in terms of plate theory. The available information is still so meager, they say, that such efforts "acquire a certain dreamlike quality." Such doubts and criticisms remind us that the progress of science, despite its pretense to objective rationality, is a very human process. So elegant and tidy is the plate theory that it has swept the scientific community like an epidemic. But despite the strong evidence for its basic elements—plate motion, growth at ridges, and descent at trenches—the more detailed applications, as described in some of the preceding chapters, are likely to undergo modification. The words of caution expressed by Vladimir Beloussov, dean of Soviet Earth scientists, in his 1968 debate with Tuzo Wilson are worth remembering. He recalled how scientists climbed on the Earth-shrinkage bandwagon as an explanation for mountains and yet, he said, "the foundations of the contraction hypothesis collapsed."

By 1973, however, the evidence for plate tectonics was so strong that only a handful of Earth scientists—notably Arthur Meyerhoff of the American Association of Petroleum Geologists—were still holding out against it. Meyerhoff cited a variety of findings that he found inconsistent with the theory. However, as Sir Edward Bullard has put it, "There are phenomena, such as ice ages and thunderstorms, for whose occurrence there is incontrovertible evidence, but for which there is no theory that is not open to substantial objections." Even Beloussov, that old warrior against the theory, was persuaded in 1972, as chairman of the seven-man Bureau of the International Upper Mantle Committee, to sign a final statement of that multinational program, hailing plate tectonics as a revolution in Earth science.

In 1964 the program had undertaken coordinated investigations in three fields: continental margins and island arcs, the world rift system, and laboratory studies of the behavior of rocks assumed to be typical of the mantle under upper mantle conditions. At Tuzo Wilson's suggestion an added goal had been set: "to prove whether or not continental drift occurred." The final statement was edited by Leon Knopoff, the UMP (Upper Mantle Project) secretary general, to reflect (particularly by emphasis on certain words) Beloussov's lingering doubts, as well as the latter's belief and that of others among his Russian colleagues that the nature of the "Second Layer" beneath the sea floor, once it had been sampled by the *Glomar Challenger,* would settle the matter. The ship's drill had penetrated repeatedly into basalt and pillow lavas beneath the sediments and most Earth scientists regarded this, essentially, as the original sea floor—the Second Layer. But there were a few, notably among the Russians, who suspected that this was merely a veneer covering a deep succession of alternating layers of lava and sediment—a counterpart of that French pastry known as a napoleon. The UMP final statement, which, in some respects, eloquently summarized what had taken place in the decade prior to 1972, said in part:

The period of the U.M.P. in fact witnessed an extraordinary development in Earth sciences that is widely considered a "revolution:" the emergence of a unifying concept of plate tectonics. It should be emphasized that data collected *at the present* testify in favor of plate tectonics, but that the final decision will come when direct data about the structure of the oceanic crust—at least the second layer—have been obtained by drilling...The main outlines of the concept of sea-floor spreading and plate tectonics have now been accepted by most scientists. This unifying model is certainly one of the most fundamental and important concepts in the history of geological investigations, far surpassing the reasonable expectations of the planners of the U.M.P. As a by-product, the question of continental drift is answered...

There was a highly increased spirit of cooperation and mutual understanding among Earth scientists of many lands...There began to emerge a remarkable degree of interdisciplinary exchange in recent years. This has now changed us from seismologists, petrologists, or other specialists in various disciplines, to Earth scientists...A century ago geology—under the impact of the theory of evolution, the statement of the principles of stratigraphy, and widespread exploration—was perhaps the most exciting area of science. With passing years, geological activities were overshadowed by dramatic discoveries in chemistry, physics, biology and astronomy. The recent new discoveries relating to the solid Earth have put Earth sciences once again in the forefront...

Of major importance, said the statement, has been the discovery that the mantle is far from uniform in composition, indicating "dynamic instabilities" there.

We now have the general idea [it continued] that what happens on the surface of the Earth is controlled (to a much larger degree than was previously supposed) by processes in the mantle. Data available *at present* indicate the youthfulness of the ocean floor. As to the continents, the connection between their development and plate tectonics is much less clear: geological processes which are not envisioned by the plate-tectonics scheme take place on continents. The problem of interrelations between oceans and continents remains the most important problem for further research.

Actually, it is but one of many questions that remain for those seeking to apply the plate theory locally. According to Charles L. Drake, professor of geology at Dartmouth College and himself a leading participant in development of the theory, "One of the problems with a successful revolution is that if the revolutionaries gain the upper hand, they must govern. Too often the main thrust is lost after the heady moment of victory in the mass of detail and contradictions that accompany the restoration of order. This is true of scientific and technological revolutions as well as political ones. The simplicity of a new concept disappears as new facts or ideas come to light, modifications are made and they disappear into a sea of complications."

As though to insure that this does not happen to the plate theory, Drake undertook to lead an international effort to seek out the deep driving forces and resolve some of the doubts, contradictions, and complications that had arisen in attempts to apply the theory universally. What came to be called the Geodynamics Project was organized in 1967–1970 by the International Council of Scientific Unions as a direct descendant of the International Geophysical Year of 1957–1958 and the subsequent Upper Mantle Project. The prime period of observation and study was to be from 1974 to 1979. Drake was named president of the commission coordinating it, as well as chairman of the American program.

The project, in which by 1974 more than 50 nations were taking part, was, by its own definition, "an international program of research on the dynamics and dynamic history of the Earth with emphasis on deep-seated foundations of geological phenomena. This includes investigations related to movements and deformations, past and present, of the lithosphere, and all relevant properties of the Earth's interior and especially any evidence for motions at depth."

In a statement of their objectives, and a more extensive prospectus of which that statement formed a part, the committee organizing the American program pointed out that the pioneers of plate theory thought they could interpret activity deep within the Earth from a straightforward reading of evidence on its surface. "The rifts and fractures, trenches and thrust belts at the surface," said the statement, " were considered to be relatively accurate reflections of movement at depth. This simple view is no longer tenable." And 25 years later most of the questions raised are still valid.

The statement cited various features of the Earth that did not seem readily explained by the theory. Why, for example, do major earthquakes, such as that in 1811 which struck New Madrid, Missouri, occur in the central region of the North American Plate? How can plate theory explain the volcanism that produced the San Juan Mountains of Colorado? Why are there swarms of sea mounts on certain parts of the Pacific floor instead of the long, southeast–northwest lines of islands one would expect to find if the plate had drifted over a hot spot? Why are there submarine ridges, like the Bermuda Rise, that seem unrelated to plate boundaries and show no earthquake activity or evidence of plume action? If the Alps arose over a descending oceanic plate, why is andesite—the volcanic rock found in the Andes and other mountains formed in this manner—"virtually absent?"

In fact, the cause of great uplifts remains one of the most puzzling problems, the committee said. The new evidence for massive horizontal thrusts has diverted attention from indications that there have also been large vertical movements. Many "are of such magnitude that they would seem to require large volumes of lateral mass transfer within or beneath the lithosphere." Not only have these affected continental areas like the Tibetan and Colorado Plateaus, but there is evidence for alternating large uplifts and subsidences of the mid-ocean ridges. Furthermore, the eastern margin of North America, as well as the Gulf Coast and parts of the interior, have subsided over the past 100 or 200 million years in a manner some believe to be typical of the rim of an expanding ocean like the Atlantic. This subsidence has led to the accumulation of deep sediment deposits that, in places, have become reservoirs of gas and oil. Some of the vertical movements within the continents are clearly related to relief of the landscape from its heavy burden after the last ice age. The north shore of Lake Michigan has risen a half meter relative to its south shore over the past century and it has been calculated that eventually the Great Lakes may drain down the Mississippi instead of down the St. Lawrence. One way to learn what changes have occurred in the tilt of the landscape, said the prospectus, would be to seek out the shorelines of ancient lakes. In some regions, notably the basin-and-range province, where ice age lakes have largely dried up, such fossil shorelines can be seen and could serve as levels, indicating changes in tilt.

Another area of uncertainty concerns the role played by changes of the Earth's spin axis relative to geographic features on its surface. To distinguish between changes in geography caused by past plate motions and changes of the spin axis it is necessary to develop a continuous record, as far back in time as possible, of the positions of the continents relative to one another and to the magnetic poles. This should be possible if enough continental rocks, lava flows, and oceanic cores are collected and analyzed for their ancient magnetic orientations.

With modern laboratory techniques it is possible to determine ancient magnetic pole positions to within five degrees, if rock samples can be found with frozen-in magnetic compasses that are sufficiently unambiguous. Where such analyses for a given period of the past show no change in the positions of continents relative to one another, but do show a drift of the pole position, this can be attributed to polar wandering. However, the information so far has been too fragmentary to determine the extent of such wandering in a convincing way.

A point of special interest will be to see if, at some periods of the Earth's history, continental areas lay at both poles. Such a situation, some believe, would have chilled the world climate enough to produce glaciation throughout the planet. To retrace accurately the plate motions over the past 200 million years it will be necessary to fill in the great gaps in the mapping of magnetic "footprints" on the ocean floors. For the earlier history, dependence will be on the "compass needles" frozen into continental rocks. When the journeyings of each land area from warm to cold latitudes and vice versa have become known, this will provide new perspectives on the factors controlling the evolution of life in those areas.

A particular challenge was to determine, from ancient orientations of the Earth's magnetic field as preserved in the rocks, what drift, if any, occurred prior to the formation of Pangaea—that is, during the Precambrian Period more than 600 million years ago.

> We know that the Earth had a field during the Precambrian [says the Geodynamics Committee prospectus], that the field was reversing, and that polar wandering occurred. However, additional studies are needed to determine the configuration of Precambrian plates and the rates of movement between them. If dynamic processes have been as active through all of geological time as they have been during the past 70 m.y., the task of paleomagnetically reassembling the Precambrian jigsaw puzzle will be formidable. Yet, because of the importance of information about the early stages of geodynamic processes, it is one worth pursuing.

The prospectus pointed out that, despite the achievements of Bullard and Elsasser in proposing a plausible explanation of the Earth's magnetic field, its generation remains one of the most poorly understood of all phenomena relating to the Earth: "We are reasonably certain that the Earth's magnetic field is produced in the Earth's fluid outer core by magnetohydrodynamic processes. However, even after 15 years of research on this complex problem, we still have not progressed beyond a demonstration that the magnetohydrodynamic mechanism is feasible." How, for example, is the field regenerated? Is it produced chiefly by convection or by spin effects? Why does its intensity over the years vary by 50 percent above and below the average? What causes the polarity reversals? Why are they irregular? It is suspected that the changes in reversal frequency may reflect changes in the boundary region between core and mantle. If plumes originate that deeply, such changes could then be related to variations in plume activity. Thus, if we can achieve a better understanding of the Earth's magnetism, we will be looking more penetratingly into the very heart of our planet.

The central goal of the Geodynamics Project, however, was to learn the nature of the force that is driving the plates of the lithosphere hither and yon. Apart from a few who think some force related to the Earth's spin is at work, the dominant belief is in some sort of convection—the upward flow of hot material in the mantle. "Most obvious," said the Geodynamics prospectus, "is the fact that the lithosphere appears to represent the upper limb of a great convection system that is described at the top by the movement of the surface plates and whose movement at depth is not yet known. Generation of new sea floor material at oceanic ridges and extinction at the trenches may be only a small part of this overall flow." The flow may originate at the bottom of the mantle, as envisioned in the plume theory, or only a few hundred kilometers down.

The best hope for finding an answer to this most fundamental of all problems relating to the Earth is through a variety of observing and measuring methods newly available. Indeed, the current explosion in our knowledge of nature, ranging from the ocean floor, outward to distant galaxies, and into the innermost sanctum of the atom, has arisen as much from new observing techniques as from the ingenuity of theorists.

It has been shown in tests that a new, two-color laser system can measure distances to within several millimeters over ranges as great as 20 kilometers (that is, to within a small fraction of an inch over 12 miles). The laser device that was used to measure stretching of the Icelandic landscape after the eruption of Hekla generated pulses at a single wavelength of light and hence was subject to errors arising from uncertainty as to light-propagating conditions along the path of the beam to the reflector and back. By using two wavelengths—that is, two colors—that respond differently to conditions along the path, it is possible to make more precise corrections. With such a device to record year-to-year distortions of the landscape in the basin-and-range province, combined with measurements of strain variations in the surface rocks at a number of locations and more extensive seismic observations, it should become clearer to what extent the residue of motion caused by drag on the California coast by the Pacific Plate and not manifest along the San Andreas Fault is being taken up by landscape distortion and mountain-building farther east. It may also become evident where the accumulation of strain is most likely to produce great earthquakes like one that, in 1872, struck Owens Valley, California. Some regard that quake, along the western margin of the basin-and-range province near the Nevada border, as even more severe than the New Madrid quake of 61 years earlier.

Far greater distances can be measured by laser systems that, as reference points, use reflectors on Earth satellites or emplaced by astronauts on the Moon, making it possible to reach beyond the limits of the Earth's curvature. The Smithsonian Astrophysical Observatory in Cambridge, Massachusetts, converted its "Satellite Tracking Program," organized in the 1950s for the International Geophysical Year, into an "Earth Dynamics Program" to use the new tools for measuring the dimensions and shape of the Earth's surface. A method known as very-long-baseline interferometry has measured intercontinental distances with great accuracy by means of radio telescopes tuned to the same signal. This has confirmed the motions of the continents and of mid-ocean islands relative to one another. Over a decade it is hoped that polar wobbles and tiny changes in rotation rate of the Earth can be recorded on scales of centimeters and fractions of a second. One project seeks to measure, to within 10 centimeters, lines about 1000 kilometers long, spanning the entire California fault system. In this way it should be possible to record strain rates across the region from the Pacific Coast to the Rocky Mountains.

While it is likely that the forces driving the plates also distort the Earth's shape, recording such distortions on a global basis seems too costly at present. They may also be evident, however, on a continental scale. Measurements of such deformation, said the Geodynamics prospectus, "may be our most important clue to the nature of the driving force."

Another new tool in the armory of Earth scientists is a system that can probe continental plates much as the sub-bottom sounder of oceanographers can probe layering beneath the sea floor. The system uses a convoy of several very heavy vehicles that are jacked up onto legs to provide firm contact with the ground. They are stationed a few dozen meters apart and set to vibrating in an electrically coordinated manner. The vehicles are positioned in such a pattern that the combined vibrations send a succession of waves straight down into the Earth, much as a radio antenna can be patterned to send waves in a preferred direction. It has been used in many parts of North America by COCORP (the Consortium for Continental Reflection Profiling) to map deeply buried structures.

To produce a wave series whose echoes off deep layers can be distinguished clearly from natural "noise," each series of vibrations progresses through a succession of frequencies ranging from a few per second to 50 per second. If this system can penetrate to the Moho, it should become possible for a convoy of these vehicles, thumping in unison, to make a transcontinental profile of the crustal layers. Of particular importance, however, is to penetrate even deeper, charting the base of the lithosphere and probing the asthenosphere. It is apparently on the shoulders of flow within the asthenosphere that the plates are being carried, and hence the boundary region between lithosphere and asthenosphere—that is, the bottom of the plates—has become a primary target of Earth science. It has dethroned the famous Moho from the special status that, in the days of the Mohole Project, made it the goal of one of the most ambitious (albeit abortive) scientific projects ever undertaken.

One test of the various explanations for plate motion envisioned in the Geodynamics Project was to determine the thickness of the asthenosphere. If it is thin, according to the argument, it would appear that convection, carrying heat up from the deep mantle (as in the plume theory) must be at work, for otherwise it would be difficult to explain the amount of heat flowing up through the oceanic crust. Glimpses into the asthenosphere have become possible, in part, thanks to great arrays of seismometers that feed their data continuously into a central computer facility. One of the earliest and largest of these was the Large Aperture Seismic Array, or LASA, set up in an area of Montana half again as large as Massachusetts. It originally was formed of 21 sub-arrays, each with 25 seismometers, for a total of 525 instruments sensitive to waves of various lengths moving in different directions. Another in Norway, known as NORSAR, has 22 sub-arrays. Like LASA, it was set up as part of an American Defense Department effort to achieve an ability to spot clandestine underground testing of nuclear weapons; but these arrays are also capable of detecting features buried deep within the Earth.

By making possible detailed analysis of Earth tremors that have gone through the heart of the planet they have shown that, even as deep as the boundary region between the mantle and the liquid outer core, there are structural features that, from place to place, vary the speed at which quake waves are transmitted. Ideally, it should be possible to "see" the hypothetical plumes rising beneath such features as Iceland and Hawaii, but so far it has not been possible to trace them to great depth. Instead, through global monitoring of seismic waves, only a broad picture of temperatures in the deep interior has become possible. "Cold," and therefore more rigid mantle transmits seismic waves faster than "hot," softer mantle and by the 1980s such "seismic tomography" had shown the area under the Pacific and to a depth of 300 kilometers under the continents to be relatively cold, whereas that under the fast-spreading ocean ridges was "hot." This led Don L. Anderson of Caltech to propose that a Wilson, or "Supercontinent," cycle is initiated when an amalgamation of continents impedes heat flow from the interior, the continents being poor conductors of heat. As elaborated by R. Damian Nance, Thomas R. Worsley, and Judith B. Moody in *Scientific American*, this causes the supercontinent to dome upward and finally break apart. The cycle, they said, would also explain the extreme changes in sea level during the past 570 million years, such as that which, as noted earlier, largely flooded the continents in the Cretaceous.

Nor had "seismic tomography" yet settled what happens to great slabs of ocean floor that have descended through a trench into the interior. Did they keep on descending through the boundary between the upper and lower mantle, at 670 kilometers or remain to become part of the upper mantle, in which case the upper mantle would be cluttered with the "gravestones" of past plates? Do the upper and lower mantle differ chemically, or only in their crystal structure, the lower mantle being squeezed into a different configuration, but chemically the same? If there have been repeated "Wilson cycles," what has happened to all those subducted slabs?

It was earlier shown that the asthenosphere, far from being a homogeneous, slushy "sea" across which the plates are carried, is itself a region of complexity where, some believe, fragments of plates that descended long ago still survive. Whether two such plate fragments lie under the Western United States, accounting for uplift of the Rockies and activity in the basin-and-range region, as proposed by the Denver group of geologists, could be tested by such observations, as opposed to the theory that a "hot spot," overridden by the westward movement of the continent, now lies under the basin-and-range area. One difficulty in determining the fate of a plate, after it has gone down into the asthenosphere, is that, beyond a depth of a few hundred kilometers, it no longer traces its path by generating earthquakes. Does this mean it has disintegrated and dispersed? Or do parts of it, at least, remain?

In the 1970s, in addition to intensive study of the North American Plate, American scientists, in conjunction with their Latin American colleagues, decided to concentrate on an oceanic plate particularly well suited for study—the Nazca Plate that is spreading from the East Pacific Rise and descending beneath South America. Its rate of motion is unusually fast, the ore deposits of the Andes seem clearly related to its descent, and so do the devastating earthquakes of the Chilean coast. It was hoped to map the "life history" of this plate, in as great detail as possible, conducting observations not only around its edges but across its width, with a drill ship probing its central region. In this way it was thought possible, at least within limits, to trace the path of metallic sediments from their birthplace at the ridge toward the coast in search of clues to the manner in which they finally become emplaced as ores in the continental mountains. Because of the rapid rate of spreading from the East Pacific Rise, the sea floor is less mountainous than alongside slower-spreading ridges like that in the North Atlantic, and hence these sediments should be less disturbed. It was hoped, as well, to conduct deep drilling in the area of the coastal trench. An ingenious plan for observing directly the movement of a plate down into a trench, and the effects of that motion, was the placement of numerous sonar reflectors on the sea floor in the area of the Aleutian Trench. A high-precision depth measuring system would then be used to record vertical movements of these reflectors.

Emphasis in the Geodynamics Project was also placed on laboratory studies to provide a better assessment of what is going on deep under the crust. While several institutions at the start of the program had presses capable of reproducing conditions of temperature and pressure sufficient, for example, to produce diamonds, few, if any, provided a large enough compression chamber for realistic experiments in deep-Earth processes. Hence new equipment was necessary to test such questions as the role of trace amounts of water in transforming silicates under deep-Earth conditions. It was noted that the conversion of graphite to diamond is greatly accelerated by the presence of certain substances (known as catalysts). Such phase changes, at specific levels of temperature and pressure within the Earth, are thought to play a key role in mantle behavior, not only because they affect the buoyancy of the mantle material, but because a phase change may either absorb or release energy and alter the mechanical properties of the rock. Hence, the committee said, from the lesson of diamond synthesis, "We can assume that there may be phase transitions in the Earth's interior that are important to geodynamics theories but that proceed so slowly in the laboratory as to be unobserved."

So voluminous are the findings of recent years that their interpretation becomes more and more difficult and the seeking out and synthesis of such material became an important part of the Geodynamics Project. Digestion of the old and new data would "challenge the ingenuity of theoreticians and tax the capacity of computing machines," said the Geodynamics Committee. Although a complete description of processes within the planet "is presently beyond our grasp," they said, the findings should be applicable to the discovery of natural resources and to the prediction or control of earthquakes, particularly as they relate to tectonically active zones, such as California and Alaska.

> Decisions of great societal and economic importance [said the prospectus] are now being made on the basis of geophysical and geological research in these areas. For example, sites are being selected for nuclear reactors, oil pipelines, aqueducts, dams and reservoirs, hospitals, and public transit systems in these two areas. Responsible government administrators and industrial engineers need reliable information concerning the expectable disruption along faults and the maximum horizontal force from anticipated earthquakes. Better estimates are needed of the potential of geothermal power...
>
> Economic concentrations of minerals are not randomly distributed, but are related to tectonic processes and regimes recent and ancient. Most of the mineral deposits with easily recognizable surface manifestations have been found. Future success in mineral exploration will depend increasingly on a more sophisticated approach, including a clear understanding of the genetic relationships between these deposits and geodynamic phenomena.

Some, or even most of the problems that now seem difficult to explain in terms of plate movements, such as the Rocky Mountain uplift, will eventually find their place in the theory, according to the prospectus, even though "the distinct possibility remains that some of these events will require either major overhaul of the concept or will be found to be caused by other global dynamic processes operating independently of the plate-tectonics process." Such processes might be overlooked, it added, by blind efforts to fit all aspects of Earth science into the plate theory.

Since the birth of the Geodynamics Project there have been several revolutionary developments. One has been the recognition that the closing of an ancient ocean to form a supercontinent, and the latter's dispersal into the continents of today, recognized by Tuzo Wilson, was but the most recent of an indefinite number of such "Wilson cycles." Each has added or subtracted new lands to the continental cores, or cratons, and the itinerant lands have come to be known as terranes. A recent study has identified 70 of them in the Americas. The United States west of Montana, Utah, and New Mexico, as well as western Canada, are entirely formed of such terranes, broken into fragments, dragged northwest by Pacific Plate motion, then smeared onto the continents from Mexico to Alaska. The terranes can be distinguished, one from another, by their geology, by the magnetic record of their past migrations in terms of "compasses frozen into their rocks, and by their fossils, showing that some were once near, or even south of, the Equator." Parts of the clearly distinctive "Wrangellia Terrane," for example, are now in the Oregon–Washington border area, coastal British Columbia, and Alaska.

The same is true of the East Coast, where evidence has been found for the "accretion" of new terranes long before last closure of the Atlantic and probably in several stages. Whereas the West Coast swept westward, picking up new lands in an otherwise great open sea, the East Coast seems to have been subject to a succession of continent–continent collisions. It was early recognized that some such event, about one billion years ago, produced a region, known as the Grenville Province, that has been traced from Labrador, through the Adirondack Mountains, to Arkansas. Some now argue that it reached from the Baltic region (then attached to North America) to Mexico. Paul F. Hoffman of the Geological Survey of Canada has proposed that seven "microcontinents" were assembled between 2.0 and 1.8 billion years ago to form the basic core, or craton, of North America.

The most recent acquisition was the Avalon Terrane, added by last closure of the Atlantic, reaching from the Avalon Peninsula of Newfoundland south through coastal Maine, Massachusetts, and Connecticut. Another segment in the Carolinas may be a segment of the same terrane. The rocks of the Blue Ridge show that they were pushed from far to the east, as was the surface of the Piedmont from the Carolinas south into Georgia. In fact the Piedmont seems entirely of "foreign" origin. Across southern Georgia and out on the continental shelf in the Atlantic is a band of abnormal magnetism, known as the Suwannee Suture because it seems to mark where Florida and southern Georgia were glued to North America, bringing with it fossils of African affiliation. But these now seem to have been only the most recent additions to eastern North America.

As noted earlier, it has likewise been found that at least eight cratons, or core regions, have been assembled to form Africa. Likewise Eurasia has been assembled by several Wilson cycles, more than half a dozen terranes constituting the Asiatic part. Early in the days of plate tectonics it was recognized that, after India had been carried thousands of kilometers away from Antarctica, it collided with Asia 40 to 60 million years ago. But subsequent investigation has shown that between 130 and 300 million years ago at least three slivers of land had been added to the south flank of Asia. The result of these accretions is the world's highest plateau. An area the size of France averaging five kilometers above sea level, forms the Tibetan Plateau. In recent years it has been studied intensively by American, British, Chinese, and French geologists. Although ringed by the highest mountains in the world, the plateau is remarkably flat. As pointed out by Peter Molnar, one of its explorers, an off-road vehicle can be driven for hundreds of kilometers without encountering a serious obstacle. One theory proposes that 1000 kilometers of what was northern India has been thrust under Asia, doubling thickness of the Earth's crust in that area. Another proposal now widely shared attributes the shortening to compression and folding of the south Asian landscape. In either case the result is a crust 70 kilometers thick, compared to the 35 typical of continents. Much of the evidence for such folding, Molnar believes, has been eroded away. The plateau is flat, he says, because its fringing mountains impede external drainage. Instead its rivers flow inward, depositing eroded material to form a great plain. In any case, the reason for Tibet's great height, like that of the Rocky Mountains and Colorado Plateau, remains controversial, although some link to continental collisions is assumed.

The northward press of India continues. In China sections of the squeezed landscape have slipped northeast, then east and south along several great faults (see Fig. 21.1), including those that form the long, narrow valleys of Southeast Asia. Molnar suspects one rift extends into Siberia, holding Baikal, the world's deepest lake. Many geologists also believe a succession of terranes, carried westward by motion of the sea floor from now-vanished spreading centers in the Pacific, have been added to such islands of the western Pacific as Japan.

Another feature, whose formation has been much discussed in recent years, is "back-arc basins," of which there are several notable examples in the western Pacific such as the Sea of Japan. They lie on the opposite side of the island arc from the trench and subduction zone formed by sea-floor spreading. Another example is the Mariana Basin west of the Mariana Islands and the deep trench on their eastern flank. It appears that such basins are formed by a special kind of "back-arc spreading."

An extraordinary development has been the discovery of "smokers" and other hot geysers on the mid-ocean ridges and of communities of exotic creatures dependent on them. To a large degree this began with a collaboration known as FAMOUS (the French-American Mid-Ocean Undersea Study) and the exploits of a deep-diving submersible called *Alvin*. FAMOUS was the first bold attempt to study first hand the mid-ocean activity that is tearing the Earth's surface apart at the seams. The venture, closely related to the Geodynamics Project, was evocative of the adventurous tales of Jules Verne. French and American oceanographic institutions undertook to send their deepest-diving submersibles into the central rift valley of the Mid-Atlantic Ridge and the canyons of its transverse valleys (or transform faults) to observe the volcanic activity, geysers of metal-laden water and other manifestations of the creation process.

It was agreed that three submersibles would take part: the *Alvin* of the Woods Hole Oceanographic Institution, the French bathyscaphe *Archimede*, and the French

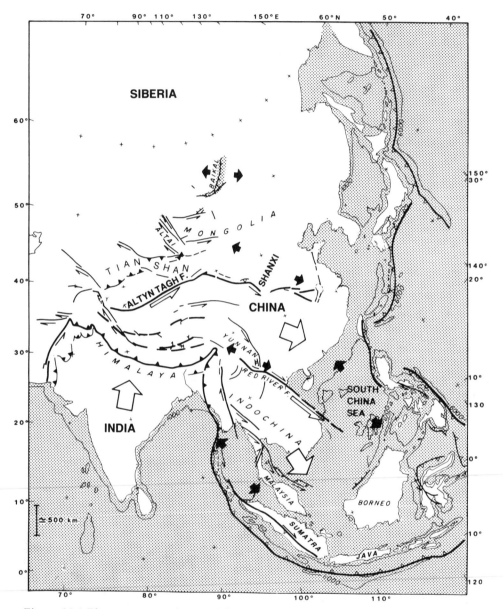

Figure 21.1 Plate movement in Asia. *This map, derived from one prepared at the Institut de Physique du Globe in Paris, illustrates schematically how eastern Asia may have evolved, driven by the impact of continental fragments from the south, culminating with the arrival of India, combined with pressure from the westward-driving Pacific and associated plates. This has created such mountain chains as the Himalayas, as well as Tian Shan, Altai, and those of Mongolia, Shanxi and Yunnan. The large white arrows indicate general plate motions with respect to Siberia. Pairs of black arrows mark crustal spreading including that which produced Lake Baikal and the South China Sea. Half arrows indicate motions along faults including the great Altyn Tagh Fault where Tibet is sliding east relative to the rest of Asia. Heavy lines represent plate boundaries with open barbs showing subduction and black barbs marking continent–continent collision. The period reaches from 50 million years ago to the present.*

"diving saucer," or *soucoupe plongeante,* known as the *SP 3000.* It was the *Alvin* that, in 1966, located the hydrogen bomb that had fallen into the sea off Palomares on the south coast of Spain after a mid-air collision. The craft, holding a pilot and two observers, consists of a manned metal sphere capable of withstanding extreme pressure, enclosed in thick panels of "syntactic foam" with sufficient buoyancy to make the craft almost neutrally buoyant. Hence, to descend, it is necessary only to flood two metal spheres not much larger than basketballs. To rise, the spheres can be blown clear of water and, if necessary, ballast can be dropped. Compared to a bathyscaphe, the vessel is relatively maneuverable. Its primary propulsion is by a hydraulically driven propeller that can be swung, like a rudder, for steering. Two smaller propellers, with adjustable directions of thrust, impart vertical motion. The main power supply is from storage batteries carried on racks in a "bomb bay" so they can be jettisoned for emergency surfacing. Before the FAMOUS project *Alvin* was fitted with a new titanium sphere whose minimum thickness was five centimeters and which was said to be the heaviest piece of that exotic (and lightweight) metal ever forged. This doubled the diving depth and doubled the payload, so that *Alvin* could operate on the floor of the mid-ocean rift valley.

The *Archimede* derived its buoyancy from an envelope filled with gasoline. It, too, was electrically driven and carried ten tons of ballast. Whereas *Alvin* is operated by a single crewman with two scientists as passengers, the *Archimede* carried a crew of two plus one observer. Its depth could be controlled much as a balloonist controls his elevation, releasing gasoline to lose buoyancy and sink, or dropping ballast to gain buoyancy and rise.

FAMOUS had many features in common with the Apollo Project for landing men on the Moon. Because human beings were being sent into an alien realm of total darkness, little-known topography, and pressure sufficient to crush an ordinary submarine, an intensive program of unmanned advance mapping and sea-floor photography was undertaken in the target area, some 350 kilometers south of the Azores. Ships of the United States Navy with special sounding systems for detailed charting of deep-sea floor crisscrossed the area. The British research ship *Discovery* towed "GLORIA," a 16-ton monster, back and forth over the site. "GLORIA" is an elaborate sonar, shaped like a gigantic bomb, that can map sea-floor topography 10 kilometers to either side of its track.

Also, as was done in Apollo, some of the divers were taken beforehand to areas on dry land thought to resemble most closely the fearsome region to be explored: Iceland, the Afar Triangle, and Hawaii. Special, remotely controlled tools were developed to enable men inside the craft to hammer chunks of rock from the canyon walls, pick up specimens, and take water samples in areas where geysering might have newly occurred.

To provide navigational reference points, four buoys were anchored at strategic locations about a kilometer apart in the dive area. Like tethered balloons, the buoys were positioned some 300 meters above the bottom so that they were above most obstructing features on the sea floor, functioning like so many super-high lighthouses. Each was equipped with a battery-powered sonar repeater, and every time the diving craft emitted a sonar "ping," each buoy, on receipt of the signal, emitted its own ping which traveled to the mother ship on the surface, far above. Thus the original signal was relayed via four routes, and, from the lengths of these routes indicated by the time lag of each signal, a computer on the mother ship could determine the position of the submarine relative to the beacons. This, in turn, was displayed automatically on a chart of the rift valley so that those on the ship, via underwater telephone, could tell the divers, in terms of a coordinate system, "you are at G-4," much as a driver on land might be told his position in terms of the coordinates on his road map.

On August 2, 1973, the *Archimede* made the first in a month-long series of dives in dress rehearsal for the more ambitious diving planned for the following summer. The scientist on board was Le Pichon who, with James Heirtzler of Woods Hole, had been a prime mover of the project. Both men, while working at the Lamont–Doherty Observatory, had helped uncover early evidence for the activity which Le Pichon and the other divers now explored on the spot. So great was the depth that it took an hour to sink into the rift valley, and the pilot, Captain Huet de Froberville, found that he was being carried by a surprisingly strong current away from the area of the sonic beacon that was their target. He estimated afterward that the flow almost equaled the maximum speed of the bathyscaphe (two knots or 3.7 kilometers per hour). By the time they reached the valley floor they were so far from the beacon that it took them some time to pick it up on the "Straza," a horizontally scanning sonar scope. As they worked their way toward the beacon, the sonar indicated the presence of a ridge, or "high," down the centerline of the valley and, Le Pichon said later, the combined output of the vertical sounder and the Straza showed the topography of this feature to be very rough. "Nearly vertical cliffs, 50 to 100 meters high, limit the central high," Le Pichon reported, "and, although the overall linearity is obvious, on a small scale—tens to hundreds of meters—lobate or semicircular forms are frequent."

Their first visual contact with the bottom was near the foot of a scarp that was nearly vertical and 50 meters high. This was the first time that human eyes gazed on the submarine rift where the world's surface is being torn asunder and born anew. By the aid of the craft's four powerful floodlights Le Pichon could see that the scarp was the terminal cliff of a fresh lava flow. Apparently the "lobate and semicircular forms" they had detected on sonar were outpourings from a series of small volcanoes rising from the centerline hump of the rift valley at intervals of about three kilometers (two miles). There was little animal life, Le Pichon said, and only a dusting of sediment:

> I must say I was very excited [he wrote]. The navigation was difficult as the pilot had to maintain the *Archimede,* which is fairly bulky, within three meters of the scarp against this fairly strong current. The pilot had to use simultaneously nearly continuously our three propellers and this used up a lot of energy. We scraped the hull of the *Archimede* several times against the rocks [nerve-wracking, for rupture of the gasoline envelope would deprive the vessel of buoyancy]. We landed a little bit downslope on a talus of broken pillow lavas, obviously fallen from the advancing flow fronts. There we sampled a fairly large pillow while fishes kept poking at our portholes....

Particularly striking were the formations of what came to be called "toothpaste lava." These long, snakelike features, thicker than a man's thigh, had apparently been extruded from small orifices and had flowed down the slopes like toothpaste forced from a tube (Fig. 21.2). The only previously known features of this sort had been seen during dives into areas of underwater eruption off Hawaii. During its two hours on the bottom the *Archimede* covered only about 300 meters horizontally, since most of the time was spent going up and down the scarps formed by two overlapping lava fronts. Le Pichon found these lofty fronts and the general roughness of the terrain "obviously very different" from volcanic features on dry land.

In keeping with the international character of the project, the second scientist to go down was Robert Ballard of Woods Hole. Whereas Le Pichon was a relative newcomer to deep diving, Ballard had made a number of descents in the *Alvin.* While his knowledge of French was meager, his two companions were able to communicate reasonably well in English. Ballard, like many of those who go deep beneath the sea or far out into space, is of a special breed—a mixture of competence and self-assurance, with a sense of humor that, in a crisis, is an antidote to panic or despair. They were qualities that served him well on this not-quite-routine dive.

In Jules Verne's Nineteenth Century account of a fictitious journey to the Moon, that in many respects anticipated the reality of a century later, two Americans were accompanied in the snug lunar projectile by a Frenchman who, to ease the tensions of the journey, produced a bottle of Chambertin wine, vintage 1853. On this descent, with two Frenchmen and a single American crowded into a spherical compartment, the French contribution was a bottle of Beaujolais.

Figure 21.2 Underwater rifts and lava formations. A camera lowered by a cord from the Woods Hole research ship Atlantis II *photographed this rift in the floor of the midocean valley in preparation for the first dives at this site. The submersibles had to approach these rifts with caution lest they became wedged in one. A compass, gleaming in the light of the camera flash, indicates the orientation of the picture. A vane extending from the compass shows the direction of the ship and camera's motion.*

When, after prolonged sinking, the vessel's sonar indicated approach to the bottom, Ballard reported afterward, the crew set the down-thrusting propeller operating at full throttle to brake the fall, just as the Apollo astronauts used their braking rocket when landing on the Moon. As in the Moon landings, this threw up great clouds of dustlike sediment. Then there was a sharp bump and Ballard knew they had arrived. Peering into a binocular device that enabled him to look through the thick compartment wall he saw that, as on the first dive, they had landed against a great scarp, or cliff.

Echo-ranging by the sonar showed that the scarp extended for a considerable distance in a remarkably straight line and, as the craft moved slowly along, Ballard could see that there was a second scarp above it, like another step in a gigantic staircase. This was the steplike pattern that has proved typical of such rifts both on land (as in the Afar Triangle) and in mid-ocean. Part of Ballard's job was to select a specimen of lava as freshly erupted as possible to be picked up by the vessel's remotely controlled claw. Looking through the binoculars and trying to coach the crew to their target, despite language problems, proved frustratingly difficult, and, to make it worse, the craft was rocking. Each time the claw seemed ready to close on the specimen, a roll lifted it away. "I wanted to stick my own arm out and grab it!" Ballard said. Once the specimen had finally been captured, they resumed their cautious journey. With visibility limited to from 10 to 30 meters it was impossible to see the opposite wall of the rift valley, but reverberation on the underwater telephone, like a voice in a darkened cavern, told them it was there.

Then something went wrong. There were indications of a drop in the electrical power supply. The bathyscaphe suddenly nosed down, throwing the three occupants of its cramped compartment off balance, and Ballard noticed that numerals on the depth indicator were flashing in rapid succession. The drop in power had apparently activated an automatic safety feature that jettisoned a ton and a half of ballast. Since the ballast was aft, its release had caused the vessel's stern to rise and its nose to drop, and now they were on their way to the surface.

This was no disaster, for they had almost completed their planned dive, but they began to smell smoke, as though electrical insulation were burning. One of the crewmen seized the telephone and began talking to the mother ship in staccato French, which Ballard could not understand but found disquieting. Since the fumes could become noxious, each man wrapped his lips around a mouthpiece attached to an emergency oxygen supply, but the more Ballard breathed through his mouthpiece, the dizzier he became. One of the crewmen saw his plight and recognized the reason. The valve on his oxygen outlet had not been opened. He had simply been rebreathing his own exhalations. As smoke haze filled the compartment their eyes burned, so they donned scuba-diving masks.

In this way they completed the long journey to the surface. The fire proved to be minor, and the electrical system was repaired at sea, making possible one more dive before returning to port in the Azores to prepare for further descents. One of the most gratifying aspects of these first three dives, according to Claude Riffaud, head of the oceanographic center in Brittany and in charge of the descents, was the demonstration that the *Archimede* could be refitted for a new dive in the open sea. Unlike *Alvin*, which could be hoisted between the catamaran pontoons of its mother ship *Lulu*, the *Archimede* could not be lifted by its tender, the *Marcel le Bihan*, and normally was towed back to port after each dive. However, it proved possible at sea to recharge its batteries, refill its gasoline envelope, and restore its ballast to full capacity. During the 1973 "campaign" there were seven dives lasting, typically, from six to eight hours and descending between 2600 and 2800 meters into what the French called "the navel of the world."

During the main part of FAMOUS, in 1974, the three submersibles descended about 50 times into the rift. I was aboard the mother ship, *Atlantis II*, making a documentary film on the project. Another participant was Kathryn Sullivan (no relation), a marine geologist who later was the first American woman to walk in space. On what became the most perilous dive, *Alvin* was swung by a deep current until it became wedged between giant rocks in the rift. Repeated efforts by the pilot to free it, using the various propellers, were futile. With only limited oxygen available for breathing, the outlook appeared grim indeed. Finally the navigator aboard the *Atlantis II*, far above on the surface, asked each occupant to describe what he could see through his saucer-sized porthole. The pilot's port looks forward and those of the two observers look more to the sides, but none looks to the rear.

From these descriptions the navigator drew a picture of what seemed the situation and, on his instructions, the pilot twisted the craft free and *Alvin* survived, not only to make more dives on that expedition, but to make many more, including those with Ballard aboard that first explored the wreck of the *Titanic.*

Probably the most important of *Alvin*'s achievements, from a scientific point of view, were its dives onto the East Pacific Rise that witnessed the eruptions of scaldingly hot water. Some, black with metallic material, came to be known as "smokers" (see Fig. 21.3). Their residue built tall, fragile towers, or "chimneys," and their chemical content, when recognized on a global scale, proved an important contributor to the composition of seawater. Surrounding them were communities of giant clams and "tube worms" each enclosed in a rigid sheath that stood as much as three meters high (Fig. 21.4). These and the clams were of great interest to ecologists, for unlike most forms of life with which we are familiar, they are totally independent of sunlight and photosynthesis. Instead their innards harbor colonies of bacteria that can derive nourishment and energy from the sulfurous geyser emissions. Such colonies have now been found by the scientists of many nations at submarine volcanic sites on both sides of the Pacific and, to a lesser degree, along the Mid-Atlantic Rift. Some have even proposed that they are examples of how life could have originated on the Earth before oxygen became a major component of the atmosphere.

If, from the many-faceted research that has been carried out so far, a certain understanding of the Earth's past history has been achieved, can we look ahead? So slow are motions within the Earth that 50 million years of the planet's internal circulation can be likened to a single day in the Earth's atmosphere. If one therefore assumes that the plate motions of today will continue for the next 50 million years, one can draw a map of the world as it might appear then. Robert Dietz and John Holden, having boldly reconstructed the history of plate movements over the past 200 million years, have done just that. They show Australia having overridden Indonesia and a new sea forming in the region of the East African rift valleys. The Atlantic is larger and the Pacific smaller.

And where will it all end? Will the Pacific close, bringing the Americas up against Asia to form a new Pangaea? After all, on a spherical surface a plate cannot drift in any direction without eventually hitting another plate. Is it, in fact, continent–continent

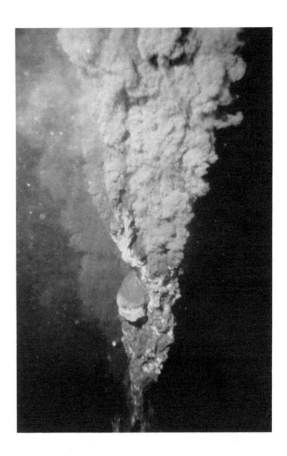

Figure 21.3 Black smoker: *hot, mineral-laden water erupting from a "chimney" of its own making on the East Pacific Rise, photographed by Robert D. Ballard.*

collisions that force the plate movements into new patterns? Or is the whole process of plate tectonics doomed to terminate, leaving the Earth a "dead" planet, with no major earthquakes, no volcanoes, no plate motions, no mountain-building?

This rather gloomy prospect has been proposed by William R. Dickinson and William C. Luth of Stanford. They said the process that has slowly extracted lighter material from the mantle to produce continental crust is running its course. As the continental lithosphere becomes thicker, the intervening asthenosphere—the zone between inner mantle and lithosphere that makes plate motions possible—becomes thinner, with the prospect that it will vanish in less than a billion years. The world, from then on, would become more and more a dull, flat sphere, possibly doomed in the end to be completely submerged under the sea.

The hypothesis makes a number of assumptions for which, at present, there is little evidence, and, if such is indeed the Earth's fate, it is a long way off. Meanwhile, we can rejoice in the slow-motion turmoil that adds to continents, thrusts up mountain ranges and plateaus, generates awesome eruptions and wreaks havoc in earthquakes. For it is this dynamism that has made our planet habitable. Its volcanic exhalations have produced the seas and rain clouds. If Mars were comparably dynamic we would see the snaking lines of its plate edges, but no such lines are visible and it has no seas. The

*Figure 21.4 **Tube worms** at a hydrothermal vent in the Galapagos Rift Valley. These semirigid worms may reach more than a meter in length.*

surface of Venus, with its eternal cloud cover, is invisible except to radar, but its history clearly has differed from that of the Earth so that it, too, is uninhabitable. As we look outward among the stars, therefore, it will be plated planets that we seek as likely abodes of life and civilization.

It is our good fortune to live in an era that has enabled mankind to understand the Earth far more clearly than ever before. Until recently, it has been said, geologists studied only one quarter of the Earth's surface (the areas not covered by water or ice), only one percent of the Earth's radius, and 10 percent of geologic time. Now they have found, in the words of the Soviet geologist Aleksandr Peyve, that "the birth of the continents was in the oceans."

What a radical change has come about in our view of the world—the ocean basins constantly recycled, the continents 20 times older than any part of the sea floor! The grandeur of the history that has set the stage for man and his works commands a new level of reverence. The events of this history, while episodic, were probably gradual. Uniformitarianism lives! But they were no less awesome, and, as our understanding of them deepens, we ourselves become grander in our perspectives. We become aware, too, of the mortality of the landscape and all of man's works. The surface features are doomed to erode away or go down into the trenches of time, only to be replaced by new mountains, valleys, and streams. What remains, as in human history, is the distillate of greatness—the continents themselves, grossly altered by time, but enduring.

References

The following references are designed primarily for those who wish to pursue further the various subjects dealt with in this book. Because some of the concepts remain controversial, an attempt has been made to include references representing diverse viewpoints, even where their description lay beyond the scope of the book itself. In most cases, where there are direct quotations in the text, the page numbers from which they were taken are cited in parentheses, following reference to the source. Where no such source is given, the quotation is from a personal communication.

Chapter 1

Bacon, Sir Francis, quoted by N. A. Rupke, *quo vide.*

Daly, Reginald Aldworth, *Our Mobile Earth*, Scribner (New York, 1926).

Du Toit, Alex. L., *Our Wandering Continents—An Hypothesis of Continental Drifting*, Oliver and Boyd (Edinburgh, 1937). Reprinted by Greenwood Press (Westport, Connecticut, 1972), p. 128.

Georgi, Johannes, "Memories of Alfred Wegener," in *Continental Drift*, S. K. Runcorn, Ed., Academic Press (New York, 1962), pp. 309–324 (317).

Georgi, Johannes, *Mid-Ice—The Story of the Wegener Expedition to Greenland*, Dutton (New York, 1935), pp. 12, 114, 115.

Holmes, Arthur, *Principles of Physical Geology*, Revised ed. The Ronald Press (New York, 1965).

Jeffreys, Sir Harold, *The Earth, Its Origin, History and Physical Constitution*, 5th Edition (Cambridge, 1970), p. 454.

Joly, John, *Radioactivity and Geology—An Account of the influence of radioactive energy on terrestrial history* (1909).

Joly, John, *The Age of the Earth*, presentation before the Royal Institution as one of its Friday Evening Discourses (1922).

Longwell, Chester, "My estimate of the Continental Drift Concept," in *Continental Drift: A Symposium, Being a Symposium on the present status of the continental drift hypothesis, held in the Geology Department of the University of Tasmania, in March, 1956.* S. W. Carey, convenor. (Hobart, 1958, revised July, 1959), pp. 1–12 (2).

Newton, Sir Isaac, letter to Thomas Burnet, Master of the Charterhouse, Jan., 1681. *The Correspondence of Isaac Newton*, vol. 2 (Cambridge, 1960) pp. 329–334 (p. 334).

Rupke, N. A., "Continental Drift Before 1900," *Nature*, vol. 227, July 25, 1970, pp. 349–350 (349).

Snider-Pellegrini, Antonio, *La Création et ses Mystères Devoilés*, Paris, 1858.

Suess, Eduard, *The Face of the Earth*, translated by H. B. C. Sollas, Oxford, 1909 (Original German Edition 1885), vol. 4, p. 621.

van Waterschoot van der Gracht, W. A. J. M., et al., *Theory of Continental Drift—A Symposium on the origin and movement of land masses both inter-continental and intra-continental, as proposed by Alfred Wegener*, held Nov. 15, 1926. American Association of Petroleum Geologists (1928). (van der Gracht, pp. 3, 4, 75; Berry, p. 194; Chamberlin, p. 83; Longwell, p. 145; Schuchert, pp. 111, 139; Willis, p. 82.)

Wegener, Alfred, *The Origin of Continents and Oceans*, translated from the Fourth Revised German Edition, published 1929, by John Biram, Dover (New York 1966), pp. vii, 29, 52, 66, 68, 77, 110, 217.

Wegener, Kurt, "Alfred Wegener," *ibid.*, pp. iii–v.

Chapter 2

Bondi, H., and T. Gold, "On the damping of the free nutation of the Earth," *Monthly Notices of the Royal Astronomical Society*, vol. 115 (1955), pp. 41–46.

Brown, Hugh A., *Cataclysms of the Earth*, Twayne (New York, 1967).

Bucher, Walter H., "Report of the Meeting of January 27, 1955," Museum Discussion Group, The American Museum of Natural History (mimeo.).

Challinor, R. A., "Variations in the Rate of Rotation of the Earth," *Science*, vol. 172 (June 4, 1971), pp. 1022–1025.

Chinnery, Michael, "The Chandler Wobble," Chapt. 6 in *Understanding the Earth*, I. G. Gass, P. J. Smith, and R. C. L. Wilson, editors, M.I.T. (Cambridge, 1971).

Cuvier, Baron G., *A Discourse on the Revolutions of the Surface of the Globe and the Changes Thereby Produced in the Animal Kingdom*. (Philadelphia, 1831), pp. 179–180.

Einstein, Albert, "Forward," in *Earth's Shifting Crust—A Key to Some Basic Problems of Earth Science*, by Charles H. Hapgood, Pantheon (New York, 1958), pp. 1–2 (1).

Gold, Thomas, "Instability of the Earth's Axis of Rotation," *Nature*, vol. 175 (March 26, 1955), pp. 526–529 (527, 529).

Goldreich, Peter, and Alar Toomre, "Some Remarks on Polar Wandering," *Journal of Geophysical Research*, vol. 74 (May 15, 1969), pp. 2555–2567.

Hapgood, Charles H., *Earth's Shifting Crust—A Key to Some Basic Problems of Earth Science*, Pantheon (New York, 1958).

Hapgood, Charles H., *The Path of the Pole*, Chilton (Philadelphia, 1970).

Irving, E., and W. A. Robertson, "Test for Polar Wandering and Some Possible Implications," *Journal of Geophysical Research*, vol. 74 (Feb. 15, 1969), pp. 1026–1036.

Kelvin, Lord (Sir William Thomson), Presidential Address to the Section of Mathematics and Physics, in "Notices and Abstracts of Miscellaneous Communications to the sections," *British Association for the Advancement of Science, Report of Meetings—1876*, pp. 1–12 (3, 11).

Mansinha, L., and D. E. Smylie, "Earthquakes and the Earth's Wobble," *Science*, vol. 161 (Sept. 13, 1968), pp. 1127–1129.

Munk, Walter H., and G. J. F. MacDonald, *The Rotation of the Earth, A Geophysical Discussion* (Cambridge, 1960).

Munk, Walter, and Roger Revelle, "On the Geophysical Interpretation of Irregularities in the Rotation of the Earth," *Monthly Notices of the Royal Astronomical Society* (Geophysical Supplement), vol. 6 (Sept., 1952), pp. 331–347 (331).

Munk, Walter, and Roger Revelle, "Sea Level and the Rotation of the Earth," *The American Journal of Science*, vol. 250 (Nov., 1952), pp. 829–833.

Murray, B. C., and M. C. Malin, "Polar Wandering on Mars?" *Science*, vol. 179 (March 9, 1973), pp. 997–1000.

Plumb, Robert, "Engineer Says Vast Polar Ice Cap Could Tip Earth Over at Any Time/Hugh A. Brown Thinks $10,000,000 Project for Atomic Blasting in Antarctic Could Avert End of World as We Know It." *The New York Times*, Aug. 30, 1948, p. 19, cols. 3 and 4.

Smylie, D.E., and L. Mansinha, "The Rotation of the Earth," *Scientific American*, vol. 225 (Dec., 1971), pp. 80–88.

Chapter 3

Agassiz, J. L. R., *Études sur les glaciers*, quoted in *Encyclopaedia Britannica*, vol. 1 (Chicago, 1963), p. 320.

Bargmann, V., and Lloyd Motz, "On the Recent Discoveries Concerning Jupiter and Venus," *Science*, vol. 138 (Dec. 21, 1962), pp. 1350–1352.

Buckland, W., *Reliquiae Diluvianae* (1923), quoted by S. J. Gould (*quo vide*).

Cohen, I. Bernard, "An Interview with Einstein," *Scientific American*, vol. 193 (July, 1955), pp. 68–73 (70).

Cotton, Richard Payne, "On the Pliocene Deposits of the Valley of the Thames at Ilford," *Annals and Magazine of Natural History*, vol. 20 (1847), pp. 164–169 (166).

Denton, George H., Richard L. Armstrong and Minze Stuiver, "Late Cenozoic Glaciation in Antarctica," in *Frozen Future, A Prophetic Report from Antarctica*, Richard S. Lewis and Philip M. Smith, eds., Quadrangle (N.Y., 1973), pp. 323–340.

Einstein, Albert, Letter to Mr. and Mrs. Velikovsky, March 17, 1955. Published in facsimile in *Pensée* (Special Issue), vol. 2, No. 2, May, 1972 (Student Academic Freedom Forum, Portland, Ore.), p. 39.

Forel, F. A., "Les ravins sous-lacustre des fleuves glaciaires," Academie des Sciences, Paris, *Comptes Rendus*, vol. 101 (1885), pp. 725–728.

Gould, S. J., "Is Uniformitarianism Necessary?" *The American Journal of Science*, vol. 263 (March, 1965), pp. 223–228.

de Grazia, Alfred, "The Scientific Reception System and Dr. Velikovsky (including 'Some Additional Examples of Correct Prognosis by Dr. Immanuel Velikovsky,')" *The American Behavioral Scientist*, vol. 7 (Sept., 1963), pp. 45–68 (*passim*) (66, 67).

Hapgood, Charles H., *The Path of the Pole, op. cit.*

Heezen, Bruce C., "The Origin of Submarine Canyons," *Scientific American*, vol. 195 (Aug., 1956), pp. 36–41.

Heezen, Bruce C., and Maurice Ewing, "Turbidity currents and submarine slumps and the 1929 Grand Banks earthquake," *American Journal of Science*, vol. 250 (Dec. 1952), pp. 849–873.

Herodotus, *Book 2*, 142, A. D. Godley, transl., Loeb Classical Library, Putnam (London, 1921).

Hess, Harry H., Correspondence with Velikovsky, 1956–1969, quoted in *Pensée*, vol. 2 (Fall, 1972), pp. 25–29 (Portland, Oregon).

Hollin, John T., "Antarctic Ice Surges," *Antarctic Journal of the U.S.*, vol. 5 (Sept.–Oct., 1970), pp. 155–156.

Hollin, John T., "Wilson's Theory of Ice Ages," *Nature*, vol. 208 (Oct. 2, 1965), pp. 12–16.

Jones, Sir Harold Spencer, *Life on Other Worlds*, English Universities Press (London, 1940), p. 168.

Kuenen, Ph. H., "Estimated Size of the Grand Banks Turbidity Current," *American Journal of Science*, vol. 250 (Dec., 1952), pp. 874–884.

Kuenen, Ph. H., "Turbidity Currents," in *The Encyclopedia of Oceanography*, R. W. Fairbridge, Editor, Reinhold (New York, 1966), pp. 943–948 (944).

Larrabee, Eric, "The Day the Sun Stood Still," *Harper's Magazine*, vol. 200 (Jan., 1950), pp. 19–26.

Larabee, Eric, "Scientists in Collision: Was Velikovsky Right?" *ibid.*, vol. 227 (August, 1963), pp. 48–55.

Latham, Harold S., *My Life in Publishing*. Dutton (New York, 1965), pp. 74, 75, 76.

Longwell, Chester R., review of *Worlds in Collision* by I. Velikovsky, *The American Journal of Science*, vol. 248 (Aug., 1950), pp. 584–589 (589).

MacDonald, Gordon J. F., "How to Wreck the Environment," in *Unless Peace Comes–A Scientific Forecast of New Weapons*, Nigel Calder, Ed., Viking (New York, 1968), pp. 181–205 (194).

Margolis, Howard, "Velikovsky Rides Again," *Bulletin of the Atomic Scientists*, vol. 20 (April, 1964), pp. 38–40 (40).

Margolis, Stanley V., and James P. Kennett, "Antarctic Glaciation during the Tertiary Recorded in Sub-Antarctic Deep-Sea Cores," *Science*, vol. 170 (Dec. 4, 1970), pp. 1085–1087.

Margolis, Stanley V., and James P. Kennett, "Paleoglacial history of Antarctica recorded in deep-sea cores," *Antarctic Journal of the U.S.*, vol. 6 (Sept.–Oct. 1971), pp. 175–176.

Martin, Paul S., "The Discovery of America," *Science*, vol. 179 (March 9, 1973), pp. 969–974.

Martin, P. S., and H. E. Wright, Jr., editors, *Pleistocene Extinctions—The Search for a Cause. Vol. 6 of the Proceedings of the VII Congress of the International Association For Quaternary Research*, Yale (New Haven, 1967).

Mason, Kenneth, "The Study of Threatening Glaciers," *Geographical Journal*, vol. 85 (Jan., 1935), pp. 24–35.

Motz, Lloyd, *Harper's Magazine*, vol. 227 (Oct., 1963), pp. 12–14 (*passim*).

O'Neill, John J., "Atomic energy charging globe held able to erupt at any time," *New York Herald Tribune*, Aug. 11, 1946, Sections II–IV, p. 10, col. 1.

Oswald, G. K. A., and G. deQ. Robin, "Lakes Beneath the Antarctic Ice Sheet," *Nature*, vol. 245 (Oct. 5, 1973), pp. 251–254.

Pettersson, H., "Exploring the Ocean Floor," *Scientific American*, vol. 183 (August, 1950), pp. 42–45 (42).

Stove, David, "The Scientific Mafia," *Pensée*, vol. 2 (May, 1972), pp. 6–49 (*passim*).

Velikovsky, Immanuel, *Ages in Chaos*, Doubleday (Garden City, N.Y., 1952).

Velikovsky, Immanuel, *Cosmos Without Gravitation—Attraction, Repulsion, and Electromagnetic Circumduction in the Solar System—Synopsis, Scientific Report IV, Scripta Academica Hiersolymitana*, Simon Velikovsky Foundation (Tel-Aviv, 1946).

Velikovsky, Immanuel, "The Dreams Freud Dreamed," *Psychoanalytic Review*, vol. 28 (1941), pp. 487–511.

Velikovsky, Immanuel, *Earth in Upheaval*, Doubleday (Garden City, N.Y., 1955), pp. 261, 262, 297.

Velikovsky, Immanuel, "Über die Energetik der Psyche und die physikalische Existenz der Gedankenwelt," *Zeitschrift für die gesamte Neurologie und Psychiatrie*, vol. 133 (1931), nos. 3 and 4, p. 437.

Velikovsky, Immanuel, *Worlds in Collision*, Macmillan (New York, 1950), pp. 329, 370, 371.

Velikovsky, I., "Zu Tolstois Kreutzersonate," *Imago*, vol. 23, pp. 363–370 (1937), published in English as "Tolstoy's Kreutzer Sonata and Unconscious Homosexuality," in *Psychoanalytic Review*, vol. 24 (1937), pp. 18–25.

Wildt, Rupert, "Note on the Surface Temperature of Venus," *Astrophysical Journal*, vol. 91 (1940), pp. 266–268.

Wilson, A.T., "Origin of Ice Ages: An Ice Shelf Theory for Pleistocene Glaciation," *Nature*, vol. 201 (Jan. 11, 1964), pp. 147–149.

Chapter 4

Carey, S. Warren, "A Tectonic Approach to Continental Drift," in *Continental Drift: A Symposium. Being a Symposium on the present status of the continental drift hypothesis, held in the Geology Department of the University of Tasmania, in March, 1956* (Hobart, 1958, revised July, 1959), pp. 177–355 (349).

Defant, Mark J. and Mark S. Drummond, "Derivation of some modern arc magmas by melting young subducted lithosphere," *Nature*, vol. 347 (Oct. 18, 1990), pp. 662–665.

Dicke, Robert H., "The Earth and Cosmology," *Science*, vol. 138 (Nov. 9, 1962), pp. 653–664.

Dicke, Robert H., "Principle of Equivalence and the Weak Interactions," *Reviews of Modern Physics*, vol. 29 (1957) pp. 355–362 (360).

Egyed, L., "The Change of the Earth's Dimensions Determined from Paleogeographical Data," *Geofisica Pura e Applicata* (Milan), vol. 33 (April, 1956), pp. 42–48.

Gamow, George, "Correspondence from an Author," *Science*, vol. 158 (Nov. 10, 1967), pp. 767–768 (767).

Heezen, Bruce C., "The Rift in the Ocean Floor," *Scientific American*, vol. 203 (Oct. 1960), pp. 98–110.

Hoyle, Fred, and J. V. Narlikar, "On the Nature of Mass," *Nature*, vol. 233 (Sept. 3, 1971), pp. 41–44.

Jordan, Pascual, *The Expanding Earth: Some Consequences of Dirac's Gravitational Hypothesis*, Translated and edited by Arthur Beer (Oxford, 1971).

Longwell, Chester, "My estimate of the Continental Drift Concept," in *Continental Drift: A Symposium... op. cit.* pp. 1–12 (3).

Rothé, J. P., "La zone séismique médiane Indo-Atlantique," *Proceedings of the Royal Society of London, Series A*, vol. 222 (March 18, 1954), pp. 387–397 (389).

Teller, Edward, "On the Change of Physical Constants," *Physical Review*, vol. 73 (April 1, 1948), pp. 801–802.

Wilson, J. T., "Some Consequences of Expansion of the Earth," *Nature*, vol. 185 (March 26, 1960), pp. 880–882.

Wiseman, J. D. H., and R. B. Seymour Sewell, "The Floor of the Arabian Sea," *Geological Magazine*, vol. 74 (May, 1937), pp. 219–230 (227).

Chapter 5

Anderson, Don L., "The Plastic Layer of the Earth's Mantle," *Scientific American*, vol. 207 (July, 1962), pp. 52–59. Reprinted in *Continents Adrift, Readings from Scientific American*, W. H. Freeman (San Francisco, 1972), pp. 28–35.

Boyle, Robert, "Of the temperatures of the subterraneal regions as to heat and cold," in *Tracts Written by the Honorable Robert Boyle*, etc. (Usually called *Cosmical Qualities*), quoted by E. C. Bullard in *Terrestrial Heat Flow*, Geophysical Monograph Series No. 8 (Publication no. 1288), American Geophysical Union (1965), pp. 1–2.

Buddington, A. F., "Memorial to Harry Hammond Hess 1906–1969," *The Geological Society of America Memorials*, vol. 1 (1973), pp. 18–26.

Bullard, Sir Edward, "The flow of heat through the floor of the Atlantic Ocean," *Proceedings of the Royal Society of London, Series A*, vol. 222 (1954), pp. 408–429 (427).

Bullard, Sir Edward (E. C. Bullard), "Heat-Flow through the floor of the ocean," *Deep-Sea Research*, vol. 1 (1954), pp. 65–66.

Bullard, Sir Edward, "Historical Introduction to Terrestrial Heat Flow," Chapt. 1 in *Terrestrial Heat Flow*, W. H. K. Lee, editor (Geophysical Monograph Series No. 8), Publication no. 1288 of the American Geophysical Union (1965).

Bullard, E. C., A. E. Maxwell and R. Revelle, "Heat Flow Through The Deep Sea Floor," *Advances in Geophysics*, vol. 3 (1956), pp. 153–181 (177).

Dietz, Robert S., "Continent and Ocean Basin Evolution by Spreading of the Sea Floor," *Nature*, vol. 190 (June 3, 1961), pp. 854–857 (854, 856).

Dietz, Robert S., "Earth's Crust, 'Eppur Si Muove,' " *Medical Opinion and Review* (Dec., 1965), pp. 82 ff.

Elsasser, Walter M., "Sea-Floor Spreading as Thermal Convection," *Journal of Geophysical Research*, vol. 76 (Feb. 10, 1971), pp. 1101–1112. Reprinted in *Plate Tectonics, Selected Papers from the Journal of Geophysical Research*, John M. Bird and Bryan Isacks, editors, American Geophysical Union (Washington, 1972), pp. 379–390.

Ewing, John, and Maurice Ewing, "Seismic-refraction measurements in the Atlantic Ocean basins, in the Mediterranean Sea, on the Mid-Atlantic Ridge and in the Norwegian Sea," *Geological Society of America Bulletin*, vol. 70 (March, 1959), pp. 291–318.

Fisher, Osmond, *Physics of the Earth's Crust*, Macmillan (London, 1889), pp. 82, 127, 129, 136, 247.

Fisher, R. L., and Roger Revelle, "The Trenches of the Pacific," *Scientific American*, vol. 193 (Nov., 1955), pp. 36–41 (36). Reprinted in *Continents Adrift, op. cit.*, pp. 10–15 (10).

Franklin, Benjamin, Letter to the Abbé Soulavie, Sept. 22, 1782, *The Writings of Benjamin Franklin, collected and edited with a life and introduction, by Alfred Henry Smyth*, Macmillan (New York, 1906), vol. 8, pp. 597–601 (598, 601).

Gutenberg, Beno, "Changes in Sea Level, Postglacial Uplift, and Mobility of the Earth's Interior," *Geological Society of America Bulletin*, vol. 52 (May 1, 1941), pp. 721–772.

Gutenberg, Beno, "Low-Velocity Layers in the Earth, Ocean and Atmosphere," *Science*, vol. 131 (April 1, 1960), pp. 959–965.

Gutenberg, Beno, "Wave Velocities in the Earth's Crust," in *Crust of the Earth (A Symposium)*, Arie Poldervaart, editor, Special Paper 62, Geological Society of America (1955, reprinted 1963), pp. 19–34.

Hess, Harry H., "Drowned Ancient Islands of the Pacific Basin," *American Journal of Science*, vol. 244 (Nov., 1946), pp. 772–791.

Hess, H. H., "History of Ocean Basins," in *Petrologic Studies—A Volume in Honor of A. F. Buddington*, A. E. J. Engle, et al., editors, Geological Society of America (1962), pp. 599–620 (599, 617, 618).

Holmes, Arthur, "Radioactivity and earth movement," *Transactions of the Geological Society of Glasgow*, vol. 18, p. 579.

Holmes, Arthur, *Principles of Physical Geology, op. cit.*

MacDonald, Gordon J. F., "Continental structure and drift," *Philosophical Transactions of the Royal Society of London, Series A*, vol. 258 (1965), pp. 215–227.

MacDonald, Gordon J. F., "The Deep Structure of Continents," *Reviews of Geophysics*, vol. 1 (Nov., 1963), pp. 587–665 (612).

Menard, H. W., "The Deep-Ocean Floor," *Scientific American*, vol. 221 (Sept., 1969), pp. 126–142. Reprinted in *Continents Adrift, op. cit.*, pp. 79–87.

Menard, H. W., "The East Pacific Rise," *Science*, vol. 132 (Dec. 9, 1960), pp. 1737–1746 (1742).

Meyerhoff, A. A., "Arthur Holmes: Originator of Spreading Ocean Floor Hypothesis," with replies by H. H. Hess and R. S. Dietz, *Journal of Geophysical Research*, vol. 73 (Oct. 15, 1968), pp. 6563–6569. Reprinted in *Plate Tectonics, Selected Papers from the Journal of Geophysical Research, op. cit.*, pp. 187–193.

Oliver, Jack, and Bryan Isacks, "Deep Earthquake Zones, Anomalous Structures in the Upper Mantle, and the Lithosphere," *ibid*, vol. 72 (Aug. 15, 1967), pp. 4259–4275. Reprinted in *Plate Tectonics, Selected Papers from the Journal of Geophysical Research, op. cit.*, pp. 29–45.

Revelle, Roger, and Arthur E. Maxwell, "Heat Flow through the Floor of the Eastern North Pacific Ocean," with a comment by Sir Edward Bullard, *Nature*, vol. 170 (Aug. 2, 1952), pp. 199–200.

Runcorn, S. K., "Changes in the convection pattern in the Earth's mantle and continental drift: evidence for a cold origin of the earth," *Philosophical Transactions of the Royal Society of London, Series A*, vol. 258 (1965), pp. 228–251.

Runcorn, S. K., "Towards a theory of continental drift," *Nature*, vol. 193 (Jan. 27, 1962), pp. 311–314.

Twain, Mark (Samuel L. Clemens) in *Mark Twain Roughing It*, 1st Ed. 1872, Signet ed. (New York, 1962), pp. 397, 398, 399.

Vening Meinesz, Felix Andries, "The Earth's Crust and Mantle," *Developments in Solid Earth Geophysics*, vol. 1 (Amsterdam, 1964).

Vening Meinesz, F. A., "Major Tectonic phenomena and the hypothesis of convection currents in the earth," *Quarterly Journal of the Geological Society of London*, vol. 103 (Jan. 31, 1948), pp. 191–207.

Wilson, J. T., "Cabot Fault, An Appalachian equivalent of the San Andreas and Great Glen Faults and some implications for continental displacement." *Nature*, vol. 195 (July 14, 1962), pp. 135–138.

Wilson, J. T., "Continental Drift," in *Scientific American*, vol. 208 (April, 1963), pp. 86–100. Reprinted in *Continents Adrift, op. cit.*, pp. 41–55.

Wilson, J. T., "Evidence from islands on the spreading of ocean floors," *Nature*, vol. 197 (Feb. 9, 1963), pp. 536–538.

Wilson, J. T., "Evidence from ocean islands suggesting movement in the Earth," *Philosophical Transactions of the Royal Society of London, Series A*, vol. 258 (1965), pp. 145–167.

Wilson, J. T., "Hypothesis of Earth's Behaviour," *Nature*, vol. 198 (June 8, 1963), pp. 925–929.

Wilson, J. T., "The Movement of Continents," Address presented at Symposium on the Upper Mantle Project, XIII General Assembly, International Union of Geodesy and Geophysics, Berkeley, 1963 (Mimeo.)

Wilson, J. T., "Did the Atlantic close and then reopen?" *Nature*, vol. 211 (Aug. 13, 1966), pp. 676–681.

Wilson, J. T., "Some Further Evidence in Support of the Cabot Fault, a Great Paleozoic Transcurrent Fault Zone in the Atlantic Provinces and New England." *Transactions of the Royal Society of Canada*, vol. 56 (June, 1962), pp. 31–36.

Wilson, J. T., "Submarine Fracture Zones, Aseismic Ridges and the International Council of Scientific Unions Line: Proposed Western Margin of the East Pacific Ridge," *Nature*, vol. 207 (Aug. 28, 1965), pp. 907–911.

Chapter 6

Blackett, P. M. S., "Lectures on Rock Magnetism" (Weizmann Memorial Lectures) December, 1954 (Jerusalem, 1956), pp. 4, 31.

Blackett, P. M. S., "A Negative Experiment Relating to Magnetism and the Earth's Rotation," *Philosophical Transactions of the Royal Society of London, Series A*, vol. 245 (Dec. 16, 1952), pp. 309–370.

Blackett, P. M. S., J. A. Clegg and P. H. S. Stubbs, "An analysis of rock magnetic data," *Proceedings of the Royal Society of London, Series A*, vol. 256 (1960), pp. 291–322 (311).

Bragg, W. L., "Trinity College, Cambridge, and Manchester University," in *Sydney Chapman, Eighty, From His Friends*, Akasofu, S. I., et al. editors, University of Alaska, University of Colorado, University Corporation for Atmospheric Research (1968), p. 50.

Bullard, Sir Edward, "The Bakerian Lecture, 1967—Reversals of the Earth's magnetic field," *Philosophical Transactions of the Royal Society of London, Series A*, vol. 263 (Dec. 12, 1968), pp. 481–524.

Bullard, E. C., "The Earth's Magnetic Field and its Origin," Chapt. 4 in *Understanding the Earth, op. cit.*

Clegg, J. A., "Rock Magnetism," *Nature*, vol. 178 (Nov. 17, 1956), pp. 1085–1087.

Clegg, J. A., Mary Almond and P. H. S. Stubbs, "The Remnant Magnetism of Some Sedimentary Rocks in Britain," *Philosophical Magazine*, vol. 45 (June, 1954), pp. 583–598.

Cox, Allan, quoted in "A perfect case of serendipity," *Mosaic*, vol. 3 (Spring, 1972), pp. 2–12 (5, 6).

Cox, Allan, G. Brent Dalrymple and Richard R. Doell, "Reversals of the Earth's Magnetic Field," *Scientific American*, vol. 216 (Feb., 1967), pp. 44–54 (53).

Cox, Allan, R. R. Doell and G. B. Dalrymple, "Geomagnetic Polarity Epochs and Pleistocene Geochronometry," *Nature*, vol. 198 (June 15, 1963), pp. 1049–1051.

Cox, Allan, Richard R. Doell and G. Brent Dalrymple, "Reversals of the Earth's Magnetic Field," *Science*, vol. 144 (June 26, 1964), pp. 1537–1543.

Dunn, J. R., Michael D. Fuller, Haro Ito and V. A. Schmidt, "Paleomagnetic Study of a Reversal of the Earth's Magnetic Field," *ibid.*, vol. 172 (May 21, 1971), pp. 840–845.

Glass, Billy P., "Australasian Microtektites in Deep-Sea Sediments," *Antarctic Research Series*, vol. 19 (D. E. Hayes, editor), American Geophysical Union (Washington, 1972), pp. 335–348.

Glass, B. P., "Bottle-Green Microtektites," *Journal of Geophysical Research*, vol. 77 (Dec. 10, 1972), pp. 7057–7064.

Glass, Billy P., "Crystalline Inclusions in a Muong Nong-Type Indochinite," *Earth and Planetary Science Letters*, vol. 16 (1972), pp. 23–26.

Glass, B. P., R. N. Baker, D. Storzer and G. A. Wagner, "North American Microtektites from the Caribbean Sea and Their Fission Track Age," *ibid.*, vol. 19 (1973), pp. 184–192.

Glass, Billy P., and Bruce C. Heezen, "Tektites and Geomagnetic Reversals," *Scientific American*, vol. 217 (July, 1967), pp. 32–38 (38).

Graham, J. W., "The Stability and Significance of Magnetism in Sedimentary Rocks," *Journal of Geophysical Research*, vol. 54 (June, 1949), pp. 131–167.

Hansteen, Christopher, quoted by S. K. Runcorn in "The Earth's Magnetism" *quo vid.*, p. 152.

Hays, James D., "Faunal Extinctions and Reversals of the Earth's Magnetic Field," *Geological Society of America Bulletin*, vol. 82 (Sept., 1971), pp. 2433–2447 (2433, 2444).

Hays, James D., and Neil D. Opdyke, "Antarctic Radiolaria, Magnetic Reversals and Climate Change," *Science*, vol. 158 (Nov. 24, 1967), pp. 1001–1011.

Heirtzler, J. R., "Evidence for Ocean Floor Spreading Across the Ocean Basins," *The History of The Earth's Crust, A Symposium*, Robert A. Phinney, Editor (Princeton, 1968), pp. 90–100.

Heirtzler, J. R., "Sea-Floor Spreading," *Scientific American*, vol. 219 (Dec., 1968), pp. 60–70 (60, 67). Reprinted in *Continents Adrift, op. cit.*, pp. 68–78 (68, 75).

Heirtzler, J. R., G. O. Dickson, E. M. Herron, W. C. Pitman III and X. LePichon, "Marine Magnetic Anomalies, Geomagnetic Field Reversals, and Motions of the Ocean Floor and Continents," *Journal of Geophysical Research*, vol. 73 (March 15, 1968), pp. 2119–2136. Reprinted in *Plate Tectonics, Selected Papers from the Journal of Geophysical Research, op. cit.*, pp. 61–78.

Heirtzler, J. R., Xavier Le Pichon and J. G. Baron, "Magnetic anomalies over the Reykjanes Ridge," *Deep Sea Research*, vol. 13 (1966), pp. 427–443.

Irving, E., and W. A. Robertson, "Test for Polar Wandering and Some Possible Implications," *Journal of Geophysical Research*, vol. 74 (Feb. 15, 1969), pp. 1026–1036. Reprinted in *Plate Tectonics, Selected Papers from the Journal of Geophysical Research, op. cit.*, pp. 194–204.

Johnson, G. Leonard, and Bruce C. Heezen, "The Arctic Mid-oceanic Ridge," *Nature*, vol. 215 (Aug. 12, 1967), pp. 724–725.

Kennett, James P., and Norman D. Watkins, "Geomagnetic Polarity Change, Volcanic Maxima and Faunal Extinction in the South Pacific," *ibid.*, vol. 227 (Aug. 29, 1970), pp. 930–934.

Le Pichon, Xavier, "Sea-Floor Spreading and Continental Drift," *Journal of Geophysical Research*, vol. 73 (June 15, 1968), pp. 3661–3697. Reprinted in *Plate Tectonics, Selected Papers from the Journal of Geophysical Research, op. cit.*, pp. 103–139.

Menard, H. W., "The East Pacific Rise," *Science*, vol. 132 (Dec. 9, 1960), pp. 1737–1746.

Menard, H. W., "Extension of Northeastern–Pacific Fracture Zones," *ibid.*, vol. 155 (Jan. 6, 1967), pp. 72–74.

Morley, L. W., and A. Larochelle, "Paleomagnetism as a means of dating geological events," *Geochronology in Canada*, F. Fitz Osborne, editor, Royal Society of Canada, Special Publication 8 (1964), University of Toronto Press.

Opdyke, N. D., "The Paleomagnetism of Oceanic Cores," in *The History of the Earth's Crust, op. cit.*, pp. 61–72.

Opdyke, N. D., B. Glass, J. D. Hays and J. Foster, "Paleomagnetic Study of Antarctic Deep-Sea Cores," *Science*, vol. 154 (Oct. 21, 1966), pp. 349–357.

Pitman, W. C. III, and J. R. Heirtzler, "Magnetic Anomalies over the Pacific–Antarctic Ridge," *ibid.*, vol. 154 (Dec. 2, 1966), pp. 1164–1171.

Raff, Arthur D., "The Magnetism of the Ocean Floor," *Scientific American*, vol. 205 (Oct., 1961), pp. 146–156 (151, 156).

Runcorn, S. K., "Continental Drift," *Research*, vol. 15 (March, 1962), pp. 103–108.

Runcorn, S. K., "The Earth's Magnetism," *Scientific American*, vol. 193 (Sept., 1955), pp. 152–162 (162).

Runcorn, S. K., "Paleomagnetic Comparisons between Europe and North America," *Philosophical Transactions of the Royal Society of London, Series A*, vol. 258 (1965), pp. 1–11.

Runcorn, S. K., "Paleomagnetic Survey in Arizona and Utah: Preliminary Results," *Geological Society of America Bulletin*, vol. 67 (March, 1956), pp. 301–316.

Runcorn, S. K., "Rock Magnetism—Geophysical Aspects," *Advances in Physics*, vol. 4 (1955), pp. 244–291.

Talwani, Manik, Xavier Le Pichon and J. R. Heirtzler, "East Pacific Rise: The Magnetic Pattern and the Fracture Zones," *Science*, vol. 150 (Nov. 26, 1965), pp. 1109–1115 (1109).

Talwani, Manik, Charles C. Windisch and Marcus G. Langseth, Jr., "Reykjanes Ridge Crest: A Detailed Geophysical Study," *Journal of Geophysical Research*, vol. 76 (Jan. 10, 1971), pp. 473–517. Reprinted in *Plate Tectonics, Selected Papers from the Journal of Geophysical Research, op. cit.*, pp. 323–367.

Vacquier, V., "Horizontal Displacements in the Earth's Crust," *Philosophical Transactions of the Royal Society of London, Series A*, vol. 258 (1965), pp. 77–81.

Vine, F. J., "Magnetic Anomalies Associated with Mid-Ocean Ridges," *The History of the Earth's Crust, op. cit.*, pp. 73–89.

Vine, F. J., "Spreading of the Ocean Floor: New Evidence," *Science*, vol. 154 (Dec. 16, 1966), pp. 1405–1415.

Vine, F. J., and D. H. Matthews, "Magnetic Anomalies over Oceanic Ridges," *Nature*, vol. 199 (Sept. 7, 1963), pp. 947–949 (948).

Vine, F. J., and J. T. Wilson, "Magnetic Anomalies over a Young Oceanic Ridge off Vancouver Island," *Science*, vol. 150 (Oct. 22, 1965), pp. 485–489 (487).

Wilson, J. T., "Transform Faults, Oceanic Ridges, and Magnetic Anomalies Southwest of Vancouver Island," *ibid.*, vol. 150 (Oct. 22, 1965), pp. 482–485.

Chapter 7

Anderson, Don, "Comment" in *The History of the Earth's Crust—A Symposium, op. cit.*, p. 147.

Auden, W. H., "Spain, 1937," in *Another Time–Poems*, Faber & Faber (London, 1940), pp. 103–106 (105).

Beloussov, V. V., "Against the Hypothesis of Ocean-Floor Spreading," *Tectonophysics*, vol. 9 (June, 1970), pp. 489–511 (505).

Beloussov, V. V., "An open letter to J. Tuzo Wilson," *Geotimes*, vol. 13 (Dec., 1968), pp. 17–19 (17, 18, 19).

Beloussov, V. V., quoted in "International Geological Congress," *ibid.*, vol. 17 (Oct., 1972), p. 21.

Beloussov, V. V., et al. (the U.M.C. Bureau) "Significance and Achievements of the Upper Mantle Project" in *The Upper Mantle—Proceedings of the final U.M.P. Review Symposium, Moscow, 9–15 August, 1971, Tectonophysics*, vol. 13 (April, 1972), pp. 1–5 (1).

Blackett, P. M. S., "Introduction," *Philosophical Transactions of the Royal Society of London, Series A*, vol. 258 (1965), pp. vii–x (viii).

Bolt, Bruce A., "The Fine Structure of the Earth's Interior," *Scientific American*, vol. 228 (March, 1973), pp. 24–33 (26).

Bullard, Sir Edward, J. E. Everett and A. Gilbert Smith, "The fit of the continents around the Atlantic," *Philosophical Transactions of the Royal Society of London, Series A*, vol. 258 (1965), pp. 41–51 (41, 42, 49).

Duffield, Wendell A., "A Naturally Occurring Model of Global Plate Tectonics," *Journal of Geophysical Research*, vol. 77 (May 10, 1972), pp. 2543–2555. See also Duffield, "Kilauea Volcano provides a model for plate tectonics," *Geotimes*, vol. 17 (April, 1972), pp. 19–21 (19).

Ewing, Maurice, George Carpenter, Charles Windisch and John Ewing, "Sediment Distribution in the Oceans: The Atlantic," *Geological Society of America Bulletin*, vol. 84 (Jan., 1973), pp. 71–88.

Ewing, Maurice, and Leonard Engel, "Seismic Shooting at Sea," *Scientific American*, vol. 206 (May 1962), pp. 116–126.

Heezen, Bruce C., "Comment," in *The History of the Earth's Crust—A Symposium, op. cit.*, p. 117.

Hurley, Patrick M., "The Confirmation of Continental Drift," *Scientific American*, vol. 218 (April, 1968), pp. 52–64. Reprinted in *Continents Adrift, op. cit.*, pp. 56–67.

Hurley, P. M., et al., "Test of Continental Drift by Comparison of Radiometric Ages—A pre-drift reconstruction shows matching geologic age provinces in West Africa and Northern Brazil," *Science*, vol. 157 (Aug. 4, 1967), pp. 495–500.

Isacks, Bryan, Jack Oliver and Lynn R. Sykes, "Seismology and the New Global Tectonics," *Journal of Geophysical Research*, vol. 73 (Sept. 15, 1968), pp. 5855–5899. Reprinted in *Plate Tectonics, Selected Papers from the Journal of Geophysical Research, op. cit.*, pp. 141–185.

Jeffreys, Sir Harold, "How Soft is the Earth?" *Quarterly Journal of the Royal Astronomical Society*, vol. 5 (1964), pp. 10–22 (16).

Menard, H. W., "Comment," in *History of the Earth's Crust—A Symposium, op. cit.*, p. 117.

Menard, H. W., "The Deep-Ocean Floor," *Scientific American*, vol. 221 (Sept., 1969), pp. 126–142 (*passim*). Reprinted in *Continents Adrift, op. cit.*, pp. 79–87.

Menard, H. W., "Extension of Northeastern–Pacific Fracture Zones," *Science*, vol. 155 (Jan. 6, 1967), pp. 72–74.

Menard, H. W., "Fracture Zones and Offsets of the East Pacific Rise," *Journal of Geophysical Research*, vol. 71 (Jan. 15, 1966), pp. 682–685.

Menard, H. W., "The World-wide oceanic Rise-Ridge System, *Philosophical Transactions of the Royal Society of London, Series A*, vol. 258 (1965), pp. 109–122.

Meyerhoff, A. A., and Howard A. Meyerhoff, " 'The New Global Tectonics': Major Inconsistencies" and " 'The New Global Tectonics': Age of Linear Magnetic Anomalies of Ocean Basins," *The American Association of Petroleum Geologists Bulletin*, vol. 56 (Feb., 1972), pp. 269–336 (269, 270) and 337–359.

Morgan, W. Jason, "Rises, Trenches, Great Faults, and Crustal Blocks," *Journal of Geophysical Research*, vol. 73 (March 15, 1968), pp. 1959–1982. Reprinted in *Plate Tectonics, Selected Papers from the Journal of Geophysical Research, op. cit.*, pp. 79–102.

Nature Editorial, "Against Sea Floor Spreading," *Nature Physical Science*, vol. 229 (Jan. 18, 1971), pp. 65–66 (65).

Oldenburg, Douglas W., and James N. Brune, "Ridge Transform Fault Spreading Pattern in Freezing Wax," *Science*, vol. 178 (Oct. 20, 1972), pp. 301–304.

Oliver, Jack, "Contributions of Seismology to Plate Tectonics," *American Association of Petroleum Geologists Bulletin*, vol. 56 (Feb., 1972), pp. 214–225.

Sykes, Lynn R., "Mechanism of Earthquakes and Nature of Faulting on the Mid-Ocean Ridges," *Journal of Geophysical Research*, vol. 72 (April 15, 1967), pp. 2131–2153. Reprinted in *Plate Tectonics, Selected Papers from the Journal of Geophysical Research, op. cit.*, pp. 5–27.

Sykes, Lynn R., "Seismological Evidence For Transform Faults, Sea Floor Spreading, and Continental Drift," in *The History of the Earth's Crust—A Symposium, op. cit.*, pp. 120–150.

Vacquier, V., "Transcurrent Faulting in the Ocean Floor," *Philosophical Transactions of the Royal Society of London, Series A*, vol. 258 (1965), pp. 77–81.

Wilson, J. T., "A New Class of Faults and Their Bearing on Continental Drift," *Nature*, vol. 207 (July 24, 1965), pp. 343–347.

Wilson, J. T., "A Revolution in earth science," *Geotimes*, vol. 13 (Dec., 1968), pp. 10–16 (10, 11, 12, 13, 16) and "A reply to V. V. Beloussov," *ibid.*, pp. 20–22 (22).

Wilson, J. T., "Transform Faults, Oceanic Ridges, and Magnetic Anomalies, Southwest of Vancouver Island," *op. cit.*

Worzel, J. Lamar, "Deep Structure of Coastal Margins and Mid-Ocean Ridges," *The Colston Papers, being the Proceedings of the Seventeenth Symposium of the Colston Research Society held in the University of Bristol, April 5th–9th*, 1965, vol. 17, pp. 335–359 (358).

Worzel, J. L., "Discussion," *Philosophical Transactions of the Royal Society of London, Series A*, vol. 258, *op. cit.*, pp. 137, 138, 139.

Chapter 8

AMSOC Committee, "Drilling Through the Earth's Crust—A Study of the Desirability and Feasibility of Drilling a Hole to the Mohorovicic Discontinuity, Conducted by the AMSOC Committee, September 1, 1959," *Publication 717*, National Academy of Sciences–National Research Council (Washington, D.C., 1959).

Bascom, Willard, *A Hole in the Bottom of the Sea*, Doubleday (Garden City, N.Y., 1961).

Bascom, Willard, "The Mohole," *Scientific American*, vol. 200 (April, 1959), pp. 41–49.

Bascom, Willard, "Penetrating Earth's Crust—Scientific Olympics Seen Developing With Soviets' Plans for Drilling," Letter to *The New York Times*, dated Sept. 24, 1961, published Sept. 18, 1961, Section 4, p. 10E, col. 5.

Darwin, Charles, Letter to Alexander Agassiz, May 5, 1881, in *The Life and Letters of Charles Darwin including an autobiographical Chapter*, edited by his son Francis. Basic Books (New York, 1959), vol. 2, p. 362.

Doyle, Sir Arthur Conan, "When the World Screamed," in *Great Stories*, by Sir Arthur Conan Doyle, John Murray (London, 1959), pp. 85, 86–87, 89, 104, 105–106.

Estabrook, Frank B., "Geophysical Research Shaft," *Science*, vol. 124 (Oct. 12, 1956), p. 686.

Greenberg, Daniel S., "Mohole: Aground on Capitol Hill," *ibid.*, vol. 153 (Aug. 26, 1966). p. 963. See also "Mohole: Senate Is Asked to Restore Funds," *ibid.* (July 1, 1966), pp. 38–39.

Greenberg, Daniel S., "Mohole: Geopolitical Fiasco," Chapt. 25 in *Understanding the Earth, op. cit.*, pp. 343–348 (347).

Hearings before the Subcommittee on Oceanography of the Committee on Merchant Marine and Fisheries—House—88th Congress, 1st Session, June 25, Oct. 29, 30, 31, Nov. 7, 12, 1963 (Serial No. 88-14). U.S. Govt. Printing Office (Washington, 1963).

Hedberg, Hollis D., in *Hearings before the Subcommittee on Oceanography, op. cit.*, pp. 39–40, 43.

Hess, H. H., "Scientific Objectives of the Mohole and a Predicted Section," Abstracts of New York Meeting, G.S.A., Dec. 27–30, 1960. *Geological Society of America Bulletin*, vol. 71 (Dec., 1960), p. 2097.

Jaggar, T. A., "Core Drilling Under the Ocean," quoted by Willard Bascom in *A Hole in the Bottom of the Sea, op. cit.*, pp. 43, 44.

Kennedy, J. F., Message to Detlev W. Bronk, president National Academy of Sciences, and Alan T. Waterman, director National Science Foundation, quoted in *The New York Times*, April 9, 1961, p. 72, col. 6.

Lill, Gordon G., Letter to D. W. Bronk, in *Hearings before the Subcommittee on Oceanography, op. cit.*, p. 172.

Lill, Gordon G., and Arthur E. Maxwell, "The Earth's Mantle," *Science*, vol. 129 (May 22, 1959), pp. 1407–1410 (1408, 1410).

Munk, Walter, Letter to Hollis D. Hedberg, quoted by Herbert Solow in "How NSF Got Lost in Mohole," *Fortune Magazine*, vol. 67 (May, 1963), pp. 138–209 (*passim*) (209).

National Academy of Sciences–National Research Council, request to National Science Foundation, in *Hearings before the Subcommittee on Oceanography, op. cit.*, p. 216.

O'Neil, Paul, "Trailbreaker of the Deeps—Willard Bascom, part Columbus and part Barnum, explores and exploits the seas," *Life Magazine*, Sept. 30, 1966, pp. 108–121 (*passim*) (108).

Solow, Herbert, "How NSF Got Lost in Mohole," *Fortune Magazine*, vol. 67 (May, 1963), pp. 138–209 (*passim*) (140).

Waterman, Alan T., in *Hearings before the Subcommittee on Oceanography, op. cit.*, pp. 10, 33.

Chapter 9

Material for Chapters 9 and 10 was drawn primarily from the following sources:

1. *Initial Reports of the Deep Sea Drilling Project*, vols. 1 ff., published by the National Science Foundation (U.S. Government Printing Office, Washington). One volume is published for each leg. Vol. 1 covers Leg 1; vol. 2 covers Leg 2, etc.

2. Logs, Operation Reports, and News Releases issued by the Deep Sea Drilling Project for each leg.

3. *Geotimes*. Preliminary accounts of each leg were published in this journal soon after completion of the leg. The references are as follows:

Leg 1 (Gulf of Mexico to New York): vol. 14 (Feb., 1969), pp. 10–12.
Leg 2 (New York to Dakar): vol. 14 (March, 1969), pp. 11–12.
Leg 3 (Dakar to Rio de Janeiro): vol. 14 (July–Aug., 1969), pp. 13–16.
Leg 4 (Rio de Janeiro to Panama): *ibid.*
Leg 5 (San Diego to Honolulu): vol. 14 (Sept., 1969), pp. 19–20.
Leg 6 (Honolulu to Guam): vol. 14 (Oct., 1969), pp. 13–17.
Leg 7 (Guam to Honolulu): vol. 14 (Dec., 1969), pp. 12–13.
Leg 8 (Hawaii to Tahiti): vol. 15 (Feb., 1970), pp. 14–15.
Leg 9 (Tahiti to Panama): vol. 15 (April, 1970), pp. 10–13.
Leg 10 (Gulf of Mexico): vol. 15 (July/Aug., 1970), pp. 11–13.
Leg 11 (Miami to Hoboken): vol. 15 (Sept., 1970), pp. 14–16.
Leg 12 (Boston to Lisbon): vol. 15 (Nov., 1970), pp. 10–14.

Leg 13 (Mediterranean): vol. 15 (Dec., 1970), pp. 12–15.
Leg 14 (Lisbon to Puerto Rico): vol. 16 (Feb., 1971), pp. 14–17.
Leg 15 (Caribbean): vol. 16 (April, 1971), pp. 12–16.
Leg 16 (Panama to Hawaii): vol. 16 (June, 1971), pp. 12–14.
Leg 17 (Hawaii to Hawaii): vol. 16 (Sept., 1971), pp. 12–14.
Leg 18 (Hawaii to Kodiak): vol. 16 (Oct., 1971), pp. 12–15.
Leg 19 (Kodiak to Yokohama): vol. 16 (Nov., 1971), pp. 12–15.
Leg 20 (Yokohama to Fiji): vol. 17 (April, 1972), pp. 10–14.
Leg 21 (Fiji to Australia): vol. 17 (May, 1972), pp. 14–16.
Leg 22 (Australia to Ceylon): vol. 17 (June, 1972), pp. 15–17.
Leg 23 (Ceylon to Djibouti via Arabian and Red Seas): vol. 17 (July, 1972), pp. 22–26.
Leg 24 (Djibouti to Mauritius): vol. 17 (Sept., 1972), pp. 17–21.
Leg 25 (Mauritius to Durban): vol. 17 (Nov., 1972), pp. 21–24.
Leg 26 (Durban to Freemantle): vol. 18 (March, 1973), pp. 16–19.
Leg 27 (Freemantle to Freemantle): vol. 18 (April, 1973), pp. 16–17.
Leg 28 (Freemantle to Lyttleton, N.Z., via Antarctic Ocean): vol. 18 (June, 1973), pp. 19–24.
Leg 29 (Lyttleton to Wellington via Antarctic Ocean): vol. 18 (July, 1973), pp. 14–17.
Leg 30 (Wellington to Guam): vol. 18 (Sept., 1973), pp. 18–21.
Leg 31 (Guam to Hakodate, Japan): vol. 18 (Oct., 1973), pp. 22–25.
Other references are as follows:
van Andle, Tjeerd H., "Deep-Sea Drilling For Scientific Purposes: A Decade of Dreams," *Science*, vol. 160 (June 28, 1968), pp. 1419–1424.
Aumento, F., R. K. Wanless and R. D. Stevens, "Potassium–Argon Ages and Spreading Rates on the Mid-Atlantic Ridge at 45° North," *ibid.*, vol. 161 (Sept. 27, 1968), pp. 1338–1339.
Bader, Richard G., et al. (Shipboard Scientific Party), "Leg 4 of the Deep Sea Drilling Project," *ibid.*, vol. 172 (June 18, 1971), pp. 1197–1205.
Ewing, John, Maurice Ewing, "Sediment Distribution on the Mid-Ocean Ridges with Respect to Spreading of the Sea Floor," *ibid.*, vol. 156 (June 23, 1967), pp. 1590–1592.
Fleischer, R. L., J. R. M. Viertl, P. B. Price and F. Aumento, "Mid-Atlantic Ridge: Age and Spreading Rates," *ibid.*, vol. 161 (Sept. 27, 1968), pp. 1339–1342.
JOIDES, "Deep Sea Drilling Project," *The American Association of Petroleum Geologists Bulletin*, vol. 51 (Sept., 1967), pp. 1787–1802.
Maxwell, A. E., *Bruun Memorial Lecture*, Intergovernmental Oceanographic Commission, Sixth Session, UNESCO, Paris, 10 Sept., 1969 (typescript).
Maxwell, A. E., et al. (The Shipboard Scientific Party), "Summary and Conclusions," *Initial Reports of the Deep Sea Drilling Project, op. cit.*, vol. 3 (1970), pp. 441–471 (462, 464).
Maxwell, Arthur E., Richard P. Von Herzen, K. Jinghwa Hsü, James E. Andrews, Tsunemasa Saito, Stephen F. Percival, Jr., E. Dean Milow and Robert E. Boyce, "Deep Sea Drilling in the South Atlantic," *Science*, vol. 168 (May 29, 1970), pp. 1047–1059.
Pautot, Guy, Jean-Marie Auzende and Xavier Le Pichon, "Continuous Deep Sea Salt Layer along North Atlantic Margins related to Early Phase of Rifting," *Nature*, vol. 227 (July 25, 1970), pp. 351–354.
Peterson, M. N. A., N. T. Edgar, C. C. von der Borch and R. W. Rex, "Cruise Leg Summary and Discussion," *Initial Reports of the Deep Sea Drilling Project, op. cit.*, vol. 2, pp. 413–427 (413, 421, 422).

Chapter 10

(See note regarding Chapter 9 references)
Baille, M. H. L. and M. A. R. Munro, "Irish tree rings, Santorini and volcanic dust veils," *Nature*, vol. 332 (March 24, 1980), pp. 344–346.
Booth, Basil, "Pyroclastic Paroxysms," *New Scientist*, vol. 59 (Sept. 20, 1973), pp. 697–700.

Chumakov, I. S., Letter to W. B. F. Ryan, quoted in preface to "Regional Distribution and Stratigraphy of Late Miocene Evaporites and Evidence of Major Depressions in the Mediterranean Sea-Level," *Initial Reports of the Deep Sea Drilling Project, op. cit.*, vol. 13, Part 2 (1973), pp. 1233–1234 (1233).

Fischer, Alfred G., et al. (Shipboard Scientific Party, Leg 6), "Geological History of the Western North Pacific," *Science*, vol. 168 (June 5, 1970), pp. 1210–1214.

Gartner, Stefan, Jr., "Sea-Floor Spreading, Carbonate Dissolution Level, and the Nature of Horizon A," *ibid.*, vol. 169 (Sept. 11, 1970), pp. 1077–1079.

Gibson, Thomas G., and Kenneth M. Towe, "Eocene Volcanism and the Origin of Horizon A," *ibid.*, vol. 172 (April 9, 1971), pp. 152–154.

Green, Charles L., diary quoted in "Odyssey of the *Glomar Challenger*," *NOAA* (Published quarterly by the National Oceanic and Atmospheric Administration), vol. 1 (Jan., 1971), pp. 52–55 (55).

Heezen, B. C., and A. G. Fischer, "Regional Problems," Chapt. 40 in *Initial Reports of the Deep Sea Drilling Project, op. cit.*, vol. 6 (Feb., 1971), pp. 1301–1305 (1301).

Hsü, Kenneth J., "When the Mediterranean Dried Up," *Scientific American*, vol. 227 (Dec., 1972), pp. 26–36 (36).

Hsü, K. J., M. B. Cita and W. B. F. Ryan, "The Origin of the Mediterranean Evaporites," *Initial Reports of the Deep Sea Drilling Project*, vol. 13, part 2, *op. cit.*, pp. 1203–1231 (1203, 1204, 1205, 1208, 1215, 1217, 1220, 1227, 1228).

Hsü, K. J., W. B. F. Ryan and M. B. Cita, "Late Miocene Desiccation of the Mediterranean," *Nature*, vol. 242 (March 23, 1973), pp. 240–244.

Jones, E. J. W., J. G. Mitchell, F. Shido and J. D. Phillips, "Igneous Rocks dredged from the Rockall Plateau," *Nature Physical Science*, vol. 237 (June 19, 1972), pp. 118–120.

Moore, Ted. C., Jr., "DSDP: successes, failures, proposals," *Geotimes*, vol. 17 (July, 1972), pp. 27–31.

Plato, *Critias*, 119, and *Timaeus*, 24 and 25, *The Dialogues of Plato*, translated into English by B. Jowett, Random House (New York, 1937), vol. 2, pp. 10, 83.

Scholl, David W., et al., "Deep Sea Drilling Project Leg 19," *Geotimes*, vol. 16 (Nov., 1971), pp. 12–15 (14, 15).

Simpson, E. S. W., and Roland Schlich, quoted in *News Release No. 188* (Sept. 19, 1972), Deep Sea Drilling Project.

Tschoegl, Nicholas, "Atlantis: Cradle of Western Civilization?" *Engineering and Science*, vol. 35 (Calif. Inst. of Tech., June, 1972), pp. 16–21.

Chapter 11

Barker, P. F., "Plate Tectonics of the Scotia Sea Region," *Nature*, vol. 228 (Dec. 26, 1970), pp. 1293–1296.

Barker, P. F., and D. H. Griffiths, "The evolution of the Scotia Ridge and Scotia Sea," *Philosophical Transactions of the Royal Society of London, Series A*, vol. 271 (1972), pp. 151–183.

Breed, William J., "Permian stromatolites from Coalsack Col," *Antarctic Journal of the U.S.*, vol. 6 (Sept.–Oct., 1971), pp. 189–190.

Colbert, Edwin H., "Antarctic Fossils and the Reconstruction of Gondwanaland," *Natural History*, vol. 81 (Jan., 1972), pp. 66–73 (68).

Colbert, Edwin H., "The Fossil Tetrapods of Coalsack Bluff," *Antarctic Journal of the United States*, vol. 5 (May–June, 1970), pp. 57–61 (58, 59).

Colbert, Edwin H., "Paleontological Investigations at Coalsack Bluff," *ibid.* (July–Aug., 1970), p. 86.

Colbert, Edwin H., quoted by John Lear in "The Bones on Coalsack Bluff, A story of Drifting Continents," *Saturday Review*, vol. 53 (Feb. 7, 1970), pp. 46–51 (48, 49).

Colbert, Edwin H., "Tetrapods and Continents," *The Quarterly Review of Biology*, vol. 46 (Sept., 1971), pp. 250–269.

Colbert, Edwin H., "Triassic tetrapods from McGregor Glacier," *Antarctic Journal of the U.S.*, vol. 6 (Sept.–Oct., 1971), pp. 188–189.

Colbert, Edwin H., *Wandering Lands and Animals*, E. P. Dutton (New York, 1973).

Dalziel, Ian W. D., and David H. Elliot, "Evolution of the Scotia Arc," *Nature*, vol. 233 (Sept. 24, 1971), pp. 246–252.

Dietz, Robert S., and John C. Holden, "The Breakup of Pangaea," *Scientific American*, vol. 223 (Oct. 1970), pp. 30–41. Reprinted in *Continents Adrift, op. cit.*, pp. 102–113.

Dietz, Robert S., and John C. Holden, "Reconstruction of Pangaea: Breakup and Dispersion of Continents, Permian to Present," *Journal of Geophysical Research*, vol. 75 (Sept. 10, 1970), pp. 4939–4956. Reprinted in *Plate Tectonics, Selected Papers from the Journal of Geophysical Research, op. cit.*, pp. 295–312.

Dietz, Robert S., John C. Holden and Walter P. Sproll, "Geotectonic Evolution and Subsidence of Bahama Platform," *Geological Society of America Bulletin*, vol. 81 (July, 1970), pp. 1915–1928.

Dietz, Robert S., and Walter P. Sproll, "Fit Between Africa and Antarctica: A Continental Drift Reconstruction," *Science*, vol. 167 (March 20, 1970), pp. 1612–1614.

DuToit, Alex. L., *Our Wandering Continents, op. cit.*, p. 128.

Elliot, David H., "Narrative and Geological Report," in "Beardmore Glacier Investigations, 1969–1970," *Antarctic Journal of the U.S.*, vol. 5 (July–Aug., 1970), pp. 83–85.

Elliot, David H., and Donald A. Coates, "Geological Investigations in the Queen Maud Mountains," *Antarctic Journal of the U.S.*, vol. 6 (July–Aug., 1971), pp. 114–118.

Elliot, David H., Edwin H. Colbert, William J. Breed, James A. Jensen and Jon S. Powell, "Triassic Tetrapods from Antarctica: Evidence for Continental Drift," *Science*, vol. 169 (Sept. 18, 1970), pp. 1197–1201 (1200).

Elliot, David H., James W. Collinson and Jon S. Powell, "Stratigraphic Setting of the Triassic Vertebrates of Antarctica," *Second Gondwana Symposium, South Africa. 1970, Proceedings and Papers*, Council for Scientific and Industrial Research, Scientia, Pretoria, South Africa, pp. 265–271 (*passim*).

Frakes, Lawrence A., and John C. Crowell, "Late Paleozoic Glacial Facies and the Origin of the South Atlantic Basin," *Nature*, vol. 217 (March 2, 1968), pp. 837–838. See also same authors in *Antarctic Journal of the U.S.*, vol. 4 (Sept.–Oct., 1969), pp. 201–202.

Gould, Laurence M., "The Geological Sledge Trip," in *Little America* by R. E. Byrd, Putnam (New York, 1930), pp. 393–412 (403).

Gould, Laurence M., and Grover Murray, cablegram, quoted in *The New York Times*, Dec. 6, 1969, p. 1, cols. 3–5.

"Great Fossil Find," (Unsigned) *Antarctic* (A News Bulletin Published Quarterly by the New Zealand Antarctic Society), Dec., 1969, pp. 330–333.

Green, A. G., "Seafloor Spreading in the Mozambique Channel," *Nature Physical Science*, vol. 236 (March 13, 1972), pp. 19–32 (*passim*).

Hurley, Patrick M., "The Confirmation of Continental Drift," *op. cit.*

Kitching, James W., "Paleontological Investigations in the McGregor Glacier Area," *Antarctic Journal of the U.S.*, vol. 6 (July–Aug., 1971), pp. 118–119.

Kitching, James W., James W. Collinson, David H. Elliot and Edwin H. Colbert, "Lystrosaurus Zone (Triassic) Fauna from Antarctica," *Science*, vol. 175 (Feb. 4, 1972), pp. 524–527 (526).

LeMasurier, Wesley E., "Spatial variation in Cenozoic volcanism of Marie Byrd Land and Ellsworth Land," *Antarctic Journal of the U.S.*, vol. 6 (Sept.–Oct., 1971), pp. 187–188.

Ramos, Victor A., "The Birth of Southern South America," *American Scientist*, vol. 77 (1989), pp. 444–450.

Ridd, M. F., "Southeast Asia as a Part of Gondwanaland," *Nature*, vol. 234 (Dec. 31, 1971), pp. 531–533.

Romer, Alfred Sherwood, "Tetrapod Vertebrates and Gondwanaland," in *Second Gondwana Symposium, South Africa, 1970, op. cit.*, pp. 111–121.

Romer, A. S., *Vertebrate Paleontology* (2nd edition), University of Chicago (1945), pp. 141, 529.

Schaeffer, Bobb, "A Jurassic Fish from Antarctica," *American Museum Novitates*, No. 2495 (The American Museum of Natural History, June 30, 1972), pp. 1–17. A similar report by Schaeffer appears in *Antarctic Journal of the U.S.*, vol. 6 (Sept.–Oct., 1971), pp. 190–191.

Scott, R. F., *Scott's Last Expedition*, Dodd Mead (New York, 1913), vol. 1, pp. 352, 387, 388, 389.

Shackleton, Sir Ernest, *The Heart of the Antarctic*, William Heinemann (London, 1932), pp. 177, 185, 193, 194.

Smith, A. Gilbert, and A. Hallam, "The Fit of the Southern Continents," *Nature*, vol. 225 (Jan. 10, 1970), pp. 139–144.

Sproll, W. P., and Robert S. Dietz, "Morphological Continental Drift Fit of Australia and Antarctica," *ibid.*, vol. 222 (April 26, 1969), pp. 345–348.

Tarling, D. H., "Another Gondwanaland," *ibid.*, vol. 238 (July 14, 1972), pp. 92–93.

Tasch, Paul, "Paleolimnology of Some Antarctic Nonmarine Deposits," *Antarctic Journal of the U.S.*, vol. 5 (July–Aug., 1970), pp. 85–86.

Chapter 12

Bonaparte, J. F., "New vertebrate evidence for a southern transAtlantic connexion during the Lower or Middle Triassic," *Paleontology*, vol. 10, Part 4 (Dec., 1967), pp. 554–563.

Colbert, Edwin H., *Evolution of the Vertebrates—A History of the Backboned Animals Through Time*, Second edition, Wiley (New York, 1969).

Colbert, Edwin H., "Tetrapods and Continents," *op. cit.*, pp. 253, 261, 267.

Colbert, Edwin H., *Wandering Lands and Animals, op. cit.*, p. 57.

Darwin, Charles, Letter to J. D. Hooker, Aug. 6, 1881, in *The Life and Letters of Charles Darwin...op. cit.*, vol. 2, pp. 422–425 (423, 424).

Dietz, Robert S., and John C. Holden, "Reconstruction of Pangaea..." *op. cit.*

Flemming, N. C., and D. G. Roberts, "Tectono-eustatic Changes in Sea Level and Seafloor Spreading," *Nature*, vol. 243 (May 4, 1973), pp. 19–22.

Fooden, Jack, "Breakup of Pangaea and Isolation of Relict Mammals in Australia, South America, and Madagascar," *Science*, vol. 175 (Feb. 25, 1972), pp. 894–898.

Fooden, Jack, "Rifting and Drift of Australia and the Migration of Mammals," *ibid.*, vol. 180 (May 18, 1973), pp. 759–761.

Hallam, A., "Continental Drift and the Fossil Record," *Scientific American*, vol. 227 (Nov., 1972), pp. 56–66 (66).

Hays, James D., and Walter C. Pitman III, "Lithospheric Plate Motion, Sea Level Changes and Climatic and Ecological Consequences," *Nature*, vol. 246 (Nov. 2, 1973), pp. 18–22.

Hooker, Sir Joseph Dalton, Letters to Charles Darwin, Aug. 4 and 11, 1881, in *Life and Letters of Sir Joseph Dalton Hooker: Based on Materials Collected and Arranged by Lady Hooker*, by Leonard Huxley, D. Appleton & Co. (New York, 1918), vol. 2, pp. 223–226 (223, 224, 226).

Hooker, Sir J. D., "On Geographical Distribution," *Report of the 51st Meeting of the British Association for the Advancement of Science, held at York in August and September, 1881*, John Murray (London, 1882), pp. 727–738.

Kurtén, Björn, "Continental Drift and Evolution," *Scientific American*, vol. 220 (March, 1969), pp. 54–64 (58). Reprinted in *Continents Adrift, op. cit.*, pp. 114–123 (118).

McKenna, Malcolm C., "Possible Biological Consequences of Plate Tectonics," *Bioscience*, vol. 22 (Sept., 1972), pp. 519–525.

Panchen, A. L., "Anthracosauria" in *Encyclopedia of Paleoherpetology*. Part 5A, Fischer (Stuttgart, 1970), pp. 1–83 (68).

Purrett, Louise, "Continental drift and the diversity of species," *Science News*, vol. 100 (Dec. 11, 1971), pp. 394–395.

Romer, Alfred Sherwood, "Fossils and Gondwanaland," *Proceedings of the American Philosophical Society*, vol. 112 (Oct., 1968), Paper read April 19, 1968, in the Symposium on "Gondwanaland Revisited: New Evidence for Continental Drift," pp. 335–343 (335).

Romer, A. S., "The Late Carboniferous Vertebrate Fauna of Kounova (Bohemia) Compared with that of the Texas Redbeds," *The American Journal of Science*, vol. 243 (Aug., 1945), pp. 417–442.

Romer, A. S., *The Procession of Life*, World (Cleveland, 1968), p. 199.

Romer, A. S., "Tetrapod Vertebrates and Gondwanaland," *op. cit.*, p. 120.

Scholl, David W., quoted in "Ocean-floor record links dinosaurs, plankton, climate," *Science News*, vol. 100 (Oct. 23, 1971), p. 279.

Sun Ai-lin, "Permo-Triassic Reptiles of Sinkiang," *Scientia Sinica* (Notes), vol. 16 (Feb., 1973), pp. 152–156 (155).

Valentine, James W., "Plate Tectonics and Shallow Marine Diversity and Endemism, An Actualistic Model," *Systematic Zoology*, vol. 20 (Sept., 1971), pp. 253–264.

Valentine, James W., and Eldridge Moores, "Global Tectonics and the Fossil Record," *Journal of Geology*, vol. 80 (1972), pp. 167–184.

Valentine, J. W., and E. M. Moores, "Plate-tectonic Regulation of Faunal Diversity and Sea Level: a Model," *Nature*, vol. 228 (Nov. 14, 1970), pp. 657–659.

Wood, Albert E., "An Eocene Hystricognathous Rodent from Texas: Its Significance in Interpretations of Continental Drift," *Science*, vol. 175 (March 17, 1972), pp. 1250–1251.

Chapter 13

Baker, B. H., "The structural pattern of the Afro-Arabian rift system in relation to plate tectonics," *Philosophical Transactions of the Royal Society of London, Series A*, vol. 267 (Oct. 29, 1970), pp. 383–391.

Baker, B. H., and J. Wohlenberg, "Structure and Evolution of the Kenya Rift Valley," *Nature*, vol. 229 (Feb. 19, 1971), pp. 538–541.

Bonatti, Enrico, Cesare Emiliani, Gote Ostlund and Harold Rydell, "Final Desiccation of the Afar Rift, Ethiopia," *Science*, vol. 172 (April 30, 1971), pp. 468–469.

Bonatti, Enrico, and Haroun Tazieff, "Exposed Guyot from the Afar Rift, Ethiopia," *Science*, vol. 168 (May 29, 1970), pp. 1087–1089.

Bradbury, Ray, Arthur C. Clarke, Bruce Murray, Carl Sagan and Walter Sullivan, *Mars and the Mind of Man*, Harper and Row (New York, 1973).

Brewer, P. G., et al., "Hydrographic Observations on the Red Sea Brines indicate a Marked Increase in Temperature," *Nature*, vol. 231 (May 7, 1971), pp. 37–38.

Brown, G. F., "Eastern margin of the Red Sea and the coastal structures in Saudi Arabia," *Philosophical Transactions of the Royal Society of London, Series A*, vol. 267 (1970), pp. 75–87.

de Chardin, P. Teilhard (1930), quoted by H. Fauré in "L'Afar, Structure Profonde Evolution Magmatique et Paleogéographique," *Rapport de la IVe Mission en Afar (1970–1971)*, Centre National de la Recherche Scientifique, R. C. P. 180 (Paris, July, 1971), p. f. Translation by the author.

Dadet, P., J. Marchesseau, R. Millon and E. Motti, "Mineral Occurrences related to stratigraphy and tectonics in Tertiary sediments near Umm Lajj, eastern Red Sea area, Saudi Arabia," *Philosophical Transactions of the Royal Society of London, Series A*, vol. 267 (1970), pp. 99–106 (105).

Davies, D., and C. Tramontini, "The deep structure of the Red Sea," *ibid.*, pp. 181–189.

Degens, Egon T., and David A. Ross, "The Red Sea Hot Brines," *Scientific American*, vol. 222 (April, 1970), pp. 32–42 (39).

Dunham, K. C., "Introduction to the general discussion," in "A Discussion on the Structure and Evolution of the Red Sea and the Nature of the Red Sea, Gulf of Aden and Ethiopia Rift Junction," *Philosophical Transactions of the Royal Society of London, Series A*, vol. 267 (1970), pp. 397–398 (398).

Fairhead, J. D., and R. W. Girdler, "Seismicity of the Red Sea," *ibid.*, pp. 49–74.

Fairhead, J. D., J. G. Mitchell and L. A. J. Williams, "New K/Ar Determinations on Rift Volcanics of S. Kenya and their Bearing on Age of Rift Faulting," *Nature Physical Science*, vol. 238 (July 31, 1972), pp. 66–69.

Fisher, Robert L., et al. (Shipboard Scientific Party), "Deep Sea Drilling Project in dodo land—Leg 24," *Geotimes*, vol. 17 (Sept., 1972), pp. 17–21.

Frazier, S. B., "Adjacent structures of Ethiopia: that portion of the Red Sea coast including Dahlak Kebir Island and the Gulf of Zula," *Philosophical Transactions of the Royal Society of London, Series A*, vol. 267 (1970), pp. 131–141.

Freund, R., Z. Garfunkel, I. Zak, M. Goldberg, T. Weissbrod and B. Derin, "The shear along the Dead Sea rift," *ibid.*, pp. 107–130.

Genesis, chapter 19, verses 24–28.

Gouin, Pierre, "Wolenchiti Fracturing," *Event Notification Report 1019* (Event 86-70), Center for Short-Lived Phenomena, Smithsonian Institution (Cambridge, Mass., Sept. 29, 1970).

Gregory, J. W., "Contributions to the Physical Geography of British East Africa," *The Geographical Journal*, vol. 4 (Oct., 1894), pp. 289 ff. (290).

Griffiths, D. H., R. F. King, M. A. Khan and D. J. Blundell, "Seismic Refraction Line in the Gregory Rift," *Nature Physical Science*, vol. 229 (Jan. 18, 1971), pp. 69–71 (71).

Isaac, Glynn Ll., Richard E. F. Leakey and Anna K. Behrensmeyer, "Archeological Traces of Early Hominid Activities, East of Lake Rudolf, Kenya," *Science*, vol. 173 (Sept. 17, 1971), pp. 1129–1134.

Leakey, R. E. F., A. K. Behrensmeyer, F. J. Fitch, J. A. Miller and M. D. Leakey, "New Hominid Remains and Early Artefacts from Northern Kenya," *Nature*, vol. 226 (April 1970), pp. 223–230.

Le Bas, M. J., "Per-alkaline Volcanism. Crustal Swelling and Rifting," *Nature Physical Science*, vol. 230 (March 22, 1971), pp. 85–87.

Lowell, James D., and Gerard J. Genik, "Sea-Floor Spreading and Structural Evolution of Southern Red Sea." *American Association of Petroleum Geologists Bulletin*, vol. 56 (Feb., 1972), pp. 247–259.

Maglio, Vincent J., "Vertebrate Faunas and Chronology of Hominid-bearing Sediments East of Lake Rudolf, Kenya," *Nature*, vol. 239 (Oct. 13, 1972), pp. 379–385.

McKenzie, D. P., D. Davies and P. Molnar, "Plate Tectonics of the Red Sea and East Africa," *ibid.*, vol. 226 (April 18, 1970), pp. 243–248.

Megrue, G. H., E. Norton and D. W. Strangway, "Tectonic History of the Ethiopian Rift as Deduced by K–Ar Ages and Paleomagnetic Measurements of Basaltic Dikes," *Journal of Geophysical Research*, vol. 77 (Oct. 10, 1972), pp. 5744–5754.

Mohr, P. A., "The Afar Triple Junction and Sea-Floor Spreading," *ibid.*, vol. 75 (Dec. 10, 1970), pp. 7340–7352.

Mohr, P. A., "Plate Tectonics of the Red Sea and East Africa," *Nature*, vol. 228 (Nov. 7, 1970), pp. 547–548.

Mohr, P. A., "Volcanic Composition in Relation to Tectonics in the Ethiopian Rift System: A Preliminary Investigation," *Bulletin Volcanologique*, vol. 34 (1970), pp. 141–157.

Ross, David A., "The Red and the Black Seas," *American Scientist*, vol. 59 (July–Aug., 1971), pp. 420–424.

Ross, David A., "Red Sea Hot Brine Area: Revisited," *Science*, vol. 175 (March 31, 1972), pp. 1455–1456.

Ross, David A., et al. (The Shipboard Scientific Party), "Deep Sea Drilling Project in the Red Sea," *Geotimes*, vol. 17 (July, 1972), pp. 24–26.

Ross, David A., et al. (Shipboard Scientific Party, Leg 23-B), "Red Sea Drillings," *Science*, vol. 179 (Jan. 26, 1973), pp. 377–380.

Schmitt, Harrison H., in *Apollo 17 Mission Commentary* (5:46 am. Central Standard Time, Dec. 7, 1972), tape 75, take 1.

Tazieff, Haroun, "The Afar Triangle," *Scientific American*, vol. 222 (Feb., 1970), pp. 32–40 (34, 37). Reprinted in *Continents Adrift, op. cit.*, pp. 133–141 (134, 137).

Tazieff, H., G. Marinelli, F. Barberi and J. Varet, "Géologie de l'Afar Septentrional," *Bulletin Volcanologique*, vol. 33 (April, 1969), pp. 1039–1072.

Tazieff, Haroun and Jacques Varet, "Signification Tectonique et Magmatique de l'Afar Septentrional (Éthiopie)," *Revue de Géographie Physique et de Géologie Dynamique* (2), vol. 11, Fasc. 4 (Paris, 1969), pp. 429–450.

Varet, Jacques, "Erta'Ale Volcanic Activity (1972)," *Event Notification Report 1363* (Event 16–72), Center for Short-Lived Phenomena, Smithsonian Institution (Cambridge, Mass., March 6, 1972).

Wegener, Alfred, *The Origin of Continents and Oceans, op. cit.*, p. 190.

Zechariah, Chapter 14, Verse 4.

Chapter 14

Andrews, James E., "Gravitational Subduction of a Western Pacific Crustal Plate," *Nature Physical Science*, vol. 233 (Sept. 27, 1971), pp. 81–83.

Bodvarsson, G., and G. P. L. Walker, "Crustal drift in Iceland," *Geophysical Journal of the Royal Astronomical Society*, vol. 8 (Feb., 1964), pp. 285–299.

Bowin, Carl, "Origin of the Ninety East Ridge from Studies near the Equator," *Journal of Geophysical Research*, vol. 78 (Sept. 10, 1973), pp. 6029–6043.

van Breemen, O., and P. Bowden, "Sequential Age Trends for some Nigerian Mesozoic Granites," *Nature Physical Science*, vol. 242 (March 5, 1973), pp. 9–11.

Casenave, Anny, Annie Souriau and Kien Dominh, "Global coupling of Earth surface topography with hotspots, geoid and mantle heterogeneities," *Nature*, vol. 340 (July 6, 1989), pp. 54–57.

Colgate, S. A., and Thorbjörn Sigurgeirsson, "Dynamic Mixing of Water and Lava," *Nature*, vol. 244 (Aug. 31, 1973), pp. 552–555 (552).

Crowell, J. C., and L. A. Frakes, "Early History of the South Atlantic," Abstract of Paper 32 in "Continental Drift Emphasizing the History of the South Atlantic Area. A UNESCO/IUGS Symposium Held at Montevideo, Uruguay, on October 16–19, 1967." *EOS* (*Transactions of the American Geophysical Union*), vol. 53 (Feb., 1972), p. 178.

Decker, R. W., Páll Einarsson and P. A. Mohr, "Rifting in Iceland: New Geodetic Data," *Science*, vol. 173 (Aug. 6, 1971), pp. 530–533.

Deffeyes, Kenneth S., "Plume Convection with an Upper Mantle Temperature Inversion," *Nature*, vol. 240 (Dec. 29, 1972), pp. 539–544.

Duncan, R. A., N. Petersen and R. B. Hargraves, "Mantle Plumes, Movement of the European Plate, and Polar Wandering," *ibid.*, vol. 239 (Sept. 8, 1972), pp. 82–86.

Eder, Richard, "The Morning the Volcano Awoke Iceland Isle: 'Grass Was Burning,' " *The New York Times*, Jan. 27, 1973, p. 3, cols. 1–3.

Ewing, Maurice, Xavier Le Pichon and John Ewing, "Crustal Structure of the Mid-Ocean Ridges," *Journal of Geophysical Research*, vol. 71 (March 15, 1966), pp. 1611–1636 (1611).

Forsyth, Donald, "Compressive Stress between Two Mid-Ocean Ridges," *Nature*, vol. 243 (May 11, 1973), pp. 78–79.

Frey, F. A., and the 121 Scientific Drilling Party...Ocean Drilling Project Leg 121: "Shipboard Results Bearing on Origin and Evolution of Ninetyeast Ridge," Abstracts 28th International Geological Congress, Washington, D.C., 1989, p. 1–511.

Gibson, I. L., and J. D. A. Piper, "Structure of the Icelandic basalt plateau and the process of drift," *Philosophical Transactions of the Royal Society of London, Series A*, vol. 271 (Jan. 27, 1972), pp. 141–149.

Gold, Thomas, cited by E. C. Bullard, A. E. Maxwell and R. Revelle, in "Heat Flow Through the Deep Sea Floor," *op. cit.*, p. 177.

Handschumacher, David, "Formation of the Emperor Seamount Chain," *Nature*, vol. 244 (July 20, 1973), pp. 150–152.

Hey, Richard N., K. S. Deffeyes, G. Leonard Johnson and Allen Lowrie, "The Galapagos Triple Junction and Plate Motions in the East Pacific," *ibid.*, vol. 237 (May 5, 1972), pp. 20–22.

Hooper, Peter R., "The Timing of Crustal Extensions and the Eruption of Continental Floor Balsalts," *Nature*, vol. 345 (May 17, 1990), pp. 246–249.

Jeffreys, Sir Harold, *The Earth...*, *op. cit.*, p. 496.

Kaula, William M., "Earth's Gravity Field: Relation to Global Tectonics," *Science*, vol. 169 (Sept. 4, 1970), pp. 982–985.

Le Pichon, Xavier, Jean Francheteau and Jean Bonnin, *Plate Tectonics*, Elsevier (Amsterdam, 1973).

Luyendyk, Bruce, and Thomas A. Davies, quoted in "Answer Near for Date of Break-Up of Gondwanaland and Birth of Indian Ocean," *D.S.D.P. Release No. 190* (on Leg 26), Scripps Institution of Oceanography (Oct. 29, 1972), p. 2.

Maack, Reinhard, *Kontinentaldrift und Geologie des südatlantischen Ozeans*, Walter de Gruyter (Berlin, 1969).

Mahoney, J. J., J. D. Macdougall, G. W. Lugmair and K. Gopalan, "Kerguelen hotspot source for Rajmahal Traps and Ninetyeast Ridge?" *Nature*, vol. 303 (June 2, 1983), pp. 385-389.

McKenzie, D. P., and W. J. Morgan, "Evolution of Triple Junctions," *Nature*, vol. 224 (Oct. 11, 1969), pp. 125-133.

Molnar, Peter, and Tanya Atwater, "Relative Motion of Hot Spots in the Mantle," *ibid.*, vol. 246 (Nov. 30, 1973), pp. 288-291.

Molnar, Peter, and Joann Stock, "Relative motions of hotspots in the Pacific, Atlantic and Indian Oceans since late Cretaceous time," *Nature*, vol. 327 (June 18, 1987), pp. 587-591, and accompanying commentary by Peter Olson, pp. 559-560.

Morgan, W. Jason, "Convection Plumes in the Lower Mantle," *Nature*, vol. 230 (March 5, 1971), pp. 42-43.

Morgan, W. Jason, "Deep Mantle Convection Plumes and Plate Motions," *The American Association of Petroleum Geologists Bulletin*, vol. 56 (Feb., 1972), pp. 203-213.

O'Hara, M. J., "Non-Primary Magmas and Dubious Mantle Plume beneath Iceland," *Nature*, vol. 243 (June 29, 1973), pp. 507-508.

Orowan, Egon, "The Origin of the Oceanic Ridges," *Scientific American*, vol. 221 (Nov., 1969), pp. 102-119.

Oversby, Virginia M., "Lead in Oceanic Islands: Faial, Azores, and Trinidade," *Earth and Planetary Science Letters*, vol. 11 (Aug., 1971), pp. 401-406.

Oversby, Virginia M., and Paul W. Gast, "Isotopic Composition of Lead from Oceanic Islands," *Journal of Geophysical Research*, vol. 75 (April 10, 1970), pp. 2097-2114.

S., P. J., "The Importance of Being Plume Conscious," News and Views, *Nature*, vol. 242 (April 27, 1973), pp. 551-552.

Schilling, J.-G., "Afar Mantle Plume: Rare Earth Evidence," *Nature Physical Science*, vol. 242 (March 5, 1973), pp. 2-5.

Schilling, J.-G., "Iceland Mantle Plume: Geochemical Study of Reykjanes Ridge," *Nature*, vol. 242 (April 27, 1973), pp. 565-571.

Schubert, Gerald, and D. L. Turcotte, "Phase Changes and Mantle Convection," *Journal of Geophysical Research*, vol. 76 (Feb. 10, 1971), pp. 1424-1432.

Schubert, Gerald, D. L. Turcotte and E. R. Oxburgh, "Phase Change Instability in the Mantle," *Science*, vol. 169 (Sept. 11, 1970), pp. 1075-1077.

"South Atlantic Islands and Ocean Floor," papers presented in "Continental Drift Emphasizing the History of the South Atlantic Area," *EOS*, vol. 53, *op. cit.*, pp. 168-170.

Storey, M., A. D. Saunders, J. Tarney, I. L. Gibson, M. J. Norry, M. F. Thirwall, P. Leat, R. N. Thompson and M. A. Menzies, "Contamination of Indian Ocean asthenosphere by the Kerguelen-Heard mantle plume," Nature, vol. 338 (April 13, 1989), pp. 574-576.

Sykes, Lynn R., Robert Kay and Orson Anderson, "Mechanical Properties and Processes in the Mantle—Report of a Symposium Held at Flagstaff, Arizona, June 24–July 3, 1970," *EOS* (*Transactions of the American Geophysical Union*), vol. 51 (Dec., 1970), pp. 874-879.

Thorarinsson, S., S. Steinthórsson, Th. Einarsson, H. Kristmannsdóttir and N. Oskarsson, "The Eruption on Heimaey, Iceland," *Nature*, vol. 241 (Feb. 9, 1973), pp. 372-375.

Verne, Jules, *Voyage au Centre de la Terre* (1864), Translation in Bascom, *A Hole in the Bottom of the Sea op. cit.*, p. 58.

Vogt, P. R., "Asthenosphere motion recorded by the ocean floor south of Iceland," *Earth and Planetary Science Letters*, vol. 13 (Dec. 11, 1971), pp. 153-160 (153).

Vogt, P. R., "Evidence for Global Synchronism in Mantle Plume Convection, and Possible Significance for Geology," *Nature*, vol. 240 (Dec. 8, 1972), pp. 338-342.

Vogt, P. R., "Subduction and Aseismic Ridges," *ibid.*, vol. 241 (Jan. 19, 1973), pp. 189-191.

Vogt, P. R., G. L. Johnson, T. L. Holcombe, J. G. Gilg and O. E. Avery, "Episodes of Sea-Floor Spreading Recorded by the North Atlantic Basement," in "Global Tectonics and Sea-Floor Spreading," Special Issue of *Tectonophysics*, vol. 12 (1971) (based on the Symposium on Global Tectonics and Sea-Floor Spreading, Tokyo, 1970), pp. 211-234.

Williams, Richard S., and James G. Moore, "Iceland Chills a Lava Flow," *Geotimes*, vol. 18 (Aug., 1973), pp. 14–17 (16, 17).

Wilson, J. Tuzo, "Evidence from Islands on the Spreading of Ocean Floors," *Nature*, vol. 197 (Feb. 9, 1963), pp. 536–538.

Wilson, J. Tuzo, "A Possible Origin of the Hawaiian Islands," *Canadian Journal of Physics*, vol. 41 (1963), pp. 863–870.

Wilson, J. Tuzo, "Submarine Fracture Zones, Aseismic Ridges and the International Council of Scientific Unions Line..." *op. cit.*

Chapter 15

Bailey, Edward B., *Tectonic Essays, Mainly Alpine*, Oxford (1935).

di Brozolo, Filippo Radicati, and Gaetano Giglia, "Further Data on the Corsica–Sardinia Rotation," *Nature*, vol. 241 (Feb. 9, 1973), pp. 389–391.

Collet, Leon W., "The Alps and Wegener's Theory," Chapt. VII in *The Structure of the Alps*, Arnold (London, 1935).

Elsasser, Walter M., "The Earth as a Planet—Origin of the Surface Features," *Encyclopedia Britannica*, vol. 7 (Chicago, 1963), p. 851.

Escher von der Linth, Arnold, quoted by Bailey *op. cit.*, p. 50.

Escher von der Linth, Arnold, quoted in "Tectonophysics," by N. Pavoni, in *The Development of Geodesy and Geophysics in Switzerland*, J. C. Thams, ed., Commemorative Book Presented to Participants in the 14th General Assembly of the International Union of Geodesy and Geophysics, Swiss Academy of Natural Sciences (Zurich, 1967), pp. 42–52 (46).

Gass, I. G., and J. D. Smewing, "Intrusion, Extrusion and Metamorphism at Constructive Margins: Evidence from the Troodos Massif, Cyprus," *Nature*, vol. 242 (March 2, 1973), pp. 26–29.

Hsü, K. J., "Origin of the Alps and Western Mediterranean," *ibid.*, vol. 233 (Sept. 3, 1971), pp. 44–48 (47).

Hsü, K. J., "Paleocurrent Structures and Paleogeography of the Ultrahelvetic Flysch Basins, Switzerland," *Geological Society of America Bulletin*, vol. 71 (May, 1960), pp. 577–610.

Hsü, K. Jinghwa, "A Preliminary Analysis of the Statics and Kinetics of the Glarus Overthrust," *Eclogae Geologicae Helvetiae*, vol. 62 (Basle, June, 1969), pp. 143–154.

Hsü, K. J., and S. O. Schlanger, "Ultrahelvetic Flysch Sedimentation and Deformation Related to Plate Tectonics," *Geological Society of America Bulletin*, vol. 82 (May, 1971), pp. 1207–1218.

Khan, M. A., C. Summers, S. A. D. Bamford, P. N. Chroston, C. K. Poster and F. J. Vine, "Reversed Seismic Refraction Line on the Troodos Massif, Cyprus," *Nature Physical Science*, vol. 238 (Aug. 28, 1972), pp. 134–136 (134).

Laubscher, H. P., "The Large-Scale Kinematics of the Western Alps and the Northern Apennines and Its Palinspastic Implications," *American Journal of Science*, vol. 271 (Oct., 1971), pp. 193–226.

Le Borgne, Eugene, Jean-Louis Le Mouël and Xavier Le Pichon, "Aeromagnetic Survey of South-Western Europe," *Earth and Planetary Science Letters*, vol. 12 (1971), pp. 287–299.

Le Pichon, Xavier, Guy Pautot and J. P. Weill, "Opening of the Alboran Sea," *Nature Physical Science*, vol. 236 (April 10, 1972), pp. 83–85.

Le Pichon, Xavier, and Jean-Claude Sibuet, "Comments on the Evolution of the North-East Atlantic," *Nature*, vol. 233 (Sept. 24, 1971), pp. 257–258.

Le Pichon, Xavier, and Jean-Claude Sibuet, "Western Extension of Boundary Between European and Iberian Plates During the Pyrenean Orogeny," *Earth and Planetary Science Letters*, vol. 12 (1971), pp. 83–88.

Longwell, Chester R., and Richard Foster Flint, *Introduction to Physical Geology*, John Wiley (New York, 1955), p. 335.

McKenzie, D. P., "Plate Tectonics of the Mediterranean Region," *Nature*, vol. 226 (April 18, 1970), pp. 239–243.

McKenzie, D. P., "Speculations on the Consequences and Causes of Plate Motions," *Geophysical Journal of the Royal Astronomical Society*, vol. 18 (1969), pp. 1–32 (1).

Moores, Eldridge, "Ultramafics and Orogeny, with Models of the US Cordillera and the Tethys," *Nature*, vol. 228 (Nov. 28, 1970), pp. 837–842.

Moores, E. M., and F. J. Vine, "The Troodos Massif, Cyprus and other ophiolites as oceanic crust: evaluation and implications," *Philosophical Transactions of the Royal Society of London, Series A*, vol. 268 (1971), pp. 443–466.

Pawley, G. S. and N. Abrahamsen, "Do the Pyramids Show Continental Drift?" *Science*, vol. 179 (March 2, 1973), pp. 892–893.

S., J., "Studying the Formation of Ocean Floors," News and Views, *Nature*, vol. 242 (March 2, 1973), pp. 9–10 (9).

Smith, A. Gilbert, "Alpine Deformation and the Oceanic Areas of the Tethys, Mediterranean, and Atlantic," *Geological Society of America Bulletin*, vol. 82 (Aug., 1971), pp. 2039–2070.

Staub, R., "Der Bau der Alpen," Beiträge zur geologischen Karte der Schweiz, N. F., no. 52 (Bern, 1924), p. 257, and *Der Bewegungsmechanismus der Erde* (Berlin, 1928), quoted by Wegener in *The Origin of Continents and Oceans, op. cit.*, pp. 10–11, 159.

Trümpy, Rudolf, "Geotectonic Evolution of the Central and Western Alps," *Geological Society of America Bulletin*, vol. 71 (June, 1960), pp. 843–908 (847).

Chapter 16

Bird, John M., and John F. Dewey, "Lithosphere Plate-Continental Margin Tectonics and the Evolution of the Appalachian Oregon," *Geological Society of America Bulletin*, vol. 81 (April, 1970), pp. 1031–1060.

Bird, John M., John F. Dewey and W. S. F. Kidd, "Proto-Atlantic Oceanic Crust and Mantle: Appalachian/Caledonian Ophiolites," *Nature Physical Science*, vol. 231 (May 10, 1971), pp. 28–31.

Burke, Kevin, and J. B. Waterhouse, "Saharan Glaciation Dated in North America," *Nature*, vol. 241 (Jan. 26, 1973), pp. 267–268.

Dewey, John F. "Evolution of the Appalachian/Caledonian Orogen," *ibid.*, vol. 222 (April 12, 1969), pp. 124–129.

Dewey, John F., and John M. Bird, "Mountain Belts and the New Global Tectonics," *Journal of Geophysical Research*, vol. 75 (May 10, 1970), pp. 2625–2647. Reprinted in *Plate Tectonics, Selected Papers from the Journal of Geophysical Research, op. cit.*, pp. 257–279.

Dewey, John F., and John M. Bird, "Origin and Emplacement of the Ophiolite Suite: Appalachian Ophiolites in Newfoundland," *ibid.*, vol. 76 (May 10, 1971), pp. 3179–3206. Reprinted in *Plate Tectonics, Selected Papers from the Journal of Geophysical Research, op. cit.*, pp. 441–470.

Dewey, John F., and John M. Bird, "Plate Tectonics and Geosynclines," *Tectonophysics*, vol. 10 (1970), pp. 625–638.

Dietz, Robert S., "Collapsing Continental Rises: An Actualistic Concept of Geosynclines and Mountain Building," *The Journal of Geology*, vol. 71 (May, 1963), pp. 314–333.

Dietz, Robert S., "Geosynclines, Mountains, and Continent-Building," *Scientific American*, vol. 226 (March, 1972), pp. 30–38. Reprinted in *Continents Adrift, op. cit.*, pp. 124–132.

Dietz, Robert S., "Origin of Continental Slopes," *American Scientist*, vol. 52 (March, 1964), pp. 50–69.

Dietz, Robert S., and John C. Holden, "Deep-sea deposits in but not on the continents," *Bulletin of the American Association of Petroleum Geologists*, vol. 50 (Feb., 1966), pp. 351–362.

Dietz, Robert S., and John C. Holden, "Miogeoclines (Miogeosynclines) in Space and Time," *The Journal of Geology*, vol. 74 (Sept., 1966), pp. 566–583 (572).

Dixon, H. Roberta, and Lawrence W. Lundgren, Jr., "Structure of Eastern Connecticut," Chapt. 16 in *Studies of Appalachian Geology—Northern and Maritime*, E-an Zen et al., editors, Interscience (New York, 1968).

Dott, Robert H., Jr., "Geosyncline concept—alive and well?" *Geotimes*, vol. 18 (Feb., 1973), pp. 16–18.

Drake, C. L., and J. E. Nafe, "Geophysics of the North Atlantic Region," Paper 26 (abstract) in "Continental Drift Emphasizing the History of the South Atlantic Area," *EOS*, vol. 53, *op. cit.*, pp. 175–176.

Fairbridge, Rhodes W., "An Ice Age in the Sahara," *Geotimes*, vol. 15 (July–Aug., 1970), pp. 18–20 (20).

Fairbridge, Rhodes W., "The Sahara Desert Ice Cap," *Natural History*, vol. 80 (June–July, 1971), pp. 66–73 (72–73).

Fairbridge, Rhodes W., "South Pole Reaches the Sahara," *Science*, vol. 168 (May 15, 1970), pp. 878–881 (*passim*). See also letter by Fairbridge, *Science*, vol. 172 (April 2, 1971), p. 86.

Gwinn, Vinton E., "Thin-Skinned Tectonics in the Plateau and Northwestern Valley and Ridge Provinces of the Central Appalachians," *Geological Society of America Bulletin*, vol. 75 (Sept., 1964), pp. 863–900.

Hurley, Patrick M., "The Confirmation of Continental Drift," *op. cit.*

Hurley, P. M., et al., "Test of Continental Drift by Comparison of Radiometric Ages," *op. cit.*

Kay, Marshall, "Geosynclines in Continental Development," *Science*, vol. 99 (June 9, 1944), pp. 461–462.

Leo, G. W., and R. W. White, "Geologic Reconnaissance in Western Liberia," Paper 21 (abstract) in "Continental Drift Emphasizing the History of the South Atlantic Area," *EOS*, vol. 53, *op. cit.*, p. 175.

Le Pichon, Xavier, and Paul J. Fox, "Marginal Offsets, Fracture Zones, and the Early Opening of the North Atlantic," *Journal of Geophysical Research*, vol. 76 (Sept. 10, 1971), pp. 6294–6308. Reprinted in *Plate Tectonics, Selected Papers from the Journal of Geophysical Research, op. cit.*, pp. 472–486. See also article by the same authors, which precedes this one (and follows it in the reprint), on early opening of the South Atlantic.

May, Paul R., "Pattern of Triassic-Jurassic Diabase Dikes around the North Atlantic in the Context of Predrift Position of the Continents," *Geological Society of America Bulletin*, vol. 82 (May, 1971), pp. 1285–1292.

Naylor, Richard S., "Acadian Orogeny: An Abrupt and Brief Event," *Science*, vol. 172 (May 7, 1971), pp. 558–560 (559).

Peyve, A. V., "Oceanic Crust of the Geologic Past," *Geotectonics* (English Edition, 1969), pp. 210–224.

Reid, P. C., and M. E. Tucker, "Probable Late Ordovician Glacial Marine Sediments from Northern Sierra Leone," *Nature Physical Science*, vol. 238 (July 17, 1972), pp. 38–40.

Rodgers, John, "Chronology of Tectonic Movements in the Appalachian Region of Eastern North America," *American Journal of Science*, vol. 265 (May, 1967), pp. 408–427.

Rodgers, John, "The Eastern Edge of the North American Continent during the Cambrian and Early Ordovician," Chapt. 10 in *Studies of Appalachian Geology: Northern and Maritime, op. cit.*

Rodgers, John, "Latest Precambrian (Post-Grenville) Rocks of the Appalachian Region," *American Journal of Science*, vol. 272 (June, 1972), pp. 507–520 (517).

Rodgers, John, "The Taconic Orogeny," *Geological Society of America Bulletin*, vol. 82 (May, 1971), pp. 1141–1177.

Rodgers, John, *The Tectonics of the Appalachians*, Wiley–Interscience (New York, 1970).

Uchupi, E., "Long lost Mytilus," *Oceanus*, vol. 14 (Oct., 1968), pp. 2–7.

Wilson, J. Tuzo, "Did the Atlantic Close and Then Re-Open?" *Nature*, vol. 211 (Aug. 13, 1966), pp. 676–681 (677, 679).

Zietz, Isidore, and E-an Zen, "Northern Appalachians," *Geotimes*, vol. 18 (Feb., 1973), pp. 24–28.

Chapter 17

Anderson, Don L., "The San Andreas Fault," *Scientific American*, vol. 225 (Nov., 1971), pp. 52–68. Reprinted in *Continents Adrift, op. cit.*, pp. 142–157.

Atwater, Tanya, "Implications of Plate Tectonics for the Cenozoic Tectonic Evolution of Western North America," *Geological Society of America Bulletin*, vol. 81 (Dec., 1970), pp. 3513–3536 (3525).

Cobbing, E. J., and W. S. Pitcher, "Plate Tectonics and the Peruvian Andes," *Nature Physical Science*, vol. 240 (Nov. 20, 1972), pp. 51–53.

Dickinson, W. R., and Trevor Hatherton, "Andesitic Volcanism and Seismicity around the Pacific," *Science*, vol. 157 (Aug. 18, 1967), pp. 801–803.

Dietz, Robert S., "Earth's Crust, 'Eppur Si Muove,'" *Medical Opinion and Review*, Dec., 1965, pp. 82 ff. (86).

Dott, R. H., Jr., "Circum-Pacific Late Cenozoic Structural Rejuvenation: Implications for Sea Floor Spreading," *Science*, vol. 166 (Nov. 14, 1969), pp. 874–876.

Ernst, W. G., "Mineral Paragenesis in Franciscan Metamorphic Rocks, Panoche Pass, California," *Geological Society of America Bulletin*, vol. 76 (Aug., 1965), pp. 879–914.

Ernst, W. G., "Tectonic Contact between the Franciscan Mélange and the Great Valley Sequence—Crustal Expression of a Late Mesozoic Benioff Zone," *Journal of Geophysical Research*, vol. 75 (Feb. 10, 1970), pp. 886–901.

Francis, T. J. G., "Effect of Earthquakes on Deep-Sea Sediments," *Nature*, vol. 233 (Sept. 10, 1971), pp. 98–102.

Geodynamics Committee (U.S.), *U.S. Program for the Geodynamics Project—Scope and Objectives*, National Academy of Sciences (Washington, 1973), pp. 61–62, 63, 99.

Gough, D. I., "Dynamic Uplift of Andean Mountains and Island Arcs," *Nature Physical Science*, vol. 242 (March 19, 1973), pp. 39–41.

Hamilton, Warren, "Mesozoic California and the Underflow of Pacific Mantle," *Geological Society of America Bulletin*, vol. 80 (Dec., 1969), pp. 2409–2430.

Hamilton, Warren, "The Uralides and the Motion of the Russian and Siberian Platforms," *ibid.*, vol. 81 (Sept., 1970), pp. 2553–2576.

Hamilton, Warren, "The Volcanic Central Andes—A Modern Model For the Cretaceous Batholiths and Tectonics of Western North America," in *Proceedings of the Andesite Conference (International Upper Mantle Project, Scientific Report 16)*; Oregon Dept. of Geology and Mineral Industries, *Bulletin 65*, A. R. McBirney, ed. (1969), pp. 175–184.

Hamilton, Warren, and W. Bradley Myers, "Cenozoic Tectonics of the Western United States," *Reviews of Geophysics*, vol. 4 (Nov., 1966), pp. 509–549.

Hsü, K. Jinghwa, "Franciscan Mélanges as a Model for Eugeosynclinal Sedimentation and Underthrusting Tectonics," *Journal of Geophysical Research*, vol. 76 (Feb. 10, 1971), pp. 1162–1170. Reprinted in *Plate Tectonics, Selected Papers from the Journal of Geophysical Research, op. cit.*, pp. 418–426.

Ingle, James C., Jr., et al. (Shipboard Scientific Party, Leg 31, D.S.D.P.), "Western Pacific floor," *Geotimes*, vol. 18 (Oct., 1973), pp. 22–25.

James, David E., "The Evolution of the Andes," *Scientific American*, vol. 229 (Aug., 1973), pp. 61–69.

Kropotkin, P. N., "Eurasia as a composite continent," in "Global Tectonics and Sea-Floor Spreading," special issue of *Tectonophysics*, vol. 12 (1971), *op. cit.*, pp. 261–266.

Kropotkin, P. N., Letter to J. D. Bernal of Birkbeck College, London, in *Philosophical Transactions of the Royal Society of London, Series A*, vol. 258 (1965), pp. 316–318 (316–317).

Kropotkin, P. N., "The Mechanism of Crust Movements," *Geotectonics* (English Edition), No. 5 (1967), pp. 276–285.

Larson, Roger L., and Clement G. Chase, "Late Mesozoic Evolution of the Western Pacific Ocean," *Geological Society of America Bulletin*, vol. 83 (Dec., 1972), pp. 3627–3644.

Larson, Roger L., and Walter C. Pitman III, "World-Wide Correlation of Mesozoic Magnetic Anomalies, and Its Implications," *ibid.*, pp. 3645–3662.

Lipman, Peter W., H. J. Prostka and R. L. Christiansen, "Cenozoic volcanism and plate-tectonic evolution of the Western United States," *Philosophical Transactions of the Royal Society of London, Series A*, vol. 271 (Jan. 27, 1972), pp. 217–284.

Lipman, Peter W., Harold J. Prostka and Robert L. Christiansen, "Evolving Subduction Zones in the Western United States, as Interpreted from Igneous Rocks," *Science*, vol. 174 (Nov. 19, 1971), pp. 821–825 (825).

Locke, Augustus, Paul Billingsley and Evans B. Mayo, "Sierra Nevada Tectonic Pattern," *Geological Society of America Bulletin*, vol. 51 (April 1, 1940), pp. 513–540.

Longwell, Chester R., "Possible Explanation of Diverse Structural Patterns in Southern Nevada," *American Journal of Science*, vol. 258-A (1960), pp. 192–203.

Luyendyk, Bruce P., "Dips of Downgoing Lithospheric Plates Beneath Island Arcs," *Geological Society of America Bulletin*, vol. 81 (Nov., 1970), pp. 3411–3416.

McKenzie, D. P., and W. J. Morgan, "Evolution of Triple Junctions," *Nature*, vol. 224 (Oct. 11, 1969), pp. 125–133.

McKenzie, D. P., and J. G. Sclater, "The Evolution of the Indian Ocean," *Scientific American*, vol. 228 (May, 1973), pp. 62–72.

Menard, H. W., and Tanya Atwater, "Changes in Direction of Sea Floor Spreading," *Nature*, vol. 219 (Aug. 3, 1968), pp. 463–467.

Moores, Eldridge, "Ultramafics and Orogeny, with Models of the US Cordillera and the Tethys," *op. cit.*

Oxburgh, E. R., "Flake Tectonics and Continental Collision," *Nature*, vol. 239 (Sept. 22, 1972), pp. 202–204.

Peyve, Aleksandr V., "Faults and Tectonic Movements," *Geotectonics* (English edition), No. 5 (1967), pp. 268–275.

Powell, C. McA., and P. J. Conaghan, "Plate Tectonics and the Himalayas," *Earth and Planetary Science Letters*, vol. 20 (Sept., 1973), pp. 1–12.

Ross, David A., "Sediments of the Northern Middle America Trench," *Geological Society of America Bulletin*, vol. 82 (Feb., 1971), pp. 303–322.

Rutland, R. W. R., "Andean Orogeny and Ocean Floor Spreading," *Nature*, vol. 233 (Sept. 24, 1971), pp. 252–255.

Sales, John K., "Crustal Mechanics of Cordilleran Foreland Deformation: A Regional and Scale-Model Approach," *American Association of Petroleum Geologists Bulletin*, vol. 52 (Oct., 1968), pp. 2016–2044.

Scholl, David W., Mark N. Christensen, Roland von Huene and Michael S. Marlow, "Peru–Chile Trench Sediments and Sea-Floor Spreading," *Geological Society of America Bulletin*, vol. 81 (May, 1970), pp. 1339–1360.

Scholl, David W., Roland von Huene and James B. Ridlon, "Spreading of the Ocean Floor: Undeformed Sediments in the Peru–Chile Trench," *Science*, vol. 159 (Feb. 23, 1968), pp. 869–871.

Scholz, Christopher H., Muawia Barazangi and Marc L. Sbar, "Late Cenozoic Evolution of the Great Basin, Western United States, as an Ensialic Interarc Basin," *Geological Society of America Bulletin*, vol. 82 (Nov., 1971), pp. 2979–2990.

Solomon, Sean C., "Seismic-Wave Attenuation and Partial Melting in the Upper Mantle of North America," *Journal of Geophysical Research*, vol. 77 (March 10, 1972), pp. 1483–1502.

Turcotte, D. L., and G. Schubert, "Frictional Heating of the Descending Lithosphere," *ibid.*, vol. 78 (Sept. 10, 1973), pp. 5876–5886.

Vogt, P. R., C. N. Anderson and D. R. Bracey, "Mesozoic Magnetic Anomalies, Sea-Floor Spreading, and Geomagnetic Reversals in the Southwestern North Atlantic," *ibid.*, vol. 76 (July 10, 1971), pp. 4796–4823.

Wilson, J. Tuzo, and Kevin Burke, "Two Types of Mountain Building," *Nature*, vol. 239 (Oct. 20, 1972), pp. 448–449.

Wise, Donald U., "An Outrageous Hypothesis for The Tectonic Pattern of the North American Cordillera," *Geological Society of America Bulletin*, vol. 74 (March, 1963), pp. 357–362.

Chapter 18

Abelson, Philip H., "Observing and Predicting Earthquakes," *Science*, vol. 180 (May 25, 1973), p. 819.

Aggarwal, Yash P., Lynn R. Sykes, John Armbruster and Marc L. Sbar, "Premonitory Changes in Seismic Velocities and Prediction of Earthquakes," *Nature*, vol. 241 (Jan. 12, 1973), pp. 101–104.

Anderson, Don L., and James H. Whitcomb, "Time Dependent Seismology" *Journal of Geophysical Research*, vol. 80 (1975), pp. 1497–1503.

Brace, W. F., and R. J. Martin III, "A Test of the Law of Effective Stress for Crystalline Rocks of Low Porosity," *International Journal of Rock Mechanics and Mining Sciences*, vol. 5 (1968), pp. 415–426.

Brace, W. F., B. W. Paulding, Jr. and C. Scholz, "Dilitancy in the Fracture of Crystalline Rocks," *Journal of Geophysical Research*, vol. 71 (Aug. 15, 1966), pp. 3939–3953.

Challinor, R. A., "Variations in the Rate of Rotation of the Earth," *Science*, vol. 172 (June 4, 1971), pp. 1022–1024.

Environmental Research Laboratories, *A Study of Earthquake Losses in the San Francisco Bay Area—Data and Analysis. A Report Prepared for the Office of Emergency Preparedness*, U.S. Department of Commerce, National Oceanic and Atmospheric Administration (1972).

Evans, David M., "The Denver Area Earthquakes and the Rocky Mountain Arsenal Disposal Well," *The Mountain Geologist*, vol. 3 (Jan., 1966), pp. 23–36.

Hammond, Allen L., "Earthquake Predictions: Breakthrough in Theoretical Insight?" *Science*, vol. 180 (May 25, 1973), pp. 851–853.

Healy, J. H., W. W. Rubey, D. T. Griggs and C. B. Raleigh, "The Denver Earthquakes," *ibid.*, vol. 161 (Sept. 27, 1968), pp. 1301–1310.

Heck, N.H., and R. A. Eppley, *Earthquake History of the United States*, Part 1, Revised Edition (through 1956), U.S. Department of Commerce, Coast and Geodetic Survey, No. 41-1 (1958), pp. 33–34.

Howell, B. F., Jr., "Earthquake Hazard in the Eastern United States," *Earth and Mineral Sciences* (Pennsylvania State University), vol. 42 (March, 1973), pp. 41–45.

Hubbert, M. King, quoted in "Quake Prevention—Messing with the Mousetrap," *Science News*, vol. 95 (Feb. 8, 1969), pp. 138–139 (139).

Kelleher, John A., "Rupture Zones of Large South American Earthquakes and Some Predictions," *Journal of Geophysical Research*, vol. 77 (April 10, 1972), pp. 2087–2103.

Kisslinger, Carl, "Earthquake Prediction Research in Japan and the USSR," paper presented at the symposium on earthquake prediction, National Academy of Sciences, April 24, 1972 (typescript).

Kisslinger, Carl, "Seismology," in *Geotimes*, vol. 18 (Jan., 1973), p. 30.

Knopoff, Leon, "Correlation of Earthquakes with Lunar Orbital Motions," in *The Moon*, vol. 2, Reidel (Dordrecht, Holland, 1970), pp. 140–143.

Mansinha, L., and D. E. Smylie, "Earthquakes and the Earth's Wobble," *op. cit.*

Nature, "How Useful is Earthquake Prediction?" (unsigned), vol. 241 (Jan. 12, 1973), p. 85.

Nur, Amos, and Gene Simmons, "The Effect of Saturation on Velocity in Low Porosity Rocks," *Earth and Planetary Science Letters*, vol. 7 (1969), pp. 183–193.

Scholz, Christopher H., Lynn R. Sykes and Yash P. Aggarwal, "Earthquake Prediction: A Physical Basis," *Science*, vol. 181 (Aug. 31, 1973), pp. 803–810.

Sykes, L. R., "Aftershock Zones of Great Earthquakes, Seismicity Gaps, and Earthquake Prediction for Alaska and the Aleutians," *Journal of Geophysical Research*, vol. 76 (Nov. 10, 1971), pp. 8021–8041.

Tamrazyan, Gurgen P., "Principal Regularities in the Distribution of Major Earthquakes Relative to Solar and Lunar Tides and Other Cosmic Forces," *Icarus*, vol. 9 (1968), pp. 574–592.

Whitcomb, James H., Jan D. Garmany and Don L. Anderson, "Earthquake Prediction: Variation of Seismic Velocities before the San Francisco [sic] Earthquake," *Science*, vol. 180 (May 11, 1973), pp. 632–635. ("San Francisco" is apparently a misprint; the report is on the San Fernando quake.)

Whitten, Charles A., "Preliminary Investigation of the Correlation of Polar Motion and Major Earthquakes," Report 62 in *Reports on Geodetic Measurements of Crustal Movement, 1906–71*, National Ocean Survey (NOAA), Department of Commerce, U.S. Govt. Printing Office stock no. 0317-00167 (Washington, July, 1973).

Wilson, J. T., "Mao's Almanac—3,000 Years of Killer Earthquakes," *Saturday Review*, vol. 55 (Feb. 19, 1972), pp. 60–64.

Chapter 19

Austin, C. F., and J. L. Moore, "Structural Interpretation of the Coso Geothermal Field," Naval Weapons Center China Lake, CA, Sept. 1987.

Banwell, C. J., "Life Expectation of Geothermal Fields," Paper presented at Seminar on Development and Use of Geothermal Energy, United Nations, Jan. 8, 1973 (typescript).

Banwell, John, and Tsvi Meidav, "Geothermal Energy for the Future," Paper presented at the Annual Meeting of the American Association for the Advancement of Science, Philadelphia, Dec., 1971 (typescript).

Barnea, Joseph, "Geothermal Power," *Scientific American*, vol. 226 (Jan., 1972), pp. 70–77 (77).

Barnea, Joseph, "Opening Statement," Seminar on Development and Use of Geothermal Energy, United Nations, Jan. 8, 1973 (typescript).

Condy, Charles T., "California Energy Company Update and Outlook," Remarks to Shareholders San Francisco, April 7, 1989.

DiPippo, Ronald, "International Developments in Geothermal Power Production." Geothermal Resources Council Bulletin, May, 1988.

Doctor, R. D., et al., *Calfornia's Electricity Quandary: III. Slowing the Growth Rate*, Report R–1116–NSF/CSA. The Rand Corporation (Sept., 1972), p. 123.

Einarsson, S. S., "Utilization of Low Enthalpy Water for Space Heating, Industrial, Agricultural and Other Uses," Rapporteur's Report, *Proceedings of the United Nations Symposium on the Development and Utilization of Geothermal Resources, Pisa, 1970*, vol. 1, Special Issue 2, *Geothermics* (in press).

Elders, Wilfred A., Robert W. Rex, Tsvi Meidav, Paul T. Robinson and Shawn Biehler, "Crustal Spreading in Southern California," *Science*, vol. 178 (Oct. 6, 1972), pp. 15–24 (22, 23).

Fenner, David, and Joseph Klarmann, "Power from the Earth," *Environment*, vol. 13 (Dec., 1971), pp. 19–34 (*passim*).

Gilmore, C. P., "Hot New Prospects for Power from the Earth," *Popular Science*, vol. 201 (Aug., 1972), pp. 56–60.

Hammond, Allen L., "Geothermal Energy: An Emerging Major Resource," *Science*, vol. 177 (Sept. 15, 1972), pp. 978–980.

Harlow, Francis H., and William E. Pracht, "A Theoretical Study of Geothermal Energy Extraction," *Journal of Geophysical Research*, vol. 77 (Dec. 10, 1972), pp. 7038–7048.

Hickel, Walter J., "Geothermal Energy—A National Proposal for Geothermal Resources Research," *Final Report of the Geothermal Resources Research Conference, Battelle Seattle Research Center, Seattle, Washington, Sept. 18–20, 1972*. The University of Alaska (undated), p. 5.

Hodgson, Susan F., Editor, "The Geothermal Hot Line," State of California Resources Agency, December, 1988.

Jones, P. H., "Geothermal Resources of the Northern Gulf of Mexico Basin," *Proceedings of the United Nations Symposium...Pisa, 1970, op. cit.*, vol. 2, part 1, pp. 14–26 (23).

Lockchine, B. A., and I. M. Dvorov, "Applications Experimentales et Industrielles de l'Energie Géothermique en URSS," *ibid.*, vol. 2, part 2, pp. 1079–1085.

Makarenko, F. A., et al., "Geothermal Resources of USSR and Prospects for Their Practical Use," *ibid.*, vol. 2, part 2, pp. 1086–1091.

Otte, Carel, quoted in "Geothermal Power Goes Commercial," *Seventy-Six* (Union Oil. Co.), Nov.–Dec., 1971.

Peet, Creighton, "Ultima Thule—A report on Iceland's National Reforestation Program and on Their Use of Geothermal Energy to Provide Nonpolluting Heat and Power," *American Forests*, vol. 77 (June, 1971), pp. 12–54 (*passim*).

Rex, Robert W., "Geothermal Energy—The Neglected Energy Option," *Science and Public Affairs* (*Bulletin of the Atomic Scientists*), vol. 27 (Oct., 1971), pp. 52–56 (52, 54).

Rex, Robert W., "Geothermal Energy—Its Potential Role in the National Energy Picture," *Contribution 72-10*, The Institute of Geophysics and Planetary Physics, University of California, Riverside, Calif. (April, 1972).

Rex, Robert W., "Investigation of Geothermal Resources in the Imperial Valley and Their Potential Value for Desalination of Water and Electricity Production," Chapt. 10 in *California Water: A Study of Resource Management*, D. Seckler, Ed., University of California Press (Berkeley, June, 1971).

Tamrazyan, G. P., "Continental Drift and Thermal Fields," *Proceedings of the United Nations Symposium...Pisa, 1970, op. cit.*, vol. 2, part 2, pp. 1212–1225.

Tikhonov, A.N., and I. M. Dvorov, "Development of Research and Utilization of Geothermal Resources in the USSR," *ibid.*, pp. 1072–1078.

United Nations Department of Economic and Social Affairs, *New Sources of Energy and Economic Development* (1957).

United States Congress, "An Act to authorize the Secretary of the Interior to make disposition of geothermal steam and associated geothermal resources, and for other purposes," *Public Law 91-581*, 91st Congress, S. 368 (Dec. 24, 1970) Sec. 2 (e).

Chapter 20

Abelson, Philip H., et al. (conference participants), "Plate tectonics and evolution of continents," (Report on conference at Airlie Conference Center, March 20–24, 1972), *Geotimes*, vol. 17 (Aug., 1972), pp. 18–19 (18, 19).

Boyd, F. R., "Progress and Problems in Understanding Kimberlites," Presidential Address to the Geochemical Society (typescript), 1973.

Brimhall, George, "The Genesis of Ores," *Scientific American,* vol. 264 (May, 1991), pp. 84–91.

Brobst, Donald A., and Walden P. Pratt, "Introduction," *United States Mineral Resources*, Geological Survey Professional Paper 820, D. A. Brobst and W. P. Pratt, editors (Washington, 1973), pp. 1–8 (7).

Brookfield, M., "Location of Ancient Mid-oceanic Rises," *Nature Physical Science*, vol. 229 (Feb. 15, 1971), pp. 204–205.

Burk, C. A., "Global Tectonics and World Resources," *The American Association of Petroleum Geologists Bulletin*, vol. 56 (Feb., 1972), pp. 196–202.

Business Week, "Vacuuming ores from the ocean floor," no. 2127 (June 6, 1970), pp. 60–69 (*passim*).

Chase, Clement G., and Todd H. Gilmer, "Precambrian Plate Tectonics: The Midcontinent Gravity High," *Earth and Planetary Science Letters*, vol. 21 (Dec., 1973), pp. 70–78.

Clifford, Tom N., "Location of Mineral Deposits," Chapt. 22 in *Understanding the Earth, op. cit.*

Cloud, Preston, "Meetings—The Precambrian," (The Third Penrose Conference of the Geological Society of America), *Science*, vol. 173 (Aug. 27, 1971), pp. 851–854 (*passim*).

Commission of the European Communities, "A Community Policy to Meet the Changes in the World Energy Market," *Industry Research and Technology*, No. 187 (April 30, 1973), Annex 1, pp. 1–4.

Corliss, J. B., "The Origin of Metal-Bearing Submarine Hydrothermal Solutions," *Journal of Geophysical Research*, vol. 76 (Nov. 20, 1971), pp. 8128–8138.

Engel, A. E. J., and Donald L. Kelm, "Pre-Permian Global Tectonics: A Tectonic Test," *Geological Society of America Bulletin*, vol. 83 (Aug., 1972), pp. 2325–2340.

Faul, Henry, "Century-old diamond hoax reexamined," *Geotimes*, vol. 17 (Oct., 1972), pp. 23–25.

Geodynamics Committee (U.S.), *U.S. Program for the Geodynamics Project—Scope and Objectives, op. cit.*, (p. 100).

Goeller, H. E., "The Ultimate Mineral Resource Situation—An Optimistic View," *Proceedings, National Academy of Sciences USA*, vol. 69 (Oct., 1972), pp. 2991–2992.

Gold, David P., "Natural and Synthetic Diamonds and the North American Outlook," *Earth and Mineral Sciences*, Pennsylvania State University, vol. 37 (Feb., 1968), pp. 37–43.

Hallam, A., "Re-evaluation of the Paleogeographic Argument for an Expanding Earth," *Nature*, vol. 232 (July 16, 1971), pp. 180–182 (181).

Hammond, Allen L., "Plate Tectonics: The Geophysics of the Earth's Surface," *Science*, vol. 173 (July 2, 1971), pp. 40–41, and "Plate Tectonics (II): Mountain Building and Continental Geology," *ibid.* (July 9, 1971), pp. 133–134.

Von Herzen, Richard P., Hartley Hoskins and Tjeerd H. Van Andel, "Geophysical Studies in the Angola Diapir Field," *Geological Society of America Bulletin*, vol. 83 (July, 1972), pp. 1901–1910.

Horsfield, Brenda, and John Dewey, "How continents are made and moved," *Science Journal*, vol. 7 (Jan., 1971), pp. 43–48.

Hurley, Patrick M., "The Confirmation of Continental Drift," *op. cit.*

Kazmin, V., "Some Aspects of Precambrian Development in East Africa," *Nature*, vol. 237 (May 19, 1972), p. 160.

King, Elizabeth R., and Isidore Zietz, "Aeromagnetic Study of the Midcontinent Gravity High of the Central United States," *Geological Society of America Bulletin*, vol. 82 (Aug., 1971), pp. 2187–2207.

Lister, C. R. B., "On the Thermal Balance of a Mid-Ocean Ridge," *Geophysical Journal of the Royal Astronomical Society*, vol. 26 (1972), pp. 515–535.

McCurry, Patricia, "Plate Tectonics and the Pan-African Orogeny in Nigeria," *Nature Physical Science*, vol. 229 (Feb. 1, 1971), pp. 154–155.

McGetchin, Thomas R. and Leon T. Silver, "Compositional Relations in Minerals from Kimberlite and Related Rocks in the Moses Rock Dike, San Juan County, Utah," *The American Mineralogist*, vol. 55 (1970), pp. 1738–1771.

McGetchin, Thomas R., and Leon T. Silver, "A Crustal-Upper-Mantle Model for the Colorado Plateau Based on Observations of Crystalline Rock Fragments in the Moses Rock Dike," *Journal of Geophysical Research*, vol. 77 (Dec. 10, 1972), pp. 7022–7037 (7022).

McGetchin, T. R., L. T. Silver and A. A. Chodos, "Titanoclinohumite: A Possible Mineralogical Site for Water in the Upper Mantle," *Journal of Geophysical Research*, vol. 75 (Jan. 10, 1970), pp. 255–259.

McGetchin, T. R., and G. Wayne Ullrich, "Xenoliths in Mars and Diatremes with Inferences for the Moon, Mars and Venus," *ibid.*, vol. 78 (April 10, 1973), pp. 1832–1853 (1842).

McKelvey, Vincent E., "Mineral Resource Estimates and Public Policy," *United States Mineral Resources*, D. A. Brobst and W.P. Pratt, editors, *op. cit.*, pp. 9–19 (18).

Melton, Charles E., C. A. Salotti and A. A. Giardini, "The observation of nitrogen, water, carbon dioxide, methane and argon as impurities in natural diamonds," *The American Mineralogist*, vol. 57 (Sept.–Oct., 1972), pp. 1518–1523.

Moorbath, S., R. K. O'Nions, R. J. Pankhurst, N. H. Gale and V. R. McGregor, "Further Rubidium–Strontium Age Determinations on the Very Early Precambrian Rocks of the Godthaab District, West Greenland," *Nature Physical Science*, vol. 240 (Nov. 27, 1972), pp. 78–82.

Moores, E. M., and F. J. Vine, "The Troodos Massif, Cyprus, and other ophiolites as oceanic crust: evaluation and implications," *Philosophical Transactions of the Royal Society of London, Series A*, vol. 268 (1971), pp. 443–466.

Noble, James A., "Metal Provinces of the Western United States," *Geological Society of America Bulletin*, vol. 81 (June, 1970), pp. 1607–1624.

Pinet, Paul R., "Diapir-Like Features Offshore Honduras: Implications Regarding Tectonic Evolution of Cayman Trough and Central America, *ibid.*, vol. 83 (July, 1972), pp. 1911–1922.

Piper, David Z., "Origin of Metalliferous Sediments from the East Pacific Rise," *Earth and Planetary Science Letters*, vol. 19 (May, 1973), pp. 75–82.

Purrett, Louise A., "Before Pangaea—What?" *Science News*, vol. 102 (Sept. 30, 1972), pp. 220–222.

Rona, Peter A., "Comparison of Continental Margins of Eastern North America at Cape Hatteras and Northwestern Africa at Cap Blanc," *The American Association of Petroleum Geologists Bulletin*, vol. 54 (Jan., 1970), pp. 129–153.

Rona, Peter A., "Plate Tectonics and Mineral Resources," *Scientific American*, vol. 229 (July, 1973), pp. 86–95 (86, 94).

Rona, Peter A., et al., "Axial Processes of the Mid-Atlantic Ridge," papers presented at the Twelfth Annual Meeting of the American Geophysical Union, April 9, 1974. Abstracted in *EOS* (*Transactions of the American Geophysical Union*), vol. 55 (April, 1974), pp. 292–294.

Sawkins, Frederick J., "Sulfide Ore Deposits in Relation to Plate Tectonics," *The Journal of Geology*, vol. 80 (July, 1972), pp. 377–397.

Sillitoe, Richard H., "Relation of Metal Provinces in Western America to Subduction of Oceanic Lithosphere," *Geological Society of America Bulletin*, vol. 83 (March, 1972), pp. 813–817.

Sillitoe, Richard H., "Tectonic segmentation of the Andes: implication for magmatism and metallogeny," *Nature*, vol. 250 (August 16, 1974), pp. 542–545.

Sutton, John, "Two billion years of global change," *New Scientist and Science Journal*, vol. 49 (March 4, 1971), pp. 472–476.

Sykes, Lynn R., "Earthquake Swarms and Sea-Floor Spreading," *Journal of Geophysical Research*, vol. 75 (Nov. 10, 1970), pp. 6598–6611.

Tarling, D. H., "Continental Drift and Reserves of Oil and Natural Gas." *Nature*, vol. 243 (June 1, 1973), pp. 277–279 (277).

"Vacuuming the Atlantic Floor" (unsigned), *Science News*, vol. 98 (Aug. 15, 1970), pp. 134–135.

Wilson, J. T., "A revolution in earth science," *op. cit.*, p. 16.

Chapter 21

Bellaiche, G., J. L. Cheminee, J. Francheteau, R. Hekinian, X. Le Pichon, H. D. Needham and R. D. Ballard, "Inner floor of the Rift Valley: first submersible study," *Nature*, vol. 250 (August 16, 1974), pp. 558–560.

Beloussov, V. V., Luis R. A. Capurro, J. M. Harrison, Leon Knopoff, A. E. Ringwood, S. K. Runcorn and Kiyoo Wadati (the U.M.C. Bureau), "Significance and Achievements of the Upper Mantle Project," *op. cit.*, pp. 1–5 (1, 2, 3).

Bostrom, R. C., "Westward Displacement of the Lithosphere," *Nature*, vol. 234 (Dec. 31, 1971), pp. 536–538.

Bullard, Sir Edward, "Concluding Remarks," *Philosophical Transactions of the Royal Society of London, Series A*, vol. 258, *op. cit.*, p. 322.

Davies, D., and R. M. Sheppard, "Lateral Heterogeneity in the Earth's Mantle," *Nature*, vol. 239 (Oct. 6, 1972), pp. 318–323.

Dietz, Robert S., and John C. Holden, "The Breakup of Pangaea," *op. cit.*

Drake, Charles L., "Future Considerations Concerning Geodynamics," *American Association of Petroleum Geologists Bulletin*, vol. 56 (Feb., 1972), pp. 260–268.

Drake, Charles L., *The Geological Revolution*, Condon Lectures, Oregon State System of Higher Education (Eugene, Ore., 1970). p. 41.

"Earth Dynamics Program Begun," *SAO News*, vol. 12 (Autumn, 1972), Smithsonian Astrophysical Observatory (Cambridge, Mass.).

Eddington, Sir Arthur S., "The Borderland of Astronomy and Geology," A lecture delivered before the Geological Society of London on Nov. 21, 1922, published in *Nature*, vol. 111 (Jan. 6, 1923), pp. 18–21.

Galileo Galilei, Recantation, quoted in *Galileo Galilei: A biography and inquiry into his philosophy of science*, by Ludovico Geymonat, McGraw-Hill (New York, 1965), pp. 153–154 (153–154).

Geodynamics Committee (U.S.), *U.S. Program for the Geodynamics Project—Scope and Objectives, op. cit.*, pp. 11, 12, 20, 25–26, 30, 31, 109, 144, 149, 174, 179.

"Geodynamics Project: Development of a U.S. Program—Preliminary Statement of the U.S. Geodynamics Committee," in *EOS (Transactions of the American Geophysical Union)*, vol. 52 (May, 1971), pp. 396–405 (405).

Geodynamics Project, Statement of objectives of the Inter-Union Commission on Geodynamics, published in *Geodynamics Project: Initial Recommendations, 1970.*

Gilluly, James, "Steady Plate Motion and Episodic Orogeny and Magmatism," *Geological Society of America Bulletin*, vol. 84 (Feb., 1973), pp. 499–513.

Hess, H. H., "Seismic Anisotropy of the Uppermost Mantle under Oceans," *Nature*, vol. 203 (Aug. 8, 1964), pp. 629–631.

Hoffman, Paul F., "Speculations on Laurentia's first gigayear (2.0 to 1.0 Ga)," *Geology*, vol. 17 (Feb., 1989), pp. 135–138.

Huxley, T. H., in *Lay Sermons, Addresses, and Reviews* (1871), D. Appleton & Co. (New York, 1910), p. 18.

Joly, John, "Shifting of the Outer Crust under Tide-generating Forces," in his book *The Surface-History of the Earth*, 2nd Edition (Oxford, 1930), pp. 106–108.

Julian, Bruce R., and Mrinal K. Sengupta, "Seismic Travel Time Evidence for Lateral Inhomogeneity in the Deep Mantle," *Nature*, vol. 242 (April 13, 1973), pp. 443–447.

Kane, Martin F., "Rotational Inertia of Continents: A Proposed Link between Polar Wandering and Plate Tectonics," *Science*, vol. 175 (March 24, 1972), pp. 1355–1357.

Karig, D., et al., "Origin of the West Philippine Basin," ("...results of the Deep Sea Drilling Project leg 31..."), *Nature*, vol. 246 (Dec. 21/28, 1973), pp. 458–461.

Knopoff, Leon and A. Leeds, "Lithospheric Momenta and the Deceleration of the Earth," *ibid.*, vol. 237 (May 12, 1972), pp. 93–95.

Knopoff, Leon, K. A. Poehls and R. C. Smith, "Drift of Continental Rafts with Asymmetric Heating," *Science*, vol. 176 (June 2, 1972), pp. 1023–1024.

Leg 127 and Leg 128 shipboard scientific parties, "Evolution of the Japan Sea," *Nature*, vol. 346 (July 5, 1990), pp. 18–20.

Le Pichon, Xavier, Jean Francheteau and Jean Bonnin, *Plate Tectonics, op. cit.*

Menard, H. W., "Epeirogeny and Plate Tectonics," *EOS* (*Transactions, American Geophysical Union*), vol. 54 (Dec., 1973), pp. 1244–1255.

Meyerhoff, A. A., and James L. Harding, "Some problems in current concepts of continental drift," in "Global Tectonics and Sea-Floor Spreading," special issue of *Tectonophysics*, vol. 12 (1971) *op. cit.*, pp. 235–260.

Molnar, Peter, "Continental tectonics in the aftermath of plate tectonics," *Nature*, vol. 335 (Sept. 8, 1988), pp. 131–137.

Molnar, Peter, "The Geologic Evolution of the Tibetan Plateau," *American Scientist*, vol. 77 (July–Aug., 1989), pp. 350–359.

Nance, R. Damian, Thomas R. Worsley and Judith B. Moody, "The Supercontinent Cycle," *Scientific American*, vol. 259 (July, 1988), pp. 72–79.

Needham, R. E., and D. Davies, "Lateral Heterogeneity in the Deep Mantle from Seismic Body Wave Amplitudes," *Nature*, vol. 244 (July 20, 1973), pp. 152–153.

Nelson, T. H., and P. G. Temple, "Mainstream Mantle Convection: A Geologic Analysis of Plate Motion," *American Association of Petroleum Geologists, Bulletin*, vol. 56 (Feb., 1972), pp. 226–246.

Oxburgh, E. R., and D. L. Turcotte, "Origin of Paired Metamorphic Belts and Crustal Dilation in Island Arc Regions," *Journal of Geophysical Research*, vol. 76 (Feb. 10, 1971), pp. 1315–1327. Reprinted in *Plate Tectonics, Selected Papers from the Journal of Geophysical Research, op. cit.*, pp. 427–439.

Oxburgh, E. R., and D. L. Turcotte, "Thermal Structure of Island Arcs," *Geological Society of America Bulletin*, vol. 81 (June, 1970), pp. 1665–1688.

Turcotte, D. L., and G. Schubert, "Frictional Heating of the Descending Lithosphere," *Journal of Geophysical Research*, vol. 78 (Sept. 10, 1973), pp. 5876–5886.

Vogt, P. R., "Asthenosphere motion recorded by the ocean floor south of Iceland," *op. cit.*

Vogt, P. R., "Evidence for Global Synchronism in Mantle Plume Convection, and Possible Significance for Geology," *op. cit.*, p. 342.

Wilson, J. T., "Submarine Fracture Zones, Aseismic Ridges and the International Council of Scientific Unions Line: Proposed Western Margin of the East Pacific Ridge," *op. cit.*

Credits and Permissions

Among passages in this book excerpted from copyrighted or privately held material are those appearing on the following pages:

Pages 33 and 39. *Earth in Upheaval* by Immanuel Velikovsky. Copyright © 1955 by Immanuel Velikovsky. Reprinted by permission of Doubleday & Co., Inc.

Page 35. *My Life in Publishing* by Harold S. Latham. Copyright © 1965 by Harold S. Latham. Reprinted by permission of E. P. Dutton & Co., Inc.

Page 42. *Worlds in Collision* by Immanuel Velikovsky. Copyright © 1950 by Immanuel Velikovsky. Reprinted by permission of Doubleday & Co., Inc.

Pages 46 and 47. "How to Wreck the Environment" by Gordon MacDonald from *Unless Peace Comes* edited by Nigel Calder. Copyright © 1968 by Nigel Calder. Reprinted by permission of The Viking Press, Inc.

Page 89. Letter from Albert Einstein to William M. Barrett, Jan. 6, 1931. Reprinted by permission of the Estate of Albert Einstein, which retains all rights thereto.

Permission has also been granted for the use of the following illustrations, some derived fully or in part from copyrighted publications.

CHAPTER 1

Figure 1.1 Antonio Snider-Pellegrini: *La Création et ses Mystères Devoilés*, Paris (1858), pp. 314-315.
Figure 1.2 Courtesy of the Alfred Wegener Institute for Polar and Marine Research.
Figure 1.3 *The Origin of Continents and Oceans*, tr. 4th revised German Edition, published 1929, by John Biram, Dover, NY. (1966).
Figure 1.4 Courtesy of the Swiss National Tourist Office.

CHAPTER 3

Figure 3.1 Paul S. Martin in *Science,* vol. 179 (1973), p. 972.
Figure 3.2 *The New York Times*, 19 Dec 1959. Copyright © 1960 by The New York Times Company. Reprinted by permission.

Chapter 4

Figure 4.1 J. R. Rothé, "La zone séismique médian Indo-Atlantique," in *Proceedings of the Royal Society of London*, Series A, vol. 222 (1954), p. 388.

Chapter 5

Figure 5.1 Harry Hammond Hess, photograph, Courtesy of the Firestone Library, Princeton University.
Figure 5.2 "Trenches of the Pacific" by Robert L. Fisher and Roger Revelle. Copyright © 1955 by *Scientific American*, Inc. All rights reserved.
Figure 5.3 Upper diagram by H. H. Hess in *American Journal of Science*, vol. 244 (1946), p. 783.
Figure 5.4 As envisioned by F. A. Vening Meinesz.
Figure 5.5 *The New York Times*, March 27, 1960. Copyright © 1960 by The New York Times Company. Reprinted by permission.
Figure 5.6 As envisioned by Harry H. Hess.
Figure 5.8 As envisioned by Robert S. Dietz and John C. Holden.
Figure 5.9 As envisioned by J. T. Wilson.
Figure 5.10 US Geological Survey, Hawaiian Volcano Observatory.

Chapter 6

Figure 6.1 "Continental Drift" by S. K. Runcorn et al., Academic Press Inc., New York, 1962.
Figure 6.2 *Science*, vol. 144, 1964, pp. 1537 *and Nature*, vol. 198, 1963, pp. 1049 +
Figure 6.3 John A. O'Keefe in *The Journal of Non-Crystalline Solids,* 84 (1986), p. 310.
Figure 6.4 Arthur D. Raff and Donald G. Mason in *Geological Society of America Bulletin,* vol 72 (1961), p. 1268.
Figure 6.5 F. J. Vine in *Science*, vol. 154 (1966), p. 1407.
Figure 6.6 F. J. Vine in *Science*, vol. 154 (1966), p. 1409.
Figure 6.7 Map prepared by Susan K. Goter of the US Geological Survey, National Earthquake Information Center.

Chapter 7

Figure 7.1 Sir Edward Bullard, E. J. Everett and A. Gilbert Smith in *Philosophical Transactions of the Royal Society of London*, Series A, vol. 258 (1965), pp. 41-51.
Figure 7.2 Adapted from "Mechanism of Earthquakes and Nature of Faulting on the Mid-Oceanic Ridges" by Lynn R. Sykes, *Journal of Geophysical Research,* Vol. 72, No. 8 (1967), p. 2137.
Figure 7.3 J. T. Wilson in *Nature*, vol. 207 (1965), p. 345.
Figure 7.4 Douglas W. Oldenburg and James N. Brune, "Ridge Transform Fault Spreading Pattern in Freezing Wax," *Science,* Vol. 178, 1972, pp. 301–304. *The New York Times,* Oct. 25, 1972. Copyright © 1973 by The New York Times Company. Reprinted by permission.
Figure 7.5 Courtesy of the American Institute of Physics, Niels Bohr Library.

Chapter 9

Figure 9.1 Courtesy of the Scripps Institute of Oceanography, University of California, San Diego.
Figure 9.2 Courtesy of the Scripps Institute of Oceanography, University of California, San Diego.
Figure 9.3 Courtesy of the Scripps Institute of Oceanography, University of California, San Diego.
Figure 9.4 The Woods Hole Oceanographic Institution.
Figure 9.5 The Deep Sea Drilling Project.

CHAPTER 10

Figure 10.1 The Deep Sea Drilling Project. Adapted by *The New York Times*, May 29, 1972. Copyright ©
1972 by The New York Times Company. Reprinted by permission.
Figure 10.2 The Deep Sea Drilling Project.
Figure 10.3 The Deep Sea Drilling Project. Adapted by *The New York Times*, March 23, 1973. Copyright
© 1973 by The New York Times Company. Reprinted by permission.
Figure 10.4 The Deep Sea Drilling Project.
Figure 10.5 The Deep Sea Drilling Project.

CHAPTER 11

Figure 11.1 Drawn by the American Geographical Society.
Figure 11.2 Courtesy The American Museum of Natural History.
Figure 11.3 US Navy Photo, courtesy of E. H. Colbert.
Figure 11.4 *The Macmillan Illustrated Encyclopedia of Dinosaurs and Prehistoric Animals,* edited by Dougal
Dixon et al., New York: Macmillan, 1988.
Figure 11.5 L. A. Frakes and J. C. Crowell in *Nature*, vol. 217 (1968), p 838.
Figure 11.6 I. W. D. Dalziel and D. H. Elliot in *Nature*, vol. 233 (1971), p. 246–252. (figure 3)
Figure 11.7a Adapted from map published in 1985 by the British Antarctic Survey, the University of
Birmingham, and the Lamont-Doherty Geological Observatory of Columbia University.
Figure 11.7b I. W. D. Dalziel and D. H. Elliot in *Nature*, vol. 233 (1971), p. 246–252. (figure 4)

CHAPTER 12

Figure 12.1 *The New York Times,* June 25, 1973. Copyright © 1973 by The New York Times Company.
Reprinted by permission.
Figure 12.2 A. S. Romer in *Proceedings of the American Philosophical Society,* vol. 112, No. 5 (1968), pp. 338–
341, figure 2 (after Osborn).
Figure 12.3 "Evolution of the Vertebrates" by E. H. Colbert (figure 48), John Wiley and Sons, Inc., 1955.
Figure 12.4 *The Macmillan Illustrated Encyclopedia of Dinosaurs and Prehistoric Animals,* edited by Dougal
Dixon et al., New York: Macmillan, 1988.
Figure 12.5 J. W. Valentine and E. M. Moores in *Nature*, vol. 228 (1971), pp. 657–659 (figure 2).

CHAPTER 13

Figure 13.2 Courtesy of Haroun Tazzieff.

CHAPTER 14

Figure 14.1 J. T. Wilson in *Nature,* vol. 207 (1965), p. 907.
Figure 14.2 *EOS* vol. 70 no. 24, June 13, 1989.
Figure 14.3 *Geotimes,* June, 1972, p. 15, and March, 1973, p. 16.

CHAPTER 15

Figure 15.1 Photo by the author.

Figure 15.2 The Encyclopedia of Oceanography, edited by Rhodes Fairbridge, p. 947. Copyright © 1966 by Reinhold Publishing Corporation. Reprinted by permission.

Figure 15.3 K. J. Hsü in *Geological Society of America Bulletin,* vol. 71 (1960), p. 579.

Figure 15.4 K. J. Hsü and S. O. Schlanger in *Geological Society of America Bulletin,* vol. 82 (1971), p. 1208.

Figure 15.5 D. P. McKenzie in *Geophysical Journal,* vol. 18 (1969), p. 27 (figure 13).

Figure 15.6 Peter Vogt et al., in *Journal of Geophysical Research,* vol. 76 (1971), p. 3211 (figure 2). Copyright © 1971 by the American Geophysical Union.

Figure 15.7 K. J. Hsü in *Nature,* vol. 233 (1971), pp. 46 and 47 (figures 2,3,4).

CHAPTER 16

Figure 16.1 John Rodgers in *Studies of Appalachian Geology—Northern and Maritime,* E-an Zen et al., editors, p. 147, John Wiley & Sons, Inc., 1968.

Figure 16.2 J. T. Wilson in *Nature,* vol. 211 (1966), pp. 667 and 678 (figures 1 and 2).

Figure 16.3 Adapted from "Geosynclines, Mountains and Continent-building" by Robert S. Dietz. Copyright © 1972 by *Scientific American,* Inc. All rights reserved.

Figure 16.4 The New York Times, Sept 12, 1970. Copyright © 1970 by The New York Times Company. Reprinted by permission.

Figure 16.5 US Geological Survey.

Figure 16.6 Geologic Map of the US, 1932, US Geological Survey, as excerpted in figure 15-2 on p. 264, J. H. Zumberge, *Elements of Geology,* 2nd Ed., John Wiley & Sons, 1965.

Figure 16.7 R. W. Fairbridge in *Geotimes,* July/August, 1970, p. 19.

CHAPTER 17

Figure 17.1 H. Kuno in *Bulletin Volcanologique,* vol. 20 (Series 2), 1959, pp. 37-76.

Figure 17.2 Adapted from Geologic Map of the United States, Geological Society of America, as excerpted in J. H. Zumberge, *Elements of Geology,* 2nd Ed., pp. 267 and 289. John Wiley & Sons, 1965.

Figure 17.3 Photo by Goodyear Aerospace Corporation, courtesy of Physics Today.

Figure 17.4 US Geological Survey. Photo by Robert M. Krimmel, Tacoma, Washington; 80S3-139.

Figure 17.5 Photo by Robert E. Wallace and Parker D. Snavely, Jr., courtesy of the US Geological Survey.

Figure 17.6 P. W. Lipman et al., in *Science,* vol. 174 (1971), pp. 821–825.

CHAPTER 18

Figure 18.1 Courtesy Public Affairs Office, US Geological Survey.

Figure 18.2 US Geological Survey Map MF-709 (1989).

Figure 18.3 The New York Times, Jan. 17, 1973. Copyright © 1973 by The New York Times Company. Reprinted by permission.

Figure 18.4 The New York Times, March 27, 1966. Copyright © 1966 by The New York Times Company. Reprinted by permission.

Figure 18.5 US Seismicity 1960–1988. Map prepared by Susan K. Goter of the US Geological Survey, National Earthquake Information Center.

CHAPTER 19

Figure 19.1 Monkmeyer Press Photo Service.
Figure 19.2 Based on map from US Department of the Interior, Geological Survey Circular 647.

CHAPTER 20

Figure 20.1 The New York Times, July 14, 1973. Copyright © 1973 by The New York Times Company. Reprinted by permission. (Adapted from C. G. Chase and T. H. Gilmer, University of Minnesota).
Figure 20.2 Adapted from National Atlas of the USA
Figure 20.3 National Atlas of the USA and J. A. Noble in *Geological Society of America Bulletin*, vol. 81 (1970).
Figure 20.4 The New York Times, March 20, 1973. Copyright © 1973 by The New York Times Company. Reprinted by permission.

CHAPTER 21

Figure 21.1 Adapted from P. Tapponnier, G. Peltzer, A. Y. Le Dain, R. Armijo and P. Cobbold, *Geology* 10 (1982) pp. 611–616.
Figure 21.2 Woods Hole Oceanographic Institution.
Figure 21.3 Photo by Robert D. Ballard, courtesy of the Woods Hole Oceanographic Institution.
Figure 21.4 Photo by Kathleen Crane, courtesy of the Woods Hole Oceanographic Institution.

COLOR PLATES

Plate 5b was produced by the US Geological Survey as a key to plate 5a. Plate 10 was prepared by the Soil Conservation Service of the US Department of Agriculture. All other plates were provided courtesy of the National Aeronautics and Space Administration.

Index

A

Abelson, Philip H., 322, 342
Absaroka volcanic field, 294, 303
Acapulco Trench, 73
Adirondacks, 268
Adriatic plate, 264
Afar Triangle, 220, 221 (Fig. 13.1), 228-235
  geological explorations, 229, 230 (Fig. 13.2), 231-232
  geothermal potential, 332
  oil reserves, 227, 347
  plume theory, 247
  rifts, 228, 229, 231-234
  volcanoes, 230 (Fig. 13.2), 231, 232
Africa, 13, 111, 115, 198-201, 207-209, 211, 214-215, 246, 249, 262-264, 266, 272, 273 (Fig. 16.2), 280-281, 283, 287, 339, 368
  Congo River, 52
  East Africa, 9, 220
  fossils, 95, 191, 195, 197, 229
  geothermal potential, 332
  ice sheet, 138, 281-283 (Fig. 16.7)
  link to Antarctica, 195-197
  link to South America, 4, 10-11, 197, 244 (Fig. 14.1)
African plate, 261, 286, 340
African rift valleys, 220, 221 (Fig. 13.1), 222, 228-229, 232-234, 294
Agassiz, Alexander, 136
Agassiz, Jean Louis Rodolphe, 21
Aggarwal, Yash P., 314
Ailin, Sun, 209

Alabama, 267, 269, 279
Alaska, 287, 300
  earthquakes, 310, 314, 319
  fossils, 39-42
Albania, 257, 264
Albatross (research ship), 48, 56, 63, 69, 176, 222
Aleutian Trench, 64, 183, 367
Aleutians, 287, 292
Algerian Petroleum Institute, 281
All American Canal, 326-327
Allegheny Front, 275 (Fig. 16.3)
Allegheny Mountains, 271, 279
Almond, Mary, 91
Alps, 3, 240-254, 344
  continent-continent collision and, 251-266
  curvature of, 261
  Limestone Alps, 258, 262
  "nappes," 9, 15, 252-254, 264
  Pennine Alps, 262
  (*See also* mountain-building)
Altiplano, 291
aluminum silicates, 270
Alvarez, Luis W., 96
Alvin (submarine), 248, 369, 371, 373, 375
Amazon, 128, 182, 338, 343
American Association of Petroleum Geologists, 77, 129
  symposium on Wegener's hypothesis, 14-16
American Behavioral Scientist, 35-36
"American" fauna, 272
American Geological Institute, 132
American Geophysical Union, 95, 110

American Institute of Aeronautics and Astronautics, 38
American Miscellaneous Society (AMSOC), 138, 140-144, 146
American Museum of Natural History, 29, 40, 189, 194, 196
American Philosophical Society, 81
amniote eggs, 207
amphibians, 191, 207
Amundsen, Roald, 187
animals and plants, drift theory and distribution of, 9-10, 204-219
  "quick-freezing" in Arctic, 22, 39-42
Anderson, Don L., 83, 121, 314, 366
Andes (*See* mountain-building)
andesite, 66, 259, 262, 293, 300, 362
anhydrite, 155, 164
Antarctic ice, 22, 24-25
Antarctic Peninsula, 201, 203, 216
Antarctic plate, 212
Antarctica, 7, 27, 29, 111, 186-203, 207, 214-217, 246, 264, 282-283, 302, 368
  absence of land animals, 188
  Byrd's expeditions, 188, 189
  coal-bearing formations, 187, 193
  deep-sea drilling, 171, 172 (Fig. 10.3), 174
  fitting into Gondwanaland, 199 (Fig. 11.5), 200 (Fig. 11.6), 201-203
  fossils, 186-190 (Fig. 11.1), 191-198, 240
  mineral deposits, 187, 195, 351
  Ordovician Period (located near Africa), 281

role in evolution, 188, 214, 217-218
sandstone formations, 188, 211
Sullivan Ridge, 197

anteaters, 213
Anthracosaurs, 205
Antilles, 175
Apennines (*See* mountain-building)
Apollo Project, 15, 56, 140, 371
Appalachian Mountains, 13, 79, 211, 251, 254, 263, 267-280, 284, 287, 291

Caledonian links, 274, 279-280
coal-forming period, 271
crumpling process theories, 271
development of eastern margin of North America, 269 (Fig. 16.1)
marble deposits, 279
nappes, 268-271
opening and closing of Paleozoic Atlantic, 273 (Fig. 16.2), 276-281
reconstruction by Tuzo Wilson, 271-274
sediments similar to Euro-African rim, 279
Taconic episode, 268, 269, 276
volcanic activity, 267-270
(*See also* mountain-building)

Arabian Plate, 220, 221 (Fig. 13.1), 228
Arabian Sea, 150, 183
Archimede (submarine), 369, 371-374
Arctic, 217
Arctic Ocean, 60, 109, 283, 287, 343
Argand, Emile, 255
Argentina, 209
Argo (research ship), 60
Arizona, 296
Armbruster, John, 314
Armenia, 309
Armero, 292
Arolla series, 253-254
Arrole Volcano, 230 (Fig. 13.2)
Artiodactyla, 214
Ascension Island, 249
Asia, 286, 370 (Fig. 21.1)
(*See also* Eurasia)
asthenosphere, 83, 131, 250, 286, 353-354, 359, 365-366, 376
"crystal slush," 82

Astoria Fan, formed by Columbia River, 182
astronauts, 56
astronomic observations of continental drift, 16, 364
Aswan High Dam, 166
Atlantic Ocean, 39, 58-59, 61, 70, 80, 84-85, 117, 165, 168, 272, 274, 278-281, 287, 357
ancient (Paleozoic Era), 271-281, original continental border, 269 (Fig. 16.1)
cleavage of continents and, 158-159
earthquake activity, 58-59
matching features on both sides, 13, 114
metallic sediments, 160-161, 347
North Atlantic, 215
salt deposits, 154, 161-162
South Atlantic, 109, 111, 121
Atlantic opening, 122, 160
Atlantis (ship), 48, 56, 127, 175-176, 222
Atlantis II, 222, 345, 375
Atlantis (legendary continent), 11, 175-176
Atlas Mountains, 182
atolls, 68, 136-137
Atomic Energy Commission, 328
Atwater, Gordon, 35
Atwater, Tanya, 248, 298, 300
Auden, W. H., 117
Australia, 9, 43, 59, 91-92, 111, 114, 172, 196, 199, 201, 214-216, 246, 283, 341, 375
fossil finds, 196
link to South America, 7
marsupials, 9–10, 214-216
northward drift, 341
split from Antarctica, 112, 171-172 (Fig. 10.3), 199-200
avalanches, 49
Avalon Terrane, 368
Azores, 7, 159, 371, 374

B

"back-arc spreading," 369
Bacon, Francis, 4
Baikal, 369
Bailey, Sir Edward, 65, 117, 255

Baja California, 329
Baker, C. P. (drill ship), 153
Baker, Howard B., 5, 32
Ballard, Robert, 373-374
ballooning, 5-6
Baltic, 368
Baltimore, 278
Bank of England, 89
Banwell, John, 333, 336-337
Barbados, 175
Barker, P. F., 203
Barnea, Joseph, 332, 337
Barr, Frank T., 167
Barre, Vermont, 270
Barrell, Joseph, 83
Barrett, Peter J., 89, 192, 208
basalt, 61, 69-70, 73, 128, 137, 139, 141, 157-159, 195, 211, 231, 233, 235, 240, 247, 250, 256, 280, 294, 300, 302, 304, 348, 360
basalt plateaus, 98, 180
Bascom, Willard, 138-140, 142-145, 147
basin-and-range province, 294, 297, 300, 362, 364-366
batholiths, 288, 290 (Fig. 17.2), 293, 302
bathyscaphe, 371
Beagle, 216
Bear Mountain, 269
Beardmore Glacier, 186-187, 189, 192
Behrensmeyer, Anna K., 234
Bell Telephone Laboratories, 59
Beloussov, Vladimir V., 82, 113, 130, 132-133, 360
Bengal, Bay of, 182, 286
Benioff, Hugo, 66
Benioff zones, 66, 288, 291
Benson, William, 138, 148
Berezovka mammoth, 40, 42
Bergell region, 262, 270
Bering land bridge, 12, 41, 43, 192, 198
Bering Sea, 182
Berkeley Hills, 305
Berkshire Mountains, 268, 270
Bermuda and Bermuda Rise, 160, 301, 362
Berry, Edward W., 15
Betic range, 264

Bible, flood and other references, 4, 13, 20-22, 32, 42, 46, 227, 235
Bird, John M., 274, 276, 277 (Fig. 16.4), 279, 285
Biscay, Bay of, 91, 115, 264
Bjerknes, Jakob A. B., 90
black smokers, 369, 375, 376 (Fig. 21.3)
Blackett, Patrick M. S., 89, 91, 94, 113-114
Blue Mountain Lake, 314
Blue Ridge, 270-271, 368
Bodvarsson, G., 240
Bohemia, 100
Bohr, Niels, 55
Bolivar Trench, 215
Bolivia, 291
Bonaparte, J. F., 209
Bonatti, Enrico, 232
Bondi, Herman, 26
Bonin trench, 179, 181
Bonnin, Jean, 360
Book of Zechariah, 227
Bora Bora, 136
Borch, C. C. von der, 161
Borhyaena, 215
Boston, 84, earthquake, 272
Boulder, Colorado, 294
Boulder batholith, 302
Bouvet Island, 249
Bowie, William, 126
Boyd, Francis R., 353-355
Boyle, Robert, 68
Brace, William F., 317, 327
brachiosaurs, 214
Bragg, W. Lawrence, 90
Bragg, William Henry, 90
Brahmaputra River, 286
Brans, Carl H., 55-56
Brazil, 9, 11, 13, 274
    objection to deep-sea drilling, 168
Brazilian coast, 179
Britain, 91, 271, 280
British Association for the Advancement of Science, 69, 218
British Columbia, 296, 302
Bronk, Detlev W., 143
Brown, Hugh Auchincloss, 22, 24, 28, 33
Brown and Root, 142, 144, 146
Brune, James N., 123

Brunhes, Bernard, 93, 95
Bucher, Walter H., 29
Buckland, William, 52
Buffon, Georges-Louis Leclerc, Comte de, 218
Bullard, Sir Edward C., 65, 69-70, 73, 90, 93, 103, 105, 113-115, 117, 126, 129-130, 174, 177, 360, 363
Bulletin of the Atomic Scientists, 36
bumper subs, 148
Bureau International de l'Heure, 25
Burke, Kevin, 290
Burmese-Indonesian Arc, 290
Byrd, Richard E., 188

C

cables (*See* submarine cables)
Cabot Fault, 84
calcium carbonate, 129, 177, 179
Caldrill (drill ship), 147
Caledonian, 13, 274, 279-280
California, 293-294, 296, 299, 324-330
    earthquakes, 298, 308-309, 319
    geothermal potential, 324-328, 330
    volcanic activity, 293-295
Campbell, James H., 28-29
Canada, 215, 278-279, 348
Canadian Arctic, 168, 287
Canadian Journal of Physics, 85
Canadian Shield, 339
Canary Islands, 215
canyons, undersea, 48
Cape Cod, 248
Cape Hatteras, 276, 279
Cape Mendocino, 329
carbon dating (*See* dating methods and radioactive elements)
carbon dioxide, 211, 238, 342
    in Venus atmosphere, 34, 38-39
Carey, S. Warren, 54-55, 60-61, 79, 114-115, 130, 299
Caribbean and Caribbean Plate, 117, 150, 174, 207
Carlsberg Ridge, 57-58, 103-106, 108, 228
Carnegie Institute of Technology, 37
Carnegie Institution of Washington, 88, 93
Carnegie Ridge, 169-170

Carnivora, 214
Carolinas, 273 (Fig. 16.2), 279, 368
Caroline Islands, 179
Cascade Mountains, 295
catastrophes and catastrophic theories, 4, 20-22, 31-53
    on ocean floor, 48-52, 165
    ice surges, 45-47
    tumbling of the Earth, 21, 28
Catskills, 271
Cedar Mountain quake, 304
Centennial Mountains, 294
Central America, 117, 169-171, 245
    "flood gate," 215
    geothermal potential, 332
Cerro de Pasco, 224
Ceylon, 201
Chain (research ship), 223
Challenger, H.M.S., 11, 57, 147
Chamberlin, Rollin T., 15
Chandler, S. C., 23
Chandler wobble, 23-24, 26, 99, 310, 364
Channel Islands, 280, 305
Chapman, Sydney, 90
Chardin, Pierre Teilhard de, 222
Chase, Clement G., 300
Chase, John, 100
chert, 156-157, 183-184
Chevron Oil Company, 315
Chief Mountain, 293
Chile, 82, 349
Chile Trench, 260
China, 207, 209, 264, 286, 369
Chinese Plate, 358
Christiansen, Robert L., 300, 303-304
Chumakov, I. S., 166-167
Churchill Province, 341
Cita, Maria B., 164
clams, giant, 375
Clegg, J. A., 91
climate changes, 13-14
    cataclysmic, 211-212
    effect of polar drifts, 20-22
    European, 167
Climax, Colorado, 348
Clipperton fracture zone, 118
Clovis points, 43
Coachella Drainage Canal, 326
coal, 33, 205, 271
    discovery in Antarctica, 13, 187
Coalsack Bluff, 192-197, 211

Coast Ranges of California, 289, 296
coastline matching, 7, 114-116 (Fig. 7.1), 117, 268-280
Coates, Donald A., 197
COCORP, the Consortium for Continental Reflection Profiling, 271, 365
Cocos Plate, 174, 245, 298, 358
Cocos Ridge, 169-170
Coe, Robert S., 98
Cohen, J. Bernard, 37
Colbert, Edwin H., 189, 191-192, 193 (Fig. 11.3), 194-196, 207, 214-215
Colgate, Stirling A., 238
Collinson, James W., 197
Colorado, 293-295
Colorado Plateau, 293, 304, 362
Colorado River, 296, 327, 329-330
Colorado River Basin Salinity Control Act, 326
Columbia Plateau and Columbia Plateau basalts, 294, 300
comets and planets, collisions with Earth, 32-34, 99-100
compass, 87-88, 90
Comstock Lode, 348
Conan Doyle, Sir Arthur, 135
Congo and Congo River, 52, 199
Congress, 143-145
Connecticut, 268, 272, 279, 368
Connecticut River and Valley, 270, 278, 280, 288
Conrad (research ship), 60
continent-continent collision, 255-266, 269-281, 312
continental drift
  astronomical proof, 16, 364
  balloon hypothesis, 54-61
  changing of Earth's spin axis, 20, 22-30
  critics of, 113-114
  evolution and, 204-219
  magnetism and, 87-112
  Royal Society symposium (1964), 113-118
continental shelf and continental slope, 12, 48, 114, 274, 342-343
continents
  balance between oceans and, 359-360

compressed by converging currents, 76
floating on material, 11
formation of
  "big bang" and "steady state" schools, 341
  timetable of breakup, 78 (Fig. 5.8)
  "underplating," 341-342
convection and convection currents, 14, 68, 72 (Fig. 5.5), 91, 118, 242-243, 340, 364-365
  Arthur Holmes and Osmond Fisher on, 77-81
  in mantle, 70-75, 91
Copernicus, 86, 356
copper, 223-226, 302, 347-349
coral, 136, 179, 283
cordillera, 7, 285, 290-291, 293, 299, 348-349
CORE (Consortium for Oceanic Research and Exploration), 146
core of Earth, 25, 26, 71-72, 81, 87, 90
core specimens (sea floor), 136-138, 363
  analysis of, 150-151, 164
  piston coring, 47, 48, 56, 226
  Red Sea, 226
Cos Cob, Connecticut, 279
cosmic rays, 89
Cosmos and Chronos, 37
Costa Rica uplifts, 169
Cotopaxi, 292
Cotton, R. P., 46
Cox, Allan, 93-95, 105-106, 109
cratons, 339-342, 368
Cretaceous Period, 96, 98
Crete and Rhodes, 261
Cripple Creek, 348
Crowell, John C., 199 (Fig. 11.5)
crust of Earth, 11, 29, 81, 83, 139, 242-243
  deep drilling, 136-137 (*See also* Mohole Project)
  formation, 4, 5, 7, 242-243
  principle of isostasy, 11
  shrinkage theory, 7, 79, 132, 254
  slippage theory, 28, 29
Cumberland Mountains, Tennessee, 271
"cuprium", 347

curie point, 88
CUSS I (drill ship), 140-141, 143, 147
Cuvier, George Léopold Chrétien Frédéric Dagobert, Baron, 21
cynodonts, 198
Cynognathus, 208-209, 210 (Fig. 12.4)
Cyprus, 256-257, 347, 355
Czechoslovakia, 206

D

Dalrymple, G. Brent, 93-95, 105-106
Dalton, John, 90
Daly, Reginald A., 14, 49-50
Dalziel, Ian W. D., 203
Dana (research ship), 57
Dana, James Dwight, 254
Danakil Alps, 228
Darwin, Charles, 21, 68, 80, 86, 216-218, 357
  atoll origins, 136-137
Darwin, George, 4, 80
Darwin Rise, 179
dating methods, 42, 94, 150-151, 157 (*See also* radioactive elements and potasium/argon method)
Davy, Sir Humphrey, 7
de Grazia, Alfred, 36
Dead Sea, 227-228
Deccan Traps, 98
Deception Island, 65
Decker, Robert W., 241
Deep Sea Drilling Project, 147-185, 211, 246, 290, 345
  Soviet participation, 184-185
  summary of results, 184-185
  (*See also* drill ships and *Glomar Challenger*)
Deepsea Miner, 350
Defense Department, 365
Deffeyes, Kenneth S., 249, 359
Degens, Edgon T., 223
dendrochronology, 94
Dent Blanche Nappe, 253-254, 262
Denver, 315
depth-measuring systems, 66-68, 367
  depth charges, 48, 125-126 (*See also* sonar)
Deutsche Seewarte, 6
Dewey, John F., 274, 276, 277 (Fig. 16.4), 285

Diablo-Mount Hamilton Range, 296
diamond hoax, 354-355
diamond pipes, 351-353
diamonds, 13, 351, 352 (Fig. 20.4), 354, 367
diapirs, (*See also* salt and salt domes), 153-156, 331
diatoms, 156, 164, 182, 324
diatremes, 351-355
Dicke, Robert H., 55-56, 61, 79
Dickinson, William R., 288, 289 (Fig. 17.1), 293, 376
Dietz, Robert S., 77-78, 103, 118, 200, 207, 249, 274-276, 286, 375-376
dikes, 93, 240, 257, 280, 352
dinosaurs, 96, 98, 208, 212, 214, 264
dip needle, 87
DiPippo, Ronald, 327
Dirac, P. A. M., 55-56
Discoverer (American research ship), 350
Discovery (British research ship), 60, 70, 222-223, 371
Discovery II, 70
Doell, Richard R., 93-95, 105-106
domes, in Connecticut River Valley, 346 (*See also* salt and salt domes)
Doubleday Book Company, 35
Drake, Charles L., 268, 361
drill ships, 147-150
    bit, 148, 156
    bumper subs, 148
    coring system, 150
    position-keeping system, 151
    (*See also* Deep Sea Drilling Project)
Du Toit, Alexander L., 13, 186, 199-200
Duffield, Wendell A., 123
Duncan, Robert A., 248
Dunham, K. C., 225
Dutton, Clarence, 254
"dynamo" theories of the Earth's magnetism, 93, 98

E

"Earth Dynamics Program," 364
earthquakes, 24, 65-66, 82, 254, 266, 307-322
    Alaska, 310

Armenia, 309
axis changes and Chandler wobble, 22, 310
Boston, 319
cables severed by, 50-52
Central America, 170
Charleston, 319
Chile, 291
Colorado, 316 (Fig. 18.4)
East Coast, 319-320
Haicheng, 318
Hebgen Lake, 304
Khait, 308, 311
Lisbon, 307
Loma Prieta, 309
magnetic precursors, 318
Middle East, 266
New Madrid, 319
Owen's Valley, 364
plate movements and, 320, 362
    descent of oceanic plates, 288
plume theory and, 358
pore theory, 254, 315-318, 322
predicting, 310-322, 367
    electrical conductivity and radon, 311-312, 318
    "gap" method, 318-319
    lifting of landscape, 311, 318
    pressure wave changes, 312-315
San Fernando, 314, 318
San Francisco, 309
Shanxi, 307
sloping zones, 66
Tangshan, 318
Tokyo, 307
zones, 58-60, 66, 75, 119, 121
    (*See also* San Andreas Fault)
Earth's spin, 291
    (*See also* spin axis)
East African rift, 375
East Coast of the United States, 269 (Fig. 16.1), 276
East Indies, 62, 216
East Pacific Rise, 60, 73-74, 106, 109-110, 117, 160, 174, 179, 224-225, 245-246, 257, 297-298, 302, 329, 349, 366
    Magnetic survey, 110
    metal deposits, 160
Easter Island Rise
    (*See* East Pacific Rise)

echo sounding, 56, 57, 66, 125-128
    (*See also* sonar)
"ecological niches," 213, 216, 219
Ecuador, 292
Eddington, Sir Arthur, 358
Edgar, N. T., 161
Egyed, Laszlo L., 59, 79, 341
Egypt, 261-262
Ehrenberg, Christian Gottfried, 175
Eifel Mountains, 248
Einarsson, Páll, 241
Einstein, Albert, 28, 31, 36-37, 56, 89
Eismitte, 17
electroencephalogram, 32
Elliot, David H., 192, 193 (Fig. 11.3), 194, 196-197, 203
Elsasser, Walter M., 90, 251
Eltanin (research ship), 109
Emiliani, Cesare, 146
Emperor Sea Mounts, 359
Eötövs, Baron Roland von, 14
Equator, 87, 205, 279, "chalk line," 182
equatorial bulge, 14, 27
Escher, Arnold, 252
Escher, Hans Conrad, von der Linth, 252
eskers, 200, 282
Estabrook, Frank B., 137
Ethiopia, 220, 230, 332
Etruscan vases, 87
eugeosyncline, 268, 275-276, 292, 296
Euler, Leonard, 23, 115
Eurasia and Eurasian Plate, 266, 368
Europe, 115, 205, 262-264, 286
"European" fauna, 272
Evans, David M., 315
Evans, Edgar, 188
Evans, Robley D., 94
Evaporites (salt deposits), 14, 162, 343
    Afar Triangle, 232
    in Mediterranean, 163-164
    Red Sea, 226
    (*See also* salt and salt domes)
evolution, theory of, 20, 21, 47, 139, 204-219, 363
    "adaptive radiation," 213-218
    continental drift and, 204-219
    effect of Earth's atmosphere, 211
    fish, 205

"parallelism," 213
  warm temperatures and, 212
Ewing, John, 74, 248
Ewing, Maurice, 38, 50, 51 (Fig. 3.2),
    56, 58, 60, 74, 108, 120, 124-
    125 (Fig. 7.5), 126-129, 137,
    143-144, 146, 152
Exodus, 32, 42, 227
expansion theory
  (balloon hypothesis), 54-61
  Carey on, 54-61, 79, 341
  relationship between gravity, elec-
    trical force, and, 55-56, 61
  ridge-and-rift network, 60
extinctions, 21, 211, 216, 219
  by hunting, 42-44
  by asteroid, 96, 98
  oceanic, 96

F

Fairbridge, Rhodes, 282
Falkland Islands, 200
FAMOUS, the French-American
    Mid-Ocean Undersea Study, 161,
    332, 369, 371-375
Farallon Plate, 298, 300, 303-304
faults, 304, 320
  "transform" vs. "transcurrent,"
    118-124, 221, 228, 236
  (See also San Andreas Fault and frac-
    ture zones)
Fernandez, Juan, 257
Field, Richard M., 126, 137
Fischer, A. G., 179
fish, 205
Fisher, Reverend Osmond, 79-80, 123
Fisher, Robert L., 64
Flagstaff, Arizona, 242
Flatiron Mountains, 294
"flipping," 261 (Fig. 15.5)
Florey, Howard W., 113
Florida, 273 (Fig. 16.2), 279-280
Florida Keys, 7
flowering plants, 217
flysch, 258-259, 264, 288
Foothills Copper Belt, 348
foraminifera, 164, 260
Forel, F. A., 49
40th Parallel Survey, 355
fossils, 2, 21-22
  Africa, 191, 195, 234
  Antarctica, 186-198

"Atlantic" and "Pacific"
    fauna, 272
  world distribution, 206 (Fig. 12.1)
  (See also dinosaurs and
    extinctions)
fossil fish, 196
foundry explosions, 239
"Four Corners," 293
fracture zones, 76, 77 (Fig. 5.7), 118
Frakes, L. A., 184, 199 (Fig. 11.5)
France, 91, 115
Francheteau, Jean, 360
Franciscan Mélange, 296
Franklin, Benjamin, 81
Franklin, Sir John, 217
Franklin Institute, 38
Frazier, A. B., 227
Fremouw Formation, 189, 193, 195
Freuchen, Peter, 17
Freud, Sigmund, 32, 38
Froberville, Captain Huet de, 372
Front Range, 294, 303
Fujiyama, 295

G

Galapagos Islands and Rift, 169 (Fig.
    10.2), 245 (Fig. 14.2), 246, 377
    (Fig. 21.4)
Galathea (research ship), 64
Galileo, 86, 356
Gamow, George, 55
Ganges River, 286
Garm District, Tadzhikstan, 311-312,
    313 (Fig. 18.3)
garnets, 353-354
Gass, I. G., 257
Gast, Paul W., 250
Gay Head, 100
Genesis, 20, 22, 227
geodimeter, 241
Geodynamics Project (See Interna-
    tional Geodynamics Project)
Geological Society of America,
    29, 126
George D. Keathley (research ship), 301
Georgi, Johannes, 5-6, 16-18
geosyncline, 254-255, 267, 276
geothermal energy, 323-337
  effect on underground water, 327
  electric power generation, 324, 328
  fracturing technique, 328-329

global use, 324, 327-328, 332-337
Iceland, 323-324
Italy, 323
mineral by-products, 224, 330
pollution problems, 326
prospecting for, 328-329
reserves of, 335-336 (Fig. 19.2),
    337
salt content of water, 326
technological challanges, 326-329
United Nations and, 332-334, 336
Geothermal Steam Act of 1970, 335
Geysers Steam Field, 324-325, 325
    (Fig. 19.1), 335-336
Ghana, 100
Giardini, A. A., 354
Gibraltar, 168-169, 207-208
  as floodgate, 164-168
  waterfall, 165
Gibson, I. L., 240
Giddings, Louis, 94-95
Gilsa Event, 95
Glacier National Park, 293
Glaciers, Kutiah, Garumbar, Spitsber-
    gen, 45
Glarus region, Switzerland, 252-253
Glass, Billy, 95, 100
Global Marine Inc., 47, 141
Glomar Challenger (drill ship), 135,
    137, 147 (Fig. 9.1), 148, 149
    (Fig. 9.2), 150-186, 225-227,
    245-246, 256, 258, 263, 301-
    302, 345, 347, 360
  Antarctic drilling, 171-174
  Caribbean drilling, 164-165, 168
  Indian Ocean drilling, 246-247
    (Fig. 14.3)
  Nazca Plate drilling, 366
  number of holes drilled by, 184
  Panamanian floodgate exploration,
    169-170, 174
  Red Sea drilling, 225-227
  Rockfall Bank drilling, 177-178
  (See also drill ships)
GLORIA sonar, 155, 371
Glossopteris flora, 10-11
gneiss, 259, 339, 342
Goeller, H. E., 350
gold, 89
Gold, Thomas, 26-27, 243
Gomorrah, 235
Gondwana flora, 188, 218

Gondwanaland, 11, 13, 111, 200, 203, 207, 209, 211, 213-216, 246, 345
  fitting together, 4, 8, 78, 115-117, 186, 199, 200-202
  near the South Pole, 282-283
Gordon, Kermit, 144
Gouin, Pierre, 230
Gould, Laurence M., 188, 196
Gould, Stephen Jay, 52-53, 216
Graham, John W., 88, 93
Grand Banks, 50, 248
Grand Canyon, 91, 293
Grand Combin, 262
Grand Tetons, 294
granite, 61, 69, 72, 128, 259, 262, 270, 288, 328-329, 339
  effect of pressure and swelling, 270, 317
Graphite Peak, 189, 192-193, 197
gravity
  cause of drift, 14
  decline in, 55-56, 61
  measurements, 12, 63, 65-66, 70, 73, 117
graywacke, 252
Grazia, Alfred de, 31
Great Glen Fault, 84
Great Lakes, 362
Great Pyramid, 266
Great Salt Lake, 304
Great Smoky Mountains, 268
Great Valley, Appalachians, 271
Great Valley, California, 296
Greece, 257, 261-264
Green, Charles L., 181
Green Mountains, 269-270
Greenberg, Daniel S., 142, 145
"greenhouse" effect, 34, 38, 211
Greenland, 3, 6-7, 16-17, 24, 115, 177, 205, 215, 280, 287, 342
Greenland ice, 46
greenstones, 339, 342
Gregory, J. W., 229
Griffths, D. H., 203
Grommé, C. S., 95
Guadalupe drilling, 141
Gulf of Aden, 58, 224, 227-228, 247
Gulf of California, 329-330
Gulf of Mexico, 46, 117, 207
  drilling, 152-156, 331
Gulf Stream, 170, 171, 359

Gunnison, Black Canyon of the, 295
Gurnigel Flysch, 259-260
"gusher" (oil well), 153
Gutenberg, Beno, 82
Guyots, 66-67 (Fig. 5.3), 68, 75, 179, 232 (*See also* sea mounts)

H

Habkern Island, 259-260 (Fig. 15.4), 262
Haicheng, 318
Haile Selassie I University, 229, 231
Haleakala, 83
Hall, James, 254
Hallam, Anthony, 214, 341-342
Hamilton, Warren, 109, 285, 287-288, 293
Hansteen, Christopher, 88
Hapgood, Charles H., 28-30, 36, 39-40, 42
Harte, Bret, 114
Hatherton, Trevor, 288
Hawaiian Islands, 65, 80, 243, 248, 359, 365, 371, 373
  formation of, 83-84 (Fig. 5.9), 85 (Fig. 5.10)
Hawaiian Volcano Observatory, 137
Hay, Richard L., 95
Hayes, Dennis, 184
Hays, James D., 96
Healy, John H., 315
heat-flow, 56, 68-71, 73, 117, 310
Heath, G. Ross, 170
Hedberg, Hollis D., 144
Heezen, Bruce C., 50, 51 (Fig. 3.2), 59-60, 62, 99-100, 113, 118, 123, 179
Heimaey eruption, 236-239
Heirtzler, James, 106, 110-111, 121, 372
Hekla volcano, 235-236, 239, 364
Helvetic Plate, 260-262
Hemerli, Felix, 251
Hercynian uplift, 280
Hersey, J. B., 146
Hess, Harry H., 37, 44, 53, 62-63 (Fig. 5.1), 66, 68, 70, 73-78, 104-106, 126, 128, 138, 140-141, 167, 179, 249
Hickel, Walter J., 334, 337
Hilpert, Lowell S., 354

Himalayas (*See* mountain-building)
hippopotamus, 9
Hoffmann, Paul F., 368
Hoggar massif, 281
Holden, John C., 200, 207, 274, 276, 375
Hollin, John T., 44-46
Hollister, California, 319
Holmes, Arthur, 77-79, 86
Holyoke Range, 280
hominid, 219, 234
Hooker, Sir Joseph Dalton, 217-218
Horizon (research ship), 60, 70
hot brines, 221-227
hot dry rock, 328-329
"hot spots" (*See* plumes)
Hoyle, Sir Fred, 26, 56
Hsü, Kenneth J., 164, 258, 265 (Fig. 15.7)
Hubbert, King, 315, 317
Hudson Canyon, 48, 52
Hudson Highlands, 269
Hudson River, 49, 280
Humboldt, Alexander von, 4
Hurley, Patrick M., 280, 357
Huxley, Thomas Henry, 205, 357
hydrofracturing, 328
hydrothermal vents, 347, 350, 369, 375-377 (Fig. 21.4)

I

ice ages and ice sheets, 42, 45-47, 331
  cycles, 282-283
  effect on length of day and Earth's spin axis, 24-25, 29
  effect on uplift and sea level, 11, 12
  relation to Panamanian Isthmus, 170-171
  vestiges in Africa, 281-283
  water beneath, 45-46
  (*See also* surges)
icebergs, 128, 173, 180
Iceland, 7, 16, 80, 159, 215, 228, 239, 257, 280, 358, 365
  dike intrusions, 240
  geothermal energy, 323-324, 333
  jökulhlaup (bursting icecap), 236
  measurements of land deformation, 241-242
  plume theory and, 246-247, 249
  rifts (*gja*), 231-232

similarities to Cyprus, 257
volcanoes, 235-240
    Skaftarjökull or Laki eruption, 235
    Surtsey eruption, 237
    Vestmannaeyjar eruption, 220, 237-239
(*See also* Reykjanes Ridge)
Ichthyostegid, 205
Idaho, 294, 296
Illinois, 254
impacts, 96, 98
Imperial Fault, 330
Imperial Valley, geothermal potential, 324, 326-327
India, 10, 13, 92, 246, 286, 359, 368
    fossils, 206, 207, 209, 214
    Gondwana puzzle, 198, 199
Indian Ocean, 11, 57-59, 103, 105, 109, 113, 165, 168, 183, 216, 246, 248, 286-287
    deep-sea drilling, 185, 246, 247 (Fig. 14.3)
    magnetic surveys, 103-104
    Mediterranean linked to, 185
    metallic deposits, 160-161
    Mid-Indian Ocean Ridge, 172
    (*See also* Carlsberg Ridge and Ninetyeast Ridge)
Indian Ocean Plate, 246, 358
Indonesia, 294, 341, 375
    ore deposits, 348
Indus River, 286
Insectivora, 214
Institute for Space Studies, 121-123
Institute of Polar Studies, Ohio State University, 89
Institute for Defense Analyses, 47
Institut Français du Pétrole, 281
Interferometry, very-long-baseline, 364
"interfingering," 286
International Council of Scientific Unions, 361
International Earth Rotation Service, 24
International Geodynamics Project, 296, 300, 361-368
International Geophysical Year of 1957-1958, 37, 60, 82, 90, 120, 133, 189, 192, 361, 364

International Indian Ocean Expedition, 103
International Latitude Service, 25
International Polar Motion Service, 23
International Union of Geodesy and Geophysics, 85, 126, 130, 139
Interstate highways, 280
Inverness, 84
Iowa, 254
Iran, 309
Ireland, Giant's Causeway, 272
iridium, 98
iron, 205, 339, 340, 350
    in ocean sediments, 160
    Red Sea deposits, 223-224
irrigation, 326, 332
Isaac, Glynn Ll., 234
Isaacs, John, 24
Isacks, Bryan, 242
Isaiah, 42
island arcs, 7, 63, 75, 268, 274, 278, 285, 287-291, 293, 314, 339, 341, 348
isostasy, 11-12
Italy, 262-264
    geothermal potential, 323, 336
Ivory Coast, 100

J

Jaggar, T. A., 137
James, David E., 292
Jan Mayen, 159
Japan, 63, 80, 245, 289-290, 369
Japan Sea, 289-290
Japan Trench, 179, 182, 245
Jaramillo Event, 95-96, 100, 106
Jean Charcot (research ship), 161
Jeffreys, Harold, 18, 114, 240
Jemez Mountain, 328
Jensen, James, 196
jet stream, 17
jökulhlaup, 236
John Murray (ship), 57
Johnston, David, 189
JOIDES (Joint Oceanographic Institutions for Deep Earth Sampling), 146-147, 150
JOIDES Resolution (drill ship), 147, 150, 184
Jones, Paul H., 331

Jones, Sir Harold Spencer, 34
Jordan, Pascual, 55, 58
Jordan Valley, 227-228
Journal of Geophysical Research, 85
Juan de Fuca Plate, 296-298, 305
Juan de Fuca Ridge, 102 (Fig. 6.4), 106, 108, 298
Juan Fernandez Island, 257
Jungfrau group, 262
Jupiter (planet), 32, 39
Jura Mountains, 271

K

Kamchatka, 294, 311
Kannemeyeria, 209, 210 (Fig. 12.4)
Kansas, 297
Karroo Basin, 195
Kaula, William M., 242
Kay, Marshall, 268
Kazakhstan, 287
Keathley sequence, 301-302
Keflavik, 108
Kelvin, Lord (William Thomson), 9, 24, 27, 81
Kennedy, John F., 141
Kenya and Mount Kenya, 229, 232, 332
Kerguelen Island, 217, 246, 248
Khait, 308
Kilauea volcano, 80, 123, 137
Kilimanjaro, 229, 231
kimberlite, 351
King, Clarence, 355
Kinsman, David J. J., 347
Kisslinger, Carl, 312, 320
Kitching, James W., 197
Klamath Mountains, 293, 296
Klemme, H. Douglas, 346
"klippen," 269
Knopoff, Leon, 357-358, 360
Known Geothermal Resources Area, or KGRA, 335, 336 (Fig. 19.2)
Koch, J. P., 7, 17
Krakatoa, 176, 239
Kropotkin, P. N., 287
Kuenen, Philip, 50-51, 127
Kullenberg, Börje, 47
Kuno, Hisashi, 289 (Fig. 17.1)
Kure, 83
Kurile Arc and Trench, 300
Kurtén, Björn, 212, 213, 215, 219

L

Labrador Sea, 128, 171
labyrinthodonts, 189, 191 (Fig. 11.2), 205-206
Ladd, Harry S., 137
Lagomorphs, 214
Lake Baikal, 233
Lake Char Fault, 272
Lake Constance, 49
Lake Geneva, 49, 52
Lake Hannington, 232
Lake of Zurich, 252
Lake Rudolf, 234
Lake Saltonsall, 280
Lake Nyasa and Lake Tanganyika, 229
Lake Tiberias, 227
Lake Walensee, 253 (Fig. 15.1)
Laki eruption, 235
Lamont, Thomas W., 58
Lamont Geological Observatory (later the Lamont-Doherty Geological Observatory), 50, 58, 95, 101, 106, 108, 124, 127, 146, 150
land bridges
    Canada-Greenland-Europe, 215
    Europe-Africa, 264
    (*See also* Bering Land Bridge)
Laramide Revolution, 293
Large Aperture Seismic Array, or LASA, 365
Larochelle, A., 104
Larrabee, Eric, 34
Larsen, Captain C. A., 202, 302
Larson, Roger L., 300, 302
Laschamp Event, 95
laser reflectors on the Moon and satellites, 16, 364
Lassen Peak, 205
laterite, 274
Latham, Harold Strong, 35
Laurasia, 13, 111, 207, 211, 213-214, 272
Laurasian plate, 248
lava, 88, 288
    fountains, 85 (Fig. 5.10), 237
    magnetic analyses, 88-89
    (*See also* magnetic reversals)
    "pillow," 257, 347
    in rift, 373 (Fig. 21.2)
    "toothpaste," 372

Le Bas, M. J., 234
Leadville, 348
Leakey, Richard E. F. and Louis S. B., 229, 234
Lehigh University, 126
Lemuria and lemurs, 9
Lewis Overthrust, 293
Libya, 167, 261-262
lightning, 310
Lilenthal, Christoph, 4
Lill, Gordon G., 137, 140, 143
limestone, 252, 254, 259, 271
Limestone Alps, 258, 262
Lipman, Peter W., 300, 303-304
Lisbon, 307
lithosphere, 83, 131, 365, 376
LOCO (Long Cores committee), 146
Loewe, Fritz, 17-18
Loihi Sea Mount, 83
Lomonossov Ridge, 60
London "brickearth," 46
Long Beach, 327
Long Island, 282
Long Island Sound, 272
Long Valley caldera, 329
Longwell, Chester R., 15, 34, 54
Los Alamos National Laboratory, 328
Los Angeles, 309, 319
Los Angeles Basin, 304, 343
Louisiana, 331
Lowell, James D., 346
Lulu (catamaran), 374
lunakhods, 16
Luth, William C., 376
Lyell, Sir Charles, 53, 167
Lyme Dome, 270
Lystrosaurus, 195-197, 207-209, 210 (Fig. 12.4), 211

M

MacDonald, Gordon J. F., 26, 46-47, 73, 113, 358
Macmillan and Company, 34-35
Madagascar, 160, 199, 201, 206
Magdalena River, 52
Magnetic Airborne Director, or M.A.D., 101
magnetic field of the Earth, 25, 37, 56, 81, 87-91, 283, 287, 363
    in outer space, 34, 36, 39
    source of, 89, 90, 93, 251, 363

magnetic reversals ("magnetic footprints"), 27, 32, 76, 87-112, 123, 128, 158, 239-240, 297, 339
    extinctions, 96, 98
    Juan de Fuca Ridge, 107 (Fig. 6.5)
    Keathley sequence, 301-302
    Mid-West gravity belt, 340
    ocean cores, 129
    Pacific floor, 296-302
    polar wandering, 91-92, 362, 363
    record in lava and clay, 88
    Reykjanes Ridge, 106, 108, 109 (Fig. 6.6), 111
    sources of, 98, 99, 363
    surveys, 76-77, 101
    timetables, 94-97 (Fig. 6.2), 101-111, 158–159 (Fig. 9.5), 297-302
magnetometers, 56, 101
Maine, 270, 272, 368
Malin, Michael C., 28
Mallet, Robert, 125
mammals, 96, 191, 208
    evolution of, 211-217, 219
    extinction of ice age, 22, 39-44
mammoths, 22, 39, 41-42
man, 212
Managua, 170
Manchester Literary and Philosophical Society, 90
manganese, 223-224, 350
Manhattan (tanker), 168
Mansinha, Lalantendu, 310
mantle, 70-71 (Fig. 5.4), 83, 144, 224, 242-243, 250, 341, 354-355, 358, 361, 364, 366-367, 376
    composition of, 73-75, 137-139, 256-257
    (*See also* convection and convection currents, plumes, ocean floor layers, and Upper Mantle Project)
marble, 279
Marcel le Bihan (ship), 374
Mariana Basin, 369
Marianas Trench, 76, 179
marine fossils, 13, 150, 151 (Fig. 9.3), 157-160, 175, 177, 179, 183, 184, 251, 254
"Mariner 9" (mission to Mars), 28, 56

Marquesas, 359
Mars, 28, 32, 56, 234, 376
Mars, canals, 133
marsupials, 9, 212, 214-216
Martha's Vineyard, 100
Martin, Paul S., 42, 42-44 (Fig. 3.1)
Martinique (volcanic eruption), 49, 175
Maryland, 267
Mason, Brian, 351
Mason, Kenneth, 45
Mason, Roland G., 101
Massachusetts, 268-269, 272, 368
Mastodons, 39-40, 212
Matterhorn, 253-255, 257, 262
Matthews, Drummond H., 104, 105, 108
Matuyama, Motonori, 93, 95
Matuyama Epoch, 95, 97 (Fig. 6.2), 106
Mauna Loa, 65
Maxwell, Arthur E., 69, 70, 73, 138, 140, 143
May, Paul R., 280
McDougall, I., 95
McGetchin, Thomas R., 353, 355
McGregor Glacier, 192, 196-197
McKelvey, Vincent E., 350
McKenzie, Dan, 253, 261, 264, 266, 285, 297-298
McMurdo Sound, 192
Mecca, 222
Mediterranean, 155, 208, 357
    Deep Sea Drilling Project, 162-169
    development of Alps and, 261-266
    effect of abrupt filling on Earth's spin, 166
    evaporite deposits, 163-165
    extensive drying of, 164-168
    link to Indian Ocean, 165
    oil resevoirs, 168
    volcanic ash blanket, 176, 260
Meidav, Tsvi, 333, 337
Meinesz, F. A. Vening (*See* Vening Meinesz)
Mélanges, 289, 296
Melton, Charles E., 354
Menard, Henry, 76, 77 (Fig. 5.7), 118, 122

Mendocino and Mendocino Fracture, 103, 118
Mesabi Range, 339
Mesosaurus, 208 (Fig. 12.2)
metals, 223-226, 232, 338-339, 347-348, 351
    depletion of, 350
Meteor (research ship), 57-58
meteorite impacts, 99
Mexico, geothermal energy, 60, 117, 300, 324-325, 327, 329-330, 368
Meyerhoff, Arthur A., 77-79, 130, 360
Meyerhoff, Howard, 130
microcontinents, 263-264, 368
microearthquakes, 314
Mid-Atlantic Ridge, 7, 11, 83, 106, 111, 117-118, 128-129, 131, 161, 177, 220, 225, 231-232, 236, 239, 323, 350
    deep-sea drilling, 157-162
    echo-sounding profile, 57-60
    fossil evidence for "rejuvenation," 177, 179
    plume theory, 243, 244
    rift, 58-59, 60, 70, 74, 158-159, 375
    sediment, 74-75, 129
    transform faults, 119, 121, 122
    Vema Fracture, 182
mid-continent gravity high, 339
Mid-Indian Ocean Ridge, 172-173
mid-ocean ridges, 58, 59 (Fig. 4.1), 60, 62, 72, 79, 103-111 (Fig. 6.7), 112, 118, 121-123, 132, 256, 347, 362
Middle America Trench, 169-170, 292
Minnesota, 269, 278
Minnesota River Valley, 339
Minoan civilization, 176
miogeosyncline, 267, 275, 276, 292, 296
Mississippi, 279, 338
Mississippian period, 271
Mohole Project, 74, 128, 185, 256, 355, 365
    costs and controversies, 132-148
    feasibility study, 139
    intermediate phase, 140-141
    (*See also* Deep Sea Drilling Project)

Mohorovičić, Andrija, 81
Mohorovičić discontinuity, or "Moho," 39, 82, 127, 139, 141, 143-144, 256, 365
Mohr, Paul A., 229-233, 241
Molnar, Peter, 242, 248, 369
molybdenum, 224
Mont Blanc massif, 262
Montana, 293-294, 302-304, 365
Moody, Judith B., 366
Moon, 32, 37, 39, 40, 56, 80, 242, 342
    reflectors on, 16, 364
Moore, James G., 238
Moores, Eldridge M., 217 (Fig. 12.5), 219, 256
Morgan, W. Jason, 242, 249, 285, 297, 298, 332, 346, 358
Morley, Lawrence W., 104
Morocco, 200, 283, 309
Mother Lode, 224, 348
Motz, Lloyd, 36
Mount Adams, 295
Mount Garibaldi, 295
Mount Hekla, 241
Mount Hood, 295
Mount Katahdin, 271
Mount Kenya, 197, 229, 231
Mount of Olives, 227
Mount Rainier, 98, 295
Mount Saint Helens, 65, 295 (Fig. 17.4), 305
mountain-building
    Alps, 13, 21, 49, 79, 251-266, 285, 288
    Andes, 63, 75, 201, 212, 285, 287, 291-293, 338, 343, 346
    Apennines, 264, 267, 289
    Appalachians, 13, 79, 211, 251, 254, 263, 267-280, 284, 287, 291
        Alleghenian upheaval, 271, 280
        Acadian phase, 270-271, 278, 280
        Grenville phase, 268, 280, 368
        ribbon limestone, 270
        ridge-and-valley province, 267, 275, 279
        Taconic episode, 268-269, 276, 280
    Carpathians, 258, 261
    crustal compression or shrinkage, 7

effect of turbidity currents, 258-260
flysch deposits, 258-259, 264, 288
Himalayas, 5, 7, 11, 13, 251, 285-286. Ophiolites, 257
layer inversions, 252, 253 (Fig. 15.1)
Pyrenees, 91, 115, 264
Urals, 257, 285, 287-288, 291, 346, 348
Western U.S., 288-290, 293-301 (Fig. 17.6), 302-306
(*See also* nappes, ophiolites, overthrusts, and under individual mountain ranges)
Mull, Isle of, 109
Munk, Walter H., 24, 73, 91, 138, 145, 358
Murfreesboro, Arkansas, 352
Murray, Bruce C., 28
Murray, Grover, 196
Murray Ridge, 58

N

Nance, R. Damian, 366
Napoleon, 252
nappes, 9, 10 (Fig. 1.4), 15, 79, 252, 253 (Fig. 15.1), 255, 270, 278, 293
National Aeronautics and Space Administration (NASA), 38, 121-122
National Center for Earthquake Research, 311
National Oceanic and Atmospheric Administration, 345
National Academy of Sciences, 138-139, 141, 296, 309, 312, 339
National Science Foundation, 138, 142-143, 146-147, 192, 196, 334
Nature (British journal), 27, 55, 70, 104, 113, 223, 239, 285, 322, 343
Nautilus (submarine), 60
navigation, 181
Navy Seabees, 192
Naylor, Richard S., 270
Nazca Plate, 245, 292, 358, 366
"Nemesis," 98
"nepheloid layer," 128
Nevada, 293-294

Nevado del Ruiz, 292
New Brunswick, 270, 272
New England, 278-279
New England Sea Mounts, 247
New Guinea, 7
New Hampshire, 272
New Haven, Connecticut, 280
New Jersey, 269
New York City, 282
New York State, 268, 271
New Zealand, 201, 289, 326, 336-337, 345
Newark Trough (Triassic rift), 211, 280, 281 (Fig. 16.6)
Newcomb, Simon, 23-24
Newfoundland, 115, 267, 270, 278-279, 368
Newton, Sir Isaac, 18
Ngorongoro, 229
Nierenberg, William A., 185
Niger River, 115, 128
Niigata, 318
Nile River, 166 (Fig. 10.1), 168
Ninetyeast Ridge, 246, 347
Nobel, James A., 349 (Fig. 20.3)
Normandy, 280
North America and North American Plate, 115, 205, 264, 297-299, 302, 303, 366, joined to Europe, 83-84
North Pole, soundings through the ice, 60
North Pole (magnetic), 91-92
Northrotheriops, 43
Norway, 13, 254, 272, 279-280
Nova Scotia, 200, 272, 279
Novaya Zemlya, 287
nuclear explosions, 47, 328
nuclear weapons and tests, 58, 82
nuée ardente, 49, 175
Nur, Amos, 317

O

Oakland, 309
Oates, L. E. G., 188
ocean currents, 128-129
ocean floor, "second" and "third" layers, 57, 128, 137, 140-141, 144, 183, 256-257, 360
ocean floors, 11-12
    balance between continental areas

and, 287, 359-360
basalt, 69
canyons, 48
depth, 214
formation of, 55
mapping of, 58-59
new equipment for, 56
recycled, 377
(*See also* Deep-Sea Drilling Project, echo-sounding, gravity (measurements), heat flow, Mid-Atlantic Ridge, ore formation, piston corer, seismic probing, subduction, and turbidity currents)
oceanography and politics, 168-169
oceans, 359
Office of Emergency Preparedness, 309
Office of Naval Research, 138
oil and gas reservoirs, 174
    Antarctica, 172 (Fig. 10.3)
    deep-sea drilling benefits, 185
    depletion of, 350-351
    formation of, 342-344 (Fig. 20.2), 345-347
    heat-flow and, 342-343, 345-346
    map, 344 (Fig. 20.2)
    Mediterranean, 168, 345-347
    in ocean-floor sediment, 150-155
    in river deltas, 343
    off-shore drilling, 161-162, 185
    Red Sea area, 227
    (*See also* salt domes and evaporites)
oil traps, 153, 154 (Fig. 9.4)
"Old Red" sandstone, 114, 205, 274
Oldenburg, Douglas W., 123
Olduvai Gorge, 95, 229
Oliver, Jack, 271
O'Neill, John J., 34
Ontario, 270, 339
Opdyke, Neil, 95-96, 110
Operation Highjump, 60
ophiolites, 256-257, 264, 287-88, 296, 339-340, 347, 353
opossums, 216
Oppenheimer, J. Robert, 28
ore formation, 224, 339, 340 (Fig. 20.1), 342, 347, 349 (Fig. 20.3), 351, 375, 376

by-products of geothermal energy, 330
deep-sea drilling, 160-161, 185, 350
depletion of, 350
plate theory and, 347-351
Red Sea hot spots, 223-226
Oregon, 183, 294, 296
Origin of Continents and Oceans, The, 3, 8 (Fig. 1.3)
origin of life, 218
Oslo, 272
ostrich, 215
Otte, Carel, 337
Oversby, Virginia M., 250
overthrust, 252, 253 (Fig. 15.1), 293
Owen (British research ship), 60, 103, 106
Owens Valley, 364
oxygen, atmospheric, 219

P

Pacific basin, lunar origin, 80
Pacific floor, 182
Pacific Gas and Electric Company, 324, 337
Pacific Ocean, 39, 59-60, 64, 111, 117, 170, 207, 264, 289
age of its floor, 179 (Fig. 10.5), 180
being squeezed smaller, 274
deep-sea drilling, 170, 179-182
echo-sounding, 59-60
missing ridge, 297-298, 300
North Pacific floor, 179
northwest movement of, 181-182
origin of Moon, 4, 32, 80
plate movements, 157, 158, 159 (Fig. 9.5), 297-305, 359
(See also East Pacific Rise, trenches, and magnetic reversals)
Pacific Plate, 174, 179, 245, 297-300 (Fig. 17.6), 310, 364
Pacific trenches, 63-64 (Fig. 5.2)
Palisades, 195, 280
Pamir Mountains, 308
Panamanian Isthmus, 169-170, 184, 215-216, 219, effect on the fossil record, 184
Panchen, A. L., 205
Pangaea, 111, 207, 255, 341, 363, 375

Panthalassa, 14
Paterson, New Jersey, 280
Paulding, B. W., Jr., 317
Pele's hair, 231
penguins, 186
Pennine Alps, 257, 262
Pennine Plate, 260-262
Pennsylvania, 267
Pennsylvanian Period, 271, 279
Pensée (magazine), 38
peridotite, 74
Perissodactyla, 214
Persian Gulf, 164
Peru, 291-292
Peterson, Melvin N. A., 161, 185
Pettersson, Hans, 48
Péwé, Troy, 41
Peyve, Aleksandr, V., 184, 287, 377
Philippine Arc, 290
Philippine Plate, 358
Philippine Sea, 289
Phoenix Islands, 300, 302
Pichon, Xavier Le, 108, 112, 248, 264, 358, 359, 360, 372-373
Pick and Hammer Club, 142
Piedmont, 275 (Fig. 16.3), 368
Piggot, Charles S., 47
pillow lavas, 347, 360, 372
Pioneer (ship), 101, 102 (Fig. 6.4), 103
Pioneer Fracture Zone, 103
Piper, J. D. A., 239-240
Piri Re'is, 171
piston corer, 47-48, 56, 74
Pitman, Walter C., III, 108, 121, 300
Pittsfield, 272
plankton, 184
plate movement theory, 302
"budgets" of plate gains and losses, 359, 360
critics of, 113-134, 360
description of, 73-75 (Fig. 5.6), 76-77, 83
diamonds and, 353-355
driving force, 241-249, 357-359
effect on Earth's history, 338-355
gravitational tugging, 242, 357
International Geodynamics Project to investigate, 361-368
list of plates, 78, 111 (Fig. 6.7), 358
observing and measuring methods, 364

phase changes and, 250
"push" versus "pull" or "drag" mechanisms, 241-242, 357-358
timetable of, 298
future trends, 375-377
(See also convection and convection currents, magnetic reversals, plume theory, subduction, and trenches)
Plato, 175-176
Plowshare program, 328
plumes ("hot spots"), 235-250, 280, 343, 346, 358-359, 364-366
plutons, 270
Poconos, 271
Poland, 249, 258
polar wandering, 26, 89, 92 (Fig. 6.1)
Poldervaart, Arie, 100
Poles: drift of, 23-30
flattening of Earth at, 73
geographic and magnetic, 25
wandering hypothesis, 20-30, 89, 363
Polfluchtkraft, 14
pore pressure and earthquakes, 254, 317-318
potassium, 288, 303-304, 348
potassium-argon dating, 94, 233
Precambrian Period, 363
Press, Frank, 120
pressure waves, 71, 312, 313 (Fig. 18.3)
Pribilof Islands, 95
primates, 212, 214
Prince Charles Mountains, Antarctica, 247
Princeton University, 37, 44, 126, 146
Proboscidea, 214
Prostka, Harold J., 300, 303-304
Prévot, Michel, 98
Psalms, 307
Pyramid, orientation of, 266
Pyrenees, 91, 115, 264 (See also mountain-building)

R

radio telescopes, 364
radioactive elements, 9, 69-70, 132, 293 (See also dating methods)
radiolarians, 151 (Fig. 9.3), 156, 170

radon, 311, 318
Raff, Arthur D., 101, 103
Raleigh, Cecil B. (Barry), 311
Ramsey Glacier, 197
Rand Corporation, 330
Rangely oil field, 315
Rangely Lakes, 270
rare earths, 247
Red Sea, 32, 58, 109, 160, 220-234,
    247, 266, 294, 345-347
    brine pools, 221 (Fig. 13.1),
        222-227
    drilling by *Glomar Challanger*,
        226-227
    metal deposits, 160, 223-226,
        345-347
    plate motions, 227-228
    salt layers, 227
    (*See also* Afar Triangle)
"redbeds," 205, 207
relativity, "general theory" of, 56
reptiles, 207, 212, 264
Revelle, Roger, 24-25, 64-65, 69-70,
    73, 138, 143, 146
Rex, Robert W., 161, 329-330, 334,
    336
Reykjanes Ridge, 106, 108, 110, 177,
    236, 239-240, 247
    magnetic survey of, 106, 108, 109
        (Fig. 6.6)
Rhine River, 233, 262
Rhone River, 167-168, 262
ridge-and-rift network:
    Afar Triangle and Africa, 228-229,
        233-234
    convection currents, 74
    deep-sea explorations, 369-377
        (Figs. 21.3 and 21.4)
    Earth-expansion theory, 54-61
    earthquake activity and, 59
        (Fig. 4.1), 60
    East Pacific Rise, 60
    faults, 118-124
    Iceland, 232-233
    (*See also* Mid-Atlantic Ridge and
        "triple junctions")
Ries Basin, 100
Riffaud, Claude, 374
rift valleys of Africa, 339
rifts, 29, 58-59, 62, 74
"ring of fire" around Pacific, 65, 174
Rio Grande Ridge, 243

Rivera Plate, 298
Rockall Bank, 115, 177, 178
    (Fig. 10.4), 205, 215, 280
Rockefeller Mountains, 201
rocks
    age determination, 9, 233, 283
    potassium content, 289, 339
Rocky Mountain Arsenal, 315
Rocky Mountains, development, 285,
    292-297, 302-303, 306, 338,
    366-367
    ore, 348, 349 (Fig. 20.8)
    two down-slopping planes, 301
        (Fig. 17.6), 303, 304
    (*See also* mountain-building)
Rodgers, John, 271, 278-279
Romer, Alfred Sherwood, 189, 206,
    208-209, 216
Rona, Peter, 345, 350
Ross, David A., 223
Ross, Sir James Clark, 217
Ross Ice Shelf, 171, 201
rotation of Earth and other bodies,
    22-30, 89
rotation poles, 115, 129, 207, 215
Rothé, J. P., 58, 59 (Fig. 4.1)
Rousseau, J. J., 307
Royal Gorge of the Arkansas River,
    295
Royal Prussian Aeronautical Observa-
    tory, 5
Royal Society of London, 123, 128,
    131
    Symposium on Continental Drift,
        113-118, 121, 124,
    Symposium on Red Sea, 225, 227
Rubey, William W., 315
Ruhr, 279
Runcorn, S. Keith, 91-93, 113-114
Russia, 207, 285, 287
Ryan, William B. F., 164-167

S

Sable Island, 346
Sahara ice sheets, 281-283 (Fig. 16.7)
Saint Bernard Pass, 262
Saint Paul Rocks, 74, 256
salt and salt domes, 154-155 (Fig.
    9.4), 161, 226, 331, 343, 345,
    346

Deep-Sea Drilling Project, 152-
    154 (Fig. 9.4), 155-156, 168
    (*See also* diapirs, evaporites)
Salton Sea, 224, 330
    geothermal energy, 324, 326, 327
San Andreas Fault, 120, 298, 299
    (Fig. 17.5), 300, 304, 305 (Fig.
    17.7), 308 (Fig. 18.1), 317-319,
    321, 329-330, 364
San Francisco, 296, 308-309
    (Fig. 18.2), 319
San Gabriel Mountains, 318
San Juan Mountains, 293, 303, 362
Sandia National Laboratories, 329
Saporta, Gaston de, 218
Sardinia, 264
Saudi Arabia, 222
satellites, 174
Sawkins, Frederick J., 348
Sbar, Marc L., 314
Scandinavia, 79, 279
Schaeffer, Bobb, 194, 196
Schilling, J. G., 247
schist, 259
Schlanger, Seymour O., 259,
    261, 262
Schlich, Roland, 183
Schlieren Flysch, 259-260
Schmitt, Harrison H., 222
Scholl, David W., 211
Scholz, Christopher H., 317
Schopf, James M., 193
Schubert, Gerald, 250
Schuchert, Charles, 15
Science, 110, 142, 223, 285, 322
Scientific American, 37, 103, 345, 350
Scotia Arc, 201-202 (Figs. 11.7A and
    B), 203
Scotland, 13, 272, 274, 279-280, 342
    Great Glen Fault, 83-84
    rifts, 231
Scott, Captain Robert Falcon, 7, 10,
    187
Scottish Highlands, 254
Scripps Institution of Oceanography,
    24, 64-65, 69, 101, 146-147, 150
sea-floor spreading (*See* plate move-
    ment theory)
sea level, 12, 25, 45, 341
sea mounts (guyots), 66, 68, 75,
    302, 362

Sea of Japan, 369
Seattle, 305
sediment, metallic, 160, 330
sedimentation, 181, 255
sediments, oceanic, 12, 33, 46-48, 64,
    74-75, 88, 137, 139-141, 143,
    146, 156-159, 161-162, 170,
    179, 184, 254, 264, 267-268,
    274, 275 (Fig. 16.3), 276, 286,
    289, 291-292, 331, 341, 345,
    347
  chalk-line evidence, 182
  charting thickness, 124-130
  effects of turbidity currents, 50-51
    (Fig. 3.2), 128
  fans of river sediment, 182
  (*See also* ocean floor, layers)
seismic networks and arrays, 58, 82,
    121, 365
seismic probing, reflection, 127
seismic probing, refraction, 126
seismic sounding, 17, 268
  of sea floor, 126-127
seismic tomography, 366
Seismological Laboratory, California
    Institute of Technology, 314
seismology, 81, 312
  detecting bomb tests, 120
  U.S. Seismicity (map), 321
    (Fig. 18.5)
serpentine, 74
Sewell, R. B. Seymour, 57-59
Seymour Island, 216
Shackleton, Sir Ernest, 10, 187
Shanxi, 307
shape of the Earth, 14, 73, 365
shear waves, 71-72, 312, 313
    (Fig. 18.3)
Shishaldin volcano, 295
shock waves, 125-127, 232
shocked quartz, 98
shrinkage (contraction) hypothesis, 7,
    79, 132
Siberia, 40-41, 59, 217, 285, 287,
    343, 354
Sicily, 264
Sierra Nevada, 288, 293, 295-296,
    299, 306
Sigsbee Knolls, 152-153, 156, 163
Sigurgeirsson, Thorbjörn, 238
Sillitoe, Richard H., 348-349

Simmons, Gene, 317
Simplon Tunnel, 253, 255
Simpson, E. S. W., 183
Simpson, George Gaylord, 198
Siple, Paul, 27-28
Skate (submarine), 60
Skye, 110
sloths, 216
Smewing, J. D., 257
Smith, A. Gilbert, 264
Smithsonian Astrophysical
    Observatory, 229, 364
"smokers" (*See* hydrothermal vents)
Smylie, Douglas, 310
Snaeffelsjökull, 235
Snake River basalts, 294
Snider, Antonio, 4 (Fig. 1.1)
Sodom, 235
solar system, 32–39
solar wind, 39
Solutré, 43
Somalis, 226
sonar (echo sounders), 57, 124-125,
    365, 372, 374
sonar beacons, 152, 158
Sonoma volcanic field, 305
Soulavie, Abbé, 81
sound waves (pressure waves), 71
South Africa, 93, 130
South America, 111, 117, 283,
    290, 302
  coastal trench and cordillera, 291
  coastlines, 276
  fossil remains, 10, 191-192, 206
    (Fig. 12.1), 207-209, 214, 216
  links to Africa, Australia, and An-
    tarctica, 3, 7, 10, 13, 199-202,
    206, 214
  ore deposits, 348-349
  fitting coastlines together, 3, 7, 13,
    115, 116 (Fig. 7.1), 129
  salt domes and match of, 161
South American Plate, 358
South Pole, 7, 10, 45, 92, 192, 283
  (*See also* Antarctica)
South Sandwich Islands, 202
Southeast Asia, 348, ore deposits, 346
Soviet Academy of Sciences, 82, 147,
    166, 184
Soviet Union, 147, 185, 342
  deep-drilling program, 139-140
  diamond search, 354

earthquake prediction, 311-313
    (Fig. 18.3), 314
  geothermal potential, 328, 333
  oil and gas deposits, 343, 346
  sea between Siberia and Russia,
    285-288
  space probes, 39

SP 3000 (*soucoupe plongeante*), 371
Spain, change in orientation, 91, 115,
    262, 263 (Fig. 15.6), 264
Spencer F. Baird (research ship), 60, 64
spin axis, 14, 20, 22, 30, 33, 88-89,
    91, 115, 215, 362
  change in rate, 24-25
  influence of ice sheets, 24-27
  (*See also* Chandler wobble)
Spindletop (oil well), 153
Spitsbergen, 58, 217, 279
Sproll, Walter, 200
Staël, Madame de, 252
Staub, Rudolf, 255
"steady state" school, 340
steel, 14, 81
Steen Mountain, 98
Stove, David, 38
Strait of Bab el Mandeb, 226-227
Strait of Gibraltar, 165
Stubbs, Peter H. S., 91
subducted slabs, 366
subduction, 66, 183, 245, 261-262,
    276, 287, 292, 300, 302-303,
    348-349, 359, 369
  "flip," 261 (Fig. 15.5), 262
submarine cables, 50, 51 (Fig. 3.2),
    57, 59
submarines, 62, 70, 117, 369-375
  (*See also* Alvin)
Sudan, 222
Suess, Eduard, 10
Suez Canal, 176
Sunda Islands, 9
surges (glacial), 45-47
Surtsey, 237
suture zones, 85, 257, 273 (Fig. 16.2)
Suwannee Suture, 368
Sweden, 254
Swedish Deep-Sea Expedition, 47-48
Switzerland, 21, 251-266
Sykes, Lynn R., 118-119 (Fig. 7.2),
    121, 123, 133, 314

T

Taiwan-Luzon Arc, 290
Talwani, Manik, 108
Tarling, Don H., 95, 343, 345
Tashkent, 311
Tasman Sea, 183
Tasmania, 216, 345
Taylor, Frank Bursley, 5
Taylor, J. H., 114
Taymyr Peninsula, 287
Tazieff, Haroun, 228-232
tektites, 99 (Fig. 6.3), 100
Teller, Edward, 55
Tennessee, 270
tephroite, 351
terranes, 368-369
Tethys Sea, 13, 208, 215, 255-258, 264, 287
Texas, 269
Texas hot brines, 331-332
Tharp, Marie, 58-60
Thecodonts, 198, 209 (Fig. 12.3)
Thera (Santorini), 175-176, 260 (*See also* Atlantis)
"thick-skinned" and "thin-skinned" geologists, 271
Thingvellir, world's first parliament, 236
Thomas, Albert, 142, 145
Thrinaxodon, 197 (Fig. 11.4), 198, 209
thrusters (for position-keeping), 143
Tibetan Plateau, 286, 338, 362, 369
tidal waves, 48
tides, 18, 80-81, 357-358
Tierra del Fuego, 171
tillites, 199, 352
time and timetables
  of Atlantic spreading, 158–159 (Fig. 9.5)
  of continental breakup, 8 (Fig. 1.3), 9, 78 (Fig. 5.8)
  of magnetic reversals, 97 (Fig. 6.2)
tin, 348-349
Titanic, 375
Tonga-Kermadec Trench, 68
Tracey, Joshua I., 137
Transantarctic Mountains, 187, 189, 211-212
transform faults, 118, 119 (Fig. 7.2), 120-122 (Fig. 7.3), 123-124

(Fig. 7.4), 228, 233-234, 236, 329, 343
tree rings, 195
trenches, ocean floor, 62-63, 64 (Fig. 5.2), 65-66, 68, 70, 74-75, 118, 121, 130-131, 257, 278, 288-292, 296-297, 302, 305, 359, 364, 369
  gravity measurements, 62-63, 70
  sea mounts on edge of, 68
  sediment in, 130
Trident (ship), 247
triple junctions, 243, 245, 297, 302
Tristan da Cunha, 80, 83, 86, 159, 243, 244 (Fig. 14.1)
Trümpy, Rudolf, 255
tsunami, 311
tube worms, 375
tungsten, 348
turbidity currents, 48-52, 128, 182, 258, 268, 289, 292, 296
  Alps, traces in, 258 (Fig. 15.2), 259 (Fig. 15.3)
  submarine cables severed, 50-51 (Fig. 3.2), 52
  volume of sediment, 51
Turcotte, D. L., 250
Turkey, 261
  Anatolian Fault, 266
Twain, Mark, 9, 80, 123
typhoon, 180-181

U

Ullrich, G. Wayne, 353
"underplating," 341
"uniformitarianism," 5, 21, 52, 167, 377
Union Oil Company of California, 335, 337
United Nations, 22, 332, 337
United States Coast and Geodetic Survey, 101, 126, 276, 330
United States Geological Survey, 93, 109, 335, 349
United States Navy, 56, 60, 106, 371, 248
University of Alaska at College, 41
University of California at Berkeley, 196
University of Cambridge, 114, 126
University of Graz, 16

University of Marburg, 6
University of Miami, 146, 232
University of Pittsburgh, 37
Upper Mantle Project (UMP), 82, 133-134, 360, 361
Urals, 257, 285, 287-288, 291, 346, 348 (*See also* mountain-building)
Urey, Harold, 96
USNS DeSteiguer, 245
USNS George D. Keathley, 301
USNS Lynch, 67 (Fig. 5.3)
U.S.S. *Barracuda*, 126
U.S.S. *Cape Johnson*, 66
U.S.S. *Staten Island*, 109
U.S.S. *Tusk*, 117
Utah, 296, 304

V

Vacquier, Victor, 103, 118
Valdivia (ship), 225
Valentine, J. W., 217 (Fig. 12.5), 218
Valpelline formation, 253-254
van Andel, Tjeerd, 70
van Waterschoot van der Gracht, W. A. J. M., 14
vanadium, 224
Vancouver Island, 296
varved clay, 88
Velikovsky, Immanuel, 31-39, 42, 44-45, 47-48
Vema (research ship), 60, 127, 137, 152, 176, 183
Vema Fracture Zone, 182
Venice, 327
Vening Meinesz, Felix Andries, 12, 74, 79, 117
  convection current hypothesis, 70-72
  gravity measurements, 62, 65, 70
Venus (planet), 5, 32-34, 38-39, 377
Vereshchagin, K. K., 43
Vermont, 268-270
Verne, Jules, 235
very-long-baseline interferometry, 364
Vine, Frederick J., 102 (Fig. 6.4), 104-106, 108-109, 122, 256, 305, 347
Virginia, 266, 270, 279
Vityaz (research vessel), 64, 222
Vogt, Peter R., 248, 301

Voisey, A. H., 114
volcanic activity, 65-66, 98, 108, 160, 170, 174-175, 177, 237, 282, 291 (Fig.17.3), 292-294, 297, 303, 310
   above descending plates, 75, 288, 291
   Afar Triangle, 220, 230 (Fig. 13.2), 231, 232
   explosive, 65
   Hawaiian, 65, 80, 85 (Fig. 5.10), 123, 137, 231
   Iceland, 235-240
   Mid-Atlantic Ridge, 159, 369, 372, 373 (Fig. 21.2), 375, 376 (Fig. 21.3)
   Pacific Ocean, 375-376 (Fig. 21.3)
   Western United States, 295 (Fig. 17.4), 298 (*See also* hydrothermal vents, island arcs, plumes, and Thera)
volcanic islands, 270, 278-279
Voltaire, 307
Vostok, 202

W

Walensee, 252
Wales, 272
Walker, G. P. L., 240
Walker Lane, 304
Wallace, Alfred Russel, 218
Walvis Ridge, 243
Wasatch Range, 304
Washington state, 147, 183, 294, 296
Waterman, Alan, 142, 153

Watkins, H. G. ("Gino"), 18
weather satellites, 171, 174
Wegener, Alfred, 3, 5-6 (Fig. 1.2), 7-9, 11-19, 27, 54, 73-74, 78, 86, 90, 131, 199, 216, 221 (Fig. 13.1), 228, 255, 356
   critics, 14-16
   expedition to Greenland, 16-18
   on Afar Triangle, 220
Wegener, Kurt, 5
Weizmann, Chaim, 31
West Africa, 279
West Indies, 62-63, 117
Westchester County, 279
Whitcomb, James H., 317
Whitten, Charles A., 310
Wigner, Eugene, 55
Wild, Frank, 187
Wildt, Rupert, 34, 38
Wilkes Land, Antarctica, 45, 172, 351
Williams, Richard S., 238
Willis, Bailey, 15
Willumsen, Rasmus, 17-18
Wilson, A. T., 45
Wilson, Edward A., 188
Wilson, J. Tuzo, 60, 102 (Fig. 6.4), 105-106, 113, 121, 126, 271-272, 273 (Fig. 16.2), 274, 280, 305, 355, 359-360, 368
   debate with Beloussov, 130-134
   origin of Hawaiian Islands, 83-85
   plume concept, 243, 248
"Wilson cycles," 366, 368
Wise, Donald U., 298
Wiseman, John D. H., 57-59
Wisconsin ice age, 42

Woods Hole Oceanographic Institution, 64, 127, 146, 223, 369
   research ships, 127, 369, 371, 373-375
Woodward, C. Don, 146
World War II, 137
World-Wide Standardized Seismograph Network, 120
Worsley, Thomas R., 211, 366
Worzel, J. Lamar, 117, 124, 152
"Wrangellia Terrane," 368
Wyoming, 293-294, 303

X

X-ray crystallography, 90
Xinjiang, 211

Y

Yangtze River, 286, 338
Yemen, 222
Yellow River, 286
Yellowstone Park, 249, 294
Young, Brigham, 304
Yugoslavia, ophiolites, 257, 264
Yuma, Arizona, 326-327
Yungay, 308

Z

Zavodovski, Lieutenant Commander, 202
Zeitz, Isadore, 278 (Fig. 16.5)
zinc, 223-224, 348, 349-350

About the Author

Walter Sullivan has been described as America's most distinguished science writer. As Science Editor of *The New York Times* for 23 years, ending in 1987, he covered many of the major science stories of that time, including the launching of Sputnik, the birth of plate tectonics, and the recognition of black holes. He witnessed many of the scientific discoveries described in this volume, having accompanied five research expeditions to Antarctica, where Sullivan's Ridge has been named in his honor. In addition to his news coverage, he has written eight books, two of them for children.

Sullivan has received numerous awards and medals, among them the Public Welfare Medal of the National Academy of Sciences, the Distinguished Public Service Award of the National Science Foundation, the 1990 John Wesley Powell Award for Citizen's Achievement by the U.S. Geological Survey and the Walter Sullivan Award from the American Geophysical Union, a new award for excellence in popularizing the earth sciences named in his honor. He was three times winner of the Westinghouse Award of the American Association for the Advancement of Science, and has been awarded six honorary doctorates, including degrees from Yale and Ohio State Universities.

Because of his remarkable ability to make complex scientific subjects understandable and exciting, he is a popular lecturer and has made frequent television appearances. He has been featured in the ABC television program *Search* and in a program on his book *We Are Not Alone*. Sullivan was also host for a federally financed documentary on Project FAMOUS, the French-American Mid-Ocean Undersea Study, in which the *Alvin* and two French deep-diving submersibles made the first direct observations of the Mid-Atlantic Ridge.

During World War II Sullivan served in the U.S. Navy, winning 12 combat stars and ending as commander of the *U.S.S. Overton*, a converted destroyer. He graduated from Yale University in 1940.